Contributions in Mathematical and Computational Sciences • Volume 2

Editors
Hans Georg Bock
Willi Jäger
Otmar Venjakob

For other titles published in this series, go to
http://www.springer.com/series/8861

Jakob Stix
Editor

The Arithmetic of Fundamental Groups

PIA 2010

MATCH HEIDELBERG UNIVERSITY

IWR

Springer

Editor
Jakob Stix
MATCH - Mathematics Center Heidelberg
Department of Mathematics
Heidelberg University
Im Neuenheimer Feld 288
69120 Heidelberg
Germany
stix@mathi.uni-heidelberg.de

ISBN 978-3-642-23904-5 e-ISBN 978-3-642-23905-2
DOI 10.1007/978-3-642-23905-2
Springer Heidelberg Dordrecht London New York

Library of Congress Control Number: 2011943310

Mathematics Subject Classification (2010): 14H30, 14F32, 14F35, 14F30, 11G55, 14G30, 14L15

Printed on acid-free paper

Springer is part of Springer Science+Business Media (www.springer.com)

to Lucie Ella Rose Stix

Preface to the Series

Contributions to Mathematical and Computational Sciences

Mathematical theories and methods and effective computational algorithms are crucial in coping with the challenges arising in the sciences and in many areas of their application. New concepts and approaches are necessary in order to overcome the complexity barriers particularly created by nonlinearity, high-dimensionality, multiple scales and uncertainty. Combining advanced mathematical and computational methods and computer technology is an essential key to achieving progress, often even in purely theoretical research.

The term mathematical sciences refers to mathematics and its genuine sub-fields, as well as to scientific disciplines that are based on mathematical concepts and methods, including sub-fields of the natural and life sciences, the engineering and social sciences and recently also of the humanities. It is a major aim of this series to integrate the different sub-fields within mathematics and the computational sciences, and to build bridges to all academic disciplines, to industry and other fields of society, where mathematical and computational methods are necessary tools for progress. Fundamental and application-oriented research will be covered in proper balance.

The series will further offer contributions on areas at the frontier of research, providing both detailed information on topical research, as well as surveys of the state-of-the-art in a manner not usually possible in standard journal publications. Its volumes are intended to cover themes involving more than just a single "spectral line" of the rich spectrum of mathematical and computational research.

The Mathematics Center Heidelberg (MATCH) and the Interdisciplinary Center for Scientific Computing (IWR) with its Heidelberg Graduate School of Mathematical and Computational Methods for the Sciences (HGS) are in charge of providing and preparing the material for publication. A substantial part of the material will be acquired in workshops and symposia organized by these institutions in topical areas of research. The resulting volumes should be more than just proceedings collecting papers submitted in advance. The exchange of information and the discussions during the meetings should also have a substantial influence on the contributions.

This series is a venture posing challenges to all partners involved. A unique style attracting a larger audience beyond the group of experts in the subject areas of specific volumes will have to be developed.

Springer Verlag deserves our special appreciation for its most efficient support in structuring and initiating this series.

Heidelberg University,
Germany

Hans Georg Bock
Willi Jäger
Otmar Venjakob

Preface

During the more than 100 years of its existence, the notion of the fundamental group has undergone a considerable evolution. It started by Henri Poincaré when topology as a subject was still in its infancy. The fundamental group in this setup measures the complexity of a pointed topological space by means of an algebraic invariant, a discrete group, composed of deformation classes of based closed loops *within* the space. In this way, for example, the monodromy of a holomorphic function on a Riemann surface could be captured in a systematic way.

It was through the work of Alexander Grothendieck that, raising into the focus the role played by the fundamental group in governing covering spaces, so spaces *over* the given space, a unification of the topological fundamental group with Galois theory of algebra and arithmetic could be achieved. In some sense the roles have been reversed in this *discrete Tannakian* approach of abstract Galois categories: first, we describe a suitable class of objects that captures *monodromy*, and then, by abstract properties of this class alone and moreover uniquely determined by it, we find a pro-finite group that describes this category completely as the category of discrete objects continuously acted upon by that group.

But the different incarnations of a fundamental group do not stop here. The concept of describing a fundamental group through its category of objects upon which the group naturally acts finds its pro-algebraic realisation in the theory of Tannakian categories that, when applied to vector bundles with flat connections, or to smooth ℓ-adic étale sheaves, or to iso-crystals or ..., gives rise to the corresponding fundamental group, each within its natural category as a habitat.

In more recent years, the influence of the fundamental group on the geometry of Kähler manifolds or algebraic varieties has become apparent. Moreover, the program of anabelian geometry as initiated by Alexander Grothendieck realised some spectacular achievements through the work of the Japanese school of Hiroaki Nakamura, Akio Tamagawa and Shinichi Mochizuki culminating in the proof that hyperbolic curves over p-adic fields are determined by the outer Galois action of the absolute Galois group of the base field on the étale fundamental group of the curve.

A natural next target for pieces of arithmetic captured by the fundamental group are rational points, the genuine object of study of Diophantine geometry. Here there are two related strands: Grothendieck's section conjecture in the realm of the étale arithmetic fundamental group, and second, more recently, Minhyong Kim's idea to use the full strength of the different (motivic) realisations of the fundamental group to obtain a nonabelian unipotent version of the classical Chabauty approach towards rational points. In this approach, one seeks for a nontrivial p-adic Coleman analytic function that finds all global rational points among its zeros, whereby in the one-dimensional case the number of zeros necessarily becomes finite. This has led to a spectacular new proof of Siegel's theorem on the finiteness of S-integral points in some cases and, moreover, raised hope for ultimately (effectively) reproving the Faltings–Mordell theorem. A truely motivic advance of Minhyong Kim's ideas due to Gerd Faltings and Majid Hadian is reported in the present volume.

This volume originates from a special activity at Heidelberg University under the sponsorship of the MAThematics Center Heidelberg (MATCH) that took place in January and February 2010 organised by myself. The aim of the activity was to bring together people working in the different strands and incarnations of the fundamental group all of whose work had a link to arithmetic applications. This was reflected in the working title PIA for our activity, which is the (not quite) acronym for π_1–arithmetic, short for *doing arithmetic with the fundamental group* as your main tool and object of study. PIA survived in the title of the workshop organised during the special activity: *PIA 2010 — The arithmetic of fundamental groups*, which in reversed order gives rise to the title of the present volume.

The workshop took place in Heidelberg, 8–12 February 2010, and the abstracts of all talks are listed at the end of this volume. Many of these accounts are mirrored in the contributions of the present volume. The special activity also comprised expository lecture series by Amnon Besser on Coleman integration, a technique used by the non-abelian Chabauty method, and by Tamás Szamuely on Grothendieck's fundamental group with a view towards anabelian geometry. Lecture notes of these two introductory courses are contained in this volume as a welcome addition to the existing literature of both subjects.

I wish to extend my sincere thanks to the contributors of this volume and to all participants of the special activity in Heidelberg on the arithmetic of fundamental groups, especially to the lecturers giving mini-courses, for the energy and time they have devoted to this event and the preparation of the present collection. Paul Seyfert receives the editor's thanks for sharing his marvelous TEX–expertise and help in typesetting this volume. Furthermore, I would like to take this opportunity to thank Dorothea Heukäufer for her efficient handling of the logistics of the special activity and Laura Croitoru for coding the website. I am very grateful to Sabine Stix for sharing her organisational skills both by providing a backbone for the *to do* list of the whole program and also in caring for our kids Antonia, Jaden and Lucie. Finally, I would like to express my gratitude to Willi Jäger, the former director of MATCH, for his enthusiastic support and for the financial support of MATCH that made PIA 2010 possible and in my opinion a true success.

Heidelberg *Jakob Stix*

Contents

Part I
Heidelberg Lecture Notes

Chapter 1
Heidelberg Lectures on Coleman Integration

Amnon Besser*

Abstract Coleman integration is a way of associating with a closed one-form on a p-adic space a certain locally analytic function, defined up to a constant, whose differential gives back the form. This theory, initially developed by Robert Coleman in the 1980s and later extended by various people including the author, has now found various applications in arithmetic geometry, most notably in the spectacular work of Kim on rational points. In this text we discuss two approaches to Coleman integration, the first is a semi-linear version of Coleman's original approach, which is better suited for computations. The second is the author's approach via unipotent isocrystals, with a simplified and essentially self-contained presentation. We also survey many applications of Coleman integration and describe a new theory of integration in families.

1.1 Introduction

In the first half of February 2010 I spent 2 weeks at the Mathematics Center Heidelberg (MATCH) at the university of Heidelberg, as part of the activity *PIA 2010 – The arithmetic of fundamental groups*. In the first week I gave 3 introductory lectures on Coleman integration theory and in the second week I gave a research lecture on new work, which was (and still is) in progress, concerning Coleman integration in families. I later gave a similar sequence of lectures at the Hebrew University in Jerusalem.

A. Besser (✉)
Department of Mathematics, Ben-Gurion, University of the Negev, Be'er-Sheva, Israel
e-mail: bessera@math.bgu.ac.il

* Part of the research described in these lectures was conducted with the support of the Israel Science Foundation, grant number: 1129/08, whose support I would like to acknowledge.

J. Stix (ed.), *The Arithmetic of Fundamental Groups*, Contributions in Mathematical and Computational Sciences 2, DOI 10.1007/978-3-642-23905-2_1,

This article gives an account of the 3 instructional lectures as well as the lecture I gave at the conference in Heidelberg with some (minimal) additions. I largely left things as they were presented in the lectures and I therefore apologize for the sometimes informal language used and the occasional proof which is only sketched. As in the lectures I made an effort to make things as self-contained as possible.

The main goal of these lectures is to introduce Coleman integration theory. The goal of this theory is (in very vague terms) to associate with a closed 1-form $\omega \in \Omega^1(\mathrm{X})$, where X is a "space" over a p-adic field K, by which we mean a finite extension of $\mathbb{Q}_p$, for a prime p fixed throughout this work, a locally analytic primitive F_ω, i.e., such that $d\mathrm{F}_\omega = \omega$, in such a way that it is unique up to a constant.

In Sect. 1.4 we introduce Coleman theory. The presentation roughly follows Coleman's original approach [Col82, CdS88]. One essential difference is that we emphasize the semi-linear point of view. This turns out to be very useful in numerical computations of Coleman integrals. The presentation we give here, which does not derive the semi-linear properties from Coleman's work, is new.

In Sect. 1.5 we give an account of the Tannakian approach to Coleman integration developed in [Bes02]. The main novelty is a more self contained and somewhat simplified proof from the one given in loc. cit. Rather than rely on the work of Chiarellotto [Chi98], relying ultimately on the thesis of Wildeshaus [Wil97], we unfold the argument and obtain some simplification by using the Lie algebra rather than its enveloping algebra.

At the advice of the referee we included a lengthy section on applications of Coleman integration. In the final section we explain a new approach to Coleman integration in families. We discuss two complementary formulations, one in terms of the Gauss-Manin connection and one in terms of differential Tannakian categories.

Acknowledgements. I would like to thank MATCH, and especially Jakob Stix, for inviting me to Heidelberg, and to thank Noam Solomon and Ehud de Shalit for organizing the sequence of lectures in Jerusalem. I also want to thank Lorenzo Ramero for a conversation crucial for the presentation of Kim's work. I would finally like to thank the referee for making many valuable comments that made this work far more readable than it originally was, and for a very careful reading of the manuscript catching a huge number of mistakes.

1.2 Overview of Coleman Theory

To appreciate the difficulty of integrating a closed form on a p-adic space, let us consider a simple example. We consider a form $\omega = dz/z$ on a space

$$\mathrm{X} = \{z \in \mathrm{K}\ ;\ |z| = 1\}.$$

Morally, the primitive F_ω should just be the logarithm function $\log(z)$. To try to find a primitive, we could pick $\alpha \in \mathrm{X}$ and expand ω in a power series around α as follows:

$$\omega = \frac{\mathrm{d}(\alpha+x)}{\alpha+x} = \frac{\mathrm{d}x}{\alpha+x} = \frac{1}{\alpha}\frac{\mathrm{d}x}{1+x/\alpha} = \frac{1}{\alpha}\sum\left(\frac{-x}{\alpha}\right)^n \mathrm{d}x$$

and integrating term by term we obtain

$$\mathrm{F}_\omega(\alpha+x) = -\sum\frac{1}{n+1}\left(\frac{-x}{\alpha}\right)^{n+1} + \mathrm{C}$$

where these expansions converge on the disc for which $|x| < 1$.

So far, we have done nothing that could not be done in the complex world. However, in the complex world we could continue as follows. Fix the constant of integration C on one of the discs. Then do analytic continuation: For each intersecting disc it is possible to fix the constant of integration on that disc uniquely so that the two expansions agree on the intersection. Going around a circle around 0 gives a non-trivial monodromy, so analytic continuation results in a multivalued function, which is the log function.

In the p-adic world, we immediately realize that such a strategy will not work because two open discs of radius 1 are either identical or completely disjoint. Thus, there is no obvious way of fixing simultaneously the constants of integration.

Starting with [Col82], Robert Coleman devises a strategy for coping with this difficulty using what he called **analytic continuation along Frobenius**. To explain this in our example, we take the map $\phi : \mathrm{X} \to \mathrm{X}$ given by $\phi(x) = x^p$ which is a lift of the p-power map. One notices immediately that $\phi^*\omega = p\omega$. Coleman's idea is that this relation should imply a corresponding relation on the integrals

$$\phi^*\mathrm{F}_\omega = p\mathrm{F}_\omega + \mathrm{C}$$

where C is a constant function. It is easy to see that by changing F_ω by a constant, which we are allowed to do, we can assume that $\mathrm{C} = 0$. The equation above now reads

$$\mathrm{F}_\omega(x^p) = p\mathrm{F}_\omega(x)\,.$$

Suppose now that α satisfies the relation $\alpha^{p^k} = \alpha$. Then we immediately obtain

$$\mathrm{F}_\omega(\alpha) = \mathrm{F}_\omega(\alpha^{p^k}) = p^k\mathrm{F}_\omega(\alpha) \Rightarrow \mathrm{F}_\omega(\alpha) = 0\,.$$

This condition, together with the assumption that $\mathrm{dF}_\omega = \omega$ fixes F_ω on the disc $|z-\alpha| < 1$. But it is well known that every $z \in \mathrm{X}$ resides in such a disc, hence F_ω is completely determined.

In [Col82] Coleman also introduces iterated integrals (only on appropriate subsets of $\mathbb{P}^1$) which have the form

$$\int(\omega_n \times \int(\omega_{n-1} \times \cdots \int(\omega_2 \times \int \omega_1)\cdots)\,,$$

and in particular defines p-adic polylogarithms $\mathrm{Li}_n(z)$ by the conditions

$$
\begin{aligned}
d\,\mathrm{Li}_1(z) &= \frac{dz}{1-z}\,, \\
d\,\mathrm{Li}_n(z) &= \mathrm{Li}_{n-1}(z)\frac{dz}{z}\,, \\
\mathrm{Li}_n(0) &= 0\,,
\end{aligned}
$$

so that locally one finds

$$
\mathrm{Li}_n(z) = \sum_{k=1}^{\infty} \frac{z^k}{k^n}\,.
$$

Then, in the paper [Col85b] he extends the theory to arbitrary dimensions but without computing iterated integrals. In [CdS88] Coleman and de Shalit extend the iterated integrals to appropriate subsets of curves with good reduction.

In [Bes02] the author gave a Tannakian point of view to Coleman integration and extended the iterated theory to arbitrary dimensions. Other approaches exist. Colmez [Col98], and independently Zarhin [Zar96], used functoriality with respect to algebraic morphisms. This approach does not need good reduction but cannot handle iterated integrals. Vologodsky has a theory for algebraic varieties, which is similar in many respects to the theory in [Bes02]. Using alterations and defining a monodromy operation on the fundamental group in a very sophisticated way he is able to define iterated Coleman integrals also in the bad reductions case. Coleman integration was later extended by Berkovich [Ber07] to his p-adic analytic spaces, again without making any reductions assumptions.

Remark 1. There are two related ways of developing Coleman integration: the linear and the semi-linear way. For a variety over a finite field κ of characteristic p the absolute Frobenius φ_a is just the p-power map and its lifts to characteristic 0 are semi-linear. A linear Frobenius is any power of the absolute Frobenius which is κ-linear.

What makes the theory work is the description of weights of a linear Frobenius on the first cohomology (crystalline, rigid, Monsky-Washnitzer) of varieties over finite fields, see Theorem 4. The theory itself can be developed by imposing an equivariance conditions with respect to a lift of the linear Frobenius, as we have done above and as done in Coleman's work, or imposing equivariance with respect to a semi-linear lift of the absolute Frobenius. Even in this approach one ultimately relies on weights for a linear Frobenius.

The two approaches are equivalent. Since a power of a semi-linear Frobenius lift is linear, equivariance for a semi-linear Frobenius implies one for a linear Frobenius. Conversely, as Coleman integration is also Galois equivariant [Col85b, Corollary 2.1e] one recovers from the linear equivariance the semi-linear one.

The linear approach is cleaner in many respects, and it is used everywhere in this text with the exception of Sects. 1.3 and 1.4. There are two main reasons for introducing the semi-linear approach:

- It appears to be computationally more efficient.
- It may be applied in some situations where the linear approach may not apply, see Remark 11.

1.3 Background

Let K be a complete discrete valuation field of characteristic 0 with ring of integers R, residue field κ of prime characteristic p, uniformizer π and algebraic closure $\bar{\mathrm{K}}$. We also fix an automorphism σ of K which reduces to the p-power map on κ, and when needed extend it to $\bar{\mathrm{K}}$ such that it continues to reduce to the p-power map. When the cardinality of κ is finite we denote it by $q = p^r$.

1.3.1 Rigid Analysis

Let us recall first a few basic facts about rigid analysis. An excellent survey can be found in [Sch98]. The Tate algebra T_n is by definition

$$\mathrm{T}_n = \mathrm{K}\langle t_1,\ldots,t_n\rangle = \left\{\sum a_{\mathrm{I}} t^{\mathrm{I}} \; ; \; a_{\mathrm{I}} \in \mathrm{K}, \lim_{\mathrm{I}\to\infty} |a_{\mathrm{I}}| = 0\right\},$$

which is the same as the algebra of power series with coefficients in K converging on the unit polydisc

$$\mathrm{B}_n = \{(z_1,\ldots,z_n) \in \bar{\mathrm{K}}^n \; ; \; |z_i| \le 1\}.$$

An affinoid algebra A is a K-algebra for which there exists a surjective map $\mathrm{T}_n \to \mathrm{A}$ for some n. One associates with A its maximal spectrum

$$\begin{aligned} \mathrm{X} = \mathrm{spm(A)} :&= \{m \subset \mathrm{A} \text{ maximal ideal }\} \\ &= \{\psi : \mathrm{A} \to \bar{\mathrm{K}} \text{ a K-homomorphism}\}/\mathrm{Gal}(\bar{\mathrm{K}}/\mathrm{K}) \end{aligned}$$

i.e., the quotient of the set of K-algebra homomorphisms from A to $\bar{\mathrm{K}}$ (no topology involved) by the Galois group of $\bar{\mathrm{K}}$ over K. The latter equality is a consequence of the Noether normalization lemma for affinoid algebras from which it follows that a field which is a homomorphic image of such an algebra is a finite extension of K. Two easy examples are

$$\begin{aligned} \mathrm{spm}(\mathrm{T}_n) &= \mathrm{B}_n/\mathrm{Gal}(\bar{\mathrm{K}}/\mathrm{K}), \\ \mathrm{spm}(\mathrm{T}_2/(t_1t_2-1)) &= \{(z_1,z_2) \in \mathrm{B}_2 \; ; \; z_1z_2 = 1\}/\mathrm{Gal}(\bar{\mathrm{K}}/\mathrm{K}) \\ &= \{z \in \bar{\mathrm{K}} \; ; \; |z| = 1\}/\mathrm{Gal}(\bar{\mathrm{K}}/\mathrm{K}) \end{aligned}$$

In what follows we will shorthand things so that the last space will simply be written $\{|z| = 1\}$ when there is no danger of confusion.

The maximal spectrum $\mathrm{X} = \mathrm{spm(A)}$ of an affinoid algebra with an appropriate Grothendieck topology and sheaf of functions will be called an affinoid space, and in a Grothendieckian style we associate with it its ring of functions $\mathcal{O}(\mathrm{X}) = \mathrm{A}$. Rigid geometry allows one to glue affinoid spaces into more complicated spaces, and obtain the ring of functions on these spaces as well. We will say nothing about this except to mention that the space $\mathrm{B}_n^\circ = \{|z_i| < 1\} \subset \mathrm{B}_n$ can be obtained as the union of the spaces $\{|z_i^k/\pi| \leq 1\}$ for $k \in \mathbb{N}$, and its ring of functions is not surprisingly

$$\mathcal{O}(\mathrm{B}_n^\circ) = \left\{\sum a_{\mathrm{I}} t^{\mathrm{I}} \; ; \; \lim_{\mathrm{I}\to\infty} |a_{\mathrm{I}}| r^{|\mathrm{I}|} = 0 \text{ for any } r < 1\right\}$$

where $|(i_1,\ldots,i_n)| = i_1 + \cdots + i_n$.

1.3.2 Dagger Algebras and Monsky-Washnitzer Cohomology

The de Rham cohomology of rigid spaces is problematic in certain respects. To see an example of this, consider the first de Rham cohomology of T_1, which is the cokernel of the map

$$\mathrm{d} : \mathrm{T}_1 \to \mathrm{T}_1 dt\,.$$

This cokernel is infinite as one can write down a power series $\sum a_i t^i$ such that the a_i converge to 0 sufficiently slowly to make the coefficients of the integral

$$\sum a_i t^{i+1}/(i+1)$$

not converge to 0. On the other hand, as B_1 can be considered a lift of the affine line, one should expect its cohomology to be trivial.

To remedy this, Monsky and Washnitzer [MW68] considered so called weakly complete finitely generated algebras. An excellent reference is the paper [vdP86].

We consider the algebra

$$\mathcal{T}_n^\dagger = \left\{\sum a_{\mathrm{I}} t^{\mathrm{I}} \; ; \; a_{\mathrm{I}} \in \mathrm{R},\ \exists r > 1 \text{ such that } \lim_{\mathrm{I}\to\infty} |a_{\mathrm{I}}| r^{\mathrm{I}} = 0\right\}.$$

In other words, these are the power series converging on something slightly bigger than the unit polydisc, hence the term **overconvergence**. Integration reduces the radius of convergence, but only slightly: if the original power series converges to radius r the integral will no longer converge to radius r but will converge to any smaller radius, hence still overconverges.

An R-algebra $\mathrm{A}^\dagger$ is called a **weakly complete finitely generated** (wcfg) algebra if there is a surjective homomorphism $\mathcal{T}_n^\dagger \to \mathrm{A}^\dagger$. Since $\mathcal{T}_n^\dagger$ is Noetherian, see [vdP86] just after (2.2), such an algebra may be presented as

$$\mathrm{A}^\dagger = \mathcal{T}_n^\dagger/(f_1,\ldots,f_m)\,. \tag{1.1}$$

The module of differentials $\Omega^1_{\mathrm{A}^\dagger}$ is given, in the presentation (1.1), as

$$\Omega^1_{\mathrm{A}^\dagger} = \bigoplus_{i=1}^{n} \mathrm{A}^\dagger \mathrm{d}t_i/(\mathrm{d}f_j\,,\ j = 1,\ldots m)\,,$$

where $\mathrm{d}f = \sum_i \frac{\partial f}{\partial t_i}\mathrm{d}t_i$ as usual, see [vdP86, (2.3)]. Be warned that this is not the algebraic module of differentials. Taking wedge powers one obtains the modules of higher differential forms $\Omega^i_{\mathrm{A}^\dagger}$ and the de Rham complex $\Omega^\bullet_{\mathrm{A}^\dagger}$.

One observes that $\mathcal{T}_n^\dagger/\pi$ is isomorphic to the polynomial algebra $\kappa[t_1,\ldots,t_n]$. Thus, if $\mathrm{A}^\dagger$ is a wcfg algebra then $\bar{\mathrm{A}} := \mathrm{A}^\dagger/\pi$ is a finitely generated κ-algebra.

Assume from now throughout the rest of this work that the κ-algebras considered are finitely generated and smooth. Any such κ-algebra can be obtained as an $\bar{\mathrm{A}}$ for an appropriate $\mathrm{A}^\dagger$ by a result of Elkik [Elk73]. In addition, we have the following results on those lifts.

Proposition 2 ([vdP86, Theorem 2.4.4]). *We have:*

(1) Any two such lifts are isomorphic.
(2) Any morphism $\bar{f} : \bar{\mathrm{A}} \to \bar{\mathrm{B}}$ can be lifted to a morphism $f^\dagger : \mathrm{A}^\dagger \to \mathrm{B}^\dagger$.
(3) Any two maps $\mathrm{A}^\dagger \to \mathrm{B}^\dagger$ with the same reduction induce homotopic maps

$$\Omega^\bullet_{\mathrm{A}^\dagger} \otimes \mathrm{K} \to \Omega^\bullet_{\mathrm{B}^\dagger} \otimes \mathrm{K}\,.$$

Thus, the following definition makes sense.

Definition 3. The **Monsky-Washnitzer cohomology** of $\bar{\mathrm{A}}$ is the cohomology of the de Rham complex $\Omega^\bullet_{\mathrm{A}^\dagger} \otimes \mathrm{K}$

$$\mathrm{H}^i_{\mathrm{MW}}(\bar{\mathrm{A}}/\mathrm{K}) = \mathrm{H}^i(\Omega^\bullet_{\mathrm{A}^\dagger} \otimes \mathrm{K})\,.$$

It is a consequence of the work of Berthelot [Ber97, Corollaire 3.2] that $\mathrm{H}^i_{\mathrm{MW}}(\bar{\mathrm{A}})$ is a finite-dimensional K-vector space.

The absolute Frobenius morphism $\varphi_a(x) = x^p$ of $\bar{\mathrm{A}}$ can be lifted, by Proposition 2, to a σ-linear morphism $\phi_a : \mathrm{A}^\dagger \to \mathrm{A}^\dagger$. Indeed, $\mathrm{A}^\dagger$ with the homomorphism

$$\mathrm{R} \xrightarrow{\sigma} \mathrm{R} \to \mathrm{A}^\dagger$$

is a lift of $\bar{\mathrm{A}}$ with the map

$$\kappa \xrightarrow{x^p} \kappa \to \bar{\mathrm{A}}$$

and φ_a induces a homomorphism between $\bar{\mathrm{A}}$ and this new twisted κ-algebra. The σ-linear ϕ_a induces a well defined σ-linear endomorphism φ_a of $\mathrm{H}^i_{\mathrm{MW}}(\bar{\mathrm{A}})$. On the other hand, if κ is a finite field with $q = p^r$ elements, then φ_a^r is already κ-linear and therefore induces an endomorphism $\varphi = \varphi_a^r$ of $\mathrm{H}^i_{\mathrm{MW}}(\bar{\mathrm{A}})$. By [Chi98, Theorem I.2.2] one knows the possible eigenvalues of φ on Monsky-Washnitzer cohomology. This result, modeled on Berthelot's proof [Ber97] of the finiteness of rigid cohomology, ultimately relies on the computation of the eigenvalues of Frobenius on crystalline cohomology by Katz and Messing [KM74], and therefore on Deligne's proof of the Weil conjectures [Del74].

Theorem 4. *The eigenvalues of the κ-linear Frobenius φ on $\mathrm{H}^1_{\mathrm{MW}}(\bar{\mathrm{A}})$ are Weil numbers of weights 1 and 2. In other words, they are algebraic integers and have absolute values q or $\sqrt{q}$ under any embedding into $\mathbb{C}$.*

1.3.3 Specialization and Locally Analytic Functions

One associates with a wcfg algebra $A^\dagger$ the K-algebra A, which is the completion $\mathcal{T}_n^\dagger$ of $A^\dagger \otimes K$ by the quotient norm induced from the Gauss norm, the maximal absolute value of the coefficients of the power series. This is easily seen to be an affinoid algebra. If $A^\dagger = \mathcal{T}_n^\dagger/I$, then $A = T_n/I$. We further associate with A the affinoid space $X = \mathrm{spm}(A)$. Letting $X_\kappa = \mathrm{Spec}(\bar{A})$ we have a specialization map

$$\mathrm{Sp} : X \to X_\kappa$$

which is defined as follows. Take a homomorphism $\psi : A \to L$, with L a finite extension of K. Then one checks by continuity that $A^\dagger$ maps to $\mathcal{O}_L$ and one associates with the kernel of ψ the kernel of its reduction mod π.

For our purposes, it will be convenient to consider the space X^{geo} of geometric points of X, which means K-linear homomorphisms $\psi : A \to \bar{K}$. This has a reduction map to the set of geometric points of X_κ obtained in the same way as above.

Definition 5. The inverse image of a geometric point $x : \mathrm{Spec}\,\bar{\kappa} \to X_\kappa$ under the reduction map will be called the **residue disc** of x, denoted $U_x \subset X^{\mathrm{geo}}$.

By Hensel's Lemma and the smoothness assumption on $\bar{A}$ it is easy to see that U_x is naturally isomorphic to the space of geometric points of a unit polydisc.

Definition 6. The K-algebra A_{loc} of **locally analytic functions** on X is defined as the space of all functions $f : X^{\mathrm{geo}} \to \bar{K}$ which satisfy the following two conditions:

(i) The function f is $\mathrm{Gal}(\bar{K}/K)$-equivariant in the sense that for any $\tau \in \mathrm{Gal}(\bar{K}/K)$ we have $f(\tau(x)) = \tau(f(x))$.

(ii) For each residue disc choose parameters z_1 to z_l identifying it with a unit polydisc over some finite field extension of K. Then restricted to such a residue disc f is defined by a power series in the z_i, which is therefore convergent on the open unit polydisc.

There is an obvious injection $A^\dagger \otimes K \subset A_{\mathrm{loc}}$. The algebra of our Coleman functions will lie in between these two K-algebras.

Another way of stating the equivariance condition for locally analytic functions, given the local expansion condition, is to say that given any $\tau \in \mathrm{Gal}(\bar{K}/K)$ transforming the geometric point x of X_κ to the geometric point y, we have that τ translates the local expansion of f near x to the local expansion near y by acting on the coefficients. This way one can similarly define the A_{loc}-module Ω^n_{loc} of **locally analytic** ***n*****-forms** on X, the obvious differential $d : \Omega^{n-1}_{\mathrm{loc}} \to \Omega^n_{\mathrm{loc}}$, and an embedding, compatible with the differential, $\Omega^n_{A^\dagger} \otimes K \hookrightarrow \Omega^n_{\mathrm{loc}}$.

We define an action of the σ-semi-linear lift of the absolute Frobenius ϕ_a defined in the previous subsection on the spaces above. We first of all define an action on X^{geo} as follows. Suppose $\psi : A \to \bar{K} \in X^{\mathrm{geo}}$ is a K-linear homomorphism. Then

$$\phi_a(\psi) = \sigma^{-1} \circ \psi \circ \phi_a, \tag{1.2}$$

recall that we have extended σ to $\bar{\mathrm{K}}$. Note that this is indeed K-linear again. We can describe this action on points concretely as follows. Suppose $\mathrm{A} = \mathrm{T}_n/(f_1, \ldots, f_k)$ and let $g_i = \phi_a(t_i)$ so that ϕ_a is given by the formula

$$\phi_a(\sum a_{\mathrm{I}} t^{\mathrm{I}}) = \sum \sigma(a_{\mathrm{I}})(g_1, \ldots, g_n)^{\mathrm{I}} .$$

Suppose that $\mathbf{z} := (z_1, \ldots, z_n) \in \mathrm{X}^{\mathrm{geo}}$, so that $f_i(\mathbf{z}) = 0$ for each i. Then we have

$$\phi_a(\mathbf{z}) = (\sigma^{-1} g_1(\mathbf{z}), \ldots, \sigma^{-1} g_1(\mathbf{z})) .$$

Having defined ϕ_a on points we now define it on functions by

$$\phi_a(f)(x) = \sigma f(\phi_a(x)) \tag{1.3}$$

From (1.2) it is quite easy to see that for $f \in \mathrm{A}$ this is just the same as $\phi_a(f)$ as previously defined. We again have a compatible action on differential forms.

1.4 Coleman Theory

We define Coleman integration in a somewhat different way than the one Coleman uses, emphasizing a semi-linear condition and stressing the Frobenius equivariance.

Theorem 7. *Suppose that* K *is a finite extension of* $\mathbb{Q}_p$. *Then there exists a unique* K-*linear integration map*

$$\int : (\Omega^1_{\mathrm{A}^\dagger} \otimes \mathrm{K})^{\mathrm{d}=0} \to \mathrm{A}_{\mathrm{loc}}/\mathrm{K}$$

satisfying the following conditions:

(i) *The map* $\mathrm{d} \circ \int$ *is the canonical map* $(\Omega^1_{\mathrm{A}^\dagger} \otimes \mathrm{K})^{\mathrm{d}=0} \to \Omega^1_{\mathrm{loc}}$.
(ii) *The map* $\int \circ \mathrm{d}$ *is the canonical map* $\mathrm{A}^\dagger_{\mathrm{K}} \to \mathrm{A}_{\mathrm{loc}}/\mathrm{K}$.
(iii) *One has* $\phi_a \circ \int = \int \circ \phi_a$.

In addition, the map is independent of the choice of ϕ_a. *Finally, in the above Theorem, equivariance with respect to the semi-linear Frobenius lift* ϕ_a *may be replaced by equivariance with respect to a linear Frobenius lift* ϕ, *and yields the same theory.*

Proof. Since $\mathrm{H}^1_{\mathrm{MW}}(\bar{\mathrm{A}})$ is finite-dimensional, we may choose $\omega_1, \ldots, \omega_n \in \Omega^1_{\mathrm{A}^\dagger} \otimes \mathrm{K}$ such that their images in $\mathrm{H}^1(\Omega^\bullet_{\mathrm{A}^\dagger} \otimes \mathrm{K})$ form a basis. If we are able to define the integrals $\mathrm{F}_{\omega_i} := \int \omega_i$ for all the ω_i's, then the second condition immediately tell us how to integrate any other form. Namely, write

$$\omega = \sum_{i=1}^{n} \alpha_i \omega_i + dg \,, \ \alpha_i \in \mathrm{K} \,, \ g \in \mathrm{A}^\dagger_{\mathrm{K}} . \tag{1.4}$$

Put all the forms above into a column vector ω. Then we have a matrix $\mathrm{M} \in \mathrm{M}_{n\times n}(\mathrm{K})$ such that

$$\phi_a \omega = \mathrm{M}\omega + \mathbf{dg}\,, \tag{1.5}$$

where $\mathbf{g} \in (\mathrm{A}_{\mathrm{K}}^{\dagger})^n$. Conditions (ii) and (iii) in the theorem tell us that (1.5) implies the relation

$$\phi_a \mathrm{F}_\omega = \mathrm{M}\mathrm{F}_\omega + \mathbf{g} + \mathbf{c}\,, \tag{1.6}$$

where $\mathbf{c} \in \mathrm{K}^n$ is some vector of constants. We first would like to show that $\mathbf{c}$ may be assumed to vanish. For this we have the following key lemma.

Lemma 8. *The map* $\sigma - \mathrm{M} : \mathrm{K}^n \to \mathrm{K}^n$ *is bijective.*

Proof. We need to show that for any $\mathbf{d} \in \mathrm{K}^n$ there is a unique solution to the system of equations $\sigma(\mathbf{x}) = \mathrm{M}\mathbf{x} + \mathbf{d}$. By repeatedly applying σ to this equation we can obtain an equation for $\sigma^i(\mathbf{x})$

$$\sigma^i(\mathbf{x}) = \mathrm{M}_i\mathbf{x} + \mathbf{d}_i$$

where

$$\mathrm{M}_i = \sigma^{i-1}(\mathrm{M})\sigma^{i-2}(\mathrm{M})\cdots\sigma(\mathrm{M})\mathrm{M}\,.$$

As $[\mathrm{K} : \mathbb{Q}_p] < \infty$ there exists some l such that σ^l is the identity on K and so we obtain the equation $\mathbf{x} = \mathrm{M}_l\mathbf{x} + \mathbf{d}_l$. Recalling that the cardinality of the residue field κ is p^r, we see that r divides l and that the matrix M_l is exactly the matrix of the l/r power of the linear Frobenius φ_a^r on $\mathrm{H}^1_{\mathrm{MW}}(\bar{\mathrm{A}}/\mathrm{K})$. It follows from Theorem 4 that the matrix $\mathrm{I} - \mathrm{M}_l$ is invertible. This shows that

$$\mathbf{x} = (\mathrm{I} - \mathrm{M}_l)^{-1}\mathbf{d}_l$$

is the unique possible solution to the equation. This shows that the map is injective, and since it is $\mathbb{Q}_p$-linear on a finite-dimensional $\mathbb{Q}_p$-vector space it is also bijective (one can also show directly that $\mathbf{x}$ above is indeed a solution). □

Remark 9. In computational applications, it is important that the modified equation $\mathbf{x} = \mathrm{M}_l\mathbf{x} + \mathbf{d}_l$ can be computed efficiently in $\mathrm{O}(\log(l))$ steps, see [LL03, LL06].

Since ϕ_a acts as σ on constant functions we immediately get from the lemma that by changing the constants in F_ω we may assume that $\mathbf{c} = 0$ in (1.6).

We claim that now the vector of functions F_ω is completely determined. Indeed, since $d\mathrm{F}_\omega = \omega$ by condition (i), we may determine F_ω on any residue disc up to a vector of constants by term by term integration of a local expansion of ω. It is therefore sufficient to determine it on a single point on each residue disc. So let x be such a point. Substituting x in (1.6) and recalling the action of ϕ_a on functions (1.3) we find

$$\sigma(\mathrm{F}_\omega(\phi_a(x))) = \mathrm{M}\mathrm{F}_\omega(x) + \mathbf{g}(x)\,.$$

Since $\phi_a(x)$ is in the same residue disc as x the difference

$$\mathbf{e} := \mathrm{F}_\omega(\phi_a(x)) - \mathrm{F}_\omega(x) = \int_x^{\phi_a(x)} \omega$$

is computable from ω alone. Substituting in the previous equation we find

$$\sigma(\mathrm{F}_\omega(x)) + \sigma(\mathbf{e}) = \mathrm{M}\mathrm{F}_\omega(x) + \mathbf{g}(x)\,,$$

and rearranging we find an equation for $\mathrm{F}_\omega(x)$ that may be solved using Lemma 8 for some finite extension of K where x is defined. This shows uniqueness and gives a method for computing the integration map.

It is fairly easy to see that the method above indeed gives an integration map satisfying all the required properties. Note that by uniqueness the integration map is independent of the choice of basis ω.

When using a linear Frobenius lift ϕ, one relies instead on the fact that any point defined over a finite field will be fixed by an appropriate power of φ. Considering equivariance with respect to that power, mapping the residue disc of the point x back to itself, we can determine the integral at x by a similar method. Thus, if ϕ is a power of ϕ_a and is linear, equivariance for ϕ_a implies one with respect for ϕ, and since the theory is determined uniquely by equivariance the converse is also true.

It remains to show that it is independent of the choice of ϕ_a. By the above, it is easy to see that it suffices to do this with respect to the equivariance property with respect to a linear Frobenius. So suppose we are given two linear Frobenii ϕ and ϕ' and that we have set up the theory for ϕ. We want to show that it also satisfies equivariance with respect to ϕ'. Let ω be a closed form and suppose we have chosen the constant in Coleman integration so that $\mathrm{F}_{\phi(\omega)} = \phi\mathrm{F}_\omega$. By Proposition 2 we have $h \in \mathrm{A}_\mathrm{K}^\dagger$ such that $\phi'(\omega) - \phi(\omega) = \mathrm{d}h$. We now compute

$$\begin{aligned}\int \phi'(\omega) - \phi' \int \omega &= \int \phi'(\omega) - \phi' \int \omega - (\int \phi(\omega) - \phi \int \omega) \\ &= \int (\phi'(\omega) - \phi(\omega)) - (\phi' \int \omega - \phi \int \omega) \\ &= h - (\phi' \int \omega - \phi \int \omega)\end{aligned}$$

and substituting at a point x we get

$$h(x) - \int_{\phi(x)}^{\phi'(x)} \omega\,.$$

We need to show that this is a constant independent of x. To do this, consider the subspace of $\mathrm{X}\times\mathrm{X}$

$$\mathrm{D} := \{(x,y) \in \mathrm{X}\times\mathrm{X}\,;\ \mathrm{Sp}(x) = \mathrm{Sp}(y)\}\,,$$

in Berthelot's language this is the tube of the points reducing to the diagonal. This is a rigid analytic space and Coleman shows [Col85b, Proposition 1.2] that there exists a rigid analytic function H on D such that $\mathrm{dH} = \pi_y^*\omega - \pi_x^*\omega$, where π_x and π_y are the projections on the two coordinates. The pullback to the diagonal of H is thus constant, and may be assumed 0. It follows that $\mathrm{H}(x,y) = \int_x^y \omega$. The two lifts ϕ and

ϕ', having the same reduction, define a map $\Phi = (\phi, \phi') : \mathrm{X} \to \mathrm{T}$ and $\phi = \pi_x \circ \Phi$, $\phi' = \pi_y \circ \Phi$ on X. Therefore, we may take $h(x) = \mathrm{H}(\Phi(x)) = \int_{\phi(x)}^{\phi'(x)} \omega$. □

Example 10. Let us demonstrate the above on the example from the introduction. Our dagger algebra is

$$\mathrm{A}^\dagger = \mathcal{T}_2^\dagger/(t_1 t_2 - 1)$$

over $\mathbb{Z}_p$. Setting $t = t_1$ and $t^{-1} = t_2$ we have

$$\mathrm{A}^\dagger \cong \left\{ \sum_{i \in \mathbb{Z}} a_i t^i \ ; \ a_i \in \mathbb{Z}_p\,, \ \lim_{|i| \to \infty} |a_i| r^{|i|} = 0 \text{ for some } r > 1 \right\}$$

and the module of 1-forms is $\Omega^1_{\mathrm{A}^\dagger} = \mathrm{A}^\dagger dt$. The associated Monsky-Washnitzer cohomology $\mathrm{H}^1(\Omega^\bullet_{\mathrm{A}^\dagger} \otimes \mathrm{K})$ is clearly one-dimensional, generated by the form $\omega = \frac{dt}{t}$. Since it is clear how to integrate exact forms, it suffices to integrate ω. The integral is to be a function on the space associated with the algebra $\mathrm{A} = \mathrm{T}_2/(t_1 t_2 - 1)$ which is just $\{z \in \bar{\mathbb{Q}}_p \ ; \ |z| = 1\}$, see Sect. 1.3.1. Finally, we may take the lift of Frobenius ϕ_a such that $\phi_a(t) = t^p$.

For the computation of the Coleman integral F_ω of ω we notice that, as in the introduction, $\phi_a \omega = p\omega$. Thus, we may pick our integral so that when evaluating at a point x we have $\sigma(\mathrm{F}_\omega(\phi_a(x))) = p\mathrm{F}_\omega(x)$, where here $\phi_a(x) = \sigma^{-1}(x^p)$. We can now either proceed with a general x as in the proof of the theorem or, as in the introduction, consider an x which is a root of unity of order prime to p. In this case it is easy to see that $\phi_a(x) = x$ and so one finds the relation $\sigma(c) = pc$ for $c = \mathrm{F}_\omega(x)$. Now, if $\sigma^l(c) = c$ we find $c = p^l c$ so $c = 0$. Thus, we again discover that our integral vanishes at all these roots of unity.

Remark 11. (1) Note that we have used the semi-linear approach here, whereas in the introduction we used the linear approach.

(2) It is interesting to note that the equation $\sigma(c) = pc$ yields $c = 0$ even without assuming a finite residue field, because it implies that $\sigma(c)$, hence c, are divisible by p and iterating we find that c is divisible by any power of p hence is 0. This suggests an interesting alternative to Coleman integration, applicable when all slopes are positive, using slopes rather than weights, that works without assuming finite residue fields. It also works for example for polylogarithms. We plan to come back to this method in future work.

To end this section, let us sketch how one may define iterated integrals using an extension of the method above. Note that this differs from the method of [Col82] and [CdS88] and is again geared towards computational applications. A similar method to the one sketched above is worked out (in progress) by Balakrishnan.

As explained in the introduction, prior to the introduction of isocrystals into Coleman integration, iterated integrals were only defined on one-dimensional spaces. This restriction means that any form is closed and can therefore be integrated. Let us explain then how one can define integrals $\int(\omega \times \int \eta)$ for ω and η in $\Omega^1_{\mathrm{A}^\dagger} \otimes \mathrm{K}$, where the space X^{geo} is one-dimensional. More complicated iterated integrals are derived in exactly the same manner.

We begin by observing that when $\eta = df$ is exact, then the above integral is just $\int f\omega$ which has already been defined. To proceed, we will impose an additional condition, which is the integration by parts formula

$$\int\left(\omega\times\int\eta\right)+\int\left(\eta\times\int\omega\right)=\left(\int\omega\right)\times\left(\int\eta\right)+C, \tag{1.7}$$

see Remark 33 for a justification of this formula. Using this formula and our knowledge of $\int\omega$ and $\int\eta$ we can also compute $\int(\omega\times\int\eta)$ when ω is exact.

Consider again a basis $\omega_1,\ldots,\omega_n \in \Omega^1_{A^\dagger}\otimes K$. Decomposing both ω and η as in (1.4) and using the above it is sufficient to compute the integrals $\int(\omega_i\times\int\omega_j)$ for all pairs (i,j). If M is the matrix satisfying (1.5), then $M\otimes M$ is the matrix describing the action of ϕ_a on the basis $\{\omega_i\otimes\omega_j\}$ of $H^1(\Omega^\bullet_{A^\dagger}\otimes K)\otimes H^1(\Omega^\bullet_{A^\dagger}\otimes K)$. Eigenvalues of (appropriately linearized) $M\otimes M$ are just products of eigenvalues of M (again linearized), and they are again Weil numbers of positive weight. Thus, the same arguments used for proving Theorem 7 may be used to obtain iterated integrals.

1.5 Coleman Integration via Isocrystals

In this section we explain the approach to Coleman integration using isocrystals introduced in [Bes02]. We comment that the approach there works globally as well, but we only explain it in the affine, more precisely, the affinoid situation, in which we described Coleman's work.

The main idea is that the iterated integral

$$\int(\omega_n\int(\omega_{n-1}\int(\cdots\int\omega_1)\cdots))$$

is the y_n coordinate of a solution of the system of differential equations

$$dy_0=0,\ dy_1=\omega_1 y_0,\ldots dy_n=\omega_n y_{n-1} \tag{1.8}$$

or, in vector notation

$$\mathbf{dy}=\Omega\mathbf{y}\ ,\ \Omega=\begin{pmatrix}0&0&0&\cdots&0\\ \omega_1&0&0&\cdots&0\\ 0&\omega_2&0&\cdots&0\\ 0&0&\omega_3&\cdots&0\\ \cdots&\cdots&\cdots&\cdots&\cdots\\ 0&\cdots&0&\omega_n&0\end{pmatrix},$$

with $y_0 = 1$. This is just a unipotent differential equation. The Frobenius equivariance condition can now be interpreted as saying that we have a system $\mathbf{y}$ of *good* local solutions for this equation, in such a way that $\phi\mathbf{y}$ is a *good* system of solutions for the equation $\mathrm{d}\mathbf{y} = \phi(\Omega)\mathbf{y}$. This, as well as the independence of the choice of the lift of Frobenius, turns out to be very nicely explained by the Tannakian formalism of unipotent isocrystals.

1.5.1 The Tannakian Theory of Unipotent Isocrystals

We assume familiarity with the basic theory of neutral Tannakian categories. The standard reference is [DM82].

Definition 12. A unipotent isocrystal on $\bar{\mathrm{A}}$ is an $\mathrm{A}_{\mathrm{K}}^{\dagger}$-module M together with an integrable connection

$$\nabla : \mathrm{M} \to \mathrm{M} \otimes_{\mathrm{A}_{\mathrm{K}}^{\dagger}} \Omega^1_{\mathrm{A}^{\dagger}} \otimes \mathrm{K}$$

which is an iterated extension of trivial connections, where trivial means the object

$$\mathbb{1} := (\mathrm{A}_{\mathrm{K}}^{\dagger}, \mathrm{d}) .$$

We first observe that the module M is in fact free, because it is an iterated extension of $\mathrm{A}_{\mathrm{K}}^{\dagger}$, which is obviously split.

A morphism of unipotent isocrystals is just a map of $\mathrm{A}_{\mathrm{K}}^{\dagger}$-modules which is horizontal, meaning that it commutes with the connection.

We denote the category of unipotent isocrystals on $\bar{\mathrm{A}}$ by $\mathcal{Un}(\bar{\mathrm{A}})$. It is a basic fact of the theory [Ber96, (2.3.6) and following paragraph] that, as the notation suggests, the category depends only on $\bar{\mathrm{A}}$ and not on the particular choice of lift $\mathrm{A}^{\dagger}$.

Example 13. Let $\mathrm{M} \in \mathcal{Un}(\bar{\mathrm{A}})$ have rank 2. Then it sits in a short exact sequence

$$0 \to \mathbb{1} \to \mathrm{M} \to \mathbb{1} \to 0$$

which is non-canonically split. It is thus isomorphic to the object having underlying module $\mathrm{A}_{\mathrm{K}}^{\dagger\,2}$ and connection

$$\nabla = \mathrm{d} - \left(\begin{smallmatrix} 0 & 0 \\ \omega & 0 \end{smallmatrix}\right) .$$

By associating with M the class of ω in $\mathrm{H}^1(\Omega^{\bullet}_{\mathrm{A}^{\dagger}} \otimes \mathrm{K}) = \mathrm{H}^1_{\mathrm{MW}}(\bar{\mathrm{A}}/\mathrm{K})$ it is easy to check that one obtains a bijection

$$\mathrm{Ext}^1_{\mathcal{Un}(\bar{\mathrm{A}})}(\mathbb{1}, \mathbb{1}) \cong \mathrm{H}^1_{\mathrm{MW}}(\bar{\mathrm{A}}/\mathrm{K}) .$$

Theorem 14. *The category $\mathcal{Un}(\bar{\mathrm{A}})$ is a rigid abelian tensor category.*

To see this, assuming the corresponding result [Cre92, p. 438] for the category of all overconvergent isocrystals, one follows the proof of [CLS99, 2.3.2] which discusses

F-isocrystals but the proof is word for word the same, to show that $\mathcal{U}n(\bar{\mathrm{A}})$ is closed under sub and quotient objects, tensor products and duals in the category of all overconvergent isocrystals.

To make $\mathcal{U}n(\bar{\mathrm{A}})$ into a neutral Tannakian category what is missing is a fiber functor, i.e., an exact faithful functor into K-vector spaces preserving the tensor structure. We can associate such a functor with each κ-rational point as follows.

Definition 15. Let $x \in \mathrm{X}_\kappa(\kappa)$ be a rational point. We associate with it the functor

$$\omega_x : \mathcal{U}n(\bar{\mathrm{A}}) \to \mathrm{Vec}_\mathrm{K}\,,\ \omega_x(\mathrm{M},\nabla) = \{v \in \mathrm{M}(\mathrm{U}_x), \nabla(v) = 0\}$$

where U_x is the residue disc of x and $\mathrm{M}(\mathrm{U}_x)$ consists of the sections of M on the rigid analytic space U_x.

The fact that ω_x is indeed a fiber functor is quite standard. The key point to observe is the following: a precondition for a functor such as ω_x to be a fiber functor is that the dimension of $\omega_x(\mathrm{M},\nabla)$ equals the rank of M. For a general differential equation there is no reason why this should be the case and one introduces a condition of overconvergence, which among other things guarantees this. A unipotent isocrystal is always overconvergent. It is, however, easy to see without knowing this that indeed $\omega_x(\mathrm{M},\nabla)$ has the right dimension for a unipotent ∇ simply because finding horizontal sections amounts to iterated integration and one can integrate power series converging on the unit open polydisc to power series with the same property as the algebra of power series converging on the open polydisc of radius 1 has trivial de Rham cohomology.

In the general theory of overconvergent isocrystals one can realize the functor ω_x as simply the pullback x^* to an isocrystal on Spec(κ), see the remark just before Lemma 1.8 in [Cre92].

The general theory of Tannakian categories [DM82] tells us that the category $\mathcal{U}n(\bar{\mathrm{A}})$ together with the fiber functor ω_x determine a fundamental group

$$\mathrm{G} = \mathrm{G}_x = \pi_1(\mathcal{U}n(\bar{\mathrm{A}}), \omega_x)$$

which is an affine proalgebraic group, and an equivalence of categories between $\mathcal{U}n(\bar{\mathrm{A}})$ and the category of finite dimensional K-algebraic representations of G. We recall that G represents the functor that sends a K-algebra F to the group

$$\begin{aligned}\mathrm{Aut}^{\otimes}(\omega_x \otimes \mathrm{F}) := \{\mathrm{M} \in \mathcal{U}n(\bar{\mathrm{A}}) \to (&\alpha_\mathrm{M} : \omega_x(\mathrm{M}) \otimes \mathrm{F} \to \omega_x(\mathrm{M}) \otimes \mathrm{F})\,, \\ &\alpha_\mathrm{M} \text{ natural isomorphism and} \\ &\alpha_{\mathrm{M}\otimes\mathrm{N}} = \alpha_\mathrm{M} \otimes \alpha_\mathrm{N}\,,\ \alpha_{\mathbb{1}} = \mathrm{id}\}\,.\end{aligned} \tag{1.9}$$

The description of the Lie algebra $\mathfrak{g}$ of G is well known. Consider the algebra $\mathrm{K}[\varepsilon]$ of dual numbers where $\varepsilon^2 = 0$. Then $\mathfrak{g}$ is just the tangent space to G at the origin and is thus given by

$$\mathfrak{g} = \mathrm{Ker}(\mathrm{G}(\mathrm{K}[\varepsilon]) \to \mathrm{G}(\mathrm{K}))\,.$$

In terms of the description (1.9) to G an element $\alpha \in \mathfrak{g}$ sends $M \in \mathcal{Un}(\bar{A})$ to

$$\alpha_M = \mathrm{id} + \epsilon\beta_M \,,\ \beta_M \in \mathrm{End}(\omega_x(M)) \,.$$

Such an element is automatically invertible. The conditions on the α_M easily translate to conditions on the β_M and we obtain

$$\begin{aligned}\mathfrak{g} = \{(M \to \beta_M \in \mathrm{End}(\omega_x(M)))\,, \\ \beta_M \text{natural}, \beta_{\mathbb{1}} = 0\,, \\ \beta_{(M\otimes N)} = \beta_M \otimes \mathrm{id}_{\omega_x(N)} + \mathrm{id}_{\omega_x(M)} \otimes \beta_N\}\,.\end{aligned}$$

The Lie bracket is given in this representation by the commutator.

Lemma 16. *The elements of* G *are unipotent and the elements of* $\mathfrak{g}$ *are nilpotent in the sense that for every* $M \in \mathcal{Un}(\bar{A})$ *the corresponding* α_M *is unipotent and the corresponding* β_M *is nilpotent.*

Proof. Choose a flag $M = M_0 \supset M_1 \supset \cdots$ with trivial consecutive quotients. Then the naturality of α and β implies that with respect to a basis compatible with the associated flag on $\omega_x(M)$ the matrices of α_M and β_M are upper triangular, with 1 respectively 0 on the diagonal. □

It follows that there is well-defined algebraic exponential map $\exp : \mathfrak{g} \to G(K)$ sending β_M to $\exp(\beta_M)$ given by the usual power series. Tensoring with an arbitrary K-algebra we can easily see, using the fact that K has characteristic 0, that exp induces an isomorphism of affine schemes from the affine space associated with $\mathfrak{g}$ to G. The product structure on G translates in $\mathfrak{g}$ to the product given by the Baker-Campbell-Hausdorff formula. It is further clear that the following holds.

Proposition 17. *The reverse operations of differentiation and exponentiation give an equivalence between the categories of algebraic representations of* G *and continuous Lie algebra representations of* $\mathfrak{g}$.

Here, continuous representation means with respect to the discrete topology on the representation space and with respect to the inverse limit topology on $\mathfrak{g}$.

1.5.2 The Frobenius Invariant Path

Consider now two κ-rational points $x, z \in X_\kappa$. Then we have a similarly defined space of paths $P_{x,z} := \mathrm{Iso}^{\otimes}(\omega_x, \omega_z)$ (same functoriality and tensor conditions) which is clearly a right principal homogeneous space for G_x (and a left one for G_z, note that in [Bes02] the directions are wrong). In concrete terms, the path space $P_{x,z}$ consists of rules for "analytic continuation" for each unipotent differential equation (M, ∇), of a solution, i.e., horizontal section, $\mathbf{y}_x \in M(U_x)^{\nabla=0}$ to $\mathbf{y}_z \in M(U_z)^{\nabla=0}$ compatible with morphisms and tensor products. Composition of paths

$$\mathrm{P}_{x,z} \times \mathrm{P}_{z,w} \to \mathrm{P}_{x,w} \tag{1.10}$$

is derived from composition of natural transformations.

Suppose now that $\bar{f} : \bar{\mathrm{B}} \to \bar{\mathrm{A}}$ is a morphism. The pullback $\bar{f}^*$ in the geometric sense is a tensor functor from $\mathcal{U}n(\bar{\mathrm{B}})$ to $\mathcal{U}n(\bar{\mathrm{A}})$. We have a natural isomorphism of functors

$$\omega_{\bar{f}(x)} \to \omega_x \circ \bar{f}^* , \tag{1.11}$$

which is compatible with the tensor structure. This is obvious from the general theory since, as one may recall, we interpreted ω_x as the pullback x^* to Spec(κ). To translate into concrete terms choose a lifting $f : \mathrm{B}^\dagger \to \mathrm{A}^\dagger$ of $\bar{f}$. Then the assumptions imply that f maps U_x to $\mathrm{U}_{\bar{f}(x)}$ and the isomorphisms is obtained by composition with f of the horizontal sections on $\mathrm{U}_{\bar{f}(x)}$.

It is easy to see that $\bar{f}$ induces a map $\bar{f} : \mathrm{P}_{x,z} \to \mathrm{P}_{\bar{f}(x),\bar{f}(z)}$. In concrete terms, suppose that $\alpha \in \mathrm{P}_{x,z}$ (over some extension algebra) is a rule for analytic continuation of solutions from U_x to U_z, then $\bar{f}(\alpha)$ is a rule for analytic continuation from $\mathrm{U}_{\bar{f}(x)}$ to $\mathrm{U}_{\bar{f}(z)}$ given as follows. Start from a horizontal section in $\mathrm{M}(\mathrm{U}_{\bar{f}(x)})$, pull back by f to obtain a horizontal section of $\bar{f}^*(\mathrm{M})$ on U_x, apply the rule α to obtain a horizontal section on U_z and finally apply the inverse of pullback by f. It is formally checked that $\bar{f}$ is compatible with composition of paths (1.10). In particular, when $x = z$, $\bar{f} : \mathrm{G}_x \to \mathrm{G}_{\bar{f}(x)}$ is a group homomorphism and in general it is compatible with the structure of $\mathrm{P}_{x,z}$ as a principal homogeneous space for G_x.

Suppose now that $\bar{f} : \bar{\mathrm{A}} \to \bar{\mathrm{A}}$ and $\bar{f}$ fixes both x and z. Then we can check what it means for a path $\alpha \in \mathrm{P}_{x,z}$ to be fixed by $\bar{f}$. The analytic continuation α has the property that the following diagram commutes,

$$\begin{array}{ccc} \omega_x(\mathrm{M}) & \xrightarrow{\alpha_{\mathrm{M}}} & \omega_z(\mathrm{M}) \\ \downarrow & & \downarrow \\ \omega_x(\bar{f}^*\mathrm{M}) & \xrightarrow{\alpha_{\bar{f}^*\mathrm{M}}} & \omega_z(\bar{f}^*\mathrm{M}) \end{array}$$

where the vertical maps are the isomorphisms of (1.11). Even more concretely, restricting to the differential equation (1.8), α translates a solution $\mathbf{y}_x$ on U_x to a solution $\mathbf{y}_z$ on U_z in such a way that it now translates the local solution $f^*\mathbf{y}_x$ to the system

$$dy_0 = 0,\ dy_1 = (f^*\omega_1)y_0, \ldots, dy_n = (f^*\omega_n)y_{n-1}$$

on U_x to the solution $f^*\mathbf{y}_z$ on U_z. In particular, if we think of a collection of solutions to $dy_0 = 0$, $dy_1 = \omega y_0$, with $y_0 = 1$, compatible under α as an integral of ω, then the path α provides such an integral for each closed one-form ω in such a way that $\int f^*\omega = f^* \int \omega$, plus a constant arising from the choice of which solutions to extend. When $\bar{f}$ is a κ-linear Frobenius this is exactly what we want our Coleman integration to do. Thus, it is clear that the following theorem provides the sought after generalization of Coleman integration.

Theorem 18 ([Bes02, Corollary 3.3]). *Suppose that φ is a κ-linear Frobenius fixing the two κ-rational points x and z. Then there exists a unique $\gamma_{x,z} \in \mathrm{P}_{x,z}(\mathrm{K})$ fixed by φ. Furthermore, these paths are compatible under raising φ to some power and under composition.*

Note that we are now denoting by φ the linear Frobenius, whereas previously it was the semi-linear Frobenius, as now we have abandoned the semi-linear point of view. The proof of Theorem 18 is more or less an immediate consequence of the following theorem.

Theorem 19 ([Bes02, Theorem 3.1]). *Let φ be as above, fixing the rational point x. Then the map $g \mapsto \varphi(g)^{-1}g$ from G_x to itself is an isomorphism of schemes.*

We first prove that Theorem 19 implies Theorem 18. Clearly, the theorem implies that $g \mapsto g\varphi(g)^{-1}$ is an isomorphisms as well. Since G_x is unipotent, there exists a K-rational point $\gamma' \in \mathrm{P}_{x,z}(\mathrm{K})$ [Ser97, Prop. III.6]. Let $g' \in \mathrm{G}_x(\mathrm{K})$ be such that $\varphi(\gamma') = \gamma' g'$ and let $g \in \mathrm{G}_x(\mathrm{K})$ be the element, whose uniqueness and existence is guaranteed by Theorem 19, such that $g' = g\varphi(g)^{-1}$. Let $\gamma = \gamma' g$. Then

$$\varphi(\gamma) = \varphi(\gamma')\varphi(g) = \gamma' g' \varphi(g) = \gamma' g = \gamma$$

proving existence. On the other hand, if both γ and γ' are fixed by φ and if $\gamma = \gamma' g$, then $\varphi(g) = g$ and by the uniqueness in Theorem 19 we have that g is the identity element and $\gamma' = \gamma$. The compatibility with respect to raising Frobenius to some power and with respect to composition are both obvious from the uniqueness.

For the proof of Theorem 19 we need to study in more detail the Lie algebra $\mathfrak{g}$. As the group G is pro-algebraic, it can be written as an inverse limit of algebraic groups $\varprojlim_\alpha \mathrm{G}_\alpha$. Its Lie algebra can thus be written as an inverse limit of finite-dimensional Lie algebras

$$\mathfrak{g} = \varprojlim_{\alpha} \mathfrak{g}/\mathfrak{g}_\alpha$$

with some indexing set of α's. We consider the lower central series of $\mathfrak{g}$ obtained as follows:

$$\mathfrak{g}_1 = [\mathfrak{g}, \mathfrak{g}]\,,\ \mathfrak{g}_{n+1} = [\mathfrak{g}, \mathfrak{g}_n]\,.$$

Here, the commutators should be taken in the topological sense, i.e., completed.

Proposition 20 (Wildeshaus [Wil97, p. 32]). *There is a canonical isomorphism*

$$\mathfrak{g}/\mathfrak{g}_1 \to \mathrm{Ext}^1_{\mathcal{U}n(\bar{\mathrm{A}})}(\mathbb{1}, \mathbb{1})^*\,.$$

Proof. We exhibit a natural pairing $\mathfrak{g} \times \mathrm{Ext}^1_{\mathcal{U}n(\bar{\mathrm{A}})}(\mathbb{1}, \mathbb{1}) \to \mathrm{K}$ as follows. Consider $\ell \in \mathfrak{g}$ and an extension

$$0 \to \mathbb{1} \to \mathrm{M} \to \mathbb{1} \to 0\,.$$

When applying ω_x we can use a compatible basis to write the matrix of ℓ on $\omega_x(\mathrm{M})$ as $\left(\begin{smallmatrix} 0 & \alpha \\ 0 & 0 \end{smallmatrix}\right)$. Then the pairing will send (ℓ, M) to α, note that this is independent of the basis chosen. Since the commutator of two matrices of the form $\left(\begin{smallmatrix} 0 & * \\ 0 & 0 \end{smallmatrix}\right)$ is 0, and since the representation of $\mathfrak{g}$ on $\omega_x(\mathrm{M})$ is continuous by Proposition 17, it is clear that the pairing factors via $(\mathfrak{g}/\mathfrak{g}_1)$.

To establish the isomorphism of the Proposition we need to use the full force of Tannakian duality, that is the part of theory implying that the category $\mathcal{U}n(\bar{\mathrm{A}})$ is equivalent to the category of continuous Lie algebra representations of $\mathfrak{g}$. Thus, if the extension M is in the kernel of the pairing, it corresponds to a trivial Lie algebra representation and is therefore trivial. In the reverse direction, suppose that $a : \mathfrak{g}/\mathfrak{g}_1 \to \mathrm{K}$ is a continuous functional. It thus extends to a functional $a : \mathfrak{g} \to \mathrm{K}$ which is continuous and which vanishes on all commutators. It follows easily that

$$\ell \to \begin{pmatrix} 0 & a(\ell) \\ 0 & 0 \end{pmatrix}$$

is a continuous Lie algebra representation of $\mathfrak{g}$, which is an extension of the required type and gives back a when pairing with it. It follows that $\mathrm{Ext}^1(\mathbb{1},\mathbb{1})$ is isomorphic to the continuous dual of $\mathfrak{g}/\mathfrak{g}_1$. As $\mathrm{Ext}^1(\mathbb{1},\mathbb{1})$ is finite dimensional, it follows that so is $\mathfrak{g}/\mathfrak{g}_1$ and they are dual as discrete vector spaces. □

Proposition 21. *The quotients $\mathfrak{g}_n/\mathfrak{g}_{n+1}$ are finite-dimensional and the commutator induces a surjective map*

$$[\,] : \mathfrak{g}/\mathfrak{g}_1 \otimes \mathfrak{g}_{n-1}/\mathfrak{g}_n \to \mathfrak{g}_n/\mathfrak{g}_{n+1}\,. \tag{1.12}$$

Proof. We prove this by induction. The case $n = 0$ for the finiteness follows from the previous Proposition. The Jacobi identity immediately implies that $[\mathfrak{g}_1,\mathfrak{g}_{n-1}] \subset \mathfrak{g}_{n+1}$ and by definition $[\mathfrak{g},\mathfrak{g}_n] = \mathfrak{g}_{n+1}$. Thus, the map (1.12) is defined. Suppose we already showed that $\mathfrak{g}_k/\mathfrak{g}_{k+1}$ is finite for $k < n$. To show surjectivity (which is not obvious because we are taking completed brackets) we can choose complementary subspaces V and W for $\mathfrak{g}_1$ in $\mathfrak{g}$ and for $\mathfrak{g}_n$ in $\mathfrak{g}_{n-1}$ respectively, which are finite-dimensional by the induction hypothesis. Surjectivity follows if we show that the inclusion $[\mathrm{V},\mathrm{W}] + \mathfrak{g}_{n+1} \subset \mathfrak{g}_n$ is an equality. But this is clearly the case after completion and so we are done because the sum of a finite-dimensional subspace and a closed subspace is closed (prove this!). Finally, the surjectivity immediately proves that $\mathfrak{g}_n/\mathfrak{g}_{n+1}$ is finite-dimensional again. □

Corollary 22. *For every n the quotient $\mathfrak{g}/\mathfrak{g}_n$ is finite-dimensional.*

Proposition 23. *The topology on $\mathfrak{g}$ induced by the $\mathfrak{g}_n$ is stronger than the $\mathfrak{g}_\alpha$ topology.*

Proof. For each α the Lie algebra $\mathfrak{g}/\mathfrak{g}_\alpha$ is a finite-dimensional nilpotent Lie algebra, implying that for a sufficiently large n its lower central series vanishes, from which it follows that $\mathfrak{g}_n \subset \mathfrak{g}_\alpha$. □

Now we use again the action of a κ-linear Frobenius φ. By functoriality it induces a continuous endomorphism of $\mathfrak{g}$. It therefore clearly preserves the filtration $\mathfrak{g}_n$ and induces an endomorphism on the quotients $\mathfrak{g}/\mathfrak{g}_n$ and $\mathfrak{g}_n/\mathfrak{g}_{n+1}$.

Proposition 24. *The eigenvalues of φ on $\mathfrak{g}/\mathfrak{g}_n$ and $\mathfrak{g}_n/\mathfrak{g}_{n+1}$ have strictly negative weights.*

Proof. This follows for $\mathfrak{g}_n/\mathfrak{g}_{n+1}$ because φ has positive weights on $\mathrm{Ext}^1(\mathbb{1},\mathbb{1}) = \mathrm{H}^1_{\mathrm{MW}}(\bar{\mathrm{A}}/\mathrm{K})$ hence negative weights on its dual $\mathfrak{g}/\mathfrak{g}_1$, and by Proposition 21 we have a surjective map $(\mathfrak{g}/\mathfrak{g}_1)^{\otimes n+1} \to \mathfrak{g}_n/\mathfrak{g}_{n+1}$, compatible with φ. Since $\mathfrak{g}/\mathfrak{g}_n$ has a filtration whose quotients are of the form $\mathfrak{g}_k/\mathfrak{g}_{k+1}$ the result follows. □

Corollary 25. *The map* $\varphi - \mathrm{id}$ *is invertible on* $\mathfrak{g}/\mathfrak{g}_n$ *and* $\mathfrak{g}_n/\mathfrak{g}_{n+1}$.

Proof of Theorem 19. For simplicity we prove bijectivity on K-rational points. Since the proof relies on the Lie algebra it will work for any extension.

We begin with injectivity. Suppose that $\varphi(g) = g$ for some $g \neq 1$. Then $g = \exp(\ell)$ for some $0 \neq \ell \in \mathfrak{g}$ and since exp is an isomorphism compatible with φ we have $\varphi(\ell) = \ell$. But, by Proposition 23, for some sufficiently large n the image of ℓ in $\mathfrak{g}/\mathfrak{g}_n$ is non-zero and is therefore an eigenvector for φ with eigenvalue 1 contradicting Corollary 25.

To prove surjectivity, let $g' = \exp(\ell') \in \mathrm{G}(\mathrm{K})$. Define a sequence $\ell_n \in \mathfrak{g}_n$ as follows. Set $\ell_0 = \ell'$. Suppose we have defined ℓ_n. Consider the function

$$f(k) = \exp^{-1}\left(\exp(\varphi(k))^{-1}\exp(\ell_n)\exp(k)\right) = \ell_n + k - \varphi(k) + \text{ commutators}$$

Since $1 - \varphi$ is invertible on $\mathfrak{g}_n/\mathfrak{g}_{n+1}$ by Corollary 25 we can find $k_n \in \mathfrak{g}_n$ such that $\ell_{n+1} := f(k_n) \in \mathfrak{g}_{n+1}$. Now let

$$g_n = \exp(k_0)\exp(k_1)\cdots\exp(k_n)\,.$$

Then

$$(\varphi(g_n))^{-1}g'g_n = \exp(\ell_{n+1})\,,$$

and by Proposition 23 the limit $g = \lim_{n\to\infty} g_n$ exists. It follows that $(\varphi(g))^{-1}g'g = 1$ or $g' = \varphi(g)g^{-1}$ as required. □

Remark 26. Note the similarity of the above proof with the theory in Sect. 1.4. The main point is that $\varphi - \mathrm{I}$ is invertible on tensor powers of the dual of $\mathrm{H}^1_{\mathrm{MW}}(\bar{\mathrm{A}}/\mathrm{K})$, just like Lemma 8 was responsible for the existence of Coleman integrals of holomorphic forms and a similar invertibility on $\mathrm{H}^1_{\mathrm{MW}}(\bar{\mathrm{A}}/\mathrm{K}) \otimes \mathrm{H}^1_{\mathrm{MW}}(\bar{\mathrm{A}}/\mathrm{K})$ was responsible for iterated integrals.

1.5.3 Coleman Functions

The work of the previous subsection explains how to analytically continue solutions of differential equations to get Coleman functions. The functions themselves are obtained as components of the solutions. The iterated integral

$$\int(\omega_n \int(\omega_{n-1}\int(\cdots\int\omega_1)\cdots))$$

is going to be the component y_n in a system of local horizontal solutions of the system (1.8), compatible with respect to Frobenius invariant paths. One can do this in a more streamlined way, which extends also to the non-affine case, by considering arbitrary functionals on the underlying vector bundle for a connection instead of just the projection on the last component. This gives rise to the following definition.

Definition 27. An **abstract Coleman function** on $\mathrm{A}^\dagger$ is a four tuple, which we write

$$(\mathrm{M}, \nabla, \mathbf{y}_x, s)$$

in which ∇ is a unipotent integrable connection on an $\mathrm{A}_\mathrm{K}^\dagger$-module M, $\mathbf{y}_x$ refers to a system of horizontal sections for each U_x, compatible with the Frobenius invariant paths, and $s \in \mathrm{Hom}(\mathrm{M}, \mathrm{A}_\mathrm{K}^\dagger)$.

We note that specifying for which points x one has the $\mathbf{y}_x$ does not matter. They are all derived from one of them by doing analytic continuation so one could instead just specify $\mathbf{y}_x$ for one x and this formulation is only done for symmetry. We further note that s is usually not horizontal, because a horizontal s produces a constant function. In fact, one can define a notion of Coleman functions with values in any sheaf by changing the target of s.

Definition 28. A Coleman function is made into an actual locally analytic function by evaluating the s on the $\mathbf{y}_x$'s.

Many abstract Coleman functions may produce the same function. One way in which this can happen is the following.

Definition 29. Two abstract Coleman functions $(\mathrm{M}, \nabla, \mathbf{y}_x, s)$ and $(\mathrm{M}', \nabla', \mathbf{y}'_x, s')$ are called **equivalent** if there exists a horizontal morphism $f : \mathrm{M} \to \mathrm{M}'$ carrying the $\mathbf{y}_x$'s to the $\mathbf{y}'_x$'s and such that $s = s' \circ f$. By the properties of the invariant paths it suffices to check this for one x. More generally they are called equivalent if they are related by the equivalence relation generated by the above relation. An equivalence class of abstract Coleman functions is called a Coleman function.

It is trivial to check that equivalent abstract Coleman functions give rise to the same locally analytic function, which is therefore associated to the Coleman function as just defined. It is not immediately clear, but turns out to be true, that a Coleman function inducing the 0 function is indeed equivalent to 0. This is a consequence of the identity principle, to be discussed below. There are some advantages to defining Coleman functions without reliance on a *physical* representation as a locally analytic function. One example is integration of meromorphic differentials on curves.

We denote the K-algebra of all Coleman functions by $\mathrm{A}_{\mathrm{Col}}$. Coleman functions with values in a sheaf $\mathcal{F}$ will be denoted $\mathrm{A}_{\mathrm{Col}}(\mathcal{F})$. In particular, we have degree n Coleman differential forms defined by $\Omega^n_{\mathrm{Col}} = \mathrm{A}_{\mathrm{Col}}(\Omega^n)$.

Example 30. Consider again the rank 2 unipotent isocrystal considered in Example 13 having underlying module $\mathrm{A}_\mathrm{K}^\dagger \oplus \mathrm{A}_\mathrm{K}^\dagger$ and connection

$$\nabla(y_1, y_2) = (\mathrm{d}y_1, \mathrm{d}y_2 - \omega y_1),$$

with $\omega \in \Omega^1_{\mathrm{A}^\dagger} \otimes \mathrm{K}$ closed. Choose one residue disc U_{x_0} and let $\mathbf{y}_{x_0}$ be a horizontal section on U_{x_0} whose first coordinate is 1 and use analytic continuation to extend this to a compatible system of horizontal sections $\mathbf{y}_x$ on each residue disc U_x. Since the projection on the first coordinate is a morphism of isocrystals to the trivial isocrystal it follows from the definition of the notion of a path as being compatible with morphisms that the first coordinate of each $\mathbf{y}_x$ is 1.

Let s be the projection on the second coordinate. This gives an abstract Coleman function. The associated Coleman function F has the property that when interpreted as a locally analytic function, which is the y_2 of a horizontal section $(1, y_2)$ of ∇, it satisfies $d\mathrm{F} = \omega$. This construction is unique up to a constant.

Another choice is obtained by choosing $\mathbf{y}_{x_0}$ differently. To still be a horizontal section with first coordinate 1 we can only add a constant to the second coordinate. But since $y_2 \to (0, y_2)$ is a morphism of isocrystals from the trivial isocrystal it follows again from the properties of paths that this will add the same constant to the second coordinate of each $\mathbf{y}_x$ and thus we just add the constant to F.

The function obtained in this way is exactly the Coleman integral of ω as defined in Sect. 1.4. Indeed, we have already explained before Theorem 18 why the invariance of the path with respect to Frobenius implies Frobenius equivariance for the collection of sections with respect to a linear Frobenius. Let us show why the integral of dg, with $g \in \mathrm{A}^\dagger_{\mathrm{K}}$ is just g. The reason is that the corresponding isocrystal is trivial. In fact, the map $(y_1, y_2) \mapsto (y_1, y_2 + g y_1)$ provides a horizontal isomorphism from the trivial two dimensional isocrystal and maps the horizontal section $(1,0)$ to $(1,g)$. Since taking $(1,0)$ in each residue disc is a compatible system of horizontal sections for the trivial isocrystal, it follows that $(1,g)$ is a compatible system, hence we get $\mathrm{F} = g$ as an integral of dg. We leave checking linearity to the interested reader. More generally, the theory reduces to the theory of Coleman iterated integrals when those are defined [Bes02, Sect. 5].

Many properties of Coleman functions can easily be derived from the description above. It is easy to define sums and products of Coleman functions, compatible with the same operations on locally analytic functions. It is also easy to define pullbacks of Coleman functions by a morphism $f : \mathrm{A}^\dagger \to \mathrm{B}^\dagger$,

$$f^* : \mathrm{A}_{\mathrm{Col}} \to \mathrm{B}_{\mathrm{Col}} \tag{1.13}$$

compatible with the corresponding operation on locally analytic functions.

To give an example of the properties of Coleman functions we discuss the identity principle. This was proved by Coleman for $\mathbb{P}^1$ in [Col82] and for curves by Coleman and de Shalit [CdS88]. It says the following.

Proposition 31. *Suppose that the Coleman function* F *is* 0 *on one residue disc. Then it is identically* 0.

The proof of this result is based on the following construction: We recall that part of the data for a Coleman function is a section $s : \mathrm{M} \to \mathrm{A}^\dagger_{\mathrm{K}}$. One can construct M_s, which is the maximal subconnection contained in $\mathrm{Ker}(s)$. The point is to construct it concretely as the intersection of the kernels of the section s and its derivatives of all orders with respect to the dual connection. If F vanishes on the residue disc U_{x_0}, then the local horizontal section $\mathbf{y}_{x_0}$ in the definition of F is contained in Ker(s) and since $\nabla \mathbf{y}_{x_0} = 0$ the construction of M_s implies that $\mathbf{y}_{x_0} \in M_s(U_{x_0})$. From the compatibility of analytic continuation along paths with morphisms of isocrystals it now follows that on any residue disc U_x we have $\mathbf{y}_x \in M_s(U_x)$, with $\mathbf{y}_x$ the corresponding local horizontal section. We find F to be equivalent with $(M_s, \nabla, \mathbf{y}_x, 0)$, and this is clearly equivalent to 0.

Corollary 32. *If* $dF = 0$ *then* F *is a constant function.*

Proof. The function F is a constant on some residue disc. Subtracting that constant we may assume $F = 0$ on one residue disc, hence $F = 0$ by the identity principle. □

Remark 33. The above Corollary, together with the fact that the product of Coleman functions is again a Coleman function, immediately gives the integration by parts formula (1.7).

The main result about Coleman functions is the following Theorem.

Theorem 34. *The sequence*

$$0 \to K \to A_{Col} \xrightarrow{d} \Omega^1_{Col} \xrightarrow{d} \Omega^2_{Col}$$

is exact.

Everything is already proved except for the fact that we may integrate a closed Coleman form. The idea is roughly that having a closed Coleman form ω, the condition $dF = \omega$ can be written as a new unipotent differential equation. The closedness of ω is used to find a subconnection which is integrable in addition to being unipotent, from which F can be constructed. For full details see the proof of [Bes02, Theorem 4.15], which is a more general result.

1.5.4 Tangential Base Points

One of the advantages of the Tannakian approach to Coleman integration is that new fiber functors are integrated in the theory with no extra cost. The prime example of this so far are fiber functors coming from Deligne's *tangential base points* [Del89]. In this subsection we sketch this extension. Full details may be found in the paper [BF06].

The de Rham version of Deligne's tangential base point is defined as follows [Del89, 15.28–15.30]. Suppose C is a curve over a field K of characteristic 0, smooth at a point P, with a local parameter t at P, and suppose

$$\nabla : M \to M \otimes \Omega^1_C(\log P)$$

is a connection with logarithmic singularities at P, so that locally $\nabla = d + \Gamma$ with Γ is a section of $\mathrm{End}(M) \otimes \Omega^1_C(\log P)$. One defines the residue connection on the constant vector bundle, with fiber the fiber of M at P, on the complement of 0 in the tangent line $T_P(C)$, with log singularities at $0, \infty$, by

$$\mathrm{Res}_P(\nabla) := d + (\mathrm{Res}_P \Gamma) d\log(\bar{t})$$

where $\bar{t}$ is the induced coordinate on the tangent space. Here, the residue is defined in the usual way and since we are assuming that Γ has log-singularities may simply be defined as the value of $t\Gamma$ at P. While the definition of the residue connection looks like it depends on the parameter it is in fact not the case, up to a canonical isomorphism, and Deligne gives a coordinate free description.

There is no difficulty in replacing the algebraic curve by a p-adic analytic one. Since the action of a lift of Frobenius, assumed to fix P, extends to an action on the tangent space, one can analytically continue horizontal sections of ∇ along Frobenius to horizontal section of $\mathrm{Res}_P \nabla$ on residue discs in $T_P(C) - \{0\}$. One can set up a theory of Coleman functions *of algebraic origin* where the underlying bundle and connection are algebraic with logarithmic singularities at P, in such a way that these functions now have values at the points of $T_P(C) - \{0\}$.

This turns out to be far less mysterious than one might expect. Consider a unipotent differential equation with logarithmic singularities near P. It terms of the parameter t one easily sees that it has a full set of solutions in the ring $K[[t]][\log(t)]$. Define the constant term (with respect to t) of an element in $K[[t]][\log(t)]$ by formally setting $\log(t) = 0$ and then evaluating at 0. In [BF06, Proposition 4.5] we showed that taking the constant term of a Coleman function corresponds to analytically continuing to the tangent space and evaluating at the tangent point $\bar{t} = 1$.

This is already useful for p-adic polylogarithms. Recall from the introduction that these were defined to be Coleman functions that satisfy the unipotent system of differential equations:

$$\begin{aligned} d\mathrm{Li}_1(z) &= \frac{dz}{1-z} \\ d\mathrm{Li}_n(z) &= \mathrm{Li}_{n-1}(z)\frac{dz}{z} \\ \mathrm{Li}_n(0) &= 0\,. \end{aligned}$$

The problem with this definition is that the equations have singularities at 0 and 1 and the boundary conditions are made at the singular point 0. In practice there is no problem because things are arranged in such a way that the Li_n are holomorphic at 0. Deligne pointed out in the complex case that one should interpret the boundary conditions at the singular point 0 to mean analytic continuation from the tangent vector $\bar{t} = 1$ at 0, and this holds true in the p-adic case as well. One replaces the condition $\mathrm{Li}_n(0) = 0$ by the equivalent condition that the constant term there is 0. One can use the same method to assign values to p-adic polylogarithms and multiple polylogarithms at 1.

For multiple polylogarithms, one has to consider a generalization of the notion of a tangential base point, which is also due to Deligne [Del89, 15.1–15.2]. Given a smooth variety X and a divisor $D = \sum_{i\in I} D_i$ with normal crossings and smooth components, set, for $J \subset I$, $D_J = \cap_{j\in J} D_j$. Let N_J be the normal bundle to D_J and let N_J^0 be the complement in N_J of $N_{J'}|_{D_J}$ for $J' \subset J$, and let N_J^{00} be the restriction of N_J^0 to $D_J^0 := D_J - \cup_{j\notin J} D_j$. Note that when $|I| = \dim(X) = 1$ so D is just one point P, we have $N_J^{00} = T_P(X) - \{0\}$. Deligne associates to a connection on X with logarithmic singularities along D, residue connections on every N_J^{00} with logarithmic singularities *at infinity*.

Thus we again obtain new fiber functors on the category of unipotent connections by taking the fiber of the residue at points of the spaces N_J^{00}.

Remark 35. An important observation is that some of these constructions provide naturally isomorphic fiber functors. A typical example which captures the essence of things [BF06, Prop. 3.6 and Rem. 3.7] is the following. Suppose $X = \mathbb{A}^2$ and D_i is defined by $x_i = 0$ where x_i, $i = 1,2$, are the coordinates. One can start with a connection with log singularities along $D_1 \cup D_2$, take the residue along D_1, which can be interpreted again as a connection on $\mathbb{A}^2$ with logarithmic singularities along $\overline{x}_1 = 0, x_2 = 0$, restrict to $\overline{x}_1 = 1$, take the residue at the point $x_2 = 0$ and restrict to $\overline{x}_2 = 1$. Then this is exactly the same as taking the fiber at $(1,1)$ after taking the residue to $N_{\{1,2\}}^{00}$. Consequently, it is also the same as doing the above procedure with the roles of 1 and 2 reversed.

In [BF06, Sect. 4] we proved that if we have a Coleman function of algebraic origin on X, then one can analytically continue it to the spaces N_J^{00} and furthermore one obtains Coleman functions on these spaces. One can further deduce, essentially from the definition of the residue connection, differential relations between the Coleman functions restricted to the spaces N_J^{00} from the original differential relations. Indeed. In Proposition 4.4 there we proved, for the special case of restricting to the normal bundle of one of the components E of D, that

$$df = \sum \omega_i g_i \quad \Rightarrow \quad df^{(E)} = \sum (\mathrm{Res}_E\, \omega_i) g_i^{(E)} \tag{1.14}$$

where $f^{(E)}$ is the restriction to the normal bundle to E of f and where, if ω is locally written as $\omega' + h\,d\log(t)$, with t the defining parameter for E, then

$$\mathrm{Res}_E(\omega) = \omega'|_E + h|_E d\log(\bar{t})\,.$$

1.6 Applications of Coleman Integration

In this section we survey, giving only occasional details, several applications of Coleman integration. It is not meant as an exhaustive list and reflects the author's knowledge and personal taste.

1.6.1 The p-Adic Abel-Jacobi Map

In [Col85b, Theorem 2.3] Coleman uses his theory to define integrals of closed forms of the *second kind* on a smooth complete variety X over K with good reduction. Recall that a meromorphic differential on a complete variety is of the second kind if it can be written, locally in the Zariski topology, as the sum of a holomorphic and of an exact differential. On a curve this means that all its residues vanish.

To define the integral, Coleman covers X with affinoids, each coming from some wcfg algebra on which the form can be written as holomorphic plus exact. Since there is no problem in integrating exact differentials, and since closed holomorphic forms can be integrated by Coleman's method, one gets local integrals, and one easily observes that these can be glued to give a global integral, which is defined outside the divisor of singularity of the form. Coleman uses functoriality (1.13) and the Albanese variety to show that the integral is in fact independent of auxiliary choices and depends only on X, and is furthermore functorial with respect to arbitrary algebraic maps.

In particular, one has everywhere defined integrals for holomorphic one-forms. Let X be a smooth complete curve over K and initially assume that it has good reduction. Let $f : X \to J$ be an Albanese map of the curve into its Jacobian (extend K when needed to have a K-rational point). Any holomorphic one-form on X is f^* of an invariant differential ω on J, so by functoriality knowing F_ω suffices for computing the integral on X. Functoriality again and the invariance of ω implies that if $F_\omega(0) = 0$, then in fact F_ω is a group homomorphism from $J(\bar{K})$ to the additive group of $\bar{K}$. Locally near 0 it is the integral of ω vanishing at 0.

There is another way to obtain F_ω on J, which does not involve Frobenius at all. Namely, let $T = T_0(J)$ be the tangent space to J at 0, thought of as a vector group. The exponential map $T \to J$ can be locally inverted to provide a logarithm

$$\log_J : U_0(\bar{K}) \to T(\bar{K}),$$

which is additive, where U_0 is the residue disc of 0. However, for any $x \in J(\bar{K})$ there exists an $0 \neq n \in \mathbb{Z}$ such that $nx \in U_0(\bar{K})$ and the additivity of $\log_J$ means it can be extended to a map $\log_J : J(\bar{K}) \to T(\bar{K})$ by setting $\log_J(x) = n^{-1} \log_J(nx)$. It is now very easy to see that

$$F_\omega = \omega_0 \circ \log_J, \tag{1.15}$$

where ω_0 means the value of ω at 0 viewed as a cotangent vector.

The above procedure works even when J has bad reduction [Bou98, III, 7.6] and therefore gives a way of computing Coleman integrals in the bad reduction case as well. In fact, extended to generalized Jacobians and Albanese varieties it is the basis of the approaches of Colmez and Zarhin [Col98, Zar96] for Coleman integration. It is important to note though that even if this *geometric* method can compute Coleman integrals in greater generality, though incapable of treating iterated integrals, it gives little information and is quite hard to compute in practice.

Let X be a complete curve over K and let ω be a holomorphic form on X. Given a divisor D of degree 0 on X

$$D = \sum n_i(P_i)\,,\ n_i \in \mathbb{Z}\,,\ P_i \in X(\bar{K})\,,\ \sum n_i = 0\,,$$

we may define

$$\int_D \omega = \sum n_i F_\omega(P_i)\,,$$

which is independent of the particular choice of F_ω by the condition $\sum n_i = 0$.

Proposition 36. *If* D *is principal, then* $\int_D \omega = 0$.

Proof. This is obvious since we have shown that the integral factors via the Albanese map $X \to J$. One can also show this in two other ways:

(1) In the good reduction case it is a consequence of the reciprocity law [Bes00c, 4.10], a generalization of Coleman's reciprocity law on curves [Col89b]. Namely, one easily checks that if $D = \mathrm{div}(f)$, then $\int_D \omega$ is the global index of F_ω and $\log(f)$, and is thus the cup product in $H^1_{dR}(X/K)$ of the projections of ω and df/f, but the projection of df/f is 0. This method extends to the bad reduction case [Bes05] but it then needs Coleman integration in the bad reduction case, which either uses the methods of Colmez and Zarhin or that of Vologodsky which is far more complicated.

(2) Again in the good reduction case it follows from [Bes00a] that the integral factors via the syntomic regulator on the Chow group of zero cycles on X, see below, thus killing principal divisors. □

Let $\Omega^1(X/K)$ be the space of holomorphic forms on X and let $\left(\Omega^1(X/K)\right)^*$ be its dual. The map $\mathrm{Div}_0(X) \to \left(\Omega^1(X/K)\right)^*$ given by

$$D \mapsto (\omega \mapsto \int_D \omega)$$

is formally a p-adic analogue to the Abel-Jacobi map using p-adic integrals rather than complex integrals. This analogy can be made precise in two different but related manners:

(1) The map above is nothing but the composition

$$\mathrm{Div}_0(X) \to J(K) \xrightarrow{\log_J} T(K) \cong \left(\Omega^1(X/K)\right)^*\,, \tag{1.16}$$

and by the description of the Bloch-Kato exponential map [BK90, Example 3.10.1] one sees that the composition

$$\mathrm{Div}_0(X) \to \left(\Omega^1(X/K)\right)^* \cong H^1_{dR}(X/K)/F^1 \xrightarrow{\exp} H^1(\mathrm{Gal}(\bar{K}/K), H^1_{\acute{e}t}(X \otimes \bar{K}, \mathbb{Q}_p(1)))$$

is the p-adic étale Abel-Jacobi map [Jan88].

(2) Let $\mathcal{X}$ be a smooth and proper model of X over R. It follows from the main theorem of [Bes00a] that the map above is is also the same as the composition

$$\mathrm{Div}_0(\mathrm{X}) \to \mathrm{Div}_0(\mathcal{X}) \xrightarrow{\mathrm{reg}} \mathrm{H}^1_{\mathrm{dR}}(\mathrm{X}/\mathrm{K})/\mathrm{F}^1 \cong \left(\Omega^1(\mathrm{X}/\mathrm{K})\right)^*,$$

where the last isomorphism is by Poincaré duality. The regulator here is the syntomic regulator, see below Sect. 1.6.3, which is the p-adic analogue of the Beilinson regulater to Deligne cohomology, and for 0-divisors on curves this last regulator is nothing but the Abel-Jacobi map.

1.6.2 Torsion Points on Curves and Effective Chabauty

In [Col85b, Col86, Col87] Coleman uses the considerations in the preceding subsection to get bounds on torsion points on curves. In [Col85a] he uses them to get an effective version of an old idea of Chabauty [Cha41].

Consider a smooth complete curve C over a field L and assume it has an L-rational point P_0 which may be used to get a map of C into its Jacobian J. Let $\mathrm{L} \subset \mathrm{K}$, where K is a p-adic field. Then we have a commutative diagram

$$\begin{array}{ccc} \mathrm{C(L)} & \longrightarrow & \mathrm{C(K)} \\ \downarrow & & \downarrow \\ \mathrm{J(L)} & \longrightarrow & \mathrm{J(K)}\,. \end{array}$$

In both works the idea is to get a Coleman integral F_ω of a holomorphic form ω to vanish on $\mathrm{C(L)} \cap \mathrm{J(L)}_{\mathrm{tor}} \subset \mathrm{C(K)}$ (respectively all of $\mathrm{C(L)} \subset \mathrm{C(K)}$). This imposes strong finiteness restrictions on the respective set of L-rational points.

For torsion points it follows from the description as a logarithm that any Coleman integral F_ω vanishes on them. Coleman uses this to get for example the following result:

Theorem 37 ([Col86, Theorem A]). *Let* L *be an algebraically closed field of characteristic* 0 *and let* C *be the Fermat curve with projective equation* $\mathrm{X}^m + \mathrm{Y}^m + \mathrm{Z}^m = 0$ *over* L. *Suppose that p is a prime,* $1 \le n \le 8$, $m = (p-1)/n$ *and* $m \ge 4$. *Then, if* $\mathrm{P},\mathrm{Q} \in \mathrm{C(L)}$ *are such that* $\mathrm{P} - \mathrm{Q}$ *is torsion in the Jacobian of* C, *and one of them is a cusp (meaning that one of the projective coordinates is* 0*), then so is the other.*

For effective Chabauty, the field L is a number field and $\mathrm{K} = \mathrm{L}_v$, the completion of L with respect to some place lying above p. The idea of Chabauty, reinterpreted in terms of Coleman theory, is as follows. Suppose C has genus g but the rank of J(L) is smaller than g. The image of J(L) in T(K) under $\log_{\mathrm{J}}$ is contained in the subspace generated by the images of a basis for J(L). By the assumption on the rank of J(L) this is a proper subspace, hence there exists a non-zero functional on T(K) vanishing on it. Using (1.15) there exists an invariant differential on J whose Coleman integral, with value 0 at 0, vanishes on J(L). Since $\mathrm{C(L)} \subset \mathrm{J(L)}$, we find using functoriality that there exists a form $\omega \in \Omega^1(\mathrm{X}/\mathrm{K})$ such that for any $\mathrm{P} \in \mathrm{C(L)}$

we have $\int_{P_0}^{P} \omega = 0$. Coleman's contribution is to use his method for computing $\int_{P_0}^{P} \omega$ to get effective bounds on C(L). The result is as follows.

Theorem 38 ([Col85a, § 0 (ii)]). *Under the assumptions above suppose* C *has good reduction above v and that* $p > 2g$*, and let n be the norm of v. Then* C(L) *has less than* $n + 2g(\sqrt{n}+1)$ *points.*

The importance of this work is twofold. On the one hand, it provides a very useful tool for the explicit determination of rational points on curves [Wet97, Fly97, FW99, FW01, Bru02, Bru03, Sik09]. On the other hand, it serves as an important motivation for the work of Kim, see below Sect. 1.6.5 and also in the contribution of Majid Hadian in the present volume, which may be viewed as a non-abelian version of the Coleman-Chabauty method: it attempts to find *iterated* Coleman integrals that would vanish on the rational points.

1.6.3 Syntomic Regulators

As mentioned above, syntomic cohomology is the p-adic analogue of Deligne cohomology, in a precise sense to be recalled below. Syntomic regulators are maps from algebraic K-theory or motivic cohomology into syntomic cohomology. They are the p-adic analogue of Beilinson's regulators. The relation with Coleman integration theory is twofold. On the one hand, computation of syntomic regulators often lead to Coleman integrals. On the other hand, there is an alternative approach to Coleman integration theory which uses a cohomology theory similar to syntomic cohomology.

Let $\mathcal{X}$ be a smooth R-scheme. To explain the origin of syntomic cohomology $H^i_{\mathrm{syn}}(\mathcal{X}, n)$ with twist n of $\mathcal{X}$, we note that $\mathcal{X}$ gives rise to the following cohomological data:

(1) The de Rham cohomology of its generic fiber $H^i_{\mathrm{dR}}(\mathcal{X}_K/K)$. This is a K-vector space that comes with the Hodge filtration $F^\bullet$.
(2) The rigid cohomology of its special fiber $H^i_{\mathrm{rig}}(\mathcal{X}_s/K_0)$, where $\mathcal{X}_s$ denotes the special fiber of $\mathcal{X}$. Rigid cohomology is the global version of Monsky-Washnitzer cohomology, defined by Berthelot [Ber86]. This is a K_0-vector space, where K_0 is the maximal unramifed extension of $\mathbb{Q}_p$ inside K. It carries a σ-semi-linear map φ_a.
(3) A comparison map, known as the specialization map [BB04],

$$H^i_{\mathrm{dR}}(\mathcal{X}_K/K) \xrightarrow{\mathrm{Sp}} H^i_{\mathrm{rig}}(\mathcal{X}_s/K_0) \otimes_{K_0} K\,,$$

which need not be an isomorphism in general.

Thus, the p-adic cohomologies associated with $\mathcal{X}$ form an object $H^i(\mathcal{X})$ in the category MF^f_K consisting of triples (V_0, V, Sp) where V_0 is a K_0-vector space equipped with a σ-semilinear operator φ_a, V is a filtered K-vector space and

$\mathrm{Sp}: \mathrm{V} \to \mathrm{V}_0 \otimes_{\mathrm{K}_0} \mathrm{K}$ is a K-linear map. The category $\mathrm{MF}_{\mathrm{K}}^{\mathrm{f}}$ has twists and an identity object $\mathbb{1}$. The key property of syntomic cohomology is that it sits in a short exact sequence,

$$0 \to \mathrm{Ext}^1_{\mathrm{MF}_{\mathrm{K}}^{\mathrm{f}}}(\mathbb{1}, \mathrm{H}^{i-1}(\mathcal{X})(n)) \to \mathrm{H}^i_{\mathrm{syn}}(\mathcal{X}, n) \to \mathrm{Ext}^0_{\mathrm{MF}_{\mathrm{K}}^{\mathrm{f}}}(\mathbb{1}, \mathrm{H}^i(\mathcal{X})(n)) \to 0\,.$$

This is formally like Deligne cohomology, which sits in a similar short exact sequence where the Exts are in the category of mixed Hodge structures, in which the the Betti and de Rham cohomologies of a complex variety X, together with their comparison isomorphisms, form an object. The analogy here runs somewhat deeper, see [Ban02].

The construction of syntomic cohomology is rather involved, see [Bes00b]. However, in the affine case, a modified version of it, known as the *Gros style modified rigid syntomic cohomology*, loc. cit. Definition 9.3, can be easily described in terms of what we have already done. Namely, suppose that $\mathcal{X}$ is the spectrum of the R-algebra $\mathrm{R}[t_1,\dots,t_n]/(f_1,\dots,f_m)$. Associate with this the wcfg algebra $\mathrm{A}^\dagger = \mathcal{T}_n^\dagger/(f_1,\dots,f_m)$ and consider the complex of differentials $\Omega^\bullet_{\mathrm{A}^\dagger} \otimes \mathrm{K}$ and its stupid filtrations $\Omega^{\geq n}_{\mathrm{A}^\dagger} \otimes \mathrm{K}$. We further assume that we have a lift $\phi: \mathrm{A}^\dagger \to \mathrm{A}^\dagger$, of a linear Frobenius φ, which is of degree q.

Definition 39. The Gros style modified syntomic cohomology of $\mathcal{X}$ is

$$\tilde{\mathrm{H}}^i_{\mathrm{ms}}(\mathcal{X}, n) := \mathrm{H}^i(\mathrm{MF}(1 - \frac{\phi}{q^n} : \Omega^{\geq n}_{\mathrm{A}^\dagger} \otimes \mathrm{K} \to \Omega^\bullet_{\mathrm{A}^\dagger} \otimes \mathrm{K}))\,,$$

where MF stands for the mapping fiber, i.e., the cone shifted by -1.

Actually, one should take the direct limit of the above over all powers of ϕ, with q modified accordingly, but we shall ignore this point, as the connecting maps tend to be isomorphisms in many of the relevant cases. Taking the long exact cohomology sequence associated with the mapping fiber we obtain the following short exact sequence,

$$0 \to \frac{\mathrm{H}^{i-1}_{\mathrm{MW}}(\bar{\mathrm{A}}/\mathrm{K})}{(1 - \varphi/q^n)\mathrm{F}^n\mathrm{H}^{i-1}_{\mathrm{MW}}(\bar{\mathrm{A}}/\mathrm{K})} \to \tilde{\mathrm{H}}^i_{\mathrm{ms}}(\mathcal{X}, n) \to \mathrm{H}^i(\Omega^{\geq n}_{\mathrm{A}^\dagger} \otimes \mathrm{K})^{\varphi = q^n} \to 0\,,$$

where we set

$$\mathrm{H}^i(\Omega^{\geq n}_{\mathrm{A}^\dagger} \otimes \mathrm{K}) = \begin{cases} \mathrm{H}^i_{\mathrm{MW}}(\bar{\mathrm{A}}/\mathrm{K}) & n < i \\ (\Omega^n_{\mathrm{A}^\dagger} \otimes \mathrm{K})^{\mathrm{d}=0} & n = i \\ 0 & n > i \end{cases}, \quad \mathrm{F}^n\mathrm{H}^{i-1}_{\mathrm{MW}}(\bar{\mathrm{A}}/\mathrm{K}) = \begin{cases} \mathrm{H}^i_{\mathrm{MW}}(\bar{\mathrm{A}}/\mathrm{K}) & n \leq i \\ 0 & n > i \end{cases}$$

and on the right hand side of the sequence above the comparison $\varphi = q^n$ is done in $\mathrm{F}^n\mathrm{H}^{i-1}_{\mathrm{MW}}(\bar{\mathrm{A}}/\mathrm{K})$, to which $\mathrm{H}^i(\Omega^{\geq n}_{\mathrm{A}^\dagger} \otimes \mathrm{K})$ is mapped in the obvious way.

Remark 40. To bring up an interesting analogy with the theory of Kim to be described in Sect. 1.6.5, we comment on a certain normalization issue, compare

Remark 43. It often happens that $\mathrm{H}^i(\Omega^{\geq n}_{\mathrm{A}^\dagger} \otimes \mathrm{K})^{\varphi=q^n} = 0$ so that we have an isomorphism

$$\tilde{\mathrm{H}}^i_{\mathrm{ms}}(\mathcal{X}, n) \cong \mathrm{H}^{i-1}_{\mathrm{MW}}(\bar{\mathrm{A}}/\mathrm{K})/(1-\varphi/q^n)\mathrm{F}^n\mathrm{H}^{i-1}_{\mathrm{MW}}(\bar{\mathrm{A}}/\mathrm{K}) .$$

It turns out that it is best to identify, when possible this last group with

$$\mathrm{H}^{i-1}_{\mathrm{MW}}(\bar{\mathrm{A}}/\mathrm{K})/\mathrm{F}^n\mathrm{H}^{i-1}_{\mathrm{MW}}(\bar{\mathrm{A}}/\mathrm{K})$$

via the map $1-\varphi/q^n$, of course, only if $1-\varphi/q^n$ is invertible on $\mathrm{H}^{i-1}_{\mathrm{MW}}(\bar{\mathrm{A}}/\mathrm{K})$. Note that this is true even if $\mathrm{F}^n\mathrm{H}^{i-1}_{\mathrm{MW}}(\bar{\mathrm{A}}/\mathrm{K}) = 0$. There are several reasons for making this identification. From the point of view of the general theory of syntomic cohomology this identification is the one that exists in general [Bes00b, Prop. 8.6]. It is also the one that remains stable if we replace φ by a power of it. The most important reason, though, is that in practice it is this identification that gives the nice formulas for syntomic regulators.

In complete analogy to Beilinson's conjectures [Bei85] there is a p-adic Beilinson conjecture, first formulated in a rather different way by Perrin-Riou [PR95], see also [BBdJR09], connecting syntomic regulators (whose complicated definition we do not recall here) to special values of p-adic L-functions. Thus, there is some interest in computing syntomic regulators. We recall here two cases where this computation involves Coleman integration.

The first case in fact predates the theory of syntomic regulators and was the motivation for the extension of the theory of iterated integrals to curves. Let C be a complete curve over K with good reduction and let K(C) be its function field. We recall that the second K-group $\mathrm{K}_2(\mathrm{L})$ of a field L, is generated by Steinberg symbols $\{f,g\}$ for $f,g \in \mathrm{L}^\times$ subject to some simple relations, and that $\mathrm{K}_2(\mathrm{C})$ is the subgroup of $\mathrm{K}_2(\mathrm{K}(\mathrm{C}))$ where all *tame symbols* vanish.

In [CdS88] Coleman and de Shalit use iterated Coleman integration to define a map

$$r_{p,\mathrm{C}} : \mathrm{K}_2(\mathrm{K}(\mathrm{C})) \to \Omega^1(\mathrm{C}/\mathrm{K})^*$$
$$\{f,g\} \mapsto \left(\omega \mapsto \int_{\mathrm{div}(f)} \log(g)\omega\right)$$

to the dual of the space of holomorphic forms on C. They called the restriction of this map to $\mathrm{K}_2(\mathrm{C})$ the p-adic regulator, and for elliptic curves with complex multiplication over $\mathbb{Q}$ they showed a relation with special values of the p-adic L-function. Namely, let E be such an elliptic curve with complex multiplication by the full ring of integers of a quadratic imaginary field F and with good ordinary reduction above the prime p. Fix an embedding $\bar{\mathbb{Q}} \to \bar{\mathbb{Q}}_p$. Then $p = \mathfrak{p}\bar{\mathfrak{p}}$ in F, where $\mathfrak{p}$ is fixed to be the prime ideal coming from $\bar{\mathbb{Q}}_p$. Let ψ be the Grössencharakter of F associated with E by the theory of complex multiplication and let $\pi = \psi(\mathfrak{p})$. Let ω be an invariant differential on E defined over $\mathbb{Q}$. This differential defines a p-adic period Ω_p. Consider a symbol $\{f,g\} \in \mathrm{K}_2(\mathbb{Q}(\mathrm{E}))$, where the divisors of f and g have their support in torsion points (which implies that a multiple of the symbol extends to $\mathrm{K}_2(\mathrm{E})$) and satisfying a mild technical condition. Then Coleman and de Shalit prove, loc. cit.

Theorem 5.11, the formula

$$\left(1-\frac{1}{\pi p}\right) r_{p,\mathrm{E}}(f,g)(\omega) = c_{f,g}\Omega_p \mathrm{L}_p(\psi),$$

where $c_{f,g} \in \mathbb{Q}$ and L_p is the p-adic L-function of F [Kat76].

In [Bes00c] we showed that upon identifying $\mathrm{K}_2(\mathrm{C})$ up to tensoring with $\mathbb{Q}$ with $\mathrm{K}_2(C)$, where C is an integral model of C, the p-adic regulator of Coleman and de Shalit is exactly the syntomic regulator

$$\mathrm{K}_2(C) \to \mathrm{H}^2_{\mathrm{syn}}(C,2) \cong \mathrm{H}^1_{\mathrm{dR}}(\mathrm{C/K})/\mathrm{F}^1 \cong \Omega^1(\mathrm{C/K})^*$$

where the last identification is via Poincaré duality.

The second case involves higher K-theory of number fields. Let L be such a field. Then, for $n \geq 2$ one can write a certain motivic complex

$$\tilde{\mathcal{M}}^\bullet_{(n)}(\mathrm{L}) : \tilde{\mathrm{M}}_n \to \tilde{\mathrm{M}}_{n-1} \otimes \mathrm{L}^\times_{\mathbb{Q}} \to \tilde{\mathrm{M}}_{n-2} \otimes \bigwedge^2 \mathrm{L}^\times_{\mathbb{Q}} \to \cdots \to \tilde{\mathrm{M}}_2 \otimes \bigwedge^{n-2} \mathrm{L}^\times_{\mathbb{Q}} \to \bigwedge^n \mathrm{L}^\times_{\mathbb{Q}}$$

in degrees 1 through n, where $\mathrm{L}^\times_{\mathbb{Q}} = \mathrm{L}^\times \otimes \mathbb{Q}$, whose H^1 injects, and is conjectured to be isomorphic, to motivic cohomology

$$\mathrm{H}^1_{\mathcal{M}}(\mathrm{L}, \mathbb{Q}(n)).$$

The group $\tilde{\mathrm{M}}_n = \tilde{\mathrm{M}}_n(\mathrm{L})$ is generated by symbols $\{x\}_n$, for $x \in \mathrm{L} - \{0,1\}$ (but the relations between them are unclear) and the differential in the complex is given by

$$\{x\}_k \otimes y_1 \wedge \cdots \wedge y_{n-k} \mapsto \{x\}_{k-1} \otimes x \wedge y_1 \wedge \cdots \wedge y_{n-k}.$$

In the classical case [BD92, dJ95] show that the map induced by the Beilinson regulator and an embedding $\tau : \mathrm{L} \to \mathbb{C}$,

$$\mathrm{H}^1(\tilde{\mathcal{M}}^\bullet_{(n)}(\mathrm{L})) \to \mathrm{H}^1_{\mathcal{M}}(\mathrm{L}, \mathbb{Q}(n)) \xrightarrow{\tau} \mathrm{H}^1_{\mathcal{M}}(\mathbb{C}, \mathbb{Q}(n)) \xrightarrow{\mathrm{reg}} \mathbb{C}$$

is induced by the map $\tilde{\mathrm{M}}_n \to \mathbb{C}$ sending the symbol $\{x\}_n$ to a constant times $\mathrm{P}_n(\tau(x))$, where P_n is a certain single valued version of the complex polylogarithm. In [BdJ03] we have shown that the same result holds under some integrality assumptions for the map

$$\mathrm{H}^1(\tilde{\mathcal{M}}^\bullet_{(n)}(\mathrm{L})) \to \bar{\mathbb{Q}}_p,$$

induced by the syntomic regulator and an embedding $\tau : \mathrm{L} \to \bar{\mathbb{Q}}_p$, with the single valued version of the complex polylogarithm replaced by the function

$$\mathrm{L}_n(z) + \mathrm{L}_{n-1}(z)\frac{\log(z)}{n} \text{ with } \mathrm{L}_n(z) = \sum_{m=0}^{n-1} \frac{(-1)^m}{m!} \mathrm{Li}_{n-m}(z) \log^m(z).$$

We end this subsection by recalling another relation between syntomic cohomology and Coleman integration. Let us consider, for an affine $\mathcal{X}$ as before,

$$\tilde{\mathrm{H}}^1_{\mathrm{ms}}(\mathcal{X},1) \cong \left\{(\omega,g)\ ;\ \omega \in \Omega^1_{\mathrm{A}^\dagger} \otimes \mathrm{K}\ ,\ g \in \mathrm{A}^\dagger_{\mathrm{K}}\ ,\ \mathrm{d}\omega = 0\ ,\ \mathrm{d}g = (1-\phi/q)\omega\right\}$$

One observes that the map $\tilde{\mathrm{H}}^1_{\mathrm{ms}}(\mathcal{X},1) \to (\Omega^1_{\mathrm{A}^\dagger} \otimes \mathrm{K})^{\mathrm{d}=0}$ obtained by sending (ω,g) to ω is not surjective, because its image can contain only ω whose cohomology class is killed by $1-\varphi/q$. This can be changed if we replace $1-\phi/q$ by $\mathrm{P}(\phi)$, where P is the characteristic polynomial of P on $\mathrm{H}^1_{\mathrm{MW}}(\bar{\mathrm{A}}/\mathrm{K})$. This gives the so called **finite polynomial cohomology** of [Bes00a].

Observe that an element of this cohomology is represented by a pair (ω,g) where ω is a closed one-form and $g \in \mathrm{A}^\dagger_{\mathrm{K}}$ with $\mathrm{d}g = \mathrm{P}(\phi)\omega$. Applying Coleman integration in the linear Frobenius approach we obtain $\mathrm{P}(\phi)\mathrm{F}_\omega = g + \mathrm{C}$. This is a functional relation from which F_ω can be recovered. This is in fact Coleman's original approach. It turns out that one can simply think of the cohomology class of (ω,g) as the Coleman integral of ω, and view the value of this integral at some point as pulling back the cohomology class to the point. This approach, developed in [Bes00a], generalizes to give Coleman integrals of forms of degree greater than 1.

1.6.4 Multiple Zeta Values

In this subsection we recall the classical theory of multiple zeta values and their relations and the p-adic theory of Furusho [Fur04], and we explain how the theory of tangential base points in Coleman integration, as described in Sect. 1.5.4, was used in [BF06] to obtain some relations between p-adic multiple zeta values.

For $\mathbf{k} = (k_1,\ldots,k_m)$, $k_i > 0$, $k_m > 1$, the multiple zeta value $\zeta(\mathbf{k})$ is defined as the convergent series,

$$\zeta(\mathbf{k}) = \sum_{0<n_1<\cdots<n_m} \frac{1}{n_1^{k_1}\cdots n_m^{k_m}}\,. \tag{1.17}$$

for example, for $\mathbf{k} = (k)$, $\zeta(\mathbf{k}) = \zeta(k)$ is the usual zeta value.

These numbers, already known to Euler, are of interest because of their algebraic interrelations, which are expected to reflect deep arithmetic information. The simplest types of relations are the so called series or harmonic shuffle product formulae. The easiest example of these, which we will concentrate on in this subsection, see [BF06] for the general theory, is the formula

$$\zeta(a)\zeta(b) = \zeta(a,b) + \zeta(b,a) + \zeta(a+b) \tag{1.18}$$

which one gets by writing

$$\zeta(a)\zeta(b) = \sum_{n_1,n_2>0} \frac{1}{n_1^a}\frac{1}{n_2^b}$$

and dividing the summation over the infinite square $n_1, n_2 > 0$ into the sum over the bottom and top triangles and over the diagonal.

There is another type of relation for multiple zeta values which one obtains from an integral representation of these values. To derive it, define the **k**-th multiple polylogarithm, where the index **k** can now have $k_m = 1$, by the series

$$\mathrm{Li}_{\mathbf{k}}(z) = \sum_{0<n_1<\cdots<n_m} \frac{z^{n_m}}{n_1^{k_1}\cdots n_m^{k_m}} . \tag{1.19}$$

and observe that $\zeta(\mathbf{k}) = \mathrm{Li}_{\mathbf{k}}(1)$ when $k_m > 1$. From the power series expansion one easily arrives at the following unipotent differential equation:

$$d\,\mathrm{Li}_{k_1,\dots,k_m}(z) = \begin{cases} \mathrm{Li}_{k_1,\dots,k_m-1}(z)\frac{dz}{z} & k_m \neq 1 \\ \mathrm{Li}_{k_1,\dots,k_{m-1}}(z)\frac{dz}{1-z} & k_m = 1 . \end{cases} \tag{1.20}$$

In particular, multiple polylogarithms are iterated integrals and can be written as integrals over certain triangular domains. In fact, borrowing from the description of multiple polylogarithms in terms of the KZ differential equation, associate with **k** the word $w = \mathrm{BA}^{k_1-1}\mathrm{BA}^{k_2-1}\cdots\mathrm{BA}^{k_m-1}$, and consider the differential form

$$\omega_i^w := \begin{cases} \frac{dt_i}{t_i} & \text{if } i\text{'th place in w is A} \\ \frac{dt_i}{1-t_i} & \text{otherwise} . \end{cases}$$

Then one obtains the formula

$$\zeta(\mathbf{k}) = \int_{0\le t_1\le t_2\le\cdots\le 1} \omega_1^w(t_1)\omega_2^w(t_2)\cdots .$$

This serves as a source for the *integral shuffle product formulae*. The simplest example is:

$$\zeta(2)\zeta(2) = \left(\int_{0\le t_1\le t_2\le 1} \frac{dt_1}{1-t_1}\frac{dt_2}{t_2}\right)\left(\int_{0\le s_1\le s_2\le 1} \frac{ds_1}{1-s_1}\frac{ds_2}{s_2}\right) = \int_{\substack{0\le t_1\le t_2\le 1\\ 0\le s_1\le s_2\le 1}} \Omega$$

where $\Omega = \frac{dt_1}{1-t_1}\frac{dt_2}{t_2}\frac{ds_1}{1-s_1}\frac{ds_2}{s_2}$. We can write this as a sum of six terms, depending on the inequalities between the coordinates,

$$= \int_{t_1\le t_2\le s_1\le s_2} \Omega + \int_{s_1\le s_2\le t_1\le t_2} \Omega + \int_{t_1\le s_1\le t_2\le s_2} \Omega$$

$$+ \int_{s_1\le t_1\le s_2\le t_2} \Omega + \int_{s_1\le t_1\le t_2\le s_2} \Omega + \int_{t_1\le s_1\le s_2\le t_2} \Omega .$$

The six terms are themselves iterated integrals and one finds the formula

$$\zeta(2)^2 = 2\zeta(2,2) + 4\zeta(1,3) .$$

In [Fur04] Furusho set out to develop a p-adic theory of multiple zeta values. The immediate problem is that the series (1.17) does not converge p-adically. In order to overcome this he checked that multiple polylogarithms, defined as Coleman functions using the differential equation (1.20), have a limit when z approached 1 and this limit is then defined to be the corresponding multiple zeta value. As explained in Sect. 1.5.4, one could simplify things by using the constant term.

Since p-adic multiple zeta values were defined using the multiple polylogarithm, it is perhaps not surprising that Furusho was only able to prove the analogues of the integral shuffle product formulae. The series product formulae were established in [BF06] and further work concerning generalized multiple zeta values, covering the case $k_m = 1$ as well, was later done in [FJ07]. The strategy for proving the series shuffle relation is rather simple, but certain intricacies have to be overcome by using the tangential base points and their generalizations. Again, we only deal with the simplest case, namely, the p-adic analogue for (1.18).

The natural function to consider for proving this is the two variable p-adic multiple polylogarithm defined near $(0,0)$ by

$$\mathrm{Li}_{(a,b)}(x,y) = \sum_{0<n<m} \frac{x^n y^m}{n^a m^b} .$$

One checks easily the differential relations between these functions:

$$x\frac{\mathrm{d}}{\mathrm{d}x}\mathrm{Li}_{(a,b)}(x,y) = \begin{cases} \mathrm{Li}_{(a-1,b)}(x,y) & a>1 \\ \frac{1}{x-1}(\mathrm{Li}_b(xy) - \mathrm{Li}_b(y)) & a=1 \end{cases}$$

$$y\frac{\mathrm{d}}{\mathrm{d}y}\mathrm{Li}_{(a,b)}(x,y) = \begin{cases} \mathrm{Li}_{(a,b-1)}(x,y) & b>1 \\ \frac{y}{1-y}\mathrm{Li}_a(xy) & b=1 . \end{cases}$$

Thus, one may analytically continue $\mathrm{Li}_{(a,b)}(x,y)$ to Coleman functions in two variables. Indeed, as Li_a and Li_b are Coleman functions, so are $\mathrm{Li}_a(xy)$ and $\mathrm{Li}_b(xy)$ by functoriality (1.13) so the resulting system is easily seen to be unipotent. Now, the relation

$$\mathrm{Li}_a(x)\mathrm{Li}_b(y) = \mathrm{Li}_{(a,b)}(x,y) + \mathrm{Li}_{(b,a)}(y,x) + \mathrm{Li}_{a+b}(xy)$$

is obvious, because on the power series defining these functions near $(0,0)$ it is true by the same summation proving the series shuffle product formula, hence it is true globally by the identity principle Proposition 31. Thus, to get the required formula one only needs to substitute $x = y = 1$. This is where the main difficulty in the entire argument is. It is by no means clear that $\mathrm{Li}_{(a,b)}(1,1) = \mathrm{Li}_{a,b}(1)$. Of course, the difficulty is increased by the fact that both points are singular for the differential equations defining the two functions.

To treat this difficulty, one has to work with the generalization of the notion of a tangential base point, as in Sect. 1.5.4. We apply the results there to the functions $\mathrm{Li}_{(a,b)}$. One first observes that the differential equations defining these functions ultimately have singularities along $x = 0,1,\infty$, $y = 0,1,\infty$ and $xy = 1$, where the last divisor comes from the appearance of functions like $\mathrm{Li}(xy)$ in the expressions.

Consequently, one should blow up $\mathbb{P}^1 \times \mathbb{P}^1$ at the point $(1,1)$ to make the singular locus normal crossing. The resulting space, if one blows up further the irrelevant points $(0,\infty),(\infty,0)$ is also the Deligne-Mumford compactification for the moduli space of curves of genus 0 with 5 marked points. One gets the picture in Fig. 1.1

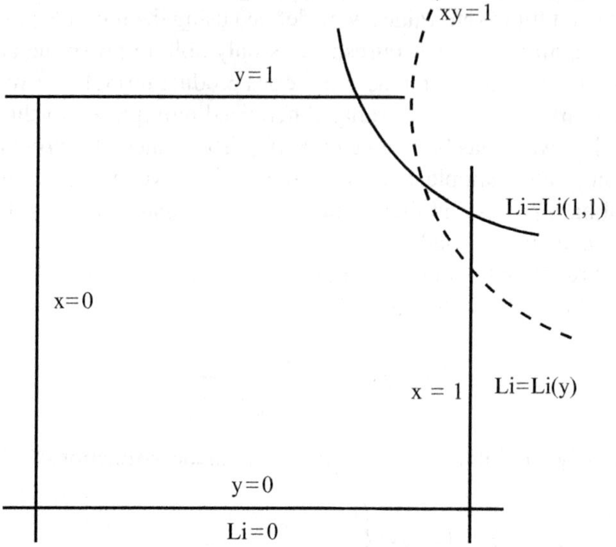

Fig. 1.1 Analytic continuation to $(1,1)$

We now try to compute $\mathrm{Li}_{(a,b)}(1,1)$. First we should interpret it as the value of $\mathrm{Li}_{(a,b)}$ on a tangent vector $(\overline{x},\overline{y}) = (1,1)$ at the point $(1,1)$. The first step in the computation of this value is to restrict to the divisor $y = 0$. By this we mean that we analytically continue to the normal bundle of $y = 0$ minus the 0 section and then restrict to the section $\overline{y} = 1$. In this case, the recipe outlined in (1.14), together with the restriction to $\overline{y} = 1$ boils down to removing the part multiplying dy and then setting $y = 0$ in the formulas. The equations are therefore going to become

$$x\frac{\mathrm{d}}{\mathrm{d}x}\mathrm{Li}_{(a,b)}(x,0) = \begin{cases}\mathrm{Li}_{(a-1,b)}(x,0) & a > 1\\ 0 & a = 1\,.\end{cases}$$

Since the boundary conditions on these functions are always set so that the constant term at 0 is 0 it follows immediately that the function $\mathrm{Li}_{(a,b)}(x,0)$ is identically 0. This is not surprising based on its expansion and the identity principle.

We now repeat the same considerations but this time restricting to the divisor $x = 1$. In the same way as before the differential equations are going to be

$$y\frac{\mathrm{d}}{\mathrm{d}y}\mathrm{Li}_{(a,b)}(1,y) = \begin{cases}\mathrm{Li}_{(a,b-1)}(1,y) & b > 1\\ \frac{y}{1-y}\mathrm{Li}_a(y) & b = 1\end{cases}$$

which are of course the same differential equations satisfied by the single variable $\mathrm{Li}_{(a,b)}$. Based on the computation at $y=0$ and on Remark 35 we see that the boundary values at $y=0$ for all of these functions are 0, which now gives the result $\mathrm{Li}_{(a,b)}(1,y)=\mathrm{Li}_{a,b}(y)$.

One cannot now just substitute at the point $(1,1)$ since that point has been blown up. However, one checks that the differential equation for restricting to the exceptional divisor formed by the blowup, with an appropriately chosen coordinate, forces $\mathrm{Li}_{(a,b)}(x,y)$ to be constant on the exceptional divisor. This makes the blowup benign and completes the proof.

It is interesting to note that in the case $b=1$ it is no longer the case that $\mathrm{Li}_{(a,b)}(x,y)$ is constant on the exceptional divisor and one therefore gets different normalizations for the multiple zeta values, depending on a chosen location on the exceptional divisor. This is worked out in [FJ07].

1.6.5 The Non-abelian Unipotent Albanese Map and the Work of Kim on Rational Points

In the last few years M. Kim developed a fascinating non-abelian analogue of the Chabauty method, and used it to obtain results, some new and some old, on rational and integral points on curves [Kim05, Kim09, Kim10a, Kim10b, CK10]. In Kim's theory, the p-adic Abel-Jacobi map is replaced by the *unipotent Albanese map*, which is intimately connected with Coleman integration theory. We describe Kim's map and the relation with Coleman theory, as well as list some results provable by Kim's method. We do not say much about the other great contribution of Kim – *Selmer varieties*, which should be viewed as the non-abelian analogues of the Jacobians in Chabauty's work. Conversations with L. Ramero contributed to our presentation.

Let $\mathcal{X}$ be a smooth R-scheme and fix on it an R-point x. We may associate with the pair $(\mathcal{X},x)$, in analogy with the situation in Sect. 1.6.3, the following data:

(1) The de Rham unipotent fundamental group $\mathrm{G}_x^{\mathrm{dR}}$ of the category of unipotent vector bundles on $\mathcal{X}_{\mathrm{K}}$ with connections with log singularities at infinity, with respect to the fiber functor of taking the fiber of the bundle at x. This is an affine group scheme over K and comes equipped with a Hodge filtration $\mathrm{F}^{\bullet}\mathrm{G}_x^{\mathrm{dR}}$, which is defined by a filtration by ideals $\mathrm{F}^{\bullet}$ on the associated Hopf algebra K[G], the filtration $\mathrm{F}^{1-i}\mathrm{K}[\mathrm{G}]$ defines $\mathrm{F}^i\mathrm{G}_x^{\mathrm{dR}}$, see [Woj93, Theorem E].

(2) The rigid fundamental group $\mathrm{G}_x^{\mathrm{rig}}$. This is the global version of the group we discussed in Sect. 1.5.1, using the definition of unipotent overconvergent isocrystals on the variety $\mathcal{X}_s$. As in Sect. 1.5.1, it is an affine group scheme over K and comes equipped with a linear Frobenius φ. One can also take the group over K_0 and use a semi-linear Frobenius but it makes little difference and for clarity we avoid this.

(3) A specialization map $\mathrm{Sp}:\mathrm{G}_x^{\mathrm{rig}}\to\mathrm{G}_x^{\mathrm{dR}}$, note the reversed direction.

Similarly, if y is another R-point, then we have spaces of paths $P^{dR}_{x,y}$ and $P^{rig}_{x,y}$, which are right torsors for the respective groups, and which have filtrations, Frobenius and specialization maps compatible with the group actions. We can make the following.

Definition 41. The **unipotent Albanese map** sends the point y to the isomorphism class of pairs of right (G^{dR}_x, G^{rig}_x) torsors with compatible additional structures represented by $(P^{dR}_{x,y}, P^{rig}_{x,y})$.

Our goal now is to make this definition concrete. Note first that the rigid side of the torsor is completely trivialized, by the Frobenius invariant path $\gamma_{x,y}$. Furthermore, the compatibility of the Hodge filtrations means that the filtration on $P^{dR}_{x,y}$ is completely determined by a path lying in F^0. The following is thus easy.

Lemma 42. *The map that sends an isomorphism class as above to the class of* $\delta_{x,y}^{-1}\mathrm{Sp}(\gamma_{x,y})$ *in* $F^0\backslash G^{dR}_x$*, where* $\delta_{x,y} \in F^0 P^{dR}_{x,y}$*, is a bijection.*

Composing with the unipotent Albanese map we obtain a map

$$\alpha : \mathcal{X}(R) \to F^0\backslash G^{dR}_x$$

which we continue to call the unipotent Albanese map. To make this even more concrete, given a function $f \in K[G^{dR}_x]$, which is invariant from the left under $F^0 G^{dR}_x$, we can evaluate f on $\alpha(y)$. This way we obtain a map $\alpha_f : \mathcal{X}(R) \to K$. In Proposition 44 we will show that $\alpha_f(y)$ is given by evaluating a certain Coleman function associated with f on y.

Remark 43. (1) The condition that f is invariant under left multiplication by F^0 is easily translated into Hopf algebra terms to mean that

$$\Delta(f) \in F^1 K[G^{dR}_x] \otimes K[G^{dR}_x],$$

with Δ the comultiplication.

(2) There is another way to classify isomorphism classes of torsors, which is suggested by Deligne [Del02]. Suppose that the specialization map is an isomorphism, as for example is the case if X is compactified such that the complement is a smooth divisor with relative normal crossings, so that we may transport φ to the de Rham torsor. Then, we may instead send the isomorphism class of $(P^{dR}_{x,y}, P^{rig}_{x,y})$ to the element $\delta_{x,y}\varphi(\delta_{x,y})^{-1}$. Different choices of $\delta_{x,y}$ mean that this element is well defined in the quotient of G^{dR}_x by the action of F^0 given by $h(g) = \varphi(h)gh^{-1}$. Let us write this quotient as $F^0\backslash_\varphi G^{dR}_x$. Clearly, the two constructions are related by the map $F^0\backslash G^{dR}_x \to F^0\backslash_\varphi G^{dR}_x$ induced by $g \mapsto \varphi(g)g^{-1}$. This should be thought of as a non-abelian analogue of the two different normalizations of syntomic cohomology, see Remark 40.

To interpret the unipotent Albanese map in terms of Coleman integration, we need to recall the construction of the the Hopf algebra $K[G^{dR}_x]$ and of the algebra of functions on path spaces [Del90]. Let G be the Tannakian fundamental group associated with the Tannakian category $\mathcal{T}$ and the fiber functor ω. The algebra K[G] may be described concretely as an *algebra of matrix coefficients*. An element in such an algebra is provided by a pair $(\mathrm{T},\mathcal{E})$ where $\mathrm{T}\in\mathcal{T}$, $\mathcal{E}\in\omega(\mathrm{T})\otimes\omega(\mathrm{T})^*$, where $\omega(\mathrm{T})^*$ is the K-dual of $\omega(\mathrm{T})$, subject to certain identifications which we do not describe. When $\mathcal{T}$ is the category of representations of an affine group scheme G and $\omega(\mathrm{T})$ is just the underlying vector space to $\mathrm{T}\in\mathcal{T}$, then a pair $(\mathrm{T},v\otimes w^*)$ is to be thought of as corresponding to the function on G given by $g\mapsto w^*(gv)$. Similarly, if ω' is another fiber functor and we replace $\omega(\mathrm{T})^*$ by $\omega'(\mathrm{T})^*$ in the above construction, we get the space of functions on the path space $\mathrm{P}_{\omega,\omega'}$.

Now, let B be a K-algebra and suppose we have a fiber functor ω' with values in locally free finite rank B-modules. Then the same construction gives a space of functions on an affine space over B, which we denote by $\mathrm{P}_{\omega,\omega'}$, whose fiber over each point $y\in\mathrm{Spec}(\mathrm{B})$ is just $\mathrm{P}_{\omega,\omega'_y}$, with ω'_y the fiber functor obtained by taking the fiber of ω' at y.

This suggests the following interpretation of Coleman functions, based on Definition 27. Let $\mathrm{A}^\dagger$ be a wcfg algebra with associated Tate algebra A and reduction $\bar{\mathrm{A}}$, and let x be a κ-rational point of X_κ. Let ω_x be the associated fiber functor on the category of unipotent $\mathrm{A}^\dagger_\mathrm{K}$-modules with an integrable connection. This is $\mathcal{U}n(\bar{\mathrm{A}})$ but we need to remember the particular $\mathrm{A}^\dagger$. and let ω' be forgetful fiber functor into $\mathrm{A}^\dagger_\mathrm{K}$-modules. This results in a path space $\mathrm{P}_{x,\omega'}$ over $\mathrm{A}^\dagger_\mathrm{K}$. Definition 27 exactly says that a Coleman function is a function on this path space and Definition 28 means that evaluating the Coleman function at a point y means evaluating the function on the associated Frobenius invariant path from x to y. More precisely, the composition of the invariant path to the reduction of y composed with the path from the reduction to y obtained by taking a horizontal section and evaluating at y.

We can state the following.

Proposition 44. *Let $f\in\mathrm{K}[\mathrm{G}_x^{\mathrm{dR}}]$ be invariant on the left by $\mathrm{F}^0\mathrm{G}_x^{\mathrm{dR}}$. Then there exists a Coleman function* F *on $\mathcal{X}$ such that $\alpha_f(y)=\mathrm{F}(y)$ for any* R*-rational point y of $\mathcal{X}$.*

Proof. This is somewhat sketched, as we have not properly defined Coleman functions on $\mathcal{X}$. Globalizing the considerations above we see that we have:

(1) A principal $\mathrm{G}_x^{\mathrm{dR}}$-bundle over X_K, P_{dR}, of *paths emminating from x*, with a filtration $\mathrm{F}^\bullet$. We denote its projection to $\mathcal{X}_\mathrm{K}$ by π.
(2) A principal $\mathrm{G}_x^{\mathrm{rig}}$-bundle $\mathrm{P}_{\mathrm{rig}}$ over an *overconvergent space* $\mathrm{X}^\dagger$ (this can either be taken literally as in [GK00] or requires an appropriate interpretation using the work of Berthelot), with a Frobenius.
(3) A specialization map Sp from $\mathrm{P}_{\mathrm{rig}}$ to the restriction of P_{dR} to $\mathrm{X}^\dagger$.

All of this compatible with the corresponding structure on $\mathrm{G}_x^{\mathrm{dR}}$ and $\mathrm{G}_x^{\mathrm{rig}}$. Now, for an affine open $\mathrm{V}\in\mathcal{X}$, we can find a section $\delta:\mathrm{V}_\mathrm{K}\to\mathrm{F}^0\mathrm{P}_{\mathrm{dR}}$, which we may reinterpret as a trivialization $v:\mathrm{P}_{\mathrm{dR}}\to\mathrm{G}_x^{\mathrm{dR}}\times\mathrm{V}$ with $v(\gamma)=(\delta(\pi(\gamma))^{-1}\gamma,\pi(\gamma))$. Starting with $f\in\mathrm{K}[\mathrm{G}_x^{\mathrm{dR}}]$, extended to a function on $\mathrm{V}\times\mathrm{G}_x^{\mathrm{dR}}$ by pulling back via the projection, we have $f\circ v\in\mathrm{K}[\mathrm{P}_{\mathrm{dR}}]$ which we may restrict to $\mathrm{X}^\dagger$ and pullback via Sp to $\mathrm{P}_{\mathrm{rig}}$ giving us, locally on $\mathrm{X}^\dagger$, a Coleman function. The left F^0 invariance of f means that these functions are independent of the choice of δ and they thus glue to give a global Coleman function F. Visibly, evaluating this on $\gamma_{x,y}$ gives $\alpha_f(y)$. □

Remark 45. For the second normalization one can also evaluate functions on G_x^{dR} on the unipotent Albanese and the results from the two normalizations are related in terms of the action of φ on G_x^{dR}. In the second normalization, it is not too hard to see that one gets certain analytic functions rather than Coleman functions. Two cases of this are worked out in [Fur07, Example 2.10].

As mentioned before, the other main ingredient in Kim's work is the Selmer variety. This is a certain pro-algebraic variety, constructed from Galois cohomology so that the unipotent Albanese for global points factors through it. If the map of the Selmer variety to $F^0\backslash G_x^{dR}$ is not surjective, more precisely, not so already on some finite level, then one can find a Coleman function vanishing on all global points. Kim shows that image of the unipotent Albanese map is Zariski dense, implying that this Coleman function does not vanish on p-adic points, giving a non-trivial condition for global points. The main difficulty in the method is to show the non-surjectivity.

We mention two results obtained by Kim. In [Kim05] he reproves a theorem of Siegel about the finiteness of integral points on $\mathbb{P}^1 - \{0, 1, \infty\}$. In [Kim10a], as corrected in [BKK11], he shows the following result.

Theorem 46. *Let* E *be an elliptic curve over* $\mathbb{Q}$ *given by the Weierstrass equation* $y^2 = x^3 + ax + b$ *and assume that it has an integral j-invariant and a Mordell-Weil group of rank* 1. *Let p be a prime of good reduction with the property that the* p^∞ *component of the Tate-Shafarevich group of* E *is finite. Suppose there is an integral solution* Q *of the Weierstrass equation above which is a point of infinite order in* $E(\mathbb{Q})$. *Then, any other integral solution is contained in the* 0*-set of the function*

$$F_\omega(Q)^2 D_2(z) - F_\omega(z)^2 D_2(Q)$$

where $\omega = dx/y$ *is the invariant differential,* $\eta = x dx/y$, *and* $D_2 = F_{\omega F_\eta}$, *and all Coleman integrals are normalized to have value* 0 *at a tangential base point at* 0, *which is an integral base for the tangent space.*

1.7 Coleman Integration in Families

An important observation to be made about Coleman integration is that it ultimately relies on the ground field having a finite residue field (by varying the field we can deal with residue fields which are algebraic over their prime fields) because we rely on the linear Frobenius. It is an interesting problem to try to remove this condition.

In this section we report on some recent work, which is still in progress. Full details will appear somewhere else. The problem that we want to address in this work is. Given an algebraic family of closed forms, how can we integrate the family in a way better than just integrating each family member separately. More precisely, suppose that $X \xrightarrow{\pi} S$ is a smooth family of overconvergent rigid spaces over the field K, which has a finite residue field as before (in this section we treat the nature of the spaces involved in a rather loose way), and $\omega \in \Omega^1_{X/S}$ is a relatively closed relative form. We would like to associate with ω a Coleman integral F_ω which, when restricted to $s \in S$ is a Coleman integral of ω_s on $\pi^{-1}(s)$, but which is more canonical then just taking a choice of integral for each fiber, with the possibility of choosing a different constant of integration for each fiber. For affine S this can be thought of as doing Coleman integration *over* $\mathcal{O}_S$, thus giving one solution to the problem posed in the previous paragraph.

One motivation for treating this problem is computational and comes from the work of Lauder [Lau03, Lau04b, Lau04a]. In this work, one sees the possibility of computing the matrix of Frobenius on a variety by putting it inside a family, deforming to a fiber where this matrix can be computed easily, and then relying on the fact that the matrix of Frobenius satisfies a differential equation, derived from the Picard-Fuchs equation and computable, to recover the matrix by solving the equation with boundary terms provided by the simple fiber. One can speculate on the possibility of doing the same with Coleman integrals, given that the computation of the matrix of Frobenius is such an important part in the computation of the Coleman integral, as we have seen in Sect. 1.4.

Following the approach to Coleman integration we presented, it is natural to attempt to look for the answer by imposing additional constraints on the association of a Coleman integral to a form. Given the type of problem it makes sense to look for a differential condition. We would like to have a condition saying roughly that the formation of Coleman integrals commutes with differentiation in the direction of the base, i.e.,

$$\int \frac{\partial}{\partial s}\omega = \frac{\partial}{\partial s} F_\omega \tag{1.21}$$

where we assume for simplicity from now onward that S is one-dimensional, and the derivative refers to some vector field on the base.

There exists a well-defined notion of differentiation of differential forms with respect to a vector field. However, there is no obvious way of lifting a vector field on S to a vector field on X, except when $X = Y \times S$, which is an interesting test case. Thus, Equation 1.21 does not quite make sense. Trying to get a meaningful statement out of it we are led to the following condition.

Lift the form ω to an absolute form $\tilde{\omega}$ on X. This can be done at least locally in the rigid topology. Since ω is relatively closed we may interpret $d\tilde{\omega}$ as an element of $\Gamma(X, \pi^*\Omega^1_S \otimes \Omega^1_{X/S})$. Note here that projecting from $\Omega^1_{X/S}$ to the first relative de Rham cohomology exactly yields the Gauss-Manin connection applied to the cohomology class of ω. Hypothesizing the existence of the theory of relative Coleman integration, we would like to integrate $d\tilde{\omega}$ to obtain a section

$$F_{d\tilde{\omega}} \in \Gamma(X, \pi^*\Omega^1_S \otimes \mathcal{O}_{Col}(X/S))$$

with a hypothetical sheaf of relative Coleman functions $\mathcal{O}_{Col}(X/S)$. Alternatively, we can integrate ω to get F_ω. We expect that $d_r F_\omega = \omega$, where d_r is the relative differential. Thus $dF_\omega - \tilde{\omega}$ is a one-form on X locally coming from the base. This suggest the following condition

$$dF_\omega - \tilde{\omega} = F_{d\tilde{\omega}}\,. \tag{1.22}$$

Our goal in the rest of this section is to show that this is indeed a meaningful condition, and gives a good theory of integration in families, from two different points of view. The first involves the Gauss-Manin connection while the second comes from fairly recent work on differential Tannakian categories. One should think of this condition as cutting the indeterminacy in solving a relative unipotent differential equation from analytic functions on some residue disc on S to just the constant ones, thus going back to working over K, where the weight arguments of the preceding sections can be used again.

1.7.1 Integration via the Gauss-Manin Connection

Let us first briefly recall the construction of the Gauss-Manin connection in [KO68]. We have a filtration on the complex of differential forms $\Omega_X^\bullet$, where

$$F^i = \mathrm{Im}(\Omega_X^{\bullet-i} \otimes \pi^* \Omega_S^i \to \Omega_X^\bullet)$$

with graded pieces $gr^i = F^i/F^{i+1} \cong \pi^*\Omega_S^i \otimes \Omega_{X/S}^{\bullet-i}$. For future use we denote the projection on the second graded piece

$$\varpi : (\mathrm{Ker} : \Omega_X^2 \to \Omega_{X/S}^2) \to \pi^*\Omega_S^1 \otimes \Omega_{X/S}^1\,. \tag{1.23}$$

The spectral sequence for the right derived functor of a filtered object, applied to the functor π_*, reads, see loc. cit. (7),

$$E_1^{p,q} = \mathbb{R}^{p+q}\pi_* gr^p \cong \Omega_S^p \otimes H_{dR}^q(X/S) \quad \Longrightarrow \quad \mathbb{R}^q \pi_* \Omega_X^\bullet\,,$$

and the Gauss-Manin connection

$$\nabla_{GM} : H_{dR}^q(X/S) \to \Omega_S^1 \otimes H_{dR}^q(X/S)$$

is simply the differential in the spectral sequence $d_1^{0,q} : E_1^{0,q} \to E_1^{1,q}$. In concrete terms this is just the composition

$$\mathbb{R}^q\pi_* gr^0 \xrightarrow{\delta} \mathbb{R}^{q+1}\pi_* F^1 \to \mathbb{R}^{q+1}\pi_* gr^1\,,$$

where the map δ is the connecting homomorphism in the long exact sequence associated with $0 \to F^1 \to F^0 \to gr^0 \to 0$, while the second map is induced from the projection $F^1 \to gr^1$. In even more concrete terms, concentrating on H_{dR}^1 for affine X and S, the Gauss-Manin connection starts with the class of $\omega \in \Omega_{X/S}^1$, closed with respect to the relative differential. One chooses a lift $\tilde{\omega} \in \Omega_X^1$, computes $\varpi d\tilde{\omega} \in \Omega_S^1 \otimes \Omega_{X/S}^1$, observes that result must lie in $\Omega_S^1 \otimes (\Omega_{X/S}^1)^{d_r=0}$ and finally projects on $\Omega_S^1 \otimes H_{dR}^1(X/S)$.

A first attempt to use the condition (1.22) to get a relative Coleman integration theory follows roughly the same line as the approach in Sect. 1.4. We assume that X and S are affine in the appropriate setting. By further localizing, we can choose a vector $\omega \in (\Omega^1_{X/S})^n$ whose entries form a basis for the relative de Rham cohomology module $H^1_{dR}(X/S)$ over $\mathcal{O}_S$ and choose a lifting $\tilde{\omega} \in (\Omega^1_X)^n$. Since $\tilde{\omega}$ consists of a basis, we find a relation of the form

$$\varpi(d\tilde{\omega}) = \Theta(s) \otimes \omega + d_r(\mathbf{g}) .$$

Here, $\Theta(s)$ in an n by n matrix with entries in Ω^1_S, $\mathbf{g}$ has entries in $\Gamma(X, \pi^*\Omega^1_S)$, d_r is the relative differential, extended Ω^1_S linearly and ϖ is the projection from (1.23). Let us first observe that upon identifying $H^1_{dR}(X/S)$ with $\mathcal{O}^n_S$ via the basis ω the Gauss-Manin connection is simply given by $d - \Theta^t$, where Θ^t is the matrix transpose to Θ. Now, applying (1.22) we get the following relation.

$$dF_\omega = \tilde{\omega} + \Theta(s)F_\omega + \mathbf{g} .$$

Rearranging terms we find

$$(d - \Theta(s))F_\omega = \tilde{\omega} + \mathbf{g} . \tag{1.24}$$

We now observe that on S the operator $d - \Theta(s)$ is nothing but the dual connection to the Gauss-Manin connection ∇^*_{GM} on the dual vector bundle $H^1_{dR}(X/S)^*$. Consequently, the equation (1.24) describes F_ω as a preimage, under $\pi^*\nabla_{GM}$ of a certain 1-form $\tilde{\omega} + \mathbf{g}$. Note that fiber by fiber $\pi^*\nabla^*_{GM}$ restricts to just ordinary derivative while $\tilde{\omega} + \mathbf{g}$ restricts to ω so fiber by fiber we indeed obtain the required integrals of our forms.

In [Col89a, Col94] Coleman extended his theory of integration to define integration (but not iterated integrals) of one-forms with values in overconvergent Frobenius isocrystals, that is, differential equations which overconverges in the appropriate sense, which have an action of Frobenius. Using this version of Coleman integration theory we obtain the required F_ω.

This method of integration can be extended to iterated integrals by using universal unipotent connections.

1.7.2 *Motivation from Differential Tannakian Categories*

In this section we explain how the condition (1.22) can be motivated by the theory of differential Tannakian categories. We only give a minimal description of this relatively new theory referring instead to the original work of Ovchinikov and Kamensky [Ovc08, Ovc09a, Ovc09b, Kam09, Kam10] as well as to our brief sketch [Bes11].

A rigid abelian tensor category $\mathcal{T}$ over the field K is given a *differential structure* by an auto-functor D, part of a short exact sequence of functors

$$0 \to \mathrm{id} \to \mathrm{D} \to \mathrm{id} \to 0\,, \tag{1.25}$$

and having a certain behavior with respect to the tensor structure.

Example 47. (1) A *differential field* is a field K together with a derivation ∂. The category Vec_K of finite-dimensional vector spaces over the differential field K is given a differential structure by sending the K-vector space V to the K-vector space D(V) which as an abelian group is $\mathrm{V}\times\mathrm{V}$ and with the K-vector space structure given by

$$\alpha(v_1,v_2) = (\alpha v_1, \alpha v_2 - \partial(\alpha)v_1)\,. \tag{1.26}$$

(2) Let $\mathrm{K} = \mathbb{C}(t)$. Consider the K-linear category of differential equations of the form

$$\nabla : \partial_x \mathbf{y} = \mathrm{A}y\,, \tag{1.27}$$

over the field $\mathbb{C}(x,t)$. Let $\mathbf{y}$ be a solution of the above equation. Using the fact that the two derivations commute we obtain

$$\partial_x(\partial_t \mathbf{y}) = \partial_t(\mathrm{A}\mathbf{y}) = (\partial_t \mathrm{A})\mathbf{y} + \mathrm{A}\partial_t \mathbf{y}\,.$$

This means that we obtain a new differential equation corresponding to the matrix $\begin{pmatrix} \mathrm{A} & 0 \\ \partial_t \mathrm{A} & \mathrm{A} \end{pmatrix}$ and a solution derived from $\mathbf{y}$ which is $\begin{pmatrix} \mathbf{y} \\ \partial_t \mathbf{y} \end{pmatrix}$. The construction of the new differential equation is clearly functorial and provides the differential structure D.

(3) A differential affine group scheme G over the differential field K is a Hopf algebra K[G] over K together with a derivation ∂ extending the one on K and commuting with multiplication and comultiplication. As an object of the category of differential K-algebras, i.e., K-algebras with a derivation ∂ compatible with its name sake on K, it represents a group valued functor. The category Rep_G of finite-dimensional representations of G is the category of K[G]-comodules which are finite-dimensional vector spaces over K. The differential structure is given by sending the K-vector space V with comodule structure $\rho : \mathrm{V} \to \mathrm{V} \otimes_\mathrm{K} \mathrm{K}[\mathrm{G}]$ to the vector space D(V) from the previous example with comodule structure $\mathrm{D}(\rho)$ given as follows: Identify the vector $v \in \mathrm{V}$ with $(0,v) \in \mathrm{D}(\mathrm{V})$ and denote the map $v \mapsto (v,0)$ by ∂. Then $\mathrm{D}(\rho)$ can be described by the two formulae

$$\mathrm{D}(\rho)(v) = \rho(v)\,,\ \mathrm{D}(\rho)\partial v = \partial(\rho(v))$$

with the rule

$$\partial(a \otimes b) = (\partial a) \otimes b + a \otimes (\partial b)$$

for the action of ∂ on tensor products. The rest of the structure is the same as in the standard theory [DM82].

Definition 48. A **differential tensor functor** $\mathcal{T}_1 \xrightarrow{\mathrm{F}} \mathcal{T}_2$ between two differential rigid abelian tensor categories, with differential structures D_1 and D_2 respectively, is a tensor functor together with a natural isomorphism $\mathrm{D}_2 \circ \mathrm{F} \cong \mathrm{F} \circ \mathrm{D}_1$ compatible in the obvious way with the short exact sequence (1.25). A **morphism of differential tensor functors** $\alpha : \mathrm{F} \to \mathrm{F}'$ is a natural transformation of tensor functors which commutes with D in the sense that the diagram

$$\begin{array}{ccc} \mathrm{F}\circ\mathrm{D}_1 & \xrightarrow{\alpha} & \mathrm{F}'\circ\mathrm{D}_1 \\ \downarrow & & \downarrow \\ \mathrm{D}_2\circ\mathrm{F} & \xrightarrow{\mathrm{D}_2(\alpha)} & \mathrm{D}_2\circ\mathrm{F}' \end{array}$$

commutes.

Example 49. Quite clearly the forgetful functor $\mathrm{Rep}_\mathrm{G} \to \mathrm{Vec}_\mathrm{K}$ is a differential tensor functor. Another example is solutions of differential equations. Consider the functor Sol that takes a differential equation ∇ as in (1.27) over the field $\mathbb{C}(x,t)$ to its space of solutions (in the differential closure of the field) considered as a vector space over $\mathbb{C}(t)$. Then, according to part (2) of Example 47 above, we can map $\mathrm{D}(\mathrm{Sol}(\nabla))$ to $\mathrm{Sol}(\mathrm{D}(\nabla))$ using the formula

$$(\mathbf{y}_1,\mathbf{y}_2) \mapsto (\mathbf{y}_1,\mathbf{y}_2+\partial_t\mathbf{y}_1)\,.$$

Note that to make this a $\mathbb{C}(t)$ linear map we exactly need to give $\mathrm{D}(\mathrm{Sol}(\nabla))$ the vector space structure (1.26).

Definition 50. A **differential fiber functor** on a differential rigid abelian tensor category is a faithful differential tensor functor ω to Vec_K. If the category has a fiber functor it is called **neutral** Tannakian.

Clearly, the category Rep_G for an affine differential group scheme G is a neutral differential Tannakian category with the obvious forgetful functor into vector spaces. The theory of differential Tannakian categories concerns itself with identifying neutral differential Tannakian categories with categories of representations of differential affine group schemes. In fact, let $\mathcal{T}$ be a differential rigid abelian tensor category. given two differential fiber functors ω_1,ω_2, consider the functor that associates to a differential K-algebra F the set of F-valued *differential paths* between ω_1 and ω_2, which is by definition the set $\mathrm{Iso}(\omega_1\otimes\mathrm{F},\omega_2\otimes\mathrm{F})$ of isomorphisms between the two functors. One can compose differential paths. In particular for a single differential fiber functor ω, the functor $\mathrm{F}\mapsto\mathrm{Aut}(\omega\otimes\mathrm{F})$ is a group valued functor on differential K-algebras.

Theorem 51 ([Ovc09b, Theorem 1]). *Let $\mathcal{T}$ be a neutral differential Tannakian category with the differential fiber functor ω. Then $\mathcal{T}$ is equivalent to the category* Rep_G *of finite-dimensional representations of an affine differential group scheme* G*. Furthermore, for a differential* K*-algebra* F *we have*

$$\mathrm{G}(\mathrm{F}) = \mathrm{Aut}(\omega\otimes\mathrm{F})\,. \tag{1.28}$$

For the present work, we are concerned only with the notion of differential paths. However, we first need to modify part (2) of Example 47, since that example is dependent on the choice of a derivation. If we want to get a theory which takes all derivations into account (like connections do) we are led, after some thought into making the following construction.

Recall that we are assuming a situation $\pi : X \to S$ and that S is one-dimensional. Suppose M is a vector bundle on X equipped with a relative connection

$$\nabla : M \to M \otimes \Omega^1_{X/S}$$

which is integrable. Suppose we can lift ∇ to an absolute connection

$$\tilde{\nabla} : M \to M \otimes \Omega^1_X .$$

Because ∇ is integrable, the curvature of $\tilde{\nabla}$, which in general lies in $\Omega^2_X \otimes \mathrm{End}(M)$, actually lies in its first filtered part $F^1 \otimes \mathrm{End}(M)$, see Sect. 1.7.1, and so we may use (1.23) to define

$$C = \varpi \tilde{\nabla}^2 \in \Omega^1_S \otimes \Omega^1_{X/S} \otimes \mathrm{End}(M) .$$

We define a new module with connection $D = D_{\tilde{\nabla}} = D(M,\nabla)_{\tilde{\nabla}}$ where $D = M \oplus \Omega^1_S \otimes M$ and the connection is defined by

$$\nabla_D(m_1, \alpha \otimes m_2) = (\nabla m_1, \alpha \otimes \nabla m_2 - C \times M) .$$

This connection is integrable. It is independent of $\tilde{\nabla}$ up to a canonical isomorphism: suppose $\tilde{\nabla}' = \tilde{\nabla} + A$ is another lift. Here $A \in \Gamma(X, \pi^{-1}\Omega^1_S \otimes \mathrm{End}(M))$ because it projects to 0 in relative forms. Then it is easy to compute that the corresponding curvature is $C' = C + \nabla(A)$ (where ∇ takes Ω^1_S as constants and acts in the induced way on End(M)). Then we get a canonical horizontal isomorphism between $D_{\tilde{\nabla}}$ and $D_{\tilde{\nabla}'}$ given by

$$(m_1, \alpha \otimes m_2) \mapsto (m_1, \alpha \otimes m_2 + A m_1) .$$

Consequently we can glue these objects, coming from different local liftings of ∇, to obtain a global object $D(M,\nabla)$. Clearly, there is a short exact sequence of vector bundles with relative connections,

$$0 \to \Omega^1_S \otimes M \to D(M,\nabla) \to M \to 0$$

because all the horizontal isomorphisms constructed commute with these short exact sequences. Clearly, the construction D is functorial.

For vector bundles over S we can make an analogous functorial construction. For such a vector bundle M define $D(M) = M \oplus \Omega^1_S \otimes M$ with the $\mathcal{O}_S$-module structure

$$s \times (m_1, \alpha \otimes m_2) = (s m_1, s\alpha \otimes m_2 + ds \otimes m_1) .$$

Suppose now that X and S are residue discs. Then, mimicking the constructions in Example 49 we have a well behaved solutions functor

$$\mathrm{Sol} : \{\text{Relative connections } (M, \nabla : M \to \Omega^1_{X/S})\} \to \{\text{Vector bundles on } S\}$$

given by taking horizontal sections, and a map

$$\mathrm{D} \circ \mathrm{Sol} \to \mathrm{Sol} \circ \mathrm{D}\,,\ (m_1, \alpha \otimes m_2) \mapsto (m_1, \alpha \otimes m_2 + \tilde{\nabla} m_1) \tag{1.29}$$

with $\tilde{\nabla}$ the local lifting of ∇ used for the construction of D(M), where m_1 and m_2 are horizontal sections for ∇, implying that $\tilde{\nabla} m_1 \in \Omega^1_{\mathrm{S}} \otimes \mathrm{M}$.

We now show that, using this modified version of differential Tannakian theory, condition (1.22) may be interpreted as coming from analytic continuation along a differential path invariant under Frobenius between two residue discs. Suppose we have a closed $\omega \in \Omega^1_{\mathrm{X/S}}$ and we lift it to a $\tilde{\omega} \in \Omega_{\mathrm{X}}$. correspondingly we have the connection ∇ and its lift $\tilde{\nabla}$ given by

$$\nabla = \mathrm{d}_r - \begin{pmatrix} 0 & 0 \\ \omega & 0 \end{pmatrix}, \quad \tilde{\nabla} = \mathrm{d} - \begin{pmatrix} 0 & 0 \\ \tilde{\omega} & 0 \end{pmatrix}.$$

The curvature of $\tilde{\nabla}$ is going to be $\left(\begin{smallmatrix} 0 & 0 \\ -d\tilde{\omega} & 0 \end{smallmatrix}\right)$. We can now compute that the connection ∇_{D} is going to be given by the following formula,

$$\nabla_{\mathrm{D}}\left(\begin{pmatrix} y_1 \\ y_2 \end{pmatrix}, \alpha \otimes \begin{pmatrix} y_3 \\ y_4 \end{pmatrix}\right) = \left(\nabla \begin{pmatrix} y_1 \\ y_2 \end{pmatrix}, \alpha \otimes \nabla \begin{pmatrix} y_3 \\ y_4 \end{pmatrix} + \begin{pmatrix} 0 & 0 \\ \varpi \mathrm{d}\tilde{\omega} & 0 \end{pmatrix}\begin{pmatrix} y_1 \\ y_2 \end{pmatrix}\right).$$

Now we check what it means for $\left(\begin{smallmatrix} 1 \\ \mathrm{F}_\omega \end{smallmatrix}\right)$ to be a horizontal section at the residue discs of x and z, say, which is compatible with respect to translation by a differential path. Appropriately translating the condition in Definition 48 we find that it simply means that another horizontal section that translates under the same path is the image of $\left(\begin{smallmatrix} 1 \\ \mathrm{F}_\omega \end{smallmatrix}\right)$ under (1.29) which is

$$\left(\begin{pmatrix} 1 \\ \mathrm{F}_\omega \end{pmatrix}, \begin{pmatrix} 0 \\ \mathrm{dF}_\omega - \tilde{\omega} \end{pmatrix}\right).$$

In other words, $\mathrm{dF}_\omega - \tilde{\omega}$ is a Coleman integral, which is just (1.22).

References

[Ban02] K. Bannai. Syntomic cohomology as a p-adic absolute Hodge cohomology. *Math. Z.*, 242(3):443–480, 2002.

[BB04] F. Baldassarri and P. Berthelot. On Dwork cohomology for singular hypersurfaces. In *Geometric aspects of Dwork theory. Vol. I, II*, pages 177–244. Walter de Gruyter GmbH & Co. KG, Berlin, 2004.

[BBdJR09] A. Besser, P. Buckingham, R. de Jeu, and X.-F. Roblot. On the p-adic Beilinson conjecture for number fields. *Pure Appl. Math. Q.*, 5(1, part 2):375–434, 2009.

[BD92] A. Beilinson and P. Deligne. Motivic polylogarithms and Zagier's conjecture. Unpublished manuscript, 1992.

[BdJ03] A. Besser and R. de Jeu. The syntomic regulator for the K-theory of fields. *Ann. Sci. École Norm. Sup. (4)*, 36(6):867–924, 2003.

[Bei85] A. A. Beilinson. Higher regulators and values of L-functions. *J. Sov. Math.*, 30:2036–2070, 1985.

[Ber86] P. Berthelot. Géométrie rigide et cohomologie des variétés algébriques de caractéristique p. *Mém. Soc. Math. France (N.S.)*, (23):3, 7–32, 1986. Introductions aux cohomologies p-adiques (Luminy, 1984).

[Ber96] P. Berthelot. Cohomologie rigide et cohomologie rigide a supports propres, premièr partie. Preprint 96-03 of IRMAR, the university of Rennes I, available online at http://www.maths.univ-rennes1.fr/~berthelo/, 1996.

[Ber97] P. Berthelot. Finitude et pureté cohomologique en cohomologie rigide. *Invent. Math.*, 128(2):329–377, 1997. With an appendix in English by A. J. de Jong.

[Ber07] V. G. Berkovich. *Integration of one-forms on p-adic analytic spaces*, volume 162 of *Annals of Mathematics Studies*. Princeton University Press, Princeton, NJ, 2007.

[Bes00a] A. Besser. A generalization of Coleman's p-adic integration theory. *Inv. Math.*, 142(2):397–434, 2000.

[Bes00b] A. Besser. Syntomic regulators and p-adic integration I: rigid syntomic regulators. *Israel Journal of Math.*, 120:291–334, 2000.

[Bes00c] A. Besser. Syntomic regulators and p-adic integration II: K_2 of curves. *Israel Journal of Math.*, 120:335–360, 2000.

[Bes02] A. Besser. Coleman integration using the Tannakian formalism. *Math. Ann.*, 322(1):19–48, 2002.

[Bes05] A. Besser. p-adic Arakelov theory. *J. Number Theory*, 111(2):318–371, 2005.

[Bes11] A. Besser. On differential Tannakian categories and Coleman integration. Preprint, available online at http://www.math.bgu.ac.il/~bessera/difftan.pdf, 2011.

[BF06] A. Besser and H. Furusho. The double shuffle relations for p-adic multiple zeta values. In *Primes and knots*, volume 416 of *Contemp. Math.*, pages 9–29. Amer. Math. Soc., Providence, RI, 2006.

[BK90] S. Bloch and K. Kato. L-functions and Tamagawa numbers of motives. In *The Grothendieck Festschrift I*, volume 86 of *Prog. in Math.*, pages 333–400, Boston, 1990. Birkhäuser.

[BKK11] J. Balakrishnan, K. Kedlaya, and M. Kim. Appendix and erratum to "Massey products for elliptic curves of rank 1". *J. Amer. Math. Soc.*, 24(1):281–291, 2011.

[Bou98] N. Bourbaki. *Lie groups and Lie algebras. Chapters 1–3*. Elements of Mathematics (Berlin). Springer-Verlag, Berlin, 1998. Translated from the French, Reprint of the 1989 English translation.

[Bru02] N. Bruin. *Chabauty methods and covering techniques applied to generalized Fermat equations*, volume 133 of *CWI Tract*. Stichting Mathematisch Centrum Centrum voor Wiskunde en Informatica, Amsterdam, 2002. Dissertation, University of Leiden, Leiden, 1999.

[Bru03] N. Bruin. Chabauty methods using elliptic curves. *J. Reine Angew. Math.*, 562:27–49, 2003.

[CdS88] R. Coleman and E. de Shalit. p-adic regulators on curves and special values of p-adic L-functions. *Invent. Math.*, 93(2):239–266, 1988.

[Cha41] C. Chabauty. Sur les points rationnels des courbes algébriques de genre supérieur à l'unité. *C. R. Acad. Sci. Paris*, 212:882–885, 1941.

[Chi98] B. Chiarellotto. Weights in rigid cohomology applications to unipotent F-isocrystals. *Ann. Sci. École Norm. Sup. (4)*, 31(5):683–715, 1998.

[CK10] J. Coates and M. Kim. Selmer varieties for curves with CM Jacobians. *Kyoto J. Math.*, 50(4):827–852, 2010.

[CLS99] B. Chiarellotto and B. Le Stum. F-isocristaux unipotents. *Compositio Math.*, 116(1):81–110, 1999.

[Col82] R. Coleman. Dilogarithms, regulators, and p-adic L-functions. *Invent. Math.*, 69:171–208, 1982.

[Col85a] R. Coleman. Effective Chabauty. *Duke Math. J.*, 52(3):765–770, 1985.

[Col85b] R. Coleman. Torsion points on curves and p-adic abelian integrals. *Annals of Math.*, 121:111–168, 1985.

[Col86] R. Coleman. Torsion points on Fermat curves. *Compositio Math.*, 58(2):191–208, 1986.

[Col87] R. Coleman. Ramified torsion points on curves. *Duke Math. J.*, 54(2):615–640, 1987.

[Col89a] R. Coleman. p-adic integration. Notes from lectures at the University of Minnesota, 1989.

[Col89b] R. Coleman. Reciprocity laws on curves. *Compositio Math.*, 72(2):205–235, 1989.

[Col94] R. Coleman. A p-adic Shimura isomorphism and p-adic periods of modular forms. *Contemp. math.*, 165:21–51, 1994.

[Col98] P. Colmez. Intégration sur les variétés p-adiques. *Astérisque*, (248):viii+155, 1998.

[Cre92] R. Crew. F-isocrystals and their monodromy groups. *Ann. Sci. École Norm. Sup. (4)*, 25(4):429–464, 1992.

[Del74] P. Deligne. La conjecture de Weil. I. *Inst. Hautes Études Sci. Publ. Math.*, (43):273–307, 1974.

[Del89] P. Deligne. Le groupe fondamental de la droite projective moins trois points. In *Galois groups over* **Q** *(Berkeley, CA, 1987)*, volume 16 of *Math. Sci. Res. Inst. Publ.*, pages 79–297. Springer, New York, 1989.

[Del90] P. Deligne. Catégories tannakiennes. In *The Grothendieck Festschrift, Vol. II*, volume 87 of *Progr. Math.*, pages 111–195. Birkhäuser Boston, Boston, MA, 1990.

[Del02] P. Deligne. Periods for the fundamental group. A short note on Arizona Winter School, 2002.

[dJ95] R. de Jeu. Zagier's conjecture and wedge complexes in algebraic K-theory. *Compositio Math.*, 96(2):197–247, 1995.

[DM82] P. Deligne and J. S. Milne. Tannakian categories. In *Hodge cycles, motives, and Shimura varieties*, volume 900 of *Lect. Notes in Math.*, pages 101–228. Springer, 1982.

[Elk73] R. Elkik. Solutions d'équations à coefficients dans un anneau hensélien. *Ann. Sci. École Norm. Sup. (4)*, 6:553–603 (1974), 1973.

[FJ07] H. Furusho and A. Jafari. Regularization and generalized double shuffle relations for p-adic multiple zeta values. *Compos. Math.*, 143(5):1089–1107, 2007.

[Fly97] E. V. Flynn. A flexible method for applying Chabauty's theorem. *Compositio Math.*, 105(1):79–94, 1997.

[Fur04] H. Furusho. p-adic multiple zeta values. I. p-adic multiple polylogarithms and the p-adic KZ equation. *Invent. Math.*, 155(2):253–286, 2004.

[Fur07] H. Furusho. p-adic multiple zeta values. II. Tannakian interpretations. *Amer. J. Math.*, 129(4):1105–1144, 2007.

[FW99] E. V. Flynn and J. L. Wetherell. Finding rational points on bielliptic genus 2 curves. *Manuscripta Math.*, 100(4):519–533, 1999.

[FW01] E. V. Flynn and J. L. Wetherell. Covering collections and a challenge problem of Serre. *Acta Arith.*, 98(2):197–205, 2001.

[GK00] E. Grosse-Klönne. Rigid analytic spaces with overconvergent structure sheaf. *J. Reine Angew. Math.*, 519:73–95, 2000.

[Jan88] U. Jannsen. *Mixed motives and algebraic K-theory*, volume 1400 of *Lect. Notes in Math.* Springer, Berlin Heidelberg New York, 1988.

[Kam09] M. Kamensky. Differential tensor categories. Unpublished lecture notes, 2009.

[Kam10] M. Kamensky. Model theory and the Tannakian formalism. Preprint, arXiv:math.LO/0908.0604v3, 2010.

[Kat76] N. M. Katz. p-adic interpolation of real analytic Eisenstein series. *Ann. of Math. (2)*, 104(3):459–571, 1976.

[Kim05] M. Kim. The motivic fundamental group of $\mathbb{P}^1 \setminus \{0,1,\infty\}$ and the theorem of Siegel. *Invent. Math.*, 161(3):629–656, 2005.

[Kim09] M. Kim. The unipotent Albanese map and Selmer varieties for curves. *Publ. Res. Inst. Math. Sci.*, 45(1):89–133, 2009.

[Kim10a] M. Kim. Massey products for elliptic curves of rank 1. *J. Amer. Math. Soc.*, 23(3):725–747, 2010.

[Kim10b] M. Kim. p-adic L-functions and Selmer varieties associated to elliptic curves with complex multiplication. *Ann. of Math. (2)*, 172(1):751–759, 2010.

[KM74] N. M. Katz and W. Messing. Some consequences of the Riemann hypothesis for varieties over finite fields. *Invent. Math.*, 23:73–77, 1974.

[KO68] N. M. Katz and T. Oda. On the differentiation of de Rham cohomology classes with respect to parameters. *J. Math. Kyoto Univ.*, 8:199–213, 1968.

[Lau03] A. G. B. Lauder. Homotopy methods for equations over finite fields. In *Applied algebra, algebraic algorithms and error-correcting codes (Toulouse, 2003)*, volume 2643 of *Lecture Notes in Comput. Sci.*, pages 18–23. Springer, Berlin, 2003.

[Lau04a] A. G. B. Lauder. Counting solutions to equations in many variables over finite fields. *Found. Comput. Math.*, 4(3):221–267, 2004.

[Lau04b] A. G. B. Lauder. Deformation theory and the computation of zeta functions. *Proc. London Math. Soc. (3)*, 88(3):565–602, 2004.

[LL03] R. Lercier and D. Lubicz. Counting points in elliptic curves over finite fields of small characteristic in quasi quadratic time. In *Advances in cryptology – EUROCRYPT 2003*, volume 2656 of *Lecture Notes in Comput. Sci.*, pages 360–373. Springer, Berlin, 2003.

[LL06] R. Lercier and D. Lubicz. A quasi quadratic time algorithm for hyperelliptic curve point counting. *Ramanujan J.*, 12(3):399–423, 2006.

[MW68] P. Monsky and G. Washnitzer. Formal cohomology. I. *Ann. of Math. (2)*, 88:181–217, 1968.

[Ovc08] A. Ovchinnikov. Tannakian approach to linear differential algebraic groups. *Transform. Groups*, 13(2):413–446, 2008.

[Ovc09a] A. Ovchinnikov. Differential Tannakian categories. *J. Algebra*, 321(10):3043–3062, 2009.

[Ovc09b] A. Ovchinnikov. Tannakian categories, linear differential algebraic groups, and parametrized linear differential equations. *Transform. Groups*, 14(1):195–223, 2009.

[PR95] B. Perrin-Riou. Fonctions L p-adiques des représentations p-adiques. *Astérisque*, (229):198pp, 1995.

[Sch98] P. Schneider. Basic notions of rigid analytic geometry. In *Galois representations in arithmetic algebraic geometry (Durham, 1996)*, volume 254 of *London Math. Soc. Lecture Note Ser.*, pages 369–378. Cambridge Univ. Press, Cambridge, 1998.

[Ser97] J.-P. Serre. *Galois cohomology*. Springer-Verlag, Berlin, 1997. Translated from the French by Patrick Ion and revised by the author.

[Sik09] S. Siksek. Chabauty for symmetric powers of curves. *Algebra Number Theory*, 3(2):209–236, 2009.

[vdP86] M. van der Put. The cohomology of Monsky and Washnitzer. *Mém. Soc. Math. France (N.S.)*, 23:33–59, 1986. Introductions aux cohomologies p-adiques (Luminy, 1984).

[Wet97] J. L. Wetherell. *Bounding the number of rational points on certain curves of high rank*. ProQuest LLC, Ann Arbor, MI, 1997. Thesis (Ph.D.)–University of California, Berkeley.

[Wil97] J. Wildeshaus. *Realizations of polylogarithms*. Springer-Verlag, Berlin, 1997.

[Woj93] Z. Wojtkowiak. Cosimplicial objects in algebraic geometry. In *Algebraic* K*-theory and algebraic topology (Lake Louise, AB, 1991)*, volume 407 of *NATO Adv. Sci. Inst. Ser. C Math. Phys. Sci.*, pages 287–327. Kluwer Acad. Publ., Dordrecht, 1993.

[Zar96] Yu. G. Zarhin. p-adic abelian integrals and commutative Lie groups. *J. Math. Sci.*, 81(3):2744–2750, 1996.

Chapter 2
Heidelberg Lectures on Fundamental Groups

Tamás Szamuely*

Abstract We survey topics related to étale fundamental groups, with emphasis on Grothendieck's anabelian program, the Section Conjecture and Parshin's proof of the geometric case of Mordell's conjecture.

As a prelude to the PIA conference, in February 2010 Amnon Besser and I gave introductory lecture series at Universität Heidelberg, following the kind request of Jakob Stix. These notes constitute a revised version of the ones I distributed during the lectures. They begin with a quick introduction to Grothendieck's concept of the algebraic fundamental group. After a reminder on basic results concerning fundamental groups of curves, we move on to discuss what is arguably the most famous open problem in the area, Grothendieck's Section Conjecture. The next section presents in detail a beautiful application of the ideas involved in the conjecture: Parshin's *hyperbolic* proof of the geometric case of Mordell's conjecture. The final section gives an overview of the most important features of anabelian geometry.

I thank Jakob Stix for giving me an opportunity to deliver the lectures and his warm hospitality at Heidelberg, as well as the referee for a very careful reading of the text.

2.1 Grothendieck's Fundamental Group

Grothendieck's theory of the algebraic fundamental group is a common generalization of Galois theory and the theory of covers in topology. Let us briefly recall both. The proofs of all statements in this section can be found in [Sza09].

T. Szamuely (✉)
Alfréd Rényi Institute of Mathematics, Hungarian Academy of Sciences, PO Box 127, 1364 Budapest, Hungary
e-mail: szamuely@renyi.hu

* The author acknowledges partial support from OTKA grant No. NK81203.

J. Stix (ed.), *The Arithmetic of Fundamental Groups*, Contributions in Mathematical and Computational Sciences 2, DOI 10.1007/978-3-642-23905-2_2,

Let k be a field. Recall that a finite dimensional k-algebra A is **étale** over k if it is isomorphic to a finite direct product of separable extensions of k. Fix a separable closure $k_s|k$. The $\mathrm{Gal}(k_s|k)$-action on k_s induces a left action on the set of k-algebra homomorphisms $\mathrm{Hom}_k(\mathrm{A}, k_s)$. The rule $\mathrm{A} \mapsto \mathrm{Hom}(\mathrm{A}, k_s)$ is a contravariant functor. The main theorem of Galois theory in Grothendieck's version is the following statement.

Theorem 1. *The contravariant functor* $\mathrm{F} : \mathrm{A} \mapsto \mathrm{Hom}_k(\mathrm{A}, k_s)$ *gives an anti-equivalence between the category of finite étale k-algebras and the category of finite sets with continuous left* $\mathrm{Gal}(k_s|k)$*-action.*

Note that the functor F depends on the choice of the separable closure k_s. The latter is not a finite étale k-algebra but a *direct limit* of such. Also, one checks that $\mathrm{Gal}(k_s|k)$ is naturally isomorphic to the **automorphism group** of the functor F, i.e., the group of natural isomorphisms $\mathrm{F} \xrightarrow{\sim} \mathrm{F}$.

Now to the topological situation. Let X be a connected, locally connected and locally simply connected topological space. Recall that a **cover** of X is a space Y equipped with a continuous map $p : \mathrm{Y} \to \mathrm{X}$ subject to the following condition: each point of X has an open neighbourhood V for which $p^{-1}(\mathrm{V})$ decomposes as a disjoint union of open subsets U_i of Y such that the restriction of p to each U_i induces a homeomorphism of U_i with V.

Given a point $x \in \mathrm{X}$, the fundamental group $\pi_1(\mathrm{X}, x)$ has a natural left action on the fibre $p^{-1}(x)$ defined as follows: given $\alpha \in \pi_1(\mathrm{X}, x)$ represented by a closed path $f : [0,1] \to \mathrm{X}$ with $f(0) = f(1) = x$ as well as a point $y \in p^{-1}(x)$, we define $\alpha y := \tilde{f}(1)$, where $\tilde{f}$ is the unique lifting of the path f to Y with $\tilde{f}(0) = y$. One checks that this indeed gives a well-defined left action of $\pi_1(\mathrm{X}, x)$, once we make the convention that the product fg of two paths f, g is given by going through g first and then through f. It is called the **monodromy action.**

Theorem 2. *The functor* Fib_x *sending a cover* $p : \mathrm{Y} \to \mathrm{X}$ *to the fibre* $p^{-1}(x)$ *equipped with the monodromy action induces an equivalence of the category of covers of* X *with the category of left* $\pi_1(\mathrm{X}, x)$*-sets.*

Here again, the functor Fib_x depends on the choice of the point x. It is in fact **representable** by a cover $\pi : \tilde{\mathrm{X}}_x \to \mathrm{X}$, i.e., we have an isomorphism of functors

$$\mathrm{Fib}_x \cong \mathrm{Hom}(\tilde{\mathrm{X}}_x, -)\,.$$

The space $\tilde{\mathrm{X}}_x$ can be constructed as the space of homotopy classes of paths starting from x, the projection π mapping the class of a path to its other endpoint. As a consequence, we have isomorphisms

$$\mathrm{Aut}(\tilde{\mathrm{X}}_x)^{\mathrm{op}} \cong \mathrm{Aut}(\mathrm{Fib}_x) \cong \pi_1(\mathrm{X}, x)\,.$$

Here $\mathrm{Aut}(\tilde{\mathrm{X}}_x)^{\mathrm{op}}$ denotes the opposite group to $\mathrm{Aut}(\tilde{\mathrm{X}}_x)$, i.e., the group with the same underlying set but with multiplication given by $(x, y) \mapsto yx$. We have to pass to the opposite group because the natural left action of $\mathrm{Aut}(\widetilde{X}_x)$ on $\widetilde{X}$ induces a right action after applying the functor $\mathrm{Hom}(\widetilde{X}_x, -)$.

Here is an important consequence. Call a cover $Y \to X$ **finite** if it has finite fibres; for connected X these have the same cardinality, called the **degree** of X.

Corollary 3. *For* X *and* x *as in Theorem 2, the functor* Fib_x *induces an equivalence of the category of* finite *covers of* X *with the category of finite continuous left* $\widehat{\pi_1(X,x)}$*-sets.*

Here $\widehat{\pi_1(X,x)}$ denotes the **profinite completion** of $\pi_1(X,x)$, i.e., the inverse limit of the natural inverse system of its finite quotients.

We can now come to Grothendieck's common generalization in algebraic geometry. Let S be a connected scheme. By a **finite étale cover** of S we mean a finite étale map $X \to S$. In particular, it is surjective and each fibre at a point $s \in S$ is the spectrum of a finite étale $\kappa(s)$-algebra. Fix a geometric point $\bar{s} : \mathrm{Spec}(\Omega) \to S$. For a finite étale cover $X \to S$ we consider the geometric fibre $X \times_S \mathrm{Spec}(\Omega)$ over $\bar{s}$, and denote by $\mathrm{Fib}_{\bar{s}}(X)$ its underlying set. This gives a set-valued functor on the category of finite étale covers of X.

We *define* the **algebraic fundamental group** $\pi_1(S,\bar{s})$ as the automorphism group of this functor. By definition an automorphism of $\mathrm{Fib}_{\bar{s}}$ induces an automorphism of the set $\mathrm{Fib}_{\bar{s}}(X)$ for each finite étale cover X. In this way we obtain a natural left action of $\pi_1(S,\bar{s})$ on the set $\mathrm{Fib}_{\bar{s}}(X)$.

Theorem 4. (Grothendieck) *Let* S *be a connected scheme, and* $\bar{s} : \mathrm{Spec}(\Omega) \to S$ *a geometric point.*

(1) The group $\pi_1(S,\bar{s})$ *is profinite, and its action on* $\mathrm{Fib}_{\bar{s}}(X)$ *is continuous for every finite étale cover* $X \to S$.
(2) The functor $\mathrm{Fib}_{\bar{s}}$ *induces an equivalence of the category of finite étale covers of* S *with the category of finite continuous left* $\pi_1(S,\bar{s})$*-sets.*

Here the functor $\mathrm{Fib}_{\bar{s}}$ is **pro-representable**, which means that there exists a filtered inverse system $P = (P_\alpha, \phi_{\alpha\beta})$ of finite étale covers and a functorial isomorphism

$$\varinjlim \mathrm{Hom}(P_\alpha, X) \cong \mathrm{Fib}_{\bar{s}}(X)\,.$$

The automorphism group of each finite étale cover $P_\alpha \to S$ is finite, and

$$\pi_1(S,\bar{s}) = \varprojlim_{\alpha} \mathrm{Aut}(P_\alpha)^{\mathrm{op}}\,,$$

which explains its profiniteness. In fact, Grothendieck showed that one may choose as a pro-representing system the system $P_\alpha \to X$ of all **Galois covers**, i.e., those connected finite étale covers for which $\mathrm{Aut}(P_\alpha|S)$ acts transitively on geometric fibres. These are turned into an inverse system by choosing a distinguished point $p_\alpha \in \mathrm{Fib}_{\bar{s}}(P_\alpha)$ for each α. For each pair α,β there is then at most one S-morphism $P_\beta \to P_\alpha$ sending p_β to p_α. We define this map to be $\phi_{\alpha\beta}$, if it exists.

Remark 5. Any two fibre functors on the category of finite étale S-schemes are (non-canonically) isomorphic. One way to prove this is to use pro-representability of the fibre functor which reduces the construction of an isomorphism between functors to the construction of a compatible system of automorphisms of the Galois objects P_α transforming one system of maps $\phi_{\alpha\beta}$ to another. This can be done by means of a compactness argument.

An isomorphism between two fibre functors

$$\gamma : \mathrm{Fib}_{\bar{s}} \xrightarrow{\sim} \mathrm{Fib}_{\bar{s}'}$$

is called a **path** from $\bar{s}$ to $\bar{s}'$. It induces an isomorphism of fundamental groups

$$\gamma(-)\gamma^{-1} : \pi_1(S, \bar{s}) \xrightarrow{\sim} \pi_1(S, \bar{s}') .$$

In the topological situation such an isomorphism is induced by the choice of a usual path between base points, whence the name in the algebraic situation. As in topology, the two isomorphisms induced by different paths differ by an inner automorphism of $\pi_1(S, \bar{s})$.

Remark 6. Historically, the case of a normal scheme was known earlier. If S is an integral normal Noetherian scheme, denote by K_s a fixed separable closure of the function field K of S, and by K_S the composite of all finite subextensions L|K of K_s such that the normalization of S in L is étale over S. Then $K_S|K$ is a Galois extension, and it can be shown that the Galois group $\mathrm{Gal}(K_S|K)$ is canonically isomorphic to the fundamental group $\pi_1(S, \bar{s})$ for the geometric point $\bar{s} : \mathrm{Spec}(\overline{K}) \to S$, where $\overline{K}$ is an algebraic closure of K containing K_s.

The following examples show that the algebraic fundamental group indeed yields a common generalization of the algebraic and topological cases.

Example 7. (1) For $X = \mathrm{Spec}(k)$, and $\bar{x} : \mathrm{Spec}(\bar{k}) \to \mathrm{Spec}(k)$ we have

$$\pi_1(X, \bar{x}) \cong \mathrm{Gal}(k_s|k) .$$

This holds basically because finite étale Spec(k)-schemes are spectra of finite étale k-algebras.

(2) For X of finite type over $\mathbb{C}$ and $\bar{x} : \mathrm{Spec}(\mathbb{C}) \to X$ there is a canonical isomorphism

$$\widehat{\pi_1^{\mathrm{top}}(X^{\mathrm{an}}, \bar{x})} \xrightarrow{\sim} \pi_1(X, \bar{x})$$

where on the left hand side we have the profinite completion of the topological fundamental group of X^{an} with base point the image of $\bar{x}$, and X^{an} denotes the complex analytic space associated with X.

This isomorphism relies on a deep algebraization theorem for finite topological covers of schemes of finite type over $\mathbb{C}$.

A base point preserving morphism of schemes induces a continuous homomorphism of fundamental groups. To construct it, let S and S′ be connected schemes, equipped with geometric points $\bar{s} : \mathrm{Spec}(\Omega) \to \mathrm{S}$ and $\bar{s}' : \mathrm{Spec}(\Omega) \to \mathrm{S}'$, respectively. Assume given a morphism $\phi : \mathrm{S}' \to \mathrm{S}$ with $\phi \circ \bar{s}' = \bar{s}$. For a finite étale cover $\mathrm{X} \to \mathrm{S}$ consider the base change $\mathrm{X} \times_{\mathrm{S}} \mathrm{S}' \to \mathrm{S}'$. The condition $\phi \circ \bar{s}' = \bar{s}$ implies that $\mathrm{Fib}_{\bar{s}}(\mathrm{X}) = \mathrm{Fib}_{\bar{s}'}(\mathrm{X} \times_{\mathrm{S}} \mathrm{S}')$. This construction is functorial in X, and thus every automorphism of the functor $\mathrm{Fib}_{\bar{s}'}$ induces an automorphism of $\mathrm{Fib}_{\bar{s}}$, which defines the required map $\phi_* : \pi_1(\mathrm{S}', \bar{s}') \to \pi_1(\mathrm{S}, \bar{s})$.

The above functoriality, together with Example 7 (1), defines the maps in the following exact sequence which is fundamental not only because it involves fundamental groups.

Proposition 8. *Let* X *be a quasi-compact and geometrically connected scheme over a field k. Fix an algebraic closure* $\bar{k}$ *of k, and let* $k_s|k$ *be the corresponding separable closure. Write* $\overline{\mathrm{X}} := \mathrm{X} \times_{\mathrm{Spec}(k)} \mathrm{Spec}(k_s)$*, and let* $\bar{x}$ *be a geometric point of* $\overline{\mathrm{X}}$ *with values in* $\bar{k}$*. The sequence of profinite groups*

$$1 \to \pi_1(\overline{\mathrm{X}}, \bar{x}) \to \pi_1(\mathrm{X}, \bar{x}) \to \mathrm{Gal}(k_s|k) \to 1 \tag{2.1}$$

induced by the maps $\overline{\mathrm{X}} \to \mathrm{X}$ *and* $\mathrm{X} \to \mathrm{Spec}(k)$ *is exact.*

The group $\pi_1(\mathrm{X}, \bar{x})$ acts on its normal subgroup $\pi_1(\overline{\mathrm{X}}, \bar{x})$ via conjugation, whence a map

$$\phi_{\mathrm{X}} : \pi_1(\mathrm{X}, \bar{x}) \to \mathrm{Aut}(\pi_1(\overline{\mathrm{X}}, \bar{x}))\,.$$

Inside $\mathrm{Aut}(\pi_1(\overline{\mathrm{X}}, \bar{x}))$ we have the normal subgroup $\mathrm{Inn}(\pi_1(\overline{\mathrm{X}}, \bar{x}))$ of inner automorphisms; the quotient is the group $\mathrm{Out}(\pi_1(\overline{\mathrm{X}}, \bar{x}))$ of **outer automorphisms**. By the commutative diagram

$$\begin{array}{ccccccccc}
1 & \longrightarrow & \pi_1(\overline{\mathrm{X}}, \bar{x}) & \longrightarrow & \pi_1(\mathrm{X}, \bar{x}) & \longrightarrow & \mathrm{Gal}(k_s|k) & \longrightarrow & 1 \\
 & & \downarrow & & \downarrow & & \downarrow & & \\
1 & \longrightarrow & \mathrm{Inn}(\pi_1(\overline{\mathrm{X}}, \bar{x})) & \longrightarrow & \mathrm{Aut}(\pi_1(\overline{\mathrm{X}}, \bar{x})) & \longrightarrow & \mathrm{Out}(\pi_1(\overline{\mathrm{X}}, \bar{x})) & \longrightarrow & 1
\end{array}$$

we get an important representation

$$\rho_{\mathrm{X}} : \mathrm{Gal}(k_s|k) \to \mathrm{Out}(\pi_1(\overline{\mathrm{X}}, \bar{x}))$$

called the **outer Galois representation**. It will appear several times in subsequent sections.

Example 9. Assume X is a smooth proper curve of genus g, and fix a prime number ℓ different from the characteristic of k. As we shall see in Remark 11, the maximal abelian pro-ℓ-quotient of $\pi_1(\mathrm{X} \times_k \bar{k}, \bar{x})$ is isomorphic to the Tate module $\mathrm{T}_\ell(\bar{\mathrm{J}})$ of the Jacobian $\bar{\mathrm{J}}$ of $\mathrm{X} \times_k \bar{k}$. Taking the pushout of the sequence (2.1) by the natural map $\pi_1(\mathrm{X} \times_k \bar{k}) \to \mathrm{T}_\ell(\bar{\mathrm{J}})$ we obtain an extension of $\mathrm{Gal}(\bar{k}|k)$ by $\mathrm{T}_\ell(\bar{\mathrm{J}})$. By the same argument as above it gives rise to a Galois representation

$$\mathrm{Gal}(k_s|k) \to \mathrm{Aut}(\mathrm{T}_\ell(\bar{\mathrm{J}})) \cong \mathrm{GL}_{2g}(\mathbb{Z}_\ell)\,.$$

It is none but the usual Galois representation on torsion points of the Jacobian, a central object of study in number theory. The outer Galois representation can thus be viewed as a non-abelian generalization.

2.2 Fundamental Groups of Curves

In the first part of this section k denotes an algebraically closed field of characteristic $p \geqslant 0$ and X a proper smooth curve over k. We recall some basic structure results about the fundamental group of X and of its open subschemes. As they concern the groups up to isomorphism, we drop base points from the notation.

Theorem 10. (Grothendieck [SGA1]) *Let* $U \subset X$ *be an open subcurve possibly equal to* X, *and* $n \geqslant 0$ *the number of closed points in* $X \setminus U$. *Then* $\pi_1(U)^{(p')}$ *is isomorphic to the profinite* p'*-completion of the group*

$$\Pi_{g,n} := \langle a_1, b_1, \ldots, a_g, b_g, \gamma_1, \ldots, \gamma_n \mid [a_1, b_1] \ldots [a_g, b_g] \gamma_1 \ldots \gamma_n = 1 \rangle .$$

Here $G^{(p')}$ denotes the **maximal prime-to-p quotient** of the profinite group G, i.e., the inverse limit of its finite continuous quotients of order prime to p. For $p = 0$ we define it to be G itself.

For $k = \mathbb{C}$ the theorem follows via Example 7 (2) from the well-known structure of the topological fundamental group. In this particular case the underlying algebraization theorem is just the Riemann existence theorem of complex analysis. One deduces the result for k of characteristic 0 using a rigidity theorem (see e.g. [Sza09, Proposition 5.6.7 and Remark 5.7.8]) which says that the fundamental group of a smooth curve does not change under extensions of algebraically closed fields of characteristic 0. This also holds in positive characteristic, but only for proper curves.

In positive characteristic Grothendieck proved the result by first lifting the curve to characteristic 0 and then proving a specialization theorem establishing an isomorphism between maximal prime-to-p quotients of the fundamental groups of the curve and its lifting. Thus this case also relies on the topological result over $\mathbb{C}$. However, Wingberg [Win84] was able to prove using delicate group-theoretic arguments that for $\ell \neq p$ a prime the maximal pro-ℓ quotients of $\pi_1(U)$ have the above structure (in the paper this is stated only for $k = \overline{\mathbb{F}}_p$ but the argument works in general).

Remark 11. For X proper and $\ell \neq p$ a prime, the theorem implies that the maximal abelian pro-ℓ-quotient of $\pi_1(X)$ is isomorphic to $\mathbb{Z}_\ell^{2g}$. On the other hand, for J the Jacobian of X, the Tate module $T_\ell(J)$ has the same structure. This is not a coincidence: by a theorem of Serre and Lang [LS57] (see also [Sza09, Theorem 5.6.10]) every finite étale cover of J of ℓ-power degree is a quotient of some cover given by

$$0 \to {}_{\ell^n}J \to J \xrightarrow{\ell^n} J \to 0 .$$

On the other hand, given some embedding $X \to J$ obtained by sending a point x to the divisor class of $x-\xi$ for a fixed base point ξ, the induced map on fundamental groups

$$\pi_1(X,\xi) \to \pi_1(J,0)$$

becomes an isomorphism on the maximal prime-to-p abelian quotient: abelian prime-to-p covers are obtained via pullback from the Jacobian.

There is also a generalization to open curves: if $U \subset X$ is an open subcurve, one still identifies the maximal abelian pro-ℓ-quotient of $\pi_1(U)$ with the ℓ-adic Tate module of the Jacobian $\widetilde{J}$ of U. The latter is a commutative group variety which is an extension of J by the $(n-1)$-st power of the multiplicative group $\mathbb{G}_m$, in accordance with the theorem, see e.g. [KL81, (2.7)] .

The **maximal pro-p quotient** $G^{(p)}$ of G is defined as the inverse limit of finite quotients of p-power order, and we have the following result.

Theorem 12. (Shafarevich [Sha47]) *Assume $p > 0$. Then $\pi_1(X)^{(p)}$ is a free pro-p group of finite rank equal to the p-rank of the Jacobian variety of* X.

For an open subcurve $U \neq X$ the group $\pi_1(U)^{(p)}$ is a free pro-p group of infinite rank equal to the cardinality of k.

Here recall that the p-**rank** of an abelian variety A over an algebraically closed field k of characteristic $p > 0$ is the dimension of the $\mathbf{F}_p$-vector space given by the kernel of the multiplication-by-p map on the k-points of A. It is a nonnegative integer bounded by $\dim A$.

Using methods of étale cohomology one can give a quick proof of this theorem, see e.g. [Gil00] for details. It is based on the group-theoretic fact that a pro-p-group G is free if and only if the Galois cohomology groups $H^i(G,\mathbb{Z}/p\mathbb{Z})$ vanish for $i > 1$. In the case $G = \pi_1(X)^{(p)}$ they can be identified with the étale cohomology groups $H^i_{\text{ét}}(X,\mathbb{Z}/p\mathbb{Z})$ of X using arguments of cohomological dimension. The latter groups are known to vanish for $i > 1$. The rank is then equal to that of the maximal abelian quotient, i.e., the dual of $H^1_{\text{ét}}(X,\mathbb{Z}/p\mathbb{Z})$, and thus can be determined using Artin–Schreier theory.

Remark 13. Observe that Theorems 10 and 12 do not elucidate completely the structure of the fundamental group of an integral normal curve over an algebraically closed field of positive characteristic. This is still unknown at the present day. The theorems give, however, a good description of its maximal abelian quotient: this group is the direct sum of its maximal prime-to-p and pro-p quotients, and hence the previous two theorems together suffice to describe it.

Concerning curves over non-algebraically closed fields, a much-studied object is the outer Galois representation

$$\rho_X : \mathrm{Gal}(k_s|k) \to \mathrm{Out}(\pi_1(\overline{X},\bar{x}))$$

over fields of arithmetic interest. One of the basic results is the following.

Theorem 14. (Matsumoto [Mat96]) *If k is a number field and X is affine such that $\overline{X}$ has non-commutative fundamental group, then ρ_X is injective.*

Recently, Hoshi and Mochizuki [HM09] proved that the result holds for proper curves of genus > 1 as well. One can easily decide using Theorem 10 which curves have noncommutative geometric fundamental group: those for which

$$(g,n) \neq (0,0),(0,1),(0,2),(1,0).$$

These are the **hyperbolic** curves: their fundamental groups are center-free and, for $n > 0$, even free.

The case $(g,n) = (0,3)$ is due to Belyi [Bel79] and is a consequence of his famous theorem stating that every smooth proper curve definable over a number field can be realized as a finite cover of $\mathbb{P}^1$ branched above at most 3 points. The proof of the general case uses different methods.

2.3 Grothendieck's Section Conjecture

Arguably the most famous open question concerning fundamental groups of curves is Grothendieck's Section Conjecture, stated in [Gro83]. It concerns the exact sequence

$$1 \to \pi_1(\overline{X},\bar{x}) \to \pi_1(X,\bar{x}) \xrightarrow{p_*} \mathrm{Gal}(k_s|k) \to 1 \tag{2.2}$$

of Proposition 8, where $p : X \to \mathrm{Spec}\, k$ is the structure map.

Any k-rational point $y : \mathrm{Spec}\, k \to X$ induces by functoriality a map

$$\sigma_y : \mathrm{Gal}(k_s|k) \to \pi_1(X,\bar{y})$$

for a geometric point $\bar{y}$ lying above y. This is not quite a splitting of the exact sequence above because of the difference of base points. But the choice of a path (see Remark 5) from $\bar{y}$ to $\bar{x}$ on $\overline{X}$ induces an isomorphism

$$\lambda : \pi_1(X,\bar{y}) \xrightarrow{\sim} \pi_1(X,\bar{x}).$$

Changing the path on $\overline{X}$ is reflected by an inner automorphism of $\pi_1(X,\bar{x})$, more precisely by an element of $\pi_1(\overline{X},\bar{x})$. The composite $\lambda \circ \sigma_y$ is then a section of the exact sequence uniquely determined up to conjugation by elements of $\pi_1(\overline{X},\bar{x})$. We thus obtain a map

$$X(k) \to \{\pi_1(\overline{X},\bar{x})\text{-conjugacy classes of sections of } p_*\}. \tag{2.3}$$

The Section Conjecture now states:

Conjecture 15. (Grothendieck [Gro83]) If k is finitely generated over $\mathbb{Q}$ and X is a smooth projective curve of genus $g \geqslant 2$, then the above map is a bijection.

Remark 16. Grothendieck also formulated a variant of the conjecture for open curves. Its formulation is, however, more complicated than the above because one has to circumvent the fact that for an affine curve U with smooth compactification X there are tons of sections coming from rational points of X \ U (see [EH08]).

Injectivity is not hard to prove and was known to Grothendieck. As the argument works over other base fields as well, we include a slightly more general statement.

Proposition 17. *Let* X *be a smooth projective curve of genus* $g > 0$ *over a field* k*. Assume any of the following:*

(i) k *is finite.*
(ii) k *is p-adic.*
(iii) k *is finitely generated over* $\mathbb{Q}$*.*

Then the map (2.3) is injective.

Proof. The proof will show that even the sections of the exact sequence

$$0 \to \pi_1^{ab}(\overline{X}) \to \Pi \to \mathrm{Gal}(k_s|k) \to 1$$

obtained from (2.2) by pushout via the abelianization map $\pi_1(\overline{X},\bar{x}) \to \pi_1^{ab}(\overline{X})$ separate the k-points of X.

Set $\Gamma := \mathrm{Gal}(\bar{k}|k)$. Fix a k-point y_0 of X and denote by s_0 the corresponding section $\Gamma \to \pi_1(X,\bar{x})$. Given another k-point y of X with corresponding section s, the composite map

$$\Gamma \to \pi_1(X,\bar{x}) \to \pi_1^{ab}(X)$$

induced by $s_0 s^{-1}$ has image in $\pi_1^{ab}(\overline{X})$ and is a continuous 1-cocycle. We thus get compatible classes in

$$H^1(\Gamma, \pi_1^{ab}(\overline{X})/m)$$

for all $m > 0$. Denoting by J the Jacobian of X we have a Galois-equivariant isomorphism $\pi_1^{ab}(\overline{X})/m \cong {}_m\overline{J}$ (see Remark 11) so we actually get maps

$$\mathrm{Div}^0(X) \to H^1(\Gamma, {}_m\overline{J})$$

for all m, where $\mathrm{Div}^0(X)$ is the group of degree 0 divisors on X. Moreover, it is an exercise to check the commutativity of the diagram

$$\begin{array}{ccc} \mathrm{Div}^0(X) & \longrightarrow & H^1(\Gamma, {}_m\overline{J}) \\ \downarrow & & \uparrow \\ J(k) & \xrightarrow{\cong} & J(\bar{k})^{\Gamma} \end{array}$$

where the right vertical map comes from the Kummer sequence

$$J(\bar{k})^{\Gamma} \xrightarrow{m} J(\bar{k})^{\Gamma} \to H^1(\Gamma, {}_m\overline{J}).$$

By this commutativity, if we assume $s = s_0$, the class of the divisor $y - y_0$ lies in the kernel of the Kummer map $J(\bar{k})^{\Gamma} \to H^1(\Gamma, {}_m\bar{J})$ for all m, i.e., it is divisible in $J(k)$. But if k satisfies any of the assumptions above, the group $J(k)$ has trivial divisible subgroup: over a finite k it is finite, over a p-adic k it has a finite index subgroup isomorphic to a finite direct power of $\mathbb{Z}_p$ by Mattuck [Mat55], and for k finitely generated over $\mathbb{Q}$ it is finitely generated by the Mordell–Weil–Lang–Néron theorem [Lan91, Chap. I, Corollary 4.3]. Therefore the class of the divisor $y - y_0$ is trivial in $J(k)$. As X has positive genus, we conclude that $y = y_0$. □

Remark 18. The injectivity result of the proposition holds for open hyperbolic curves U as well, with basically the same proof. The role of J is played by the generalized Jacobian $\widetilde{J}$ encountered in the second paragraph of Remark 11.

Remark 19. Assume k is a subfield of a finitely generated extension of $\mathbb{Q}_p$ for some prime p, for example a p-adic field or a finitely generated field over $\mathbb{Q}$. Consider the exact sequence

$$1 \to \pi_1(\overline{X}, \bar{x})^{(p)} \to \Pi^p \to \mathrm{Gal}(k_s|k) \to 1 \tag{2.4}$$

obtained from (2.2) by pushout via the map $\pi_1(\overline{X}, \bar{x}) \to \pi_1(\overline{X}, \bar{x})^{(p)}$. Mochizuki has shown in [Moc99, Theorem 19.1] as a consequence of his anabelian characterization of hyperbolic curves (see Theorem 33 below) that the injectivity result of the proposition (and also its generalization as in the previous remark) remains valid for splittings of (2.4) that come from k-points of X. On the other hand, Hoshi [Hos10] recently gave examples of curves over number fields where not all splittings of (2.4) come from k-points. Therefore the *pro-p-version* of the section conjecture is false.

So much about injectivity in the section conjecture. As for surjectivity, it is widely open: at the time of writing not a single curve is known over a number field that has a rational point and for which the map (2.3) is proven to be bijective, or at least for which the finiteness of conjugacy classes of sections is known. The latter statement would yield another proof of Mordell's conjecture. In the next section we shall see that over function fields over $\mathbb{C}$ statements of this type can actually be proven.

Let us mention, however, a nice observation that goes back to Tamagawa [Tam97] that was first stated in [Koe05].

Proposition 20. *Conjecture 15 is equivalent to the following seemingly weaker statement: if k is finitely generated over $\mathbb{Q}$ and X is a smooth projective curve of genus $g \geqslant 2$, then X has a k-rational point if and only if the sequence (2.2) splits.*

The proposition does *not* claim that for a given curve the splitting of (2.2) implies the bijectivity of (2.3); one has to consider *all* curves.

The proof is based on the following lemma which has many other applications.

Lemma 21. (Tamagawa) *Let X be a smooth curve over a field k. Assume k satisfies one of the assumptions as in Proposition 17; in the third case assume moreover that the smooth compactification of X has genus $\geqslant 2$.*

A section $s : \mathrm{Gal}(k_s|k) \to \pi_1(X, \bar{x})$ comes from a k-point if and only if for each open subgroup $H \subset \pi_1(X, \bar{x})$ containing $s(\mathrm{Gal}(k_s|k))$ the corresponding cover X_H has a k-point.

Proof. Let $\widetilde{X}$ be the "universal cover" of X, i.e., the normalization of X in the extension $K_X|K$ of Remark 6. If s is a section coming from a k-rational point P, the image $s(\mathrm{Gal}(k_s|k))$ is the stabilizer in $\pi_1(X,\bar{x})$ of a closed point Q of $\widetilde{X}$ above P. For $H \supset s(\mathrm{Gal}(k_s|k))$ it is also the stabilizer of Q under the action of $\pi_1(X_H,\bar{x})$. In particular, it maps onto $\mathrm{Gal}(k_s|k)$ under the projection

$$\pi_1(X_H,\bar{x}) \to \mathrm{Gal}(k_s|k)\,.$$

This means that the image of Q by the projection $\widetilde{X} \to X_H$ is k-rational.

For the converse we choose $\bar{x}$ to be a geometric generic point. The sets $X_H(k)$ form a natural inverse system indexed by the subgroups $H \subset s(\mathrm{Gal}(k_s|k))$. Under any of the three assumptions on k these sets are compact in their natural topology. In the first two cases this is immediate but the third case is based on a highly nontrivial input: Faltings's theorem on the finiteness of $X_H(k)$. Hence the inverse limit of the $X_H(k)$ is nonempty. An element of the inverse limit defines a point of $\widetilde{X}$ whose image in X induces s. □

The proposition follows from the lemma because the splitting of (2.2) for X by a section $s : \mathrm{Gal}(k_s|k) \to \pi_1(X,\bar{x})$ implies its splitting for X_H when $H \supset s(\mathrm{Gal}(k_s|k))$.

Remark 22. In recent years a birational analogue of the Section Conjecture was also studied. If K is the function field of a smooth proper curve X over a field k of characteristic 0, a k-rational point P of X induces a conjugacy class of sections of the exact sequence of Galois groups

$$1 \to \mathrm{Gal}(\overline{K}|K\bar{k}) \to \mathrm{Gal}(\overline{K}|K) \to \mathrm{Gal}(\bar{k}|k) \to 1\,.$$

Indeed, the local ring of P is a discrete valuation ring with fraction field K and residue field k. A decomposition group $D_P \subset \mathrm{Gal}(\overline{K}|K)$ of this valuation is isomorphic to $\mathrm{Gal}(\overline{k((t))}|k((t)))$. The natural projection

$$\mathrm{Gal}(\overline{k((t))}|k((t))) \to \mathrm{Gal}(\bar{k}|k)$$

has a section. Its image is the subgroup of $\mathrm{Gal}(\overline{k((t))}|k((t)))$ fixing the extension of $k((t))$ obtained by adjoining the n-th roots of t for all $n > 1$. The composite map

$$\mathrm{Gal}(\bar{k}|k) \to \mathrm{Gal}(\overline{k((t))}|k((t))) \xrightarrow{\sim} D_P \to \mathrm{Gal}(\overline{K}|K)$$

is a section as required.

Each k-point P is uniquely determined by the conjugacy classes of sections it induces because the arising D_P pairwise intersect trivially by an old theorem of F. K. Schmidt [Sch33]. One may then ask whether the analogue of the Section Conjecture holds over arithmetically interesting fields. Koenigsmann [Koe05] observed that the answer is yes if k is a p-adic field (see also [Pop10b] for a sharpened version). Over a global field there are only partial results, but in contrast to the original conjecture of Grothendieck at least examples are known of curves having rational points where the answer is positive [HS12, Sto07].

2.4 Parshin's Proof of Mordell's Conjecture over Function Fields

Let B be a smooth projective connected curve over the field $\mathbb{C}$ of complex numbers, and let C be a smooth projective geometrically connected curve defined over the function field $\mathbb{C}(\mathrm{B})$ of B. The following statement is usually called the geometric case of Mordell's Conjecture or the Mordell Conjecture for function fields of characteristic 0.

Theorem 23. *Assume that there is no finite extension* $\mathrm{K}|\mathbb{C}(\mathrm{B})$ *for which the base changed curve* $\mathrm{C}\times_{\mathbb{C}(\mathrm{B})}\mathrm{K}$ *can be defined over* $\mathbb{C}$. *Then* C *has only finitely many* $\mathbb{C}(\mathrm{B})$*-rational points.*

As a consequence, one gets the same result over finitely generated base fields of characteristic 0, assuming B geometrically integral.

This famous theorem has several proofs. The first one was given by Manin [Man63]. Coleman later discovered that it contained a gap which he was able to fill in [Col90]. The first complete published proof seems to be that of Grauert [Gra65]. Parshin himself gave two proofs, in [Par68] and [Par90]. It is the second one that we are going to explain now. As we shall see, it is partly inspired by the ideas explained in the previous section. We prove the following equivalent statement.

Theorem 24. *Let* V *be a smooth projective surface equipped with a proper flat morphism* $p: \mathrm{V}\to\mathrm{B}$ *with generic fibre* C *as above. If* V *as a family over* B *is non-isotrivial, then the projection* p *has only finitely many sections.*

Recall that the family $p: \mathrm{V}\to\mathrm{B}$ is **isotrivial** if there is a finite flat base change $\mathrm{B}'\to\mathrm{B}$ such that $\mathrm{V}\times_{\mathrm{B}}\mathrm{B}'\to\mathrm{B}'$ is a trivial family, i.e., isomorphic to $\mathrm{C}'\times\mathrm{B}'\to\mathrm{B}'$.

To see the equivalence of the two statements, note that one may find a smooth projective surface $\widetilde{\mathrm{V}}$ over $\mathbb{C}$ whose function field is that of the curve C of Theorem 23, by resolution of singularities for surfaces. The inclusion $\mathbb{C}(\mathrm{B})\to\mathbb{C}(\widetilde{\mathrm{V}})$ induces a rational map $\widetilde{\mathrm{V}}\to\mathrm{B}$ with generic fibre C. By elimination of indeterminacy we find a blowup V of $\widetilde{\mathrm{V}}$ in finitely many points equipped with a morphism $p: \mathrm{V}\to\mathrm{B}$ as required. A section of p induces a section on the generic fibre. On the other hand, by properness of V any section of the projection $\mathrm{C}\to\operatorname{Spec}\mathbb{C}(\mathrm{B})$ extends uniquely to a section of p.

Strategy of the proof of Theorem 24. Choose a Zariski open subset $\mathrm{B}_0\subset\mathrm{B}$ such that p is smooth over B_0. Fix a point $b_0\in\mathrm{B}_0$, and denote by F the fibre $p^{-1}(b_0)$. Fixing a base point $v_0\in\mathrm{F}$, we have a homotopy exact sequence of topological fundamental groups

$$1\to\pi_1^{\mathrm{top}}(\mathrm{F},v_0)\to\pi_1^{\mathrm{top}}(\mathrm{V}_0,v_0)\xrightarrow{p_*}\pi_1^{\mathrm{top}}(\mathrm{B}_0,b_0)\to 1$$

where $\mathrm{V}_0=p^{-1}(\mathrm{B}_0)$. A section $s_0: \mathrm{B}_0\to\mathrm{V}_0$ of p over B_0 meets F in a point v_1, whence a map

$$s_{0*}: \pi_1^{\mathrm{top}}(\mathrm{B}_0,b_0)\to\pi_1^{\mathrm{top}}(\mathrm{V}_0,v_1)\,.$$

Fixing a path from v_0 to v_1 induces an isomorphism

$$\pi_1^{\mathrm{top}}(\mathrm{V}_0,v_1)\xrightarrow{\sim}\pi_1^{\mathrm{top}}(\mathrm{V}_0,v_0)$$

unique up to inner automorphism. By composition s_{0*} induces a section of the map p_* above. Therefore we obtain a map

$$S : \{\text{sections of } p|_{V_0} : V_0 \to B_0\} \to \{\text{conjugacy classes of sections of } p_*\}\,.$$

As any section of p is determined by its restriction to B_0, the theorem follows from the two claims below. □

Claim 25. The map S has finite fibres.

Claim 26. The map S has finite image.

We begin with the proof of Claim 25. First we recall the notion of **K|*k*-trace** for abelian varieties. Given a field extension K|*k* and an abelian variety A over K, the K|*k*-trace $\mathrm{tr}_{K|k}(A)$ is the *k*-abelian variety characterized by the property that

$$\mathrm{Hom}(B_K, A) \xrightarrow{\sim} \mathrm{Hom}(B, \mathrm{tr}_{K|k}(A))$$

for all *k*-abelian varieties B. Its existence is a theorem of Chow, see [Kah06, Appendix A] or [Con06] for modern proofs. Applying the defining property with $B = \mathrm{Spec}(k)$ we obtain a bijection $A(K) \xrightarrow{\sim} \mathrm{tr}_{K|k}(A)(k)$. Its inverse is induced by the map $\tau : \mathrm{tr}_{K|k}(A)_K \to A$ obtained by setting $B = \mathrm{tr}_{K|k}(A)$ and taking the map corresponding to the identity. The image of τ is the maximal abelian subvariety of A defined over *k*.

Proof of Claim 25. The diagram

$$\begin{array}{ccc} C & \longrightarrow & \mathrm{Spec}\,\mathbb{C}(B) \\ \downarrow & & \downarrow \\ V_0 & \longrightarrow & B_0 \end{array} \qquad (2.5)$$

is Cartesian, so a section $s_0 : B_0 \to V_0$ induces a section $s : \mathrm{Spec}\,\mathbb{C}(B) \to C$; moreover, s_0 is uniquely determined by s. On the other hand, a section

$$\pi_1^{\mathrm{top}}(B_0, b_0) \to \pi_1^{\mathrm{top}}(V_0, v_0)$$

induces a map on profinite completions, i.e., a map $\pi_1(B_0, b_0) \to \pi_1(V_0, v_0)$ of *algebraic* fundamental groups. For some geometric point c_0 of C above v_0 the diagram of groups

$$\begin{array}{ccc} \pi_1(C, c_0) & \longrightarrow & \mathrm{Gal}(\overline{\mathbb{C}(B)}|\mathbb{C}(B)) \\ \downarrow & & \downarrow \\ \pi_1(V_0, v_0) & \longrightarrow & \pi_1(B_0, b_0) \end{array}$$

coming from diagram (2.5) commutes, the sections s and s_0 inducing compatible sections of the horizontal maps. Hence it is enough to show that the map

$$\mathrm{C}(\mathbb{C}(\mathrm{B})) \to \{\text{conjugacy classes of sections of } \pi_1(\mathrm{C}, c_0) \to \mathrm{Gal}(\overline{\mathbb{C}(\mathrm{B})}|\mathbb{C}(\mathrm{B}))\}$$

has finite fibres. This is done as in the injectivity part of the section conjecture. If y_0 is a a $\mathbb{C}(\mathrm{B})$-point of C and y another $\mathbb{C}(\mathrm{B})$-point inducing the same section

$$\mathrm{Gal}(\overline{\mathbb{C}(\mathrm{B})}|\mathbb{C}(\mathrm{B})) \to \pi_1(\mathrm{C}, c_0),$$

then the argument given there shows that the class of the divisor $y - y_0$ is divisible in $\mathrm{J}(\mathbb{C}(\mathrm{B}))$. But by the Lang–Néron theorem (see [Kah09] for a beautiful short proof) the group

$$\mathrm{J}(\mathbb{C}(\mathrm{B}))/\tau(\mathrm{tr}_{\mathbb{C}(\mathrm{B})|\mathbb{C}}(\mathrm{J})(\mathbb{C}))$$

is finitely generated and as such has no nontrivial divisible element. Therefore the image of y by the embedding $\mathrm{C} \to \mathrm{J}$ with base point y_0 lies in the image of the trace $\mathrm{tr}_{\mathbb{C}(\mathrm{B})|\mathbb{C}}(\mathrm{J})$. But if C is non-isotrivial, the whole of C cannot lie in the trace. This can be checked using the explicit construction of the trace in [Kah06]. Their intersection is thus a proper closed, hence finite subset of C, which shows that there can be only finitely many points y inducing the same section as y_0. □

The proof of Claim 26 is entirely topological. The idea is to bound the *size* of sections of p_* in a suitable way. This is accomplished using ideas of complex hyperbolic geometry, of which we summarize here some basic facts. See [Kob76] and [Lan87] for proofs and much more.

Equip the complex unit disc D with the Poincaré metric given by $z \mapsto (1-|z|^2)^{-1}$. It defines a distance function d_{hyp} on D which we may use to define the **Kobayashi pseudo-distance** on any complex manifold X:

$$d_{\mathrm{X}}(x,y) = \inf\left(\sum_{i=1}^{r} d_{\mathrm{hyp}}(p_i, q_i)\right)$$

where the infimum is taken over systems of points $p_i, q_i \in \mathrm{D}$ $(1 \leqslant i \leqslant r)$ for which there exist holomorphic maps $f_1, \ldots, f_r : \mathrm{D} \to \mathrm{X}$ with $f_1(p_1) = x$, $f_r(q_r) = y$ and $f_i(q_i) = f_{i+1}(p_{i+1})$. Holomorphic maps are distance-decreasing: if $\phi : \mathrm{X} \to \mathrm{Y}$ is a holomorphic map, then $d_{\mathrm{Y}}(\phi(x), \phi(y)) \leqslant d_{\mathrm{X}}(x,y)$. This follows from the case $\mathrm{X} = \mathrm{Y} = \mathrm{D}$, where it is a consequence of the Schwarz lemma.

The pseudo-distance d_{D} is identically 0, so d_{X} does not satisfy $d_{\mathrm{X}}(x,y) \neq 0$ for $x \neq y$ in general. The manifold X is said to be **hyperbolic** if $d_{\mathrm{X}}(x,y) \neq 0$ for $x \neq y$, and in this case we get a distance function that can be used to define the length of a path in X. Given a holomorphic map $\mathrm{X} \to \mathrm{Y}$ which is topologically a cover, it is known that the hyperbolicity of Y implies that of X.

By a classical theorem of Brody, a compact manifold X is hyperbolic if and only if there is no non-constant holomorphic map $\mathbb{C} \to \mathrm{X}$. In particular, a compact Riemann surface of genus $g > 1$ is hyperbolic and we may obtain a hyperbolic manifold from any compact Riemann surface after removing finitely may open discs. Also, a fibred complex manifold with base and fibre of this type is again hyperbolic.

So in our case we can make a hyperbolic manifold V' out of V_0 by removing the preimage of finitely many open discs in B, which we may assume to contain the finitely many deleted points of B. Write $B' = p(V')$ and assume that the fixed fibre F, and in particular the base point v_0, lies in V'. The inclusion $V' \hookrightarrow V_0$ then induces an isomorphism $\pi_1^{top}(V', v_0) \xrightarrow{\sim} \pi_1^{top}(V_0, v_0)$ since V' is a deformation retract of V_0. Similarly, we have a canonical isomorphism $\pi_1^{top}(B', b_0) \xrightarrow{\sim} \pi_1^{top}(B_0, b_0)$.

Lemma 27. *For each* $C > 0$ *there are only finitely many elements of* $\pi_1^{top}(V_0, v_0) \cong \pi_1^{top}(V', v_0)$ *that can be represented by paths lying in* V' *that have length at most* C *in the hyperbolic metric of* V'.

Proof. Consider the universal cover $\widetilde{V}' \to V'$. It carries a canonical holomorphic structure. Any holomorphic map $D \to V'$ lifts to $\widetilde{V}'$, therefore the definition of the pseudo-distance implies that liftings of paths of length $\leqslant C$ starting at v_0 stay inside a closed ball of radius C. As V' is hyperbolic, so is the cover $\widetilde{V}'$, and therefore the ball is compact. Closed paths around v_0 lift to paths with endpoints contained in a fixed orbit of $\pi_1^{top}(V', v_0)$. As these orbits are discrete, they intersect the compact ball in finitely many points. □

Now fix generators $x_1, \ldots, x_r$ of the finitely generated group $\pi_1^{top}(B_0, b_0)$. In view of the lemma, Claim 26 is a consequence of:

Proposition 28. *There exists a constant* $C > 0$ *such that for any section* $s : B_0 \to V_0$ *the images of* $x_1, \ldots, x_n$ *under the induced map*

$$\pi_1^{top}(B_0, b_0) \to \pi_1^{top}(V_0, v_0)$$

can be represented by paths lying in V' *that have length at most* C.

Proof. Let $s : B_0 \to V_0$ be a section, and let $s' : B' \to V'$ be its restriction to B'. We may identify the map $\pi_1^{top}(B_0, b_0) \to \pi_1^{top}(V_0, v_0)$ induced by s with the map

$$\pi_1^{top}(B', b_0) \to \pi_1^{top}(V', v_0)$$

induced by s' and hence we may assume that the x_i are represented by closed paths γ_i lying inside B'. As holomorphic maps are distance-decreasing, we have for points $x, y \in s'(B')$ a sequence of inequalities

$$d_{s'(B')}(x,y) \geqslant d_{V'}(x,y) \geqslant d_{B'}(p(x), p(y)) \geqslant d_{s'(B')}(x,y)$$

induced by the maps

$$s'(B') \hookrightarrow V' \xrightarrow{p} B' \xrightarrow{s'} s'(B').$$

Thus we have equality throughout, which shows that for each i the length of $s(\gamma_i)$ calculated with respect to $d_{V'}$ is the same as that of γ_i with respect to $d_{B'}$. This gives a uniform bound on the V'-length of the $s(\gamma_i)$. A representative of $s_*(x_i)$ in $\pi_1^{top}(V_0, v_0)$ is given by $\gamma s(\gamma_i)\gamma^{-1}$, where γ is a path lying in $F \subset V'$ joining v_0 to $s(b_0)$. But F is a compact hyperbolic Riemann surface, so we may join v_0 to any point by a path of length bounded by an absolute constant, e.g. a geodesic. This proves the proposition, and thereby Claim 26. □

2.5 Anabelian Geometry

By *anabelian geometry* one refers to a sheaf of conjectures formulated by Grothendieck in a famous letter to Faltings [Gro83]. The rough idea is that a certain category of schemes defined over finitely generated fields should be determined by their geometric fundamental groups together with its outer Galois action. There are two kinds of motivation for the conjectures. The first one comes from topology.

Fact 29. Recall that for a smooth proper curve X of genus $\geqslant 2$ over $\mathbb{C}$ the topological fundamental group has a presentation

$$\Pi = \langle a_1, b_1, \ldots, a_g, b_g \,|\, [a_1, b_1] \ldots [a_g, b_g] = 1 \rangle.$$

This group is non-commutative, and moreover, it has trivial center. The universal cover of X is the unit disc D which is contractible. Therefore the higher homotopy groups $\pi_q(\mathrm{X})$ are trivial for $q \geqslant 2$, and so X is the *Eilenberg-MacLane space* $\mathrm{K}(\Pi, 1)$. As such it is determined up to homotopy by $\Pi = \pi_1(\mathrm{K}(\Pi, 1))$.

As an algebraic curve, X may be defined over a finitely generated extension $k/\mathbb{Q}$. The hope therefore arises that the extra structure given by Galois action on Π may determine X up to algebraic isomorphism, not just up to homotopy.

The second motivation comes from the Tate conjecture.

Fact 30. Let k now be a number field, and let $\mathrm{X}_1, \mathrm{X}_2$ be smooth proper curves of genus $\geqslant 2$ over k. Assume for simplicity that both have a k-point. These k-points can be used to embed X_i in its Jacobian J_i. Write $\overline{\mathrm{X}}_i := \mathrm{X}_i \times_k \bar{k}$ and similarly for J_i. We know that for each prime ℓ and $i = 1, 2$

$$\mathrm{T}_\ell(\bar{\mathrm{J}}_i) \cong \pi_1^{\mathrm{ab}}(\bar{\mathrm{J}}_i)^{(\ell)} \cong \pi_1^{\mathrm{ab}}(\overline{\mathrm{X}}_i)^{(\ell)}$$

where T_ℓ stands for the ℓ-adic Tate module as in Remark 11.

By a fundamental theorem of Faltings [Fal83], the ex Tate conjecture, the natural map

$$\mathrm{Hom}(\mathrm{J}_1, \mathrm{J}_2) \otimes_{\mathbb{Z}} \mathbb{Z}_\ell \to \mathrm{Hom}_{\mathbb{Z}_\ell}(\mathrm{T}_\ell(\bar{\mathrm{J}}_1), \mathrm{T}_\ell(\bar{\mathrm{J}}_2))^{\mathrm{Gal}(\bar{k}|k)}$$

is bijective. In other words, Galois-equivariant homomorphisms $\mathrm{T}_\ell(\bar{\mathrm{J}}_1) \to \mathrm{T}_\ell(\bar{\mathrm{J}}_2)$ can be *approximated ℓ-adically* by morphisms $\mathrm{J}_1 \to \mathrm{J}_2$.

One can ask here whether working with the whole geometric fundamental group instead of its abelian quotient can give a stronger result: does a Galois-invariant outer homomorphism $\pi_1(\overline{\mathrm{X}}_1) \to \pi_1(\overline{\mathrm{X}}_2)$ come from a k-morphism $\mathrm{X}_1 \to \mathrm{X}_2$? Or, even more economically, does a Galois-invariant outer homomorphism

$$\pi_1(\overline{\mathrm{X}}_1)^{(\ell)} \to \pi_1(\overline{\mathrm{X}}_2)^{(\ell)}$$

between maximal pro-ℓ-quotients come from a map of curves?

Before formulating precise statements, let us elucidate the role of center-freeness. Recall that the representation

$$\rho_{\mathrm{X}} : \mathrm{Gal}(\bar{k}|k) \to \mathrm{Out}(\pi_1(\overline{\mathrm{X}}, \bar{x}))$$

is defined using the exact commutative diagram

$$\begin{array}{ccccccccc} 1 & \longrightarrow & \pi_1(\overline{\mathrm{X}}, \bar{x}) & \longrightarrow & \pi_1(\mathrm{X}, \bar{x}) & \longrightarrow & \mathrm{Gal}(\bar{k}|k) & \longrightarrow & 1 \\ & & \downarrow & & \downarrow & & \downarrow & & \\ 1 & \longrightarrow & \mathrm{Inn}(\pi_1(\overline{\mathrm{X}}, \bar{x})) & \longrightarrow & \mathrm{Aut}(\pi_1(\overline{\mathrm{X}}, \bar{x})) & \longrightarrow & \mathrm{Out}(\pi_1(\overline{\mathrm{X}}, \bar{x})) & \longrightarrow & 1 . \end{array}$$

Observe that, when the center of $\pi_1(\overline{\mathrm{X}}, \bar{x})$ is trivial, this becomes a pushout diagram. Therefore

$$\pi_1(\mathrm{X}, \bar{x}) \cong \mathrm{Aut}(\pi_1(\overline{\mathrm{X}}, \bar{x})) \times_{\mathrm{Out}(\overline{\mathrm{X}}, \bar{x})} \mathrm{Gal}(\bar{k}|k) ,$$

and $\pi_1(\mathrm{X}, \bar{x})$ is determined by $\pi_1(\overline{\mathrm{X}}, \bar{x})$ and ρ_{X}. When $k \subset \mathbb{C}$, it thus appears as a *transcendental object* endowed with a Galois action.

We now define a category of profinite groups as follows. Given two profinite groups G_1, G_2 together with morphisms $p_i : \mathrm{G}_i \to \mathrm{G}$, define $\mathrm{Hom}^*_{\mathrm{G}}(\mathrm{G}_1, \mathrm{G}_2)$ as the set of morphisms $\mathrm{G}_1 \to \mathrm{G}_2$ compatible with the p_i up to conjugation by an element of G. This set carries an action of G_1 from the left and of G_2 from the right. The latter defines a finer equivalence, so put

$$\mathrm{Hom}^{ext}_{\mathrm{G}}(\mathrm{G}_1, \mathrm{G}_2) = \mathrm{Hom}^*_{\mathrm{G}}(\mathrm{G}_1, \mathrm{G}_2)$$

modulo action of G_2. Fixing G we thus get a category $\mathbf{Prof}^{\mathrm{ext}}_{\mathrm{G}}$ with objects profinite groups with projections onto G and Hom-sets the $\mathrm{Hom}^{ext}_{\mathrm{G}}(\mathrm{G}_1, \mathrm{G}_2)$. Denote by $\mathbf{Prof}^{\mathrm{ext,open}}_{\mathrm{G}}$ the full subcategory with the same objects but with morphisms having open image. Sending a variety over a field k to its algebraic fundamental group gives a functor

$$\pi_1 : \{k\text{-varieties}\} \to \mathbf{Prof}^{\mathrm{ext}}_{\mathrm{Gal}(k)}$$

where base points do not play a role any more, so we drop them from now on. Similarly, sending a field to its absolute Galois group yields a contravariant functor

$$\mathrm{Gal} : \{\text{field extensions of } k\} \to \mathbf{Prof}^{ext}_{\mathrm{Gal}(k)} .$$

In his letter to Faltings, Grothendieck formulated the following conjecture.

Conjecture 31. Let k be a finitely generated extension of $\mathbb{Q}$. Denote by $\mathbf{Hyp}_k$ the category of hyperbolic k-curves equipped with *dominating* k-morphisms. Then

$$\pi_1 : \mathbf{Hyp}_k \to \mathbf{Prof}^{\mathrm{ext,open}}_{\mathrm{Gal}(k)}$$

is a fully faithful functor.

Recall that hyperbolic k-curves are the smooth k-curves of genus g with at least $2-2g+1$ geometric points at infinity. These are precisely the smooth curves with non-trivial center-free geometric π_1. Grothendieck also speculated about extending $\mathbf{Hyp}_k$ by including some higher-dimensional varieties called *anabelian varieties*. At present there is no precise conjectural characterization of anabelian varieties in dimensions > 1. However, there is a precisely formulated birational analogue:

Conjecture 32. Let k be finitely generated over $\mathbb{Q}$. Denote by $\mathbf{Bir}_k^{\mathrm{dom}}$ the category of fields finitely generated over k together with k-morphisms. Then

$$\mathrm{Gal} : \mathbf{Bir}_k^{\mathrm{dom}} \to \mathbf{Prof}_{\mathrm{Gal}(k)}^{ext,open}$$

is a fully faithful contravariant functor.

Here are the most important known results about these conjectures.

Theorem 33. (Mochizuki [Moc99]) *Conjecture 31 is true more generally for k sub-p-adic, i.e., a subfield of some finitely generated extension of* $\mathbb{Q}_p$. *In fact, over such fields the following holds: for a hyperbolic k-curve* X *and an arbitrary smooth k-variety* V *the map*

$$\mathrm{Hom}_k^{dom}(\mathrm{V},\mathrm{X}) \to \mathrm{Hom}_{\mathrm{Gal}(k)}^{\mathrm{ext,open}}(\pi_1(\mathrm{V}),\pi_1(\mathrm{X}))$$

is bijective. Here π_1 *may be replaced by its quotient* π_1^p *classifying covers whose base change to* $\bar{k}$ *is of p-power degree.*

This is all the more remarkable as the Tate conjecture does not hold over $\mathbb{Q}_p$. Concerning the birational version, we have:

Theorem 34. *(1)* (Pop, [Pop00, Sza04]) *The isomorphism version of conjecture 32 is true, even in positive characteristic. More precisely, if* K, L *are finitely generated fields over the prime field, the natural map*

$$\mathrm{Isom}^i(\mathrm{K},\mathrm{L}) \to \mathrm{Isom}^{ext}(\mathrm{Gal}(\mathrm{L}),\mathrm{Gal}(\mathrm{K}))$$

is bijective, where on the left Isom^i *means isomorphisms between some purely inseparable extensions* K′|K *and* L′|L.

(2) (Mochizuki [Moc99]) *Conjecture 32 is true more generally for k sub-p-adic.*

Here part (2) has been recently improved by Corry and Pop [CP09]: one can replace Gal(K) and similarly Gal(L) by its natural quotient obtained as an extension of Gal(k) by the maximal pro-p quotient of the subgroup Gal($K\bar{k}$). Thus one has a birational result that is completely analogous to Theorem 33. However, the positive characteristic analogue is not known at present.

Remark 35. The statements of Theorem 34 are of arithmetic nature. In remarkable contrast to this, Bogomolov [Bog91] initiated a program according to which finitely generated fields of transcendence degree at least 2 over *algebraically closed* fields should be characterized up to isomorphism by their absolute Galois group, and even by its maximal pro-ℓ nilpotent quotient of class 2, for a prime ℓ different from the characteristic. There has been important recent progress in this direction, by Bogomolov–Tschinkel [BT08], and in a series of papers by Pop including [Pop10a].

Observe that Pop's result does not use the augmentation $\mathrm{Gal}(K) \to \mathrm{Gal}(k)$. This hints at the possibility that *absolute* forms of Grothendieck's conjecture hold true. And indeed, Mochizuki proved by combining Theorem 31 and Theorem 32 (1):

Theorem 36. (Mochizuki [Moc04]) *Let* X *and* Y *be hyperbolic curves defined over some finitely generated extension of* $\mathbb{Q}$ *(not necessarily the same). Then the natural map*

$$\mathrm{Isom}(X, Y) \to \mathrm{Isom}^{ext}(\pi_1(X), \pi_1(Y))$$

is bijective.

Even more surprisingly, *absolute* results hold over a finite base field:

Theorem 37. *(1)* (Tamagawa [Tam97]) *Let* X *and* Y *be smooth affine curves defined over some finite field (not necessarily the same) with profinite universal covers* $\widetilde{X}$*,* $\widetilde{Y}$*, respectively. Then the natural map*

$$\mathrm{Isom}(\widetilde{X}|X, \widetilde{Y}|Y) \to \mathrm{Isom}(\pi_1(X), \pi_1(Y))$$

is bijective. Here on the left hand side we have the set of commutative diagrams of isomorphisms

$$\begin{array}{ccc} \widetilde{X} & \xrightarrow{\cong} & \widetilde{Y} \\ \downarrow & & \downarrow \\ X & \xrightarrow{\cong} & Y. \end{array}$$

(2) (Mochizuki [Moc07]) *The same statement holds for proper smooth curves of genus* $\geqslant 2$ *over a finite field.*

Here the profinite universal cover of an normal integral scheme S means its normalization in the field K_S of Proposition 6.

Remark 38. (1) Recently Saïdi and Tamagawa [ST09] proved that in the theorem above one may replace fundamental groups by their maximal prime-to-p quotients where p is the characteristic of the base field. They also proved results with even smaller quotients but it is not known whether the statement holds for the maximal pro-ℓ quotients of the fundamental groups where $\ell \neq p$ is a prime.

(2) Before the full statement of Theorem 33 was proven, by specialisation arguments Tamagawa and Mochizuki derived the statement of Theorem 33 for isomorphisms of hyperbolic curves over number fields from Theorem 37 (1). Stix [Sti02] used a similar method to prove an isomorphism statement for hyperbolic curves over global fields of positive characteristic.

Although in the oral lectures I gave a sketch of some of the ideas involved in the proofs of the above results, I feel that one cannot do them justice in just a couple of pages. On the other hand, besides the mostly well-written original papers there are quite a few detailed surveys that the interested reader may consult with profit. So let me conclude with some bibliographic indications.

The first results that can be associated with anabelian geometry, though they actually predate the formulation of the conjectures, are the theorems of Neukirch [Neu77] and Uchida [Uch77] concerning Galois characterization of global fields. They can now be viewed as special cases of Theorem 34 (1) of Pop, but in fact their methods have been highly inspirational for the proof of the general result. A nice exposition can be found in the last chapter of the book [NSW08] by Neukirch, Schmidt and Wingberg. As for Pop's theorem, the reader may consult my Bourbaki exposé [Sza04].

The impact of the Neukirch-Uchida techniques can also be seen in Tamagawa's proof of Theorem 37 (1) and the recent results of Saïdi–Tamagawa mentioned above refine this method further. An introduction to these ideas can be found in [Sza00]. But the best introduction to the contributions of Tamagawa and Mochizuki is the survey paper [NTM01] which also includes an overview of the work of Nakamura that contains germs of many of the ideas that were developed later. Mochizuki's methods are completely different from the approach initiated by Neukirch and Uchida and are based on constructions in p-adic Hodge theory. A succinct survey can be found in the Bourbaki exposé [Fal98] by Faltings.

References

[Bel79] G. V. Belyi. Galois extensions of a maximal cyclotomic field (Russian). *Izv. Akad. Nauk SSSR Ser. Mat.*, 43:267–276, 479, 1979.

[Bog91] F. Bogomolov. On two conjectures in birational algebraic geometry. In *Algebraic geometry and analytic geometry (Tokyo, 1990), ICM-90 Satell. Conf. Proc.*, pages 26–52. Springer, Tokyo, 1991.

[BT08] F. Bogomolov and Yu. Tschinkel. Reconstruction of function fields. *Geom. Funct. Anal.*, 18:400–462, 2008.

[Col90] R. F. Coleman. Manin's proof of the Mordell conjecture over function fields. *Enseign. Math.*, 36:393–427, 1990.

[Con06] B. Conrad. Chow's K/k-image and K/k-trace, and the Lang-Néron theorem. *Enseign. Math.*, 52:37–108, 2006.

[CP09] S. Corry and F. Pop. The pro-p hom-form of the birational anabelian conjecture. *J. reine angew. Math.*, 628:121–127, 2009.

[EH08] H. Esnault and Ph. H. Hai. Packets in Grothendieck's section conjecture. *Adv. Math.*, 218(2):395–416, 2008.

[Fal83] G. Faltings. Endlichkeitssätze für abelsche Varietäten über Zahlkörpern. *Invent. Math.*, 73:349–366, 1983.

[Fal98] G. Faltings. Curves and their fundamental groups (following Grothendieck, Tamagawa and Mochizuki), Séminaire Bourbaki, exposé 840. *Astérisque*, 252:131–150, 1998.

[Gil00] P. Gille. Le groupe fondamental sauvage d'une courbe affine en caractéristique $p > 0$. In J-B. Bost et al., editors, *Courbes semi-stables et groupe fondamental en géométrie algébrique*, volume 187 of *Progress in Mathematics*, pages 217–231. Birkhäuser, Basel, 2000.

[Gra65] H. Grauert. Mordells Vermutung über rationale Punkte auf algebraischen Kurven und Funktionenkörper. *Publ. Math. IHES*, 25:131–149, 1965.

[Gro83] A. Grothendieck. Brief an Faltings (27/06/1983). In L. Schneps and P. Lochak, editors, *Geometric Galois Actions 1*, volume 242 of *LMS Lecture Notes*, pages 49–58. Cambridge, 1997.

[HM09] Y. Hoshi and S. Mochizuki. On the combinatorial anabelian geometry of nodally nondegenerate outer representations. preprint RIMS-1677. 2009.

[Hos10] Y. Hoshi. Existence of nongeometric pro-p Galois sections of hyperbolic curves. *Publ. Res. Inst. Math. Sci.*, 46(4):829–848, 2010.

[HS12] D. Harari and J. Stix. Descent obstruction and fundamental exact sequence. In J. Stix, editor, *The Arithmetic of Fundamental Groups - PIA 2010*, volume 2 of *Contributions in Mathematical and Computational Sciences*. Heidelberg, 2012.

[Kah06] B. Kahn. Sur le groupe des classes d'un schéma arithmétique (avec un appendice de Marc Hindry). *Bull. Soc. Math. France*, 134:395–415, 2006.

[Kah09] B. Kahn. Démonstration géométrique du théorème de Lang-Néron et formules de Shioda-Tate. In *Motives and algebraic cycles: a celebration in honour of Spencer J. Bloch*, volume 56, pages 149–155. Fields Institute Communications, Amer. Math. Soc., 2009.

[KL81] N. M. Katz and S. Lang. Finiteness theorems in geometric class field theory. *L'Enseign. Math.*, 27:285–314, 1981.

[Kob76] S. Kobayashi. Intrinsic distances, measures and geometric function theory. *Bull. Amer. Math. Soc.*, 82:357–416, 1976.

[Koe05] J. Koenigsmann. On the "section conjecture" in anabelian geometry. *J. reine angew. Math.*, 588:221–235, 2005.

[Lan87] S. Lang. *Introduction to complex hyperbolic spaces*. Springer, Berlin, 1987.

[Lan91] S. Lang. *Number theory III*, volume 60 of *Encyclopaedia of Mathematical Sciences*. Springer, Berlin, 1991.

[LS57] S. Lang and J-P. Serre. Sur les revêtements non ramifiés des variétés algébriques. *Amer. J. Math.*, 79:319–330, 1957.

[Man63] Yu. I. Manin. Rational points on algebraic curves over function fields (Russian). *Izv. Akad. Nauk SSSR Ser. Mat.*, 27:1395–1440, 1963.

[Mat55] A. Mattuck. Abelian varieties over p-adic ground fields. *Ann. of Math.*, 62:92–119, 1955.

[Mat96] M. Matsumoto. Galois representations on profinite braid groups on curves. *J. reine angew. Math.*, 474:169–219, 1996.

[Moc99] S. Mochizuki. The local pro-p anabelian geometry of curves. *Invent. Math.*, 138:319–423, 1999.

[Moc04] S. Mochizuki. *The absolute anabelian geometry of hyperbolic curves*, pages 77–122. Kluwer, Boston, 2004.

[Moc07] S. Mochizuki. Absolute anabelian cuspidalizations of proper hyperbolic curves. *J. Math. Kyoto Univ.*, 47:451–539, 2007.

[Neu77] J. Neukirch. Über die absoluten Galoisgruppen algebraischer Zahlkörper. In *Journées Arithmétiques de Caen*, Astérisque 41-42:67–79, 1977.

[NSW08] J. Neukirch, A. Schmidt, and K. Wingberg. *Cohomology of number fields*. Volume 323 in Grundlehren der Mathematischen Wissenschaften. Springer, Berlin, 2nd edition, 2008.

[NTM01] H. Nakamura, A. Tamagawa, and S. Mochizuki. The Grothendieck conjecture on the fundamental groups of algebraic curves. *Sugaku Expositions*, 14:31–53, 2001.

[Par68] A. N. Parshin. Algebraic curves over function fields I (Russian). *Izv. Akad. Nauk SSSR Ser. Mat.*, 32:1191–1219, 1968.

[Par90] A. N. Parshin. Finiteness theorems and hyperbolic manifolds. In P. Cartier et al., editors, *The Grothendieck Festschrift, vol. III*, volume 88 of *Progress in Mathematics*, pages 163–178. Birkhäuser, Boston, MA, 1990.

[Pop00] F. Pop. Alterations and birational anabelian geometry. In H. Hauser et al., editor, *Resolution of singularities*, volume 181 of *Progress in Mathematics*, pages 519–532. Birkhäuser, Basel, 2000.

[Pop10a] F. Pop. On the birational anabelian program initiated by Bogomolov I. Preprint. 2010. to appear in Inventiones Mathematicae

[Pop10b] F. Pop. On the birational p-adic section conjecture. *Compositio Math.*, 146:621–637, 2010.

[Sch33] F. K. Schmidt. Mehrfach perfekte Körper. *Math. Ann.*, 108:1–25, 1933.

[SGA1] A. Grothendieck. *Revêtements étale et groupe fondamental (SGA 1)*. Séminaire de géométrie algébrique du Bois Marie 1960-61, directed by A. Grothendieck, augmented by two papers by Mme M. Raynaud, *Lecture Notes in Math.* 224, Springer-Verlag, Berlin-New York, 1971. Updated and annotated new edition: *Documents Mathématiques* 3, Société Mathématique de France, Paris, 2003.

[Sha47] I. R. Shafarevich. On p-extensions (Russian). *Mat. Sbornik*, 20(62):351–363, 1947. English translation: *Amer. Math. Soc. Translation Series* 4 (1956), 59-72.

[ST09] M. Saïdi and A. Tamagawa. A prime-to-p version of Grothendieck's anabelian conjecture for hyperbolic curves over finite fields of characteristic $p > 0$. *Publ. Res. Inst. Math. Sci.*, 45:135–186, 2009.

[Sti02] J. Stix. Projective anabelian curves in positive characteristic and descent theory for log étale covers. *Bonner Mathematische Schriften*, 354, 2002.

[Sto07] M. Stoll. Finite descent obstructions and rational points on curves. *Algebra and Number Theory*, 1:349–391, 2007.

[Sza00] T. Szamuely. Le théorème de Tamagawa I. In J.-B. Bost et al., editors, *Courbes semistables et groupe fondamental en géométrie algébrique (Luminy, 1998)*, volume 187 of *Progress in Mathematics*, pages 185–201. Birkhäuser, Basel, 2000.

[Sza04] T. Szamuely. Groupes de Galois de corps de type fini [d'après Pop], Séminaire Bourbaki, exposé 923. *Astérisque*, 294:403–431, 2004.

[Sza09] T. Szamuely. *Galois Groups and Fundamental Groups*. Volume 117 of Cambridge Studies in Advanced Mathematics, University Press, 2009.

[Tam97] A. Tamagawa. The Grothendieck conjecture for affine curves. *Compositio Math.*, 109:135–194, 1997.

[Uch77] K. Uchida. Isomorphisms of Galois groups of algebraic function fields. *Ann. of Math.*, 106:589–598, 1977.

[Win84] K. Wingberg. Ein Analogon zur Fundamentalgruppe einer Riemannschen Fläche im Zahlkörperfall. *Invent. Math.*, 77:557–584, 1984.

Part II
The Arithmetic of Fundamental Groups

Chapter 3
Vector Bundles Trivialized by Proper Morphisms and the Fundamental Group Scheme, II

Indranil Biswas and João Pedro P. dos Santos

Abstract Let X be a projective and smooth variety over an algebraically closed field k. Let $f : Y \longrightarrow X$ be a proper and surjective morphism of k–varieties. Assuming that f is separable, we prove that the Tannakian category associated to the vector bundles E on X such that f^*E is trivial is equivalent to the category of representations of a finite and étale group scheme. We give a counterexample to this conclusion in the absence of separability.

3.1 Introduction

The present work is a continuation of [BdS10], giving some applications of the main result in [BdS10] which throw light on the nature of the fundamental group scheme of Nori [Nor76] for a smooth projective variety.

Let X be a smooth projective variety over an algebraically closed field k. The fundamental group scheme of X is the affine group scheme obtained from the Tannakian category of essentially finite vector bundles on X, see Definition 3. The main theorem of [BdS10] says that a vector bundle E over X is essentially finite if and only if there is a proper k–scheme Y and a surjective morphism $f : Y \longrightarrow X$ such that f^*E is trivial. As an application of this theorem, we prove the following.

Theorem 1. *Let* X *be a smooth and projective variety over the algebraically closed field* k*, let* $x_0 : \mathrm{Spec}(k) \longrightarrow X$ *be a point, and* $f : Y \longrightarrow X$ *a proper and surjective morphism of varieties:*

I. Biswas
School of Mathematics, Tata Institute of Fundamental Research, Homi Bhabha Road, Bombay 400005, India
e-mail: indranil@math.tifr.res.in

J.P.P. dos Santos (✉)
Faculté de Mathématiques, Université de Paris VI. 4, Place Jussieu, Paris 75005, France
e-mail: dos-santos@math.jussieu.fr

J. Stix (ed.), *The Arithmetic of Fundamental Groups*, Contributions in Mathematical and Computational Sciences 2, DOI 10.1007/978-3-642-23905-2_3,

(i) *The full subcategory of* **VB**(X)

$$\mathcal{T}_{\mathrm{Y}}(\mathrm{X}) = \{\mathrm{V} \in \mathbf{VB}(\mathrm{X})\ ;\ f^*\mathrm{V}\ \textit{is trivial}\}$$

is Tannakian. The functor $x_0^* : \mathcal{T}_{\mathrm{Y}}(\mathrm{X}) \longrightarrow$ *(k–mod) is a fibre functor.*

(ii) *Assume that* f *is separable. Let* G(Y/X) *denote the affine group scheme obtained from* $\mathcal{T}_{\mathrm{Y}}(\mathrm{X})$ *and* x_0^*. *Then* G(Y/X) *is finite and étale.*

(iii) *If the separability assumption on* f *is removed, then there exists a counterexample to the conclusion in* (ii) *in which* G(Y/X) *is not a finite group scheme.*

Part (i) of the above theorem is routine, see Lemma 7. Part (ii) is the subject of Theorem 8, while the counterexample alluded to in (iii) is produced in Sect. 3.4.1.

Acknowledgements. We thank the referee for pertinent remarks which made the present text much clearer.

3.2 Preliminaries

Throughout k will stand for an algebraically closed field. By a variety we mean an integral scheme of finite type over k.

3.2.1 Notation and Terminology

Let V be a normal variety. Its field of rational functions will be denoted by R(V). We will let Val(V) denote the set of discrete valuations of R(V) associated to V. More precisely, a discrete valuation $v : \mathrm{R(V)} \longrightarrow \mathbb{Z} \cup \{\infty\}$ belongs to Val(V) if and only if there exists a point ξ of codimension one in V such that

$$\mathcal{O}_{\mathrm{V},\xi} = \{\varphi \in \mathrm{R(V)}\ ;\ v(\varphi) \geq 0\}\ .$$

Given a finite extension of fields L/K and a set of discrete valuations S of K, we say that L is **unramified** above S if for each discrete valuation v of S and each prolongation w of v to L, the ramification index $e(w/v) = 1$ and the extension of residue fields is separable.

A dominant morphism $f : \mathrm{W} \longrightarrow \mathrm{V}$ between two varieties is **separable** if the extension of function fields R(W)/R(V) is separable. This differs from the homonymous notion defined in [SGA1, X, Definition 1.1].

A vector bundle over a scheme is a locally free coherent sheaf. The category of all vector bundles on X will be denoted by **VB**(X). If E is a vector bundle over the k–scheme X, we will say that E comes from a representation of the étale fundamental group if there exists a finite group Γ, a representation

$$\rho : \Gamma \longrightarrow \mathrm{GL}_m(k)$$

and an étale Galois cover $Y \longrightarrow X$ of Galois group Γ, such that

$$E \cong Y \times^{\Gamma} k^{\oplus m} .$$

For the general definition of the contracted product of a torsor and a representation, see e.g. [Jan87, I 5.8, 5.14].

Given an affine group scheme G over k, we will let Rep(G) denote the category of all finite dimensional representations of G [Wat79, Chap. 3]. A morphism of affine group schemes $f : G \longrightarrow H$ is a quotient morphism if it is faithfully flat, or, equivalently, if the morphism induced on the function rings is injective [Wat79, Chap. 14].

If $\mathcal{T}$ is a Tannakian category over k [DM82] and $V \in \mathcal{T}$, we define the **monodromy category** of V to be the smallest Tannakian sub–category of $\mathcal{T}$ containing V. It will be denoted by $\langle V; \mathcal{T} \rangle_{\otimes}$.

Consider the neutral Tannakian category Rep(G) over k. For any $V \in \mathrm{Rep}(G)$, the category $\langle V; \mathrm{Rep}(G) \rangle_{\otimes}$ is equivalent to the category of representations of the image of the tautological homomorphism $\rho_V : G \longrightarrow GL(V)$. This image will be called the **monodromy group** of V, see Definition 2.5 and the remark after it in [BdS10] for more information.

3.2.2 Vector Bundles Trivialized by Proper and Surjective Morphisms

Let X be a smooth and projective variety over k. Recall that k is algebraically closed.

Definition 2 (Property T). A vector bundle E over X is said to have **property (T)** if there exists a proper k–scheme Y together with a surjective proper morphism $f : Y \longrightarrow X$ such that the pull–back f^*E is trivial.

The main result of [BdS10] relates property (T) to the more sophisticated notion of essential finiteness.

Definition 3. Following Nori [Nor76], we say that a vector bundle over X is **essentially finite** if there exists a finite group scheme G, a G–torsor $P \longrightarrow X$ and a representation $\rho : G \longrightarrow GL(V)$, such that

$$P \times^{G} V \cong E .$$

The category of all essentially finite vector bundles over X will be denoted by

$$\mathbf{EF}(X) .$$

Remark 4. Every essentially finite vector bundle enjoys property (T) as these are trivialized by a torsor under a finite group scheme.

The category **EF**(X) is Tannakian [Nor76]. The above definition of essential finiteness is not the one presented in [Nor76], but a consequence of the results of that work.

Theorem 5. *[BdS10, Theorem 1.1] Let* X *be a smooth and projective variety over the algebraically closed field k. Then a vector bundle* E *over* X *is essentially finite if and only if it satisfies property (T).*

The reader is urged to read Remark 16 at the end of this text to be directed to another proof of the case where $\dim X = 1$. This proof was suggested to us by Parameswaran and is based on [BP11, §6] which contains very interesting conceptual advancements. Also, we indicate that recently Antei and Mehta put forward a generalisation of Theorem 5 in the case where X is only normal [AM10].

It should be clarified that the smoothness condition on Theorem 5 cannot be dropped. This is shown by the following example.

Example 6. Let $X \subset \mathbb{P}^2_k$ be the nodal cubic defined by $(y^2z = x^3 + x^2z)$. Let

$$f : \mathbb{P}^1 \longrightarrow X, \quad (s:t) \mapsto (s^2t - t^3 : s^3 - st^2 : t^3),$$

be the birational morphism which identifies the points $(1:1)$ and $(-1:1)$. It is well–known that $\mathrm{Pic}^0(X) = k^*$, so that any line bundle L of infinite order over X gives a counterexample to the generalization of Theorem 5 to the case where X is not normal.

3.2.3 The Fundamental Group Scheme

Fix a k–rational point $x_0 : \mathrm{Spec}(k) \longrightarrow X$. The essentially finite vector bundles with the fibre functor defined by sending any essentially finite vector bundle E to its fibre x_0^*E over x_0 form a neutral Tannakian category [DM82, Definition 2.19]. The corresponding affine group scheme over k [DM82, Theorem 2.11] is called the **fundamental group scheme** [Nor76, Nor82]. This group scheme will be denoted by

$$\Pi^{\mathrm{EF}}(X, x_0).$$

3.3 Vector Bundles Trivialized by Separable Proper Morphisms

Throughout this section, we let X stand for a projective and smooth variety and $f : Y \longrightarrow X$ for a proper surjective morphism from a proper variety Y. We also choose a k–rational point $x_0 : \mathrm{Spec}(k) \longrightarrow X$.

3.3.1 The Object of Our Study

For general terminology on Tannakian categories the reader should consult [DM82].

Lemma 7. *The full subcategory of* $\mathbf{VB}(X)$

$$\mathcal{T}_Y(X) = \{V \in \mathbf{VB}(X)\ ;\ f^*V \text{ is trivial}\}$$

is Tannakian. The functor $x_0^* : \mathcal{T}_Y(X) \longrightarrow$ *(k–mod) is a fibre functor.*

Proof. That $\mathcal{T}_Y(X)$ is stable by tensor products and direct sums is clear. That it is an abelian category is a consequence of the fact that all vector bundles in $\mathcal{T}_Y(X)$ are Nori–semistable, so that kernels and cokernels are always vector bundles, see [BdS10, Corollary 2.3] and [Nor76, Lemma 3.6]. Using this last remark, it is easy to understand why the functor x_0^* is exact and faithful. As $\mathcal{T}_Y(X)$ has only vector bundles as objects, the rigidity axiom for a Tannakian category is satisfied. □

The affine group scheme obtained from $\mathcal{T}_Y(X)$ and the fibre functor x_0^* via the main theorem of Tannakian categories [DM82, Theorem 2.11] will be denoted by $G(Y/X)$ in the sequel.

3.3.2 Finiteness of G(Y/X) *for Separable Morphisms*

Theorem 8. *We assume that* $f : Y \longrightarrow X$ *is separable.*

(1) If the vector bundle E *is such that* f^*E *is trivial, then* E *is essentially finite and in fact comes from a representation of the étale fundamental group. Moreover, the monodromy group of* E *in the category* $\mathbf{EF}(X)$ *at the point* $x_0 \in X(k)$ *is a quotient of a fixed finite étale group scheme* Γ^{nr}.

(2) The group scheme $G(Y/X)$ *is finite and étale.*

The first step towards a proof of Theorem 8, and also of [BdS10, Theorem 1.1], is to consider the Stein factorization of f:

$$\begin{array}{ccc} Y & \xrightarrow{h} & Y' \\ & {\scriptstyle f}\searrow & \downarrow{\scriptstyle g} \\ & & X, \end{array}$$

where g is finite and $h_*(\mathcal{O}_Y) = \mathcal{O}_{Y'}$. The latter equality implies that the morphism $h^* : \mathbf{VB}(Y') \longrightarrow \mathbf{VB}(Y)$ is full and faithful, so that g^*E is already trivial.

Definition 9. Let $\varphi : V \longrightarrow X$ be a finite, surjective and separable morphism of varieties. By $R(V)^{nr}$ we denote the maximal unramified intermediate extension of $R(V)/R(X)$, which is the compositum of all sub-extensions R of $R(V)/R(X)$ which are unramified over Val(V). We let

$$\varphi^{\mathrm{nr}} : \mathrm{V}^{\mathrm{nr}} \longrightarrow \mathrm{X}$$

denote the normalization of X in $\mathrm{R(V)}^{\mathrm{nr}}$. If R(V)/R(X) is Galois of group Γ, then $\mathrm{R(V)}^{\mathrm{nr}}/\mathrm{R(X)}$ is Galois. Let Γ^{nr} denote the Galois group of the extension $\mathrm{R(V)}^{\mathrm{nr}}/\mathrm{R(X)}$.

Proof of Theorem 8. (1) That E is essentially finite is the content of Theorem 5. For the remainder, it is enough to prove statement (1) in the theorem under the assumption that f is finite. There is also no loss of generality in assuming that the field extension R(Y)/R(X) is Galois. Let Γ be its Galois group.

We first prove that if Γ^{nr} is trivial, i.e., $f^{\mathrm{nr}} = \mathrm{id}_{\mathrm{X}}$, then E is likewise. Let G be the finite group scheme associated, by Tannakian duality, to the category $\langle \mathrm{E}; \mathbf{EF}(\mathrm{X})\rangle_{\otimes}$ via the point $x_0 \in \mathrm{X}(k)$, see Sect. 3.2.1. Let P be the G–torsor associated to E [Nor76, §2]. The functor

$$\mathrm{P}\times^{\mathrm{G}}(\bullet) : \mathrm{Rep(G)} \longrightarrow \langle \mathrm{E}; \mathbf{EF}(\mathrm{X})\rangle_{\otimes}$$

induces an equivalence of monoidal categories. We denote by G^{et} the finite étale group scheme of connected components of G [Wat79, Chap. 6]. As P is connected [Nor82, Proposition 3, p. 87], so is

$$\mathrm{P}^{\mathrm{et}} := \mathrm{P}/\ker\left(\mathrm{G} \longrightarrow \mathrm{G}^{\mathrm{et}}\right) = \mathrm{P}\times^{\mathrm{G}}\mathrm{G}^{\mathrm{et}}\,.$$

Since $\mathrm{P}^{\mathrm{et}} \longrightarrow \mathrm{X}$ is an étale morphism, it follows that P^{et} is a normal variety.

Claim 10. The triviality of $f^*\mathrm{E}$ implies the triviality of the G–torsor

$$\mathrm{P_Y} := \mathrm{P}\times_{\mathrm{X}}\mathrm{Y} \longrightarrow \mathrm{Y}\,.$$

Let $\rho : \mathrm{G} \longrightarrow \mathrm{GL(V)}$ be a representation of G such that $\mathrm{E} = \mathrm{P}\times^{\mathrm{G}}\mathrm{V}$. It follows that ρ is a closed embedding and we are able to deduce the triviality of $\mathrm{P_Y}$ by using the triviality of $\mathrm{P_Y}\times^{\mathrm{G}}\mathrm{GL(V)}$ together with the fact that the natural map

$$\mathrm{H}^1_{\mathrm{fppf}}(\mathrm{Y},\mathrm{G}) \longrightarrow \mathrm{H}^1_{\mathrm{fppf}}(\mathrm{Y},\mathrm{GL(V)})$$

is injective. Indeed, the kernel is the set of all morphisms from Y to the affine scheme GL(V)/G, see [DG70, p. 373, III, §4, 4.6].

Let $h : \mathrm{Y} \longrightarrow \mathrm{P}$ be the X–morphism derived from an isomorphism $\mathrm{P_Y} \cong \mathrm{Y}\times\mathrm{G}$ and let $j : \mathrm{Y} \longrightarrow \mathrm{P}^{\mathrm{et}}$ be the morphism of X–schemes obtained from h. It is not hard to see that j takes the generic point of Y to the generic point of P^{et}, so j gives rise to a homomorphism of R(X)–fields $\mathrm{R}(\mathrm{P}^{\mathrm{et}}) \longrightarrow \mathrm{R(Y)}$. Since $\mathrm{R}(\mathrm{P}^{\mathrm{et}})/\mathrm{R(X)}$ is unramified above Val(X), we must have $\mathrm{R}(\mathrm{P}^{\mathrm{et}}) = \mathrm{R(X)}$. As a consequence, $\mathrm{P}^{\mathrm{et}} = \mathrm{X}$ and thus G^{et} is trivial. This means that G is a local group scheme. We will now prove the following.

Claim 11. If G is local, then the existence of an X–morphism $h : \mathrm{Y} \longrightarrow \mathrm{P}$ implies the triviality of P.

Let $\mathrm{Spec(A)} \subseteq \mathrm{X}$ be an affine open and let $\mathrm{Spec(B)} \subseteq \mathrm{Y}$ (respectively, $\mathrm{Spec(S)} \subseteq \mathrm{P}$) be its pre–image in Y (respectively, in P). We then have a homomorphism of A–algebras $\eta : \mathrm{S} \longrightarrow \mathrm{B}$; let $\mathrm{S}' \subseteq \mathrm{B}$ be its image. Since

$$\mathrm{Spec(S)} \longrightarrow \mathrm{Spec(A)}$$

is a G–torsor, above any maximal ideal $\mathfrak{m} \subseteq \mathrm{A}$, there exists only one maximal ideal of S. The same property is valid if we replace S by S′. Hence, the extension of fields defined by $\mathrm{S}' \supseteq \mathrm{A}$ must be purely inseparable. Because R(Y)/R(X) is a separable extension, and A is a normal ring, it follows that $\mathrm{S}' = \mathrm{A}$. This allows one to construct a section $\sigma : \mathrm{X} \longrightarrow \mathrm{P}$. Therefore E is trivial. This proves Claim 11.

Now we treat the general case. Since $f^{\mathrm{nr}} : \mathrm{Y}^{\mathrm{nr}} \longrightarrow \mathrm{X}$ is unramified above Val(X), the Zariski–Nagata purity Theorem, see [SGA1, X, 3.1], permits us to conclude that f^{nr} is étale. In particular Y^{nr} a smooth projective variety over k. Moreover, the map $f^{\mathrm{nr}} : \mathrm{Y}^{\mathrm{nr}} \longrightarrow \mathrm{X}$ is an étale Galois covering of group Γ^{nr}. Let

$$g : \mathrm{Y} \longrightarrow \mathrm{Y}^{\mathrm{nr}}$$

denote the obvious morphism, we have $g^{\mathrm{nr}} = \mathrm{id}_{\mathrm{Y}^{\mathrm{nr}}}$. Applying what was proved above to Y^{nr}, we conclude that $f^{\mathrm{nr}*}\mathrm{E}$ is trivial. By [LS77, Proposition 1.2], we conclude that

$$\mathrm{E} \cong \mathrm{Y}^{\mathrm{nr}} \times^{\Gamma^{\mathrm{nr}}} \mathrm{V},$$

where V is a representation of Γ^{nr}. This proves that the monodromy group of E in **EF**(X) is a quotient of Γ^{nr}.

(2) The proof rests on the same sort of argument used for the proof of (1). As in (1), we assume that f is finite. Let

$$\mathrm{G(Y/X)} := \mathrm{G} = \varprojlim \mathrm{G}_i$$

be the *profinite* group scheme associated to $\mathcal{T}_{\mathrm{Y}}(\mathrm{X})$ via x_0^*. Here each group G_i is finite and the transition morphisms $\mathrm{G}_j \longrightarrow \mathrm{G}_i$ are all faithfully flat. (The reader unfamiliar with this sort of structure argument will profit from [Wat79, 3.3] and [Wat79, 14.1].) Write $\mathrm{P} \longrightarrow \mathrm{X}$ for the universal G–torsor [Nor76, §2] and P_i for $\mathrm{P} \times^{\mathrm{G}} \mathrm{G}_i$. We remark that Proposition 3 on p. 87 of [Nor82] proves that $\Gamma(\mathrm{P}_i, \mathcal{O}_{\mathrm{P}_i}) = k$. In this situation, we can find X–morphisms

$$h_i : \mathrm{Y} \longrightarrow \mathrm{P}_i\,.$$

(The details of the argument are given in the proof of (1) above.) Let $\mathrm{G}_i^{\mathrm{et}}$ be the largest étale quotient of G_i [Wat79, Chap. 6]. The morphism

$$\mathrm{P}_i^{\mathrm{et}} := \mathrm{P} \times^{\mathrm{G}} \mathrm{G}_i^{\mathrm{et}} \longrightarrow \mathrm{X}$$

is finite and étale and the number of k–rational points on a fiber equals $\mathrm{rank}\, \mathrm{G}_i^{\mathrm{et}}$. From the surjectivity of the composition

$$Y \longrightarrow P_i \longrightarrow P_i^{\text{et}},$$

the integers $\operatorname{rank} G_i^{\text{et}}$ are bounded from above, so

$$G^{\text{et}} := \varprojlim G_i^{\text{et}} = G_{i_0}^{\text{et}}$$

for some i_0. Then $X' := P_{i_0}^{\text{et}} = P_i^{\text{et}}$ is a smooth and projective variety and the obvious morphism

$$P_i \longrightarrow P_i/G_i^0 = P_i/(\ker G_i \longrightarrow G_{i_0}^{\text{et}}) = X', \quad i \geq i_0$$

gives $P_i \longrightarrow X'$ the structure of a torsor over X' under the structure group G_i^0. Moreover, since $\Gamma(P_i, \mathcal{O}_{P_i}) = k$, the torsor P_i cannot be trivial over X' unless $G_i^0 = \{e\}$. Employing the X'–morphisms $Y \longrightarrow P_i$, we see, using Claim 11 proved in part (1), that $G_i^0 = \{e\}$. This means that $G = G_{i_0}^{\text{et}}$. □

3.4 Finiteness of G(Y/X), Reducedness of the Universal Torsor and Base Change Properties

As in Sect. 3.3, we let X stand for a projective and smooth variety and $f : Y \longrightarrow X$ for a proper surjective morphism from a (proper) variety Y. We also choose a k–rational point $x_0 : \operatorname{Spec}(k) \longrightarrow X$.

3.4.1 *An Instance Where* G(Y/X) *Is Not Finite and the Universal Torsor is not Reduced*

Let G(Y/X) be the affine fundamental group scheme associated to the Tannakian category

$$\mathcal{T}_Y(X)$$

by means of the fiber functor $x_0^* : \mathcal{T}_Y(X) \longrightarrow (k\text{–mod})$. If V is an object of $\mathcal{T}_Y(X)$ which is stable as a vector bundle (all vector bundles in $\mathcal{T}_Y(X)$ are semistable of slope zero [BdS10, Proposition 2.2]), the representation of G(Y/X) obtained from V must be irreducible. Since a finite group scheme only has finitely many isomorphism classes of irreducible representations – these are all Jordan–Hölder components of the right regular representation [Wat79, 3.5] – we have a proved the following lemma.

Lemma 12. *If there are infinitely many non–isomorphic stable vector bundles in* $\mathcal{T}_Y(X)$*, then the group scheme* G(Y/X) *is not finite.*

The existence of infinitely many stable bundles in $\mathcal{T}_Y(X)$ also causes the following particularity.

Proposition 13. *Assume that there are infinitely many non–isomorphic stable vector bundles in* $\mathcal{T}_{\mathrm{Y}}(\mathrm{X})$. *Then there exists a finite quotient* G_0 *of* $\mathrm{G}(\mathrm{Y}/\mathrm{X})$ *and a* G_0*–torsor over* X, *call it* P_0, *such that:*

(1) $\Gamma(\mathrm{P}_0, \mathcal{O}_{\mathrm{P}_0}) = k$ *and*
(2) The scheme P_0 *is not reduced.*

Moreover, in this case, the universal torsor $\widetilde{\mathrm{X}} \longrightarrow \mathrm{X}$ *for the fundamental group scheme* $\Pi^{\mathrm{EF}}(\mathrm{X}, x_0)$ *is not reduced as a scheme.*

Proof. Let
$$\mathrm{G}(\mathrm{Y}/\mathrm{X}) := \mathrm{G} = \varprojlim \mathrm{G}_i\,,$$
where each G_i is a finite group–scheme and the transition morphisms $\mathrm{G}_j \longrightarrow \mathrm{G}_i$ are faithfully flat, just as in the proof of Theorem 8. By Lemma 12 and the assumption, G is not a finite group scheme. We will show that the conclusion of the statement holds under the extra assumption that the group schemes G_i are all local. The general case can be obtained from this one as in the proof of Theorem 8. Let $\mathrm{P} \longrightarrow \mathrm{X}$ be the universal G–torsor associated to $\mathcal{T}_{\mathrm{Y}}(\mathrm{X}) \subset \mathbf{EF}(\mathrm{X})$ via the constructions in [Nor76, §2]. The torsor P gives rise to G_i–torsors
$$\psi_i\,:\; \mathrm{P}_i = \mathrm{P} \times^{\mathrm{G}} \mathrm{G}_i \longrightarrow \mathrm{X}\,.$$
Due to [Nor82, Proposition 3, p. 87], we have $\Gamma(\mathrm{P}_i, \mathcal{O}_{\mathrm{P}_i}) = k$. Since G_i is a local group scheme, for any field extension K/k, the map
$$\psi_i(\mathrm{K})\,:\; \mathrm{P}_i(\mathrm{K}) \longrightarrow \mathrm{X}(\mathrm{K})$$
is bijective, by [EGAI-IV, I, 3.5.10, p. 116] the map ψ_i induces a bijection on the corresponding topological spaces. Hence, ψ_i is a homeomorphism and it follows that P_i is irreducible for each i. We assume that each P_i is also reduced. Proceeding as in the proof of Theorem 8, see Claim 10, there exists a X–morphism $h\,:\,\mathrm{Y} \longrightarrow \mathrm{P}_i$ for each i. This bounds $\deg \psi_i = \operatorname{rank} \mathrm{G}_i$ by above and leads to a contradiction with the assumption that G is not finite.

The proof of the last statement is a direct consequence of what we just proved together with [Nor82, Proposition 3] and [EGAI-IV, IV_3, 8.7.2]. □

In view of Lemma 12 and Proposition 13, we can use [Pau07] to give an example of a smooth curve X having two extraordinary features: (1) there exists a finite morphism $\mathrm{Y} \longrightarrow \mathrm{X}$ such that $\mathrm{G}(\mathrm{Y}/\mathrm{X})$ is *not* finite and (2) the universal torsor $\widetilde{\mathrm{X}}$ for the fundamental group scheme $\Pi^{\mathrm{EF}}(\mathrm{X}, x_0)$ is *not* reduced. Indeed, let X be the smooth curve constructed in [Pau07, (3.1) and Proposition 4.1], a particular smooth projective curve defined by a single explicit equation in $\mathbb{P}^2_k$. Here k is a field of characteristic two.

Let $f\,:\,\mathrm{Y} \longrightarrow \mathrm{X}$ be the fourth power of the Frobenius morphism, so Y is isomorphic to X as a scheme. Pauly [Pau07, Proposition 4.1] constructs a locally free coherent sheaf over $\mathrm{X} \times \mathrm{S}$, where S is a positive dimensional k–scheme, such that for every $s \in \mathrm{S}(k)$, the vector bundle $\mathcal{E}|\mathrm{X} \times \{s\}$ is stable and $f^*(\mathcal{E}|\mathrm{X} \times \{s\})$ is trivial.

Furthermore, for two different points $s,t \in \mathrm{S}(k)$, the sheaves $\mathcal{E}|\mathrm{X}\times\{s\}$ and $\mathcal{E}|\mathrm{X}\times\{t\}$ are not isomorphic. In other words, there are infinitely many isomorphism classes of stable vector bundles of fixed rank satisfying the condition that the pullback by f is trivial. By Lemma 12, the affine group scheme $\mathrm{G}(\mathrm{Y}/\mathrm{X})$ is not finite. From Proposition 13, it follows also that the universal torsor $\widetilde{\mathrm{X}} \longrightarrow \mathrm{X}$ is not reduced.

Remark 14. In [EHS08, Remark 2.4] the reader can find an example of an α_p–torsor over a reduced variety which is not reduced. The example we have just given shows that the situation can be bad even if the ambient variety is smooth.

3.4.2 A Link Between the Quantity of Vector Bundles Trivialized by the Frobenius Morphism and the Universal Torsor

We assume that k is of positive characteristic, and let $\mathrm{F} : \mathrm{X} \longrightarrow \mathrm{X}$ be the absolute Frobenius morphism. Define

$$\mathrm{S}(\mathrm{X},r,t) = \left\{\begin{array}{l}\text{isomorphism classes of stable vector bundles of rank } r\\ \text{on X, whose pull–back by } \mathrm{F}^t \text{ is trivial}\end{array}\right\}$$

Here we refrain from using the terminology F–trivial, since there is a question of stability which is not constant in the literature [Pau07, MS08]. In their study of base change for the local fundamental group scheme and these bundles, Mehta and Subramanian [MS08] showed the following.

Theorem 15 ([MS08, Theorem, p. 208]). *Let* X *be a smooth projective variety over k. The following are equivalent:*

(*a*) *For any algebraically closed extension k'/k, any pair $r,t \in \mathbb{N}$ and any* E' *in* $\mathrm{S}(\mathrm{X}\otimes_k k';r,t)$, *there exists a vector bundle over* X *and an isomorphism*

$$\mathrm{E}\otimes_k k' \cong \mathrm{E}' .$$

(*b*) *For any two given $r,t \in \mathbb{N}$, the set* $\mathrm{S}(\mathrm{X};r,t)$ *is finite.*
(*c*) *The local fundamental group scheme of* $\mathrm{X}\otimes_k k'$ *is obtained from the local fundamental group scheme of* X *by base change.*

For the definition of the local fundamental group scheme, the reader should consult [MS08]. In Proposition 13 we have shown that

$$\left\{\begin{array}{l}\text{The universal torsor for the}\\ \text{fundamental group}\\ \text{scheme is a reduced scheme}\end{array}\right\} \Longrightarrow \{\text{Condition (b) in the above theorem holds.}\}$$

As Vikram Mehta made us realize, the reverse implication need not be true and the arguments to follow are due to him. To construct a counter–example, we consider an abelian threefold A and $\iota : \mathrm{X} \hookrightarrow \mathrm{A}$ a closed smooth surface defined by intersecting A with a hyperplane section of high degree in some projective embedding $A \hookrightarrow \mathbb{P}^N$. By the Lefschetz Theorem [BH07, Theorem 1.1], we have an isomorphism

$$\Pi^{EF}(\iota) : \Pi^{EF}(X, x_0) \xrightarrow{\cong} \Pi^{EF}(A, x_0)$$

so that, if $B \longrightarrow A$ is a pointed torsor under a finite group scheme with the property of being *Nori reduced* [Nor82, Proposition 3, p. 87], i.e.,

$$H^0(B, \mathcal{O}_B) = k ,$$

then the same can then be said about the restriction of B to X. Using the torsors

$$[p] : A \longrightarrow A \quad ([p] \text{ is multiplication by } p) ,$$

we see that X admits a Nori reduced torsor under a finite group scheme which is not reduced *as a scheme*. This follows from the factorization $[p] = VF$ and the fact that $F^{-1}(Z)$ is never reduced if $Z \subsetneq A$ is a proper closed sub–scheme. By another application of the Lefschetz Theorem [BH07, Theorem 1.1], we obtain a bijection

$$S(A, r, t) \overset{\sim}{\longleftrightarrow} S(X, r, t) .$$

Since the iteration of the Frobenius morphism $F^t_A : A \longrightarrow A$ sits in a commutative diagram

$$\begin{array}{ccc} A & \xrightarrow{[p^t]} & A \\ & \searrow & \uparrow F_A \\ & & A \end{array}$$

if ${F^t_A}^*E$ is trivial, then $[p^t]^*E$ is likewise. Consequently, we obtain an injection

$$S(A, r, t) \hookrightarrow \left\{ \begin{array}{l} \text{isomorphism classes of simple} \\ \text{representations of rank } r \text{ of } \ker[p] \end{array} \right\} .$$

This entails that $S(X, r, t)$ is always a finite set and we arrive at the desired counter–example to the above highlighted implication.

Remark 16 (Made after completion). In a recent discussion, Parameswaran called our attention to a simpler proof of the fact that a vector bundle E on X which becomes trivial after being pulled back by a finite morphism from a smooth and projective variety $f : Y \longrightarrow X$ in fact comes from a representation of the étale fundamental group of X, compare Theorem 8. The main idea is to use the coherent sheaf of algebras

$$f_*(\mathcal{O}_Y)_{max}$$

associated to a separable and finite morphism $f : Y \longrightarrow X$ from a smooth projective variety Y to X. Here the subscript *max* stands for the maximal semistable subsheaf. That this is in fact an algebra requires a proof and the reader is directed to [BP11, Lemma 6.4]. One of the consequences of [BP11] (which Parameswaran was kind enough to explain to the second author) is that $f_*(O_Y)_{max}$ is the maximal *étale* extension of O_X inside f_*O_Y. Together with [BP11, Proposition 6.8], the triviality of f^*E implies the triviality of the pull–back of E to the finite étale X–scheme $Y_{max} = \operatorname{Spec} f_*(O_Y)_{max}$ and this enough to show that E is essentially finite. The reader should also note that in [BP11, §6], the framework is such that the domain variety is smooth, which is not sufficient to obtain Theorem 8 directly. But it is possible that the methods in [BP11] can be extended (for example, to a normal domain variety) to give another proof of Theorem 8.

References

[AM10] M. Antei and V. Mehta. Vector bundles over normal varieties trivialized by finite morphisms. 2010, arXiv:1009.5234. *Archiv der Mathematik*, DOI: 10.1007/s00013-011-0327-1.

[BdS10] I. Biswas and J. P. dos Santos. Vector bundles trivialized by proper morphisms and the fundamental group scheme. *Jour. Inst. Math. Jussieu*, 10(02):225 – 234, 2010.

[BH07] I. Biswas and Y. Holla. Comparison of fundamental group schemes of a projective variety and an ample hypersurface. *J. Algebraic Geom.*, 16(3):547–597, 2007.

[BP11] V. Balaji and A. J. Parameswaran. An analogue of the Narasimhan–Seshadri theorem and some applications. *Journal of Topology*, 4(1):105–140, 2011.

[DG70] M. Demazure and P. Gabriel. *Groupes algébriques*. Masson & Cie, Paris; North-Holland Publishing Co., Amsterdam, 1970.

[DM82] P. Deligne and J. Milne. Tannakian categories. Lecture Notes in Mathematics 900, pages 101–228. Springer-Verlag, Berlin, New York, 1982.

[EGAI-IV] A. Grothendieck. Éléments de Géométrie Algébrique. *Publ. Math. IHÉS*, 8, 11 (1961), 17 (1963), 20, (1964), 24 (1965), 28 (1966), 32 (1967).

[EHS08] H. Esnault, P. H. Hai, and X. Sun. On Nori's fundamental group scheme. In *Geometry and dynamics of groups and spaces*, Progr. Math. 265, pages 377–398. Birkhäuser, Basel, 2008.

[Jan87] J. C. Jantzen. Representations of algebraic groups. *Pure and Applied Mathematics*, 131, 1987.

[LS77] H. Lange and U. Stuhler. Vektorbündel auf Kurven und Darstellungen der algebraischen Fundamentalgruppe. *Math. Zeit.*, 156:73–84, 1977.

[MS08] V. B. Mehta and S. Subramanian. Some remarks on the local fundamental group scheme. *Proc. Indian Acad. Sci. (Math. Sci.)*, 118:207–211, 2008.

[Nor76] M. V. Nori. On the representations of the fundamental group. *Compos. Math.*, 33:29–41, 1976.

[Nor82] M. V. Nori. PhD thesis. *Proc. Indian Acad. Sci. (Math. Sci.)*, 91:73–122, 1982.

[Pau07] C. Pauly. A smooth counter–example to Nori's conjecture on the fundamental group scheme. *Proceedings of the American Mathematical Society*, 135:2707–2711, 2007.

[SGA1] A. Grothendieck. *Revêtements étale et groupe fondamental (SGA 1)*. Séminaire de géométrie algébrique du Bois Marie 1960-61, directed by A. Grothendieck, augmented by two papers by Mme M. Raynaud, *Lecture Notes in Math.* 224, Springer-Verlag, Berlin-New York, 1971. Updated and annotated new edition: *Documents Mathématiques* 3, Société Mathématique de France, Paris, 2003.

[Wat79] W. C. Waterhouse. *Introduction to affine group schemes*. Number 66 in Graduate Texts in Mathematics. Springer, New York-Berlin, 1979.

Chapter 4
Note on the Gonality of Abstract Modular Curves

Anna Cadoret

Abstract Let S be a curve over an algebraically closed field k of characteristic $p \geqslant 0$. To any family of representations $\rho = (\rho_\ell : \pi_1(\mathrm{S}) \to \mathrm{GL}_n(\mathbb{F}_\ell))$ indexed by primes $\ell \gg 0$ one can associate **abstract modular curves** $\mathrm{S}_{\rho,1}(\ell)$ and $\mathrm{S}_\rho(\ell)$ which, in this setting, are the modular analogues of the classical modular curves $\mathrm{Y}_1(\ell)$ and $\mathrm{Y}(\ell)$. The main result of this paper is that, under some technical assumptions, the gonality of $\mathrm{S}_\rho(\ell)$ goes to $+\infty$ with ℓ. These technical assumptions are satisfied by $\mathbb{F}_\ell$-linear representations arising from the action of $\pi_1(\mathrm{S})$ on the étale cohomology groups with coefficients in $\mathbb{F}_\ell$ of the geometric generic fiber of a smooth proper scheme over S. From this, we deduce a new and purely algebraic proof of the fact that the gonality of $\mathrm{Y}_1(\ell)$, for $p \nmid \ell(\ell^2-1)$, goes to $+\infty$ with ℓ.

4.1 Introduction

Let k be an algebraically closed field of characteristic $p \geqslant 0$ and S a smooth, separated and connected curve over k with generic point η. Let $\pi_1(\mathrm{S})$ denote its étale fundamental group. Fix an integer $n \geqslant 1$. For each prime $\ell \gg 0$, let H_ℓ be an $\mathbb{F}_\ell$ vector space of dimension n on which $\pi_1(\mathrm{S})$ acts continuously. We will write ρ for the family of the resulting $\mathbb{F}_\ell$-linear representations

$$\rho_\ell : \pi_1(\mathrm{S}) \to \mathrm{GL}(\mathrm{H}_\ell) \simeq \mathrm{GL}_n(\mathbb{F}_\ell)\,.$$

To such data, one can associate families of **abstract modular curves** $\mathrm{S}_{\rho,1}(\ell) \to \mathrm{S}$ and $\mathrm{S}_\rho(\ell) \to \mathrm{S}$, see Sect. 4.2, which, in this setting, are the modular analogues of the classical modular curves $\mathrm{Y}_1(\ell) \to \mathrm{Y}(0)$ and $\mathrm{Y}(\ell) \to \mathrm{Y}(0)$ classifying ℓ-torsion points and full level-ℓ structures of elliptic curves respectively.

A. Cadoret (✉)
Centre Mathématiques Laurent Schwarz, École Polytechnique, 91128 Palaiseau, France
e-mail: anna.cadoret@math.polytechnique.fr

J. Stix (ed.), *The Arithmetic of Fundamental Groups*, Contributions in Mathematical and Computational Sciences 2, DOI 10.1007/978-3-642-23905-2_4,

The main examples of such representations we have in mind are the $\mathbb{F}_\ell$-linear representations arising from the action of $\pi_1(S)$ on the étale cohomology groups with coefficients in $\mathbb{F}_\ell$ of the geometric generic fiber of a smooth proper scheme over S. In particular, this includes those representations arising from the action of $\pi_1(S)$ on the group of ℓ-torsion points of the geometric generic fiber of an abelian scheme over S, see Sect. 4.2.3.

The properties satisfied by these representations motivated, in [CT11b], the introduction of technical conditions on ρ, denoted by (A), (WA) and (AWA) for *abelianization, weak abelianization* and *alternating weak abelianization* respectively, (I) for *isotriviality*, (T) for *tame* and (U) for *unipotent*, see Sect. 4.2.2 for a precise formulation of these conditions.

Let $g_{\rho,1}(\ell)$ and $g_\rho(\ell)$ (resp. $\gamma_{\rho,1}(\ell)$ and $\gamma_\rho(\ell)$) denote the genus (resp. the k-gonality) of the abstract modular curves $S_{\rho,1}(\ell)$ and $S_\rho(\ell)$ respectively. The main result of [CT11b, Thm. 2.1] asserts that, if conditions (AWA), (I), (U) are satisfied then

$$\lim_{\ell\to+\infty} g_{\rho,1}(\ell) = +\infty\,.$$

An intermediate step in the proof of this result is that, if conditions (WA), (I), (T) are satisfied then

$$\lim_{\ell\to+\infty} g_{\rho}(\ell) = +\infty\,.$$

In this note, we prove that the same holds with gonality replacing genus.

Theorem 1. *If conditions (WA), (I), (T) are satisfied then*

$$\lim_{\ell\to+\infty} \gamma_{\rho}(\ell) = +\infty\,.$$

The proof of Theorem 1 is purely algebraic and based on the equivariant-primitive decompositions introduced by A. Tamagawa in [Tam04] to estimate the gonality of Galois covers. The method, however, fails to prove the analogue for $S_{\rho,1}$.

Conjecture 2. Assume that conditions (WA), (T), (U) are satisfied. Then

$$\lim_{\ell\to+\infty} \gamma_{\rho,1}(\ell) = +\infty\,.$$

Our method shows Conjecture 2 only when we restrict to $n=2$ and primes ℓ with $p\nmid \ell(\ell^2-1)$, or, more generally, for the variant of $S_{\rho,1}(\ell)$ classifying points $v\in H_\ell$ whose $\pi_1(S)$-orbit generates a subspace of rank 2, see Proposition 17. This provides in particular an algebraic proof of the following well-known fact, cf. [Abr96], [Poo07].

Corollary 3.

$$\lim_{\substack{\ell\to+\infty \\ p\nmid \ell(\ell^2-1)}} \gamma_{Y_1(\ell)} = +\infty\,.$$

When $p=0$, it seems that variants of Theorem 1 can be proved by the techniques from differential geometry and Cayley-Schreier graph theory generalizing [Abr96] and developed in [EHK10].

Apart from their intrinsic geometric interest, statements as Theorem 1 and Conjecture 2 also have arithmetic consequences. In characteristic 0, this follows from the following corollary of [Fal91].

Corollary 4. *([Fre94]) Let k be a finitely generated field of characteristic 0 and let S be a smooth, proper, geometrically connected curve over k with k-gonality γ. Then, for any integer $1 \leqslant d \leqslant \left[\frac{\gamma-1}{2}\right]$, the set of all closed points s of S with residue field $k(s)$ of degree $[k(s):k] \leqslant d$ is finite.*

So, for instance, Conjecture 2 for $p = 0$ combined with [CT11a, Prop. 3.18], to rule out the $\overline{k}$-isotrivial torsion points of $\mathrm{A}_{\overline{\eta}}$, would imply:

For any finitely generated field k of characteristic 0, smooth, separated and geometrically connected curve S over k, abelian scheme $\mathrm{A} \to \mathrm{S}$ and integer $d \geqslant 1$ the set of closed points s of S with degree $[k(s):k] \leqslant d$ and such that A_s carries a $k(s)$-rational torsion point of order ℓ is finite for $\ell \gg 0$.

Acknowledgements. I am very indebted to Jakob Stix for his impressive editorial work (from which the exposition of this paper gained a lot) and for pointing out mathematical gaps in the last part of the proof of Theorem 1 and in the proof of Corollary 3. I am also grateful to Akio Tamagawa for his careful reading of the first version of this text as well as to the referee for his detailed and constructive report.

4.2 Abstract Modular Curves

We fix once and for all an algebraically closed field k of characteristic $p \geqslant 0$. By a curve over k we mean a connected, smooth and separated k-scheme of dimension 1.

4.2.1 Notation

Let S be a curve over k with a geometric generic point $\overline{\eta}$ above its generic point $\eta \in \mathrm{S}$. We will write $\mathrm{S} \hookrightarrow \mathrm{S}^{\mathrm{cpt}}$ for the smooth compactification of S and $\pi_1(\mathrm{S})$ for its étale fundamental group with base point $\overline{\eta}$. Fix an integer $n \geqslant 1$, and, for each prime $\ell \gg 0$, let H_ℓ be an $\mathbb{F}_\ell$-module of rank n on which $\pi_1(\mathrm{S})$ acts. We will write ρ for the family of the resulting $\mathbb{F}_\ell$-linear representations

$$\rho_\ell : \pi_1(\mathrm{S}) \to \mathrm{GL}(\mathrm{H}_\ell) \simeq \mathrm{GL}_n(\mathbb{F}_\ell)\,.$$

For every prime $\ell \gg 0$, set $\mathrm{G}_\ell = \mathrm{im}(\rho_\ell)$ and for any subgroup $\mathrm{U} \subset \mathrm{G}_\ell$, the *abstract modular curve* associated to U is the connected étale cover $\mathrm{S}_\mathrm{U} \to \mathrm{S}$ corresponding to the open subgroup $\rho_\ell^{-1}(\mathrm{U}) \subset \pi_1(\mathrm{S})$. We write $g_{\mathrm{S}_\mathrm{U}}$ and $\gamma_{\mathrm{S}_\mathrm{U}}$ for the genus and the gonality of S_U respectively.

Remark 5. As we are only interested in the asymptotic behaviour of abstract modular curves, it is enough to consider only *big enough* primes ℓ. Furthermore, in practice, H_ℓ will be an étale cohomology group $H^i(X_{\overline{\eta}},\mathbb{F}_\ell)$ for some smooth proper morphism $X \to S$ with connected geometric generic fibre $X_{\overline{\eta}}$. In particular, the dimension of $H^i(X_{\overline{\eta}},\mathbb{F}_\ell)$ may become constant only for $\ell \gg 0$, see Sect. 4.2.2.

In the following, we will consider only specific classes of abstract modular curves of two kinds. First, for $v \in H_\ell$ we denote by $S_v \to S$ the abstract modular curve associated to the stabilizer of $G_{\ell,v} \subset G_\ell$ of v, and let g_v and γ_v denote its genus and gonality respectively.

Secondly, for a $\pi_1(S)$-submodule $M \subset H_\ell$, we denote by $S_M \to S$ the abstract modular curve associated to

$$\mathrm{Fix}(M) := \{g \in G_\ell \ ; \ g|_M = \mathrm{Id}_M\},$$

and let g_M and γ_M denote its genus and gonality respectively. The connected étale cover $S_M \to S$ is Galois with Galois group $G_M = G_\ell/Fix(M)$, which is the image of the induced representation $\rho_M : \pi_1(S) \to GL(M)$.

For $v \in H_\ell$ and the $\pi_1(S)$-submodule $M(v) := \mathbb{F}_\ell[G_\ell \cdot v] \subset H_\ell$ generated by v, the cover $S_{M(v)} \to S$ is the Galois closure of $S_v \to S$.

Let $\mathcal{F} = (\mathcal{F}_\ell)$ denote a sequence of non-empty families of subgroups of G_ℓ. We will say that

$$S_{\rho,\mathcal{F}}(\ell) := \bigsqcup_{U\in\mathcal{F}_\ell} S_U \to S$$

is the **abstract modular curve associated with** $\mathcal{F}_\ell$ and define

$$d_{\rho,\mathcal{F}}(\ell) := \min\{[G_\ell : U] \ ; \ U \in \mathcal{F}_\ell\}$$
$$g_{\rho,\mathcal{F}}(\ell) := \min\{g_{S_U} \ ; \ U \in \mathcal{F}_\ell\}$$
$$\gamma_{\rho,\mathcal{F}}(\ell) := \min\{\gamma_{S_U} \ ; \ U \in \mathcal{F}_\ell\},$$

which we call the **degree, genus** and **gonality** of the abstract modular curve $S_{\rho,\mathcal{F}}(\ell)$. Following the notation for the usual modular curves, we will write

$$S_{\rho,1}(\ell),\ d_{\rho,1}(\ell),\ g_{\rho,1}(\ell),\ \gamma_{\rho,1}(\ell)$$

when $\mathcal{F}_\ell$ is the family of all stabilizers $G_{\ell,v}$ for $0 \neq v \in H_\ell$, and

$$S_\rho(\ell),\ d_\rho(\ell),\ g_\rho(\ell),\ \gamma_\rho(\ell)$$

when $\mathcal{F}_\ell$ is the family of all Fix(M), for $0 \neq M \subset H_\ell$. Note that by construction

$$d_\rho(\ell) \geqslant d_{\rho,1}(\ell),\ g_\rho(\ell) \geqslant g_{\rho,1}(\ell) \ \text{and} \ \gamma_\rho(\ell) \geqslant \gamma_{\rho,1}(\ell)\,.$$

4.2.2 Conditions (WA), (I), (T)

Given an integer $1 \leqslant m \leqslant n$ and a $\pi_1(\mathrm{S})$-submodule $\mathrm{M} \subset \Lambda^m \mathrm{H}_\ell$, write again

$$\rho_{\mathrm{M}} : \pi_1(\mathrm{S}) \to \mathrm{GL(M)}$$

for the induced representation. We consider the following technical conditions:

(WA) For any open subgroup $\Pi \subset \pi_1(\mathrm{S})$, there exists an integer $\mathrm{B}_\Pi \geqslant 1$ such that, for every prime ℓ, integer $1 \leqslant m \leqslant n$ and Π-submodule $\mathrm{M} \subset \Lambda^m \mathrm{H}_\ell$, one has

$$\rho_{\mathrm{M}}(\Pi) \text{ abelian of prime-to-}\ell \text{ order} \Rightarrow |\rho_{\mathrm{M}}(\Pi)| \leqslant \mathrm{B}_\Pi .$$

(WA)' For any open subgroup $\Pi \subset \pi_1(\mathrm{S})$, there exists an integer $\mathrm{B}_\Pi \geqslant 1$ such that, for every prime ℓ, integer $1 \leqslant m \leqslant n$ and Π-submodule $\mathrm{M} \subset \Lambda^m \mathrm{H}_\ell$, one has

$$\rho_{\mathrm{M}}(\Pi) \text{ abelian} \Rightarrow |\rho_{\mathrm{M}}(\Pi)| \leqslant \mathrm{B}_\Pi .$$

(I) For any open subgroup $\Pi \subset \pi_1(\mathrm{S})$ the $\mathbb{F}_\ell$-submodule H_ℓ^Π of fixed vectors under Π is trivial for $\ell \gg 0$.

(T) For any $\mathrm{P} \in \mathrm{S}^{\mathrm{cpt}} \setminus \mathrm{S}$ there exists an open subgroup $\mathrm{T_P}$ of the inertia group $\mathrm{I_P} \subset \pi_1(\mathrm{S})$ at P such that $\rho_\ell(\mathrm{T_P})$ is tame for $\ell \gg 0$.

In [CT11b], we introduce an additional condition (U), which asserts that for any $\mathrm{P} \in \mathrm{S}^{\mathrm{cpt}} \setminus \mathrm{S}$ there exists an open subgroup $\mathrm{U_P}$ of the inertia group $\mathrm{I_P} \subset \pi_1(\mathrm{S})$ at P such that $\rho_\ell(\mathrm{U_P})$ is unipotent for $\ell \gg 0$. Condition (U) is stronger than condition (T), but we will not use it in the following. See [CT11b, §2.3] for more details, in particular for the following lemma.

Lemma 6 ([CT11b, Lem. 2.2, 2.3 and 2.4]).

(1) Assume that condition (T) is satisfied. Set $\mathrm{K} := \bigcap_\ell \ker(\rho_\ell)$. *Then* $\pi_1(\mathrm{S})/\mathrm{K}$ *is topologically finitely generated.*

(2) Conditions (I) and (T) imply $\lim\limits_{\ell \to +\infty} d_{\rho,1}(\ell) = +\infty$.

(3) Conditions (I), (T) and (WA) imply condition (WA)'.

Assume that conditions (I), (T) and (WA) are satisfied. Since $d_\rho(\ell) \geqslant d_{\rho,1}(\ell)$, it follows from Lemma 6 parts (2) and (3) that for $\ell \gg 0$ and any $\pi_1(\mathrm{S})$-submodule $0 \neq \mathrm{M} \subset \mathrm{H}_\ell$ the group $\mathrm{G_M}$ cannot be abelian.

Corollary 7. *Assume that conditions (I), (T) and (WA) hold. Then, for any integer* $\mathrm{B} \geqslant 1$, *for every* $\pi_1(\mathrm{S})$*-submodule* $0 \neq \mathrm{M} \subset \mathrm{H}_\ell$ *and for every abelian subgroup* A *of* $\mathrm{G_M}$ *one has* $[\mathrm{G_M} : \mathrm{A}] \geqslant \mathrm{B}$ *for* $\ell \gg 0$.

Proof. Otherwise, there exists an integer $\mathrm{B} \geqslant 1$ and an infinite set of primes $\mathcal{S}$ such that, for every $\ell \in \mathcal{S}$, there exists a $\pi_1(\mathrm{S})$-submodule $0 \neq \mathrm{M}_\ell \subset \mathrm{H}_\ell$ and an abelian subgroup A_ℓ of $\mathrm{G}_{\mathrm{M}_\ell}$ with $[\mathrm{G}_{\mathrm{M}_\ell} : \mathrm{A}_\ell] \leqslant \mathrm{B}$. But, since it follows from Lemma 6 (1) that $\pi_1(\mathrm{S})$ acts through a topologically finitely generated quotient, there are only finitely many isomorphism classes of connected étale covers of S corresponding to the $\rho_{M_\ell}^{-1}(A_\ell) \subset \pi_1(S)$ for $\ell \in \mathcal{S}$. Hence at least one of them, say $S' \to S$, appears infinitely many times. Up to base-changing by $S' \to S$, we may assume that G_{M_ℓ} is abelian for infinitely many $\ell \in \mathcal{S}$, which contradicts Lemma 6 (2) and (3). □

4.2.3 Etale Cohomology

Let $X \to S$ be a smooth, proper morphism with geometrically connected fibers. For every integer $i \geqslant 0$ the $\mathbb{F}_\ell$-rank $n_{i,\ell}$ of $H^i_\ell := H^i(X_{\overline{\eta}}, \mathbb{F}_\ell)$ is finite and independent of ℓ for $\ell \gg 0$. Indeed, when $p = 0$, this follows from the comparison isomorphism between Betti and étale cohomology with finite coefficients and the fact that Betti cohomology with coefficient in $\mathbb{Z}$ is finitely generated. More generally, when $p \geqslant 0$, this follows from the fact that ℓ-adic cohomology with coefficients in $\mathbb{Z}_\ell$ is torsion free for $\ell \gg 0$ [Gab83] and that the $\mathbb{Q}_\ell$-rank of ℓ-adic cohomology with coefficients in $\mathbb{Q}_\ell$ is independent of ℓ. So, we will simply write n_i instead of $n_{i,\ell}$ for $\ell \gg 0$.

For each $i \geqslant 1$ and $\ell \gg 0$, the action of $\pi_1(S)$ on H^i_ℓ gives rise to a family $\rho^i = (\rho^i_\ell)$ of n_i-dimensional $\mathbb{F}_\ell$-linear representations

$$\rho^i_\ell : \pi_1(S) \to GL(H^i_\ell) \simeq GL_{n_i}(\mathbb{F}_\ell) .$$

It follows from [CT11b, Thm. 2.4] that the families ρ^i for $i \geqslant 1$ satisfy conditions (T) and (WA). As for condition (I), if X_η is projective over $k(\eta)$ then, for $i = 1$ it can be ensured by the condition:

$\mathrm{Pic}^0_{X_{\overline{\eta}}/k(\overline{\eta})}$ contains no non-trivial k-isotrivial abelian subvarieties.

4.3 Technical Preliminaries

The proof of Theorem 1 is based on a combination of Lemma 6 with the use of E-P decompositions and group-theoretic ingredients. We gather the results we will need in Sects. 4.3.1, 4.3.2 and 4.3.3 respectively.

4.3.1 E-P Decompositions

Consider a diagram of proper curves over k

$$\begin{array}{ccc} Y & \xrightarrow{f} & B \\ {\scriptstyle \pi}\downarrow & & \\ Y', & & \end{array} \qquad (4.1)$$

where $f : \mathrm{Y} \to \mathrm{B}$ is a non-constant k-morphism of proper curves and $\pi : \mathrm{Y} \to \mathrm{Y}'$ is a G-cover with group G, i.e., G acts faithfully on Y and $\pi : \mathrm{Y} \to \mathrm{Y}'$ is the quotient morphism $\mathrm{Y} \to \mathrm{Y}/\mathrm{G}$. We will say that a pair of maps (π, f) as in (4.1) is **equivariant** if for any $\sigma \in \mathrm{G}$ there exists $\sigma_{\mathrm{B}} \in \mathrm{Aut}_k(\mathrm{B})$ such that $f \circ \sigma = \sigma_{\mathrm{B}} \circ f$ and that (π, f) as in (4.1) is **primitive** if it does not have any equivariant nontrivial subdiagram that is, more precisely, if for any commutative diagram (4.2) of morphisms of proper curves over k

$$\begin{array}{ccccc} & & \overset{f}{\frown} & & \\ \mathrm{Y} & \xrightarrow[f']{} & \mathrm{B}' & \xrightarrow[f'']{} & \mathrm{B} \\ {\scriptstyle\pi}\downarrow & & & & \\ \mathrm{Y}' & & & & \end{array} \tag{4.2}$$

with f' and f'' of degree $\geqslant 2$, the pair (π, f') is not equivariant.

We will resort to the following corollary of the Castelnuovo-Severi inequality.

Lemma 8 ([Tam04, Thm. 2.4]). *If the pair of maps (π, f) as in (4.1) is primitive then*

$$\deg(f) \geqslant \sqrt{\frac{g_{\mathrm{Y}}+1}{g_{\mathrm{B}}+1}}\,.$$

For a pair (π, f) as in diagram (4.1), among all equivariant decompositions, i.e., diagrams as (4.2) with the pair (π, f') equivariant, we choose a pair $(\pi, f' : \mathrm{Y} \to \mathrm{C})$ with $\deg(f')$ maximal. This exists as (π, id) is equivariant and $\deg(f') \leqslant \deg(f)$ is bounded. By definition, the action of G on Y induces an action on C, hence we obtain a homomorphism $\mathrm{G} \to \mathrm{Aut}_k(\mathrm{C})$. We set $\overline{\mathrm{G}} = \mathrm{G}/\mathrm{K}$ where

$$\mathrm{K} := \mathrm{Ker}(\mathrm{G} \to \mathrm{Aut}_k(\mathrm{C}))\,.$$

Then diagram (4.1) for (π, f) can be enriched to a commutative diagram with respect to the maximal equivariant decomposition (π, f') as follows

$$\begin{array}{ccccc} \mathrm{Y} & & \overset{f}{\searrow} & & \\ \downarrow & & & & \\ \mathrm{Z} & \longrightarrow & \mathrm{C} & \longrightarrow & \mathrm{B} \\ \downarrow & & \downarrow & & \\ \mathrm{Y}' & \longrightarrow & \mathrm{C}' & & \end{array} \tag{4.3}$$

where the vertical maps $\mathrm{Y} \to \mathrm{Z} = \mathrm{Y}/\mathrm{K}$, $\mathrm{Z} \to \mathrm{Y}' = \mathrm{Z}/\overline{\mathrm{G}}$ and $\mathrm{C} \to \mathrm{C}' = \mathrm{C}/\overline{\mathrm{G}}$ are the quotient morphisms. By construction, the pair $(\mathrm{Z} \to \mathrm{Y}', \mathrm{Z} \to \mathrm{C})$ is equivariant and the pair $(\mathrm{C} \to \mathrm{C}', \mathrm{C} \to \mathrm{B})$ is primitive. We will call such a decomposition an **equivariant-primitive decomposition** (E-P decomposition for short).

4.3.2 Review of the Classification of Finite Subgroups of SL_2

We remind that k is a fixed algebraically closed field of characteristic $p \geqslant 0$. Then we have the following description of finite subgroups of $\mathrm{SL}_2(k)$.

Theorem 9 ([Suz82, Thm. 3.6.17]). *A finite subgroup* G *of* $\mathrm{SL}_2(k)$ *is one from the following list:*

(1) *A cyclic group,*
(2) *For some* $n \geqslant 2$ *a group with presentation*

$$\langle x,y \mid x^n = y^2,\ y^{-1}xy = x^{-1}\rangle ,$$

(3) $\mathrm{SL}_2(3)$*, or* $\mathrm{SL}_2(5)$*,*
(4) *The representation group* $\hat{S}_4$ *of the permutation group* S_4 *in which transpositions lift to elements of order* 4*,*
(5) *An extension*

$$1 \to \mathrm{A} \to \mathrm{G} \to \mathrm{Q} \to 1 ,$$

where A *is an elementary abelian p-group and* Q *is a cyclic group of prime-to-p order,*
(6) *A dihedral group,*
(7) $\mathrm{SL}_2(k_r)$*, where* k_r *denotes the subfield of k with* p^r *elements,*
(8) $\langle \mathrm{SL}_2(k_r), d_\pi\rangle$*, where* d_π *is the scalar matrix with diagonal entries given by a* $\pi \in k$ *such that* $k_r(\pi)$ *has* p^{2r} *elements and* π^2 *is a generator of* $k_r^\times$*.*

Case (6) occurs only when $p = 2$ *and cases (7) and (8) occur only when* $p > 0$*.*

We will use two corollaries of Theorem 9. Observing that $\mathrm{PGL}_2(k) = \mathrm{PSL}_2(k)$ when k is algebraically closed, we get the following well known corollary.

Corollary 10. *A finite subgroup* G *of* $\mathrm{PGL}_2(k)$ *is of the following form:*

(1) *A cyclic group,*
(2) *A dihedral group,*
(3) $\mathcal{A}_4$, S_4, $\mathcal{A}_5$*,*
(4) *An extension*

$$1 \to \mathrm{A} \to \mathrm{G} \to \mathrm{Q} \to 1 ,$$

where A *is an elementary abelian p-group and* Q *is a cyclic group of prime-to-p order,*
(5) $\mathrm{PSL}_2(k_r)$*,*
(6) $\mathrm{PGL}_2(k_r)$*.*

The last three cases occur only when $p > 0$*.*

Also, regarding $\mathrm{SL}_2(\mathbb{F}_\ell)$ as a subgroup of $\mathrm{SL}_2(\overline{\mathbb{F}}_\ell)$ and ruling out the groups that cannot lie in $\mathrm{SL}_2(\mathbb{F}_\ell)$, we get:

Corollary 11. *Assume that* $\ell \geqslant 5$*. A subgroup of* $\mathrm{SL}_2(\mathbb{F}_\ell)$ *is isomorphic to one of the following:*

(1) *A cyclic group,*
(2) *For some* $n \geqslant 2$ *a group with presentation*

$$\langle x,y \mid x^n = y^2,\ y^{-1}xy = x^{-1} \rangle ,$$

(3) $\mathrm{SL}_2(\mathbb{F}_3)$, *or* $\mathrm{SL}_2(\mathbb{F}_5)$,
(4) *The representation group* $\hat{\mathcal{S}}_4$ *of the permutation group* $\mathcal{S}_4$ *in which transpositions lift to elements of order* 4,
(5) *A semi-direct product* $\mathbb{F}_\ell \rtimes \mathrm{C}$ *contained in a Borel subgroup with* C *a cyclic group of prime-to-*ℓ *order,*
(6) $\mathrm{SL}_2(\mathbb{F}_\ell)$.

4.3.3 A Group-Theoretic Lemma

The following lemma provides a practical condition for a finite group to contain a large normal abelian subgroup.

Lemma 12. *Let* G *be a finite group and assume that* G *fits into a short exact sequence of finite groups*

$$1 \to \mathrm{N} \to \mathrm{G} \to \mathrm{Q} \to 1 \tag{4.4}$$

with Q *abelian and generated by* $\leqslant r$ *elements. Then the group* G *contains a normal abelian subgroup* A *with index*

$$[\mathrm{G} : \mathrm{A}] \leqslant \mu(\mathrm{Z(N)})^r \cdot |\mathrm{Aut(N)}| ,$$

where $\mu(\mathrm{Z(N)})$ *denotes the least common multiple of the order of the elements in the center* Z(N) *of* N.

Proof. The short exact sequence (4.4) induces by conjugation representations

$$\tilde{\phi} : \mathrm{G} \to \mathrm{Aut(N)} \quad \text{and} \quad \phi : \mathrm{Q} \to \mathrm{Out(N)}$$

and induces on the centralizer $\mathrm{Z_G(N)} = \ker(\tilde{\phi})$ of N in G the structure of a central extension

$$1 \to \mathrm{Z(N)} \to \mathrm{Z_G(N)} \to \ker(\phi) \to 1 .$$

Because the extension is central, taking the commutator of lifts to $\mathrm{Z_G(N)}$ defines an alternating bilinear form [,] on $\ker(\phi)$ with values in Z(N). The radical of [,]

$$\mathrm{R} = \{q \in \ker(\phi)\ ;\ [q,q'] = 0 \text{ for all } q' \in \ker(\phi)\} \subset \ker(\phi) ,$$

contains $\mu(\mathrm{Z(N)})\ker(\phi)$. We find an extension

$$1 \to \mathrm{Z(N)} \to \mathrm{Z(Z_G(N))} \to \mathrm{R} \to 1$$

where $A = Z(Z_G(N))$ is the center of $Z_G(N)$. Since N is normal in G, the abelian group A is also normal in G. We can estimate the index $[G : A]$ as

$$\begin{aligned}[G:A] &= \frac{|G|}{|Z_G(N)|}\cdot\frac{|Z_G(N)|}{|A|} \leqslant |\mathrm{Aut}(N)|\cdot\frac{|\ker(\phi)|}{|R|}\\ &\leqslant |\mathrm{Aut}(N)|\cdot\frac{|\ker(\phi)|}{|\mu(Z(N))\ker(\phi)|} \leqslant |\mathrm{Aut}(N)|\cdot\mu(Z(N))^r\end{aligned}$$

since $\ker(\phi) \subset Q$ is also generated by $\leqslant r$ elements. □

4.4 Proof of Theorem 1

Observe first that if $S' \to S$ is any connected finite étale cover then

$$\pi_1(S'_M) = \pi_1(S_M) \cap \pi_1(S') .$$

In particular, one has

$$\gamma_{S_M} \leqslant \gamma_{S'_M} \leqslant \gamma_{S_M}\deg(S'_M \to S_M) \leqslant \gamma_{S_M}\deg(S' \to S)$$

and, as a result, $\lim_{\ell\to+\infty} \gamma_{\rho|_{\pi_1(S')}}(\ell) = +\infty$ if and only if $\lim_{\ell\to+\infty} \gamma_\rho(\ell) = +\infty$. This allows to perform arbitrary base changes by connected étale covers. In particular, from condition (T), one may assume that $\pi_1(S)$ acts through its tame quotient $\pi_1^t(S)$.

For every prime ℓ, consider a $\pi_1(S)$-submodule $0 \neq M_\ell \subset H_\ell$ such that $\gamma_{M_\ell} = \gamma_\rho(\ell)$. We thus have a diagram of proper curves over k

$$\begin{array}{ccc} S^{cpt}_{M_\ell} & \xrightarrow{f_\ell} & \mathbb{P}^1_k \\ \downarrow & & \\ S^{cpt} & & \end{array} \tag{4.5}$$

with $\deg(f_\ell) = \gamma_\rho(\ell)$. We can consider an E-P decomposition of (4.5)

$$\begin{array}{ccccc} S^{cpt}_{M_\ell} & & & & \\ \downarrow & & \searrow^{f_\ell} & & \\ Z_\ell & \longrightarrow & C_\ell & \longrightarrow & \mathbb{P}^1_k \\ \downarrow & & \downarrow & & \\ S^{cpt} & \longrightarrow & B_\ell & & \end{array} \tag{4.6}$$

where, setting $K_\ell = \ker(G_{M_\ell} \to \mathrm{Aut}(C_\ell))$ and $\overline{G}_{M_\ell} = G_{M_\ell}/K_\ell$, $S^{cpt}_{M_\ell} \to Z_\ell = S^{cpt}_{M_\ell}/K_\ell$ and $C_\ell \to B_\ell = C_\ell/\overline{G}_{M_\ell}$ are the respective quotient maps.

If $\gamma_p(\ell)$ does not diverge, then there exists an infinite subset $\mathcal{S}$ of primes and an integer $\gamma \geqslant 1$ such that $\gamma_p(\ell) \leqslant \gamma$ for all $\ell \in \mathcal{S}$. In particular $|K_\ell| \leqslant \gamma$, hence

$$|\overline{G}_{M_\ell}| \geqslant \frac{d_p(\ell)}{|K_\ell|} \geqslant \frac{d_p(\ell)}{\gamma}.$$

So, from Lemma 6 (2) one has

$$\lim_{\substack{\ell \to +\infty \\ \ell \in \mathcal{S}}} |\overline{G}_{M_\ell}| = +\infty .$$

To get the contradiction, we distinguish between three cases. In the first case we assume that $g_{C_\ell} \geqslant 2$ for all but finitely many $\ell \in \mathcal{S}$. Since by [Sti73] the size of the automorphism group of a genus $g \geqslant 2$ curve over an algebraically closed field of characteristic p is bounded by $P_p(g)$ for a polynomial $P_p(T) \in \mathbb{Z}[T]$ depending only on p, we find for $\ell \in \mathcal{S}$ that $|\overline{G}_{M_\ell}| \leqslant P_p(g_{C_\ell})$, which forces

$$\lim_{\substack{\ell \to +\infty \\ \ell \in \mathcal{S}}} g_{C_\ell} = +\infty .$$

But from Lemma 8 applied to the primitive pair $(C_\ell \to B_\ell, C_\ell \to \mathbb{P}^1_k)$ in diagram (4.5), one has

$$\gamma_p(\ell) = \deg(f_\ell) \geqslant \deg(C_\ell \to \mathbb{P}^1_k) \geqslant \sqrt{g_{C_\ell} + 1} ,$$

which therefore also diverges for $\ell \in \mathcal{S}$ contradicting the choice of $\mathcal{S}$.

If we are not in the first case, then $g_{C_\ell} \leqslant 1$ for infinitely many $\ell \in \mathcal{S}$. In the second case, we assume that for infinitely many $\ell \in \mathcal{S}$, and in fact by replacing $\mathcal{S}$ by a subset, that for all $\ell \in \mathcal{S}$ we have $g_{C_\ell} = 1$. Then for $\ell \in \mathcal{S}$, the group $\overline{G}_{M_\ell}$ is an extension

$$1 \to A_\ell \to \overline{G}_{M_\ell} \to Q_\ell \to 1$$

with A_ℓ a finite quotient of $\hat{\mathbb{Z}}^2$ and $|Q_\ell| \leqslant 24$. Since $\pi_1(S)$ acts through a topologically finitely generated quotient by Lemma 6, there are only finitely many isomorphism classes of étale covers of S with degree $\leqslant 24$ corresponding to the inverse image of A_ℓ via

$$\pi_1(S) \overset{\rho_{M_\ell}}{\twoheadrightarrow} G_{M_\ell} \twoheadrightarrow \overline{G}_{M_\ell} .$$

So, by replacing S by the composite of all these étale covers of degree $\leqslant 24$, we may assume that $\overline{G}_{M_\ell} = A_\ell$ for all $\ell \in \mathcal{S}$. Now Lemma 12 applied to

$$1 \to K_\ell \to G_{M_\ell} \to A_\ell \to 1$$

shows, since $|K_\ell| \leqslant \gamma$, that G_{M_ℓ} has an abelian subgroup of index bounded above independently of $\ell \in \mathcal{S}$ in contradiction to Corollary 7.

In the last case we can and do assume that $g_{C_\ell} = 0$ for all $\ell \in \mathcal{S}$. As above, Corollary 7 shows that the subgroup

$$\overline{G}_{M_\ell} \subset \mathrm{Aut}(C_\ell) \cong \mathrm{PGL}_2(k)$$

can be only of type (4), (5) or (6) as in Corollary 10 for $\ell \gg 0$, and $\ell \in \mathcal{S}$. This occurs only if $p > 0$. Without loss of generality, by replacing $\mathcal{S}$ by an infinite subset, we may assume that $\overline{G}_{M_\ell}$ is of the same type for all $\ell \in \mathcal{S}$. To rule out these cases, we are going to use the following theorem.

Theorem 13 ([Nor87, Thm. C]). *For any integer $n \geqslant 1$ there exists an integer $d(n) \geqslant 1$ such that for any prime $\ell \geqslant n$, integer $m \leqslant n$ and subgroup G of $\mathrm{GL}_m(\mathbb{F}_\ell)$ the following holds. Let G^+ denote the (normal) subgroup of G generated by the elements of order ℓ in G. Then, there exists an abelian subgroup $A \subset G$ such that AG^+ is normal in G and $[G : AG^+] \leqslant d(n)$.*

Assume that $\overline{G}_{M_\ell}$ is of type (4) for all $\ell \in \mathcal{S}$, that is of the form

$$(\mathbb{Z}/p)^{r_\ell} \rtimes Z/N_\ell$$

for some integers $r_\ell, N_\ell \geqslant 1$ with $p \nmid N_\ell$.

Claim 14. There exists an integer $r(n) \geqslant 1$ such that $r_\ell \leqslant r(n)$ for $\ell \gg 0$ in $\mathcal{S}$.

Proof. Let T_ℓ denote the inverse image of $(\mathbb{Z}/p)^{r_\ell}$ in G_{M_ℓ} that is T_ℓ fits into the short exact sequence of finite groups

$$1 \to K_\ell \to T_\ell \to (\mathbb{Z}/p)^{r_\ell} \to 1 .$$

Because $|K_\ell| \leqslant \gamma$ we see that ℓ does not divide $|T_\ell|$ for $\ell \gg 0$ and, in particular, that T_ℓ^+ is trivial. Theorem 13 implies that T_ℓ fits into a short exact sequence

$$1 \to A_\ell \to T_\ell \to Q_\ell \to 1$$

with A_ℓ abelian and $|Q_\ell| \leqslant d(n)$. In turn, A_ℓ fits into the sort exact sequence

$$1 \to K_\ell \cap A_\ell \to A_\ell \to (\mathbb{Z}/p)^{s_\ell} \to 1$$

with $s_\ell \leqslant r_\ell$. In particular, A_ℓ is an abelian subgroup of $\mathrm{GL}(M_\ell)$ of prime-to-ℓ order and of $\mathbb{Z}$-rank $\geqslant s_\ell$. This implies $s_\ell \leqslant n$ since any abelian subgroup A of order prime-to-ℓ in $\mathrm{GL}_n(\mathbb{F}_\ell)$ is conjugate in $\mathrm{GL}_n(\overline{\mathbb{F}}_\ell)$ to a diagonal torus. So the claim follows from $r_\ell \leqslant s_\ell + \log_p |Q_\ell|$ and the bounds for s_ℓ and $|Q_\ell| \leqslant d(n)$. □

By Claim 14 and Lemma 12, the group $\overline{G}_{M_\ell}$ contains a normal abelian subgroup A_ℓ with index bounded by

$$[\overline{G}_{M_\ell} : A_\ell] \leqslant p \cdot |\mathrm{GL}_{r(n)}(\mathbb{F}_p)| .$$

Invoking again that $\pi_1(S)$ acts through a topologically finitely generated quotient, without loss of generality we may assume that $\overline{G}_{M_\ell} = A_\ell$ and then, as above the contradiction follows from the bound $|K_\ell| \leqslant \gamma$, Lemma 12 and Corollary 7.

Assume now that $\overline{G}_{M_\ell}$ is of type (5) or (6) for all $\ell \in \mathcal{S}$, that is either $PSL_2(k_{r_\ell})$ or $PGL_2(k_{r_\ell})$ for some integer $r_\ell \geqslant 1$.

For any non zero vector $v \in M_\ell$ the cover $S_{M(v)} \to S$ is a quotient of $S_{M_\ell} \to S$ hence $\gamma_{S_{M(v)}} \leqslant \gamma_{S_{M_\ell}}$. So, without loss of generality, we may assume that M_ℓ is a simple $\pi_1(S)$-module. In particular, there exists a non zero vector $v \in M_\ell$ such that $M_\ell = M(v)$ and $M_\ell^+ := \mathbb{F}_\ell[G_{M_\ell}^+ v] \subset M$ is a simple $G_{M_\ell}^+$-submodule.

Claim 15. The group $G_{M_\ell}^+$ is nontrivial for $\ell \gg 0$ in $\mathcal{S}$.

Proof. Theorem 13 applied to $G_{M_\ell} \subset GL(M_\ell)$ shows that one can write $G_{M_\ell}/G_{M_\ell}^+$ as an extension

$$1 \to A_\ell G_{M_\ell}^+/G_{M_\ell}^+ \to G_{M_\ell}/G_{M_\ell}^+ \to Q_\ell \to 1$$

with $A_\ell G_{M_\ell}^+/G_{M_\ell}^+$ abelian and $|Q_\ell| \leqslant d(n)$, because $\dim_{\mathbb{F}_\ell}(M_\ell) \leqslant n$. As a result, if $G_{M_\ell}^+ = 1$, we get a contradiction to Corollary 7. This proves the claim. □

Since $PSL_2(k_{r_\ell})$ is simple and the only nontrivial normal subgroups of $PGL_2(k_{r_\ell})$ are $PSL_2(k_{r_\ell})$ and $PGL_2(k_{r_\ell})$, Claim 15 implies that the normal subgroup

$$\overline{G}_{M_\ell}^+ := G_{M_\ell}^+/G_{M_\ell}^+ \cap K_\ell$$

of $\overline{G}_{M_\ell}$ contains $PSL_2(k_{r_\ell})$.

Claim 16. The group $Z_\ell := K_\ell \cap G_{M_\ell}^+$ is a central subgroup of $G_{M_\ell}^+$ for $\ell \gg 0$ in $\mathcal{S}$.

Proof. Because $|Z_\ell| \leqslant |K_\ell| \leqslant \gamma$ we see that (i) $\ell \nmid |Z_\ell|$ and (ii) $\ell \nmid |\mathrm{Aut}(Z_\ell)|$ for $\ell \gg 0$ in $\mathcal{S}$. From (i) and Schur-Zassenhauss, for any ℓ-Sylow $S_\ell \subset G_{M_\ell}$, the group $Z_\ell S_\ell$ is a semidirect product $Z_\ell \rtimes S_\ell$ and, from (ii), the semidirect product $Z_\ell \rtimes S_\ell$ is actually a direct product that is S_ℓ is contained in the centralizer $Z_{G_{M_\ell}^+}(H_\ell)$ of H_ℓ in $G_{M_\ell}^+$. But, by definition, for $\ell \gg 0$ the group $G_{M_\ell}^+$ is generated by the ℓ-Sylow subgroups S_ℓ of G_{M_ℓ} hence $G_{M_\ell}^+ = Z_{G_{M_\ell}^+}(Z_\ell)$. □

The group Z_ℓ acts semisimply on H_ℓ, because Z_ℓ is commutative and of prime-to-ℓ order. Since Z_ℓ is central in $G_{M_\ell}^+$ by Claim 16, the group $G_{M_\ell}^+$ preserves the isotypical decomposition

$$M \otimes_{\mathbb{F}_\ell} \overline{\mathbb{F}}_\ell = \bigoplus_\chi E_\chi$$

with respect to the characters χ of Z_ℓ. The induced projective representations

$$p_\chi : G_{M_\ell}^+ \to PGL(E_\chi)$$

factor over $\overline{G}_{M_\ell}^+$. As $\bigcap_\chi \ker(p_\chi)$ is of order prime-to-ℓ, this shows that the simple normal subgroup $PSL_2(k_{r_\ell}) \subseteq \overline{G}_{M_\ell}^+$ embeds into $PGL_m(\overline{\mathbb{F}}_\ell)$ for some $m \leqslant n$. By

[LS74, Thm. p. 419], this can occur only for finitely many values of r_ℓ, which, in turn, contradicts the fact that $r_\ell \to \infty$ for $\ell \in \mathcal{S}$ by Lemma 6 (2). The proof of Theorem 1 is now complete.

4.5 The Case of $S_{\rho,1}(\ell)$

Whenever it is defined, we set for $i = 1,\dots,n = \dim_{\mathbb{F}_\ell}(H_\ell)$

$$\gamma^i_{\rho,1}(\ell) := \min\{\gamma_v \ ; \ 0 \neq v \in H_\ell \text{ and } \dim_{\mathbb{F}_\ell}(M(v)) \leqslant i\}.$$

Note that, when $n = i$, one has $\gamma^n_{\rho,1}(\ell) = \gamma_{\rho,1}(\ell)$. Let $\mathcal{S}$ denote the set of all primes ℓ such that H_ℓ contains a $\pi_1(S)$-submodule of $\mathbb{F}_\ell$-rank 2. Assume that $\mathcal{S}$ is infinite. In this section, we prove the following.

Proposition 17. *Assume that conditions (WA), (I) and (T) are satisfied. Then:*

$$\lim_{\substack{\ell \to +\infty \\ p \nmid \ell(\ell^2-1)}} \gamma^2_{\rho,1}(\ell) = +\infty .$$

Proposition 17 will lead to a proof of Corollary 3. The proof of Proposition 17 needs some preparation. We first study the possible structure of the group G_M when $\dim_{\mathbb{F}_\ell}(M) = 2$ and $\ell \gg 0$.

Lemma 18. *Assume that conditions (WA), (I) and (T) are satisfied. Then, for $\ell \gg 0$ and any $\pi_1(S)$-submodule $M \subset H_\ell$ of $\mathbb{F}_\ell$-rank 2 one has $SL(M) \subset G_M$.*

Proof. We write G_M as an extension

$$1 \to G_M \cap SL(M) \to G_M \xrightarrow{\det} D_M \to 1,$$

where

$$D_M = \det(G_M) \subset \mathbb{F}_\ell^\times \simeq \mathbb{Z}/(\ell-1).$$

Let us show first that $|G_M \cap SL(M)|$ diverges with $\ell \to \infty$ in $\mathcal{S}$ and M is any $\pi_1(S)$-submodule $M \subset H_\ell$ of $\mathbb{F}_\ell$-rank 2. Otherwise, up to replacing $\mathcal{S}$ by an infinite subset, we may assume that there exists an upper bound

$$|G_M \cap SL(M)| \leqslant B$$

for all possible M. From Lemma 6 (2), one has

$$\lim_{\substack{\ell \to +\infty \\ \ell \in \mathcal{S}}} |G_{M_\ell}| = +\infty ,$$

which forces $|D_M|$ to diverge when $\ell \to \infty$ in $\mathcal{S}$. Let $o(B)$ denote the maximal order of the automorphism group of a group of order $\leqslant B$. Then, as D_M is cyclic, it follows from Lemma 12 that G_M contains a normal abelian subgroup of index $\leqslant B \cdot o(B)$, which contradicts Corollary 7 for $\ell \gg 0$ in $\mathcal{S}$.

Hence, for $\ell \gg 0$ in $\mathcal{S}$ and any $\pi_1(S)$-submodule $M \subset H_\ell$ of $\mathbb{F}_\ell$-rank 2, the only possibilities with respect to the list of Corollary 11 for $G_M \cap SL(M)$ are (1), (2), (5) or (6). The types (1) and (2) are ruled out by condition (WA)' and Lemma 6, and type (6) is exactly what the lemma claims. It remains to rule out type (5).

If $G_M \cap SL(M)$ is of type (5), then it is contained in a Borel and thus fixes a line $\mathbb{F}_\ell \cdot v \subset M$ for some $0 \neq v \in H_\ell$. The line is uniquely determined since the ℓ-Sylow of $G_M \cap SL(M)$ is nontrivial, and thus $\mathbb{F}_\ell \cdot v$ is also invariant under G_M. However, by condition (WA) and Lemma 6 (2), the group G_M cannot fix $\mathbb{F}_\ell \cdot v$, which is the desired contradiction. □

Lemma 19. *Assume that conditions (WA), (I) and (T) are satisfied. Then, there exists an integer* $D \geqslant 1$ *such that for* $\ell \gg 0$ *and any* $\pi_1(S)$*-submodule* $M \subset H_\ell$ *one has* $|\det(G_M)| \leqslant D$.

Proof. Let m denote the $\mathbb{F}_\ell$-rank of M. Then the action of G_M on the line $\Lambda^m M$ factors through a faithfull action of $D_M := \det(G_M)$. So the conclusion follows from condition (WA)'. □

Now we can prove Proposition 17. Let $\mathcal{S}$ denote the set of all primes ℓ such that there exists $v \in H_\ell$ with $M(v)$ of $\mathbb{F}_\ell$-rank 2. Assume that $\mathcal{S}$ is infinite and for every $\ell \in \mathcal{S}$, choose $v_\ell \in H_\ell$ with $M_\ell := M(v_\ell)$ of $\mathbb{F}_\ell$-rank 2 such that $\gamma_{v_\ell} = \gamma^2_{\rho,1}(\ell)$. By Lemma 18 and for $\ell \gg 0$ in $\mathcal{S}$ we write again G_{M_ℓ} as an extension

$$1 \to SL(M_\ell) \to G_{M_\ell} \xrightarrow{\det} D_\ell \to 1\,,$$

where

$$D_\ell = \det(G_{M_\ell}) \subset \mathbb{F}_\ell^\times \simeq \mathbb{Z}/(\ell-1)\,.$$

From Lemma 19, we have $|D_\ell| \leqslant D$. Consider an E-P decomposition

$$\begin{array}{ccccc} S^{cpt}_{M_\ell} & & \overset{f_\ell}{\searrow} & & \\ \downarrow & & & & \\ Z_\ell & \longrightarrow & C_\ell & \longrightarrow & \mathbb{P}^1_k \\ \downarrow & & \downarrow & & \\ S^{cpt} & \longrightarrow & B_\ell & & \end{array} \tag{4.7}$$

where, setting $K_\ell = \ker(G_{M_\ell} \to \mathrm{Aut}(C_\ell))$ and $\overline{G}_{M_\ell} = G_{M_\ell}/K_\ell$, $S^{cpt}_{M_\ell} \to Z_\ell = S^{cpt}_{M_\ell}/K_\ell$ and $C_\ell \to B_\ell = C_\ell/\overline{G}_{M_\ell}$ are the respective quotient maps, and $\deg(f_\ell) = \gamma_{M_\ell}$. We set D^K_ℓ for the image of K_ℓ in D_ℓ. Then K_ℓ fits into the short exact sequence

$$1 \to K_\ell \cap SL(M_\ell) \to K_\ell \to D_\ell^K \to 1 .$$

As the only normal subgroups of $SL_2(\mathbb{F}_\ell)$ are 1, $\mathbb{Z}/2$ and $SL_2(\mathbb{F}_\ell)$, there are only two possibilities for $K_\ell \cap SL(M_\ell)$, namely:

(1) $K_\ell \cap SL(M_\ell) = SL(M_\ell)$,
(2) $K_\ell \cap SL(M_\ell) = 1, \mathbb{Z}/2$.

In case (1), one has the estimate

$$\gamma_{M_\ell} = \deg(f_\ell) \geqslant \deg(S_{M_\ell}^{cpt} \to Z_\ell) = |K_\ell| = \ell(\ell^2-1) \cdot |D_\ell^K| = |G_{M_\ell}| \cdot \frac{|D_\ell^K|}{|D_\ell|} .$$

Since $SL(M_\ell)$ acts transitively on $M_\ell \setminus \{0\}$, the stabilizer G_{M_ℓ, v_ℓ} of v_ℓ under the action of G_{M_ℓ}, namely the Galois group of $S_{M(v)} \to S_v$, has index $\ell^2 - 1$ and so

$$\gamma_{\rho,1}^2(\ell) = \gamma_{v_\ell} \geqslant \frac{\gamma_{M_\ell}}{|G_{M_\ell, v_\ell}|} \geqslant (\ell^2-1) \cdot \frac{|D_\ell^K|}{|D_\ell|} \geqslant \frac{\ell^2-1}{D} \to +\infty .$$

In case (2), the stabilizer has size

$$|G_{M_\ell, v_\ell}| = \frac{|G_{M_\ell}|}{\ell^2-1} = \ell \cdot |D_\ell| ,$$

and thus Lemma 8 applied to the primitive pair $(C_\ell \to B_\ell, C_\ell \to \mathbb{P}_k^1)$ in diagram (4.7) yields the estimate

$$\gamma_{\rho,1}^2(\ell) = \gamma_{v_\ell} \geqslant \frac{\gamma_{M_\ell}}{|G_{M_\ell, v_\ell}|} \geqslant \frac{\deg(S_{M_\ell}^{cpt} \to Z_\ell) \cdot \deg(C_\ell \to \mathbb{P}_k^1)}{\ell \cdot |D_\ell|} \geqslant \frac{|K_\ell| \cdot \sqrt{g_{C_\ell} + 1}}{\ell \cdot |D_\ell|} \tag{4.8}$$

For $\ell \gg 0$ and in particular $\ell > p$, the group $\overline{G}_{M_\ell}$ contains $SL(M_\ell)$ or $PSL(M_\ell)$, and so it is not a subgroup of the automorphism group of a curve of genus 0 or 1 over an algebraically closed field of characteristic $p \geqslant 0$. As a result, one may assume that C_ℓ has genus $\geqslant 2$. If p does not divide $\ell(\ell^2-1)$ then p does not divide $|GL(M_\ell)|$ hence, a fortiori, does not divide $|\overline{G}_{M_\ell}|$. Consequently, the cover $C_\ell \to B_\ell$ lifts to characteristic 0 and we have the Hurwitz bound for the automorphism group

$$\frac{\ell(\ell^2-1)|D_\ell|}{|K_\ell|} = |\overline{G}_{M_\ell}| \leqslant 84(g_{C_\ell} - 1) .$$

In combination with (4.8) this yields

$$\gamma_{\rho,1}^2(\ell) \geqslant \frac{|K_\ell| \cdot \sqrt{g_{C_\ell} + 1}}{\ell \cdot |D_\ell|} \geqslant \frac{|K_\ell|}{\ell \cdot |D_\ell|} \sqrt{\frac{\ell(\ell^2-1)|D_\ell|}{84|K_\ell|} + 2}$$

Hence

$$\gamma^2_{\rho,1}(\ell) \geqslant \sqrt{\frac{(\ell^2-1)}{84\cdot\ell\cdot|\mathrm{D}|}} \to +\infty\,.$$

This completes the proof of Proposition 17.

Remark 20. When $p|\ell(\ell^2-1)$, one can assert only that $\ell(\ell^2-1) \leqslant \mathrm{P}_p(g_{\mathrm{C}_\ell})$ so the resulting lower bound for g_{C_ℓ} is too small to conclude. Also, from condition (T), one could observe that $\mathrm{Z}_\ell \to \mathrm{S}^{\mathrm{cpt}}$ is tame for $\ell \gg 0$ but, if $p \mid \ell(\ell^2-1)$ and $\mathrm{S}^{\mathrm{cpt}} \to \mathrm{B}_\ell$ is wildly ramified, it may happen that $\mathrm{C}_\ell \to \mathrm{B}_\ell$ is wildly ramified as well hence does not necessarily lift to characteristic 0.

Finally, we give a proof of Corollary 3. Let $\mathrm{Y}(0)$ and $\mathrm{Y}_1(\ell)$ denote the coarse moduli schemes of the stack $\mathcal{E}$ of elliptic curves and of the stack $\mathcal{E}_1(\ell)$ of elliptic curves with a torsion point of order exactly ℓ as stacks over k. For any nonisotrivial relative elliptic curve $\mathrm{E} \to \mathrm{S}$ and $0 \neq v \in \mathrm{E}_{\overline{\eta}}[\ell]$, one has the following commutative diagram

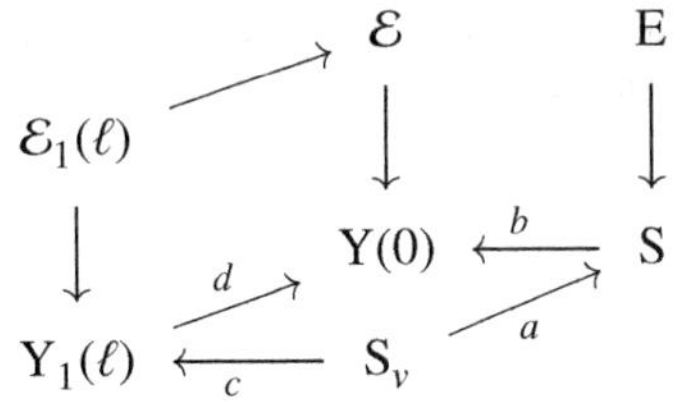

In particular we can estimate the gonality as

$$\gamma_{\mathrm{Y}_1(\ell)} \geqslant \frac{\gamma_v}{\deg(c)} = \frac{\gamma_v \deg(d)}{\deg(a)\deg(b)} = \frac{\gamma_v(\ell^2-1)/2}{|\mathrm{G}_\ell\cdot v|\deg(b)} \geqslant \frac{\gamma_v}{2\deg(b)}$$

with $\deg(b)$ independent of v and ℓ. Applying Proposition 17 to the family of rank-2 $\mathbb{F}_\ell$-linear representations

$$\rho_\ell : \pi_1(\mathrm{S}) \to \mathrm{GL}(\mathrm{E}_{\overline{\eta}}[\ell])$$

gives the conclusion of Corollary 3.

References

[Abr96] D. Abramovich. A linear lower bound on the gonality of modular curves. *Internat. Math. Res. Notices*, 20:1005–1011, 1996.

[CT11a] A. Cadoret and A. Tamagawa. On a weak variant of the geometric torsion conjecture. *Journal of Algebra*, 346:227–247, 2011.

[CT11b] A. Cadoret and A. Tamagawa. On a weak variant of the geometric torsion conjecture II. Preprint, available online at `http://www.math.polytechnique.fr/~cadoret/Travaux.html`, 2011.

[EHK10] J. Ellenberg, C. Hall, and E. Kowalski. Expander graphs, gonality and variation of Galois representations. Preprint, available online at `http://www.math.wisc.edu/~ellenber/papers.html`, 2010.

[Fal91] G. Faltings. Diophantine approximation on abelian varieties. *Annals of Math.*, 133:549–576, 1991.

[Fre94] G. Frey. Curves with infinitely many points of fixed degree. *Israel J. Math.*, 85:79–83, 1994.

[Gab83] O. Gabber. Sur la torsion dans la cohomologie ℓ-adique d'une variété. *C. R. Acad. Sci. Paris Ser. I Math.*, 297:179–182, 1983.

[LS74] V. Landazuri and G. M. Seitz. On the minimal degree of projective representations of the finite Chevalley groups. *J. Algebra*, 32:418–443, 1974.

[Nor87] M. V. Nori. On subgroups of $\mathrm{GL}_n(\mathbb{F}_p)$. *Inventiones Math.*, 88:257–275, 1987.

[Poo07] B. Poonen. Gonality of modular curves in characteristic p. *Math. Res. Letters*, 14:691–701, 2007.

[Sti73] H. Stichtenoth. Über die Automorphismengruppe eines algebraischen Funktionenkörpers von Primzahlcharakteristik I, II. *Arch. der Math. (Basel)*, 24:524–544 and 615–631, 1973.

[Suz82] M. Suzuki. *Group theory I.* Number 247 in Grundlehren der Mathematischen Wissenschaften. Springer, 1982.

[Tam04] A. Tamagawa. Finiteness of isomorphism classes of curves in positive characteristic with prescribed fundamental group. *J. Algebraic Geometry*, 13:675–724, 2004.

Chapter 5
The Motivic Logarithm for Curves

Gerd Faltings

Abstract The paper explains how in Kim's approach to diophantine equations étale cohomology can be replaced by motivic cohomology. For this Beilinson's construction of the motivic logarithm suffices, and it is not necessary to construct a category of mixed motives as it is done by Deligne–Goncharov for rational curves.

5.1 Introduction

The purpose of this note is to exhibit the definition of a motivic logarithm for smooth curves, following Beilinson. I personally prefer the name polylogarithm but I have learned that there is some opposition to this because the name polylogarithm is already in use for different (although in my opinion related) objects. Many finer properties of motivic categories are only shown over fields, but our definition makes sense over any base. We show that its different realisations give the logarithm in étale and crystalline cohomology. For rational curves this has been done in Deligne–Goncharov. They in fact achieve much more by also defining a category of mixed Tate-motives which contains the motivic logarithm. Here we are more modest and only construct the logarithm itself, without exhibiting it as an object in a category of mixed motives. It is something like the free tensor-algebra in the reduced motive of the curve, and the motivic fundamental group has as its Lie-algebra the free Lie-algebra. In fact we work with $\mathbb{Q}$-coefficients where nilpotent groups and nilpotent Lie-algebras correspond via the Hausdorff series, see [Bou75, Chap. 2, 6]. At the end we try to define the notion of a motivic torsor. Unfortunately the desired properties need some additional vanishing assumptions. For the moment these are known only for Tate-motives over a field, where however we already can cite Deligne–Goncharov [DG05].

G. Faltings (✉)
MPI für Mathematik, Vivatsgasse 7, 53111 Bonn, Germany
e-mail: gerd@mpim-bonn.mpg.de

J. Stix (ed.), *The Arithmetic of Fundamental Groups*, Contributions in Mathematical and Computational Sciences 2, DOI 10.1007/978-3-642-23905-2_5,

Our main motivation for this work was the new method of Kim [Kim05] in diophantine geometry, and its improvement by Hadian-Jazi [HJ10] which involves motivic cohomology. However so far the results are not really useful in that context because of our general inability to compute motivic cohomology groups. The few known cases concern Tate-motives mainly of number fields and use the connection with algebraic K-theory. In this context it might be useful to define some regulator with values in K-theory. For example this might work over the integers while the most essential properties of motivic cohomology are only shown over a field.

Acknowledgements. Thanks for enlightening discussions go to A. Beilinson, H. Esnault, and M. Levine, and also to the referee for his comments.

We assume that S is an arbitrary base-scheme and

$$X \to S$$

a relative smooth curve with geometrically irreducible fibres. Furthermore we assume that X admits an embedding

$$X \subseteq \bar{X}$$

into a smooth projective curve $\bar{X}$ such that the complement

$$D = \bar{X} - X$$

is a divisor which is finite étale over S. We denote by $\mathcal{M}$ the Karoubian hull of the category of smooth S-schemes, with maps given on connected components by $\mathbb{Z}$-linear combinations of actual maps of schemes. That is an object of $\mathcal{M}$ is defined by a smooth S-scheme T together with a projector in the locally constant combinations of elements of $\mathrm{End}_S(T)$. $\mathcal{M}$ is an additive category (sums are disjoint unions) with tensor product given by the fibered product of smooth S-schemes. The same is true for the category $K(\mathcal{M})$ of finite complexes with entries in $\mathcal{M}$, where the tensor product is the usual tensorproduct of complexes which is symmetric with the usual sign-rule. Objects of $K(\mathcal{M})$ have well defined cohomology for any reasonable cohomological functor on the category of smooth S-schemes, and $\mathcal{M}$ is sufficiently big to allow our basic construction. For a smooth S-scheme X we denote by M(X) its image in $\mathcal{M}$.

Later we shall pass to coefficients $\mathbb{Q}$ but keep the same notations.

5.2 Enveloping Algebras and Symmetric Groups

This section collects some general remarks for later use. All algebras are algebras over a base field k of characteristic zero. Its purpose is to construct certain operators (linear combinations of permutations) which operate on the free Lie algebra but can also be applied to the motivic logarithm. Assume first that $\mathfrak{g}$ is a Lie-algebra over k. Define its central series $Z^n(\mathfrak{g})$ by the rule that Z^n is the subspace generated by n-fold commutators, so

$$Z^1(\mathfrak{g}) = \mathfrak{g} .$$

Furthermore $U(\mathfrak{g})$ and $S(\mathfrak{g})$ denote the enveloping algebra and the symmetric algebra. The algebra $U(\mathfrak{g})$ admits a cocommutative and coassociative coproduct

$$c : U(\mathfrak{g}) \to U(\mathfrak{g}) \otimes U(\mathfrak{g})$$

which is a homomorphism of rings and satisfies

$$c(x) = x \otimes 1 + 1 \otimes x$$

for $x \in \mathfrak{g}$. If we compose with the multiplication on $U(\mathfrak{g})$ we obtain a linear operator λ on $U(\mathfrak{g})$. It has the following alternate description:

There exists, see [Bou75, Ch. 1, 2, 7], a $\mathfrak{g}$-linear isomorphism between $S(\mathfrak{g})$ and $U(\mathfrak{g})$ which sends a monomial

$$x_1 \cdot \dots \cdot x_n \in S(\mathfrak{g})$$

to the average over the symmetric group S_n of all permuted products in $U(\mathfrak{g})$. One easily checks that λ operates on the image of $S^n(\mathfrak{g})$ by multiplication by 2^n. Thus λ is a diagonalisable automorphism of $U(\mathfrak{g})$ with eigenvalues 2^n. If

$$I \subset U(\mathfrak{g})$$

denotes the augmentation-ideal one easily sees that I^n is the image of the subspace of weight $\geq n$ in $S(\mathfrak{g})$. Here the weight of an element $x \in \mathfrak{g}$ is the maximal n such that $x \in Z^n(\mathfrak{g})$, or ∞, and the weight of a monomial $x_1 \dots x_n$ is the sum of the weights of its components. It follows that λ respects I^n and that on the quotient

$$U(\mathfrak{g})/I^n$$

the $\lambda = 2$-eigenspace is isomorphic to

$$\mathfrak{g}/Z^n(\mathfrak{g}) .$$

As a generalisation, modify c to ($x \in U(\mathfrak{g})$)

$$\tilde{c}(x) = c(x) - x \otimes 1 - 1 \otimes x$$

and consider the complex

$$k \to U(\mathfrak{g}) \to U(\mathfrak{g}) \otimes U(\mathfrak{g}) \to \dots$$

where the differential is the alternating sum

$$d(x_1 \otimes x_2 \otimes \ldots) = \tilde{c}(x_1) \otimes x_2 \otimes \cdots - x_1 \otimes \tilde{c}(x_2) \otimes \ldots$$

The square of the differential vanishes because of the coassociativity of c. Via the symmetrisation map this complex is isomorphic to the analogeous complex for the symmetric algebra $\mathrm{S}(\mathfrak{g})$.

Lemma 1. *The cohomology of the $\tilde{c}$-complex is isomorphic to the exterior algebra $\wedge(\mathfrak{g})$.*

Proof. This is an assertion about k-vector spaces. For a finite dimensional vector space W let the dual W^t operate via the diagonal on the standard simplicial model for EW^t, that is the simplicial scheme with entries $\mathrm{W}^{t,n+1}$ in degree n. The complex of regular algebraic functions on EW^t gives a resolution of k by injective S(W)-modules. Our complex is the complex of W^t-invariants via the diagonal action whose cohomology is

$$\mathrm{Ext}^{\bullet}_{\mathrm{S(W}^t)}(k,k) = \wedge(\mathrm{W}) . \qquad \square$$

Remark 2. It is easy to see that the differential, a graded derivation, vanishes on the subspaces

$$\mathfrak{g}^{\otimes n} \subset \mathrm{U}(\mathfrak{g})^{\otimes n} ,$$

and the subspaces $\wedge^n(\mathfrak{g})$ of antisymmetric elements represent the cohomology.

Cocycles representing these classes are for example cup-products of linear functions on W^t, or their antisymmetrisations. Translated to Lie-algebras we obtain that our original complex has cohomology $\wedge(\mathfrak{g})$ and all classes are represented by antisymmetric elements. That is let the groups S_n operate on $\mathrm{U}(\mathfrak{g})^{\otimes n}$ and denote by ϵ_n the projector onto antisymmetric elements. Although the different ϵ_n's do not commute with the differentials they still annihilate their image because of the symmetry of $\tilde{c}$. As they operate as the identity on suitable representatives of the cohomology classes they induce an injection of the n-th cohomology into $\mathrm{U}(\mathfrak{g})^{\otimes n}$. So a cocycle is a coboundary if it is annihilated by ϵ_n.

Finally assume that $\mathfrak{g}$ is the free Lie-algebra on a k-vectorspace V. Then

$$\mathrm{U}(\mathfrak{g}) = \mathrm{T(V)}$$

is the free tensoralgebra in V, with the coproduct

$$c : \mathrm{T(V)} \to \mathrm{T(V)} \otimes \mathrm{T(V)}$$

induced by the diagonal on V, and thus given by the shuffle formula

$$c(v_1 \otimes v_2 \cdots \otimes v_n) = \sum v_{\mathrm{A}} \otimes v_{\mathrm{B}} ,$$

where the sum is over all disjoint decompositions

$$\{1, \ldots, n\} = \mathrm{A} \sqcup \mathrm{B} ,$$

and v_A, v_B denote the products (in natural order) of the v_i with $i \in A, B$. It follows that the operator λ respects the direct sum $V^{\otimes n}$ and acts on it via a certain element of the group-ring $\mathbb{Z}[S_n]$ which is independant of V. As the operation of $\mathbb{Q}[S_n]$ on $V^{\otimes n}$ is faithful if V has dimension at least n it follows that this element is semisimple with eigenvalues $\{1, 2, \ldots, 2^n\}$, that is its action by left-multiplication on $\mathbb{Q}[S_n]$ has this property. The projection onto the $\lambda = 2$-eigenspace is then defined by a certain universal element

$$e_n \in \mathbb{Q}[S_n]$$

which is a polynomial in λ. Its image consists of the image of the free Lie-algebra, that is of all Lie-polynomials of length n in its generators. For example for $n = 2$ we obtain all commutators, so

$$e_2 = (1 - \sigma)/2$$

with σ the transposition in S_2. The e_n have the property that for any k-vectorspace V the direct sum

$$\sum_n e_n(V^{\otimes n})$$

is a Lie-algebra, that is we have in $\mathbb{Q}[S_{m+n}]$ the identity (in hopefully suggestive notation)

$$(1 - e_{m+n})(e_m \otimes e_n - e_n \otimes e_m) = 0 .$$

A general formula for e_n can be found in [Bou75, Ch. II §3.2].

We give another argument which generalises to other cases. The actions of the algebraic group GL(V) and the finite group S_n on $V^{\otimes n}$ commute, and it is known, see [Wey46, Thm. 4.4.E], that the commutator of GL(V) in $\mathrm{End}(V^{\otimes n})$ is the image of the group ring $k[S_n]$, equal to it if $\dim(V) \geq n$. Also we know that the action of GL(V) is reductive. Thus any GL(V)-invariant subspace is the image of a projector e in $k[S_n]$, and the right ideal generated by e in $k[S_n]$ is canonical. So for the Lie-polynomials of degree n we obtain such a projector in $\mathbb{Q}[S_n]$ which can be chosen independantly of V.

We apply this as follows to our complex with entries $T(V)^{\otimes n}$ and differentials d the alternating sums of $\tilde{c}$'s: There exist universal matrices u, v, w with entries elements of $\mathbb{Q}[S_l]$ (l suitable) such that on $T(V)^{\otimes n}$ we have the identity

$$\mathrm{id} = d \circ u + v \circ \epsilon_n + w \circ d .$$

For applications we also need a superversion which amounts to the same with signs added at suitable locations: now $\mathfrak{g}$ is graded into even and odd parts, and for two odd elements the commutator has to be replaced by the supercommutator. By $S(\mathfrak{g})$ we denote the supersymmetric algebra, that is the tensorproduct of the symmetric algebra on even elements and the alternating algebra on odd elements. Again

$$S(\mathfrak{g}) \cong U(\mathfrak{g}) .$$

The enveloping algebra $U(\mathfrak{g})$ is also $\mathbb{Z}/(2)$-graded and the coproduct

$$c : \mathrm{U}(\mathfrak{g}) \to \mathrm{U}(\mathfrak{g}) \otimes \mathrm{U}(\mathfrak{g})$$

is a ring homomorphism induced by the diagonal on $\mathfrak{g}$. However the multiplication rule on the tensorproduct is "super", that is two odd elements in the factors anticommute. The definition of λ is the same as before, and it operates on the image of $\mathrm{S}^n(\mathfrak{g})$ as 2^n. Furthermore λ respects the powers I^n of the augmentation ideal $\mathrm{I} \subset \mathrm{U}(\mathfrak{g})$ and the $\lambda = 2$-eigenspace on $\mathrm{U}(\mathfrak{g})/\mathrm{I}^n$ is $\mathfrak{g}/\mathrm{Z}^n(\mathfrak{g})$.

Finally, if V is an odd k-vectorspace we can apply this to the free Lie-algebra in V. The coproduct on

$$\mathrm{U}(\mathfrak{g}) = \mathrm{T}(\mathrm{V})$$

is given by the shuffle formula

$$c(v_1 \otimes \ldots \otimes v_n) = \sum_{\mathrm{A,B}} \pm x_{\mathrm{A}} \otimes x_{\mathrm{B}}$$

where the sum is over all disjoint decompositions of $\{1, \ldots, n\}$ and the sign is that of the shuffle, i.e., that of the permutation which defines it. It follows that the $\lambda = 2$-eigenspace in $\mathrm{V}^{\otimes n}$ is again defined by a universal projector

$$f_n \in \mathbb{Q}[\mathrm{S}_n]$$

which differs from the previous e_n by an application of the sign-character. For example

$$f_2 = (1 + \sigma)/2 \, .$$

Also as before

$$(1 - f_{m+n})(f_m \otimes f_n - (-1)^{mn} f_n \otimes f_m) = 0 \, .$$

Finally, also the results about $\tilde{c}$ carry over.

5.3 The Definition of the Motivic Logarithm

Assume we are given an S-point $\mathrm{X} \in \mathrm{X}(\mathrm{S})$. The composition of projection and the inclusion

$$\mathrm{X} \to \mathrm{S} \to \mathrm{X}$$

is an idempotent e_x acting on M(X), and $\mathrm{M}(\mathrm{X})^\circ$ denotes the image of $1 - e_x$, the **reduced homology**. Similarly a product like

$$\mathrm{M}(\mathrm{X})^\circ \otimes \mathrm{M}(\mathrm{X})^\circ \otimes \mathrm{M}(\mathrm{X})$$

denotes the direct summand of $\mathrm{M}(\mathrm{X} \times \mathrm{X} \times \mathrm{X})$ where we apply the above idempotent to the first two factors.

Now define a projective system of complexes in $\mathrm{K}(\mathcal{M})$ by chosing for $\underline{\mathrm{P}}_n$ the complex

$$\mathrm{M(S)} \to \mathrm{M(X)}^{\circ} \to \mathrm{M(X)}^{\circ\otimes 2} \to \cdots \to \mathrm{M(X)}^{\circ,\otimes n}$$

where the maps are induced by the alternating sums of the adjacent diagonals $\delta_{i,i+1}$ which double the argument in position i, for $0 \le i \le n$. In particular, the first map is 0. It is obviously related to the simplicial object $Cosk_0(\mathrm{X})$ but should not be confused with the usual chain complex whose differentials have the opposite direction. By the equality

$$\delta \cdot e_x = (e_x \otimes e_x) \cdot \delta$$

for the diagonal

$$\delta : \mathrm{M(X)} \to \mathrm{M(X \times X)} = \mathrm{M(X)} \otimes \mathrm{M(X)}$$

the diagonals induce maps on the reduced quotient of $\mathrm{M(X)}^{\otimes m}$. There are compatible associative products

$$\underline{\mathrm{P}}_n \otimes \underline{\mathrm{P}}_n \to \underline{\mathrm{P}}_n$$

induced from the juxtaposition

$$\mathrm{M(X)}^{\otimes a} \otimes \mathrm{M(X)}^{\otimes b} \to \mathrm{M(X)}^{\otimes a+b}$$

which sends $(x_1,\ldots,x_a) \otimes (x_{a+1},\ldots,x_{a+b})$ to $(x_1,\ldots,x_{a+b})$. Also we have graded cocommutative and coassociative shuffle coproducts

$$\underline{\mathrm{P}}_{m+n} \to \underline{\mathrm{P}}_m \otimes \underline{\mathrm{P}}_n$$

induced from

$$(x_1,\ldots,x_{a+b}) \mapsto \sum_{\sigma} \mathrm{sign}(\sigma)(x_{\sigma(1)},\ldots,x_{\sigma(a)}) \otimes (x_{\sigma(a+1)},\ldots,x_{\sigma(a+b)}) ,$$

where the sum is over all permutions of $\{1,\ldots,a+b\}$ which are monotone on $\{1,\ldots,a\}$ and on $\{a+1,\ldots,a+b\}$. The sum of these makes the right unbounded complex

$$\underline{\mathrm{P}} = \varprojlim \underline{\mathrm{P}}_n$$

a cocommutative Hopf-algebra.

If we pass to $\mathbb{Q}$-coefficients we obtain as a direct summand a super Lie algebra as follows.

Proposition 3. *The direct summands $f_n(\mathrm{M(X)}^{\circ\otimes n})$ are preserved by the differentials (the f_n are polynomials in the composition of multiplication and comultiplication) and they form a super Lie algebra $\underline{\mathrm{L}}$ with truncations $\underline{\mathrm{L}}_n$.*

Proof. This follows from the rule

$$(1 - f_{m+n})(f_m \otimes f_n - (-1)^{mn} f_n \otimes f_m) = 0 .$$ □

If we are given another S-point $y \in \mathrm{X(S)}$ define a new complex $\underline{\mathrm{P}}(y)$ by the rule that its terms are the same as for $\underline{\mathrm{P}}$ but that we add to the differential

$$\underline{\mathrm{X}}_n \to \underline{\mathrm{X}}_{n+1}$$

the map induced by

$$(x_1, \dots, x_n) \to -(y, x_1, \dots, x_n) .$$

The same formulas as before define a cocommutative and coassociative coproduct

$$\underline{P}(y) \to \underline{P}(y) \otimes \underline{P}(y)$$

as well as an associative product

$$\underline{P}(y) \otimes \underline{P} \to \underline{P}(y) .$$

The terms of the complex for $\underline{P}(y)$ are the same as those for $\underline{P}$ but the differential differs by left-multiplication with the element

$$a = x - y$$

in degree zero. The element a satisfies the Maurer-Cartan equation

$$d(a) + a \otimes a = 0 .$$

If we pass to $\mathbb{Q}$-coefficients the element a lies in the super Lie algebra $\underline{L}$ and the sum of the differential d and the superbracket with a has square zero, therefore defines a new complex. If we vary y we can replace the base S by X and get a universal $\underline{P}(y)$ over X.

5.4 The Étale Realisation

Assume ℓ is a prime invertible in S. The étale log is the universal unipotent $\mathbb{Z}_\ell$-sheaf on X trivialised at x. One epigonal (to [Del89]) reference is [Fal07]. We claim that unless X is projective of relative genus zero it can be realised as the étale homology relative S, namely the dual of higher direct images of $\underline{P}(y)$ for the universal section y over X given by the diagonal in $X \times X$. This can be seen as follows.

Obviously there exists a spectral sequence starting with the homology of powers $M(X)^{\circ, \otimes m}$ for $0 \le m \le n$, shifted by degree $-m$, and converging to the homology of $\underline{P}_n$. If X is affine all nonzero terms have homological degree 0, so the spectral sequence degenerates and the homology is a repeated extension of the powers of the Tate-module $T_\ell(X)^{\otimes m}$. If X is projective the homology of $M(X)^{\circ, \otimes m}$ can be computed by the Künneth-formula, which gives a direct sum of tensor products where each factor is either T_ℓ (in degree 1) or $\mathbb{Z}_\ell(1)$ (in degree 2). If we shift by $-m$ this is the direct sum of powers $T_\ell(X)^{\otimes m-i}(i)$ placed in degree i, for $0 \le i \le m$. The first differential in the spectral sequence multiplies these by the homology class of the diagonal in $X \times X$ (the class of y vanishes in reduced étale homology). This class is a sum

$$c_\Delta = \sum \alpha_j \otimes \beta_j ,$$

with α_j and β_j forming a dual basis for the dual of the cupproduct on $\mathrm{T}_\ell(\mathrm{X})$. We claim that after the first differential only terms associated to $\mathrm{M(X)}^{\circ,\otimes m}$ survive. This comes down to the following.

Lemma 4. *Suppose R is a commutative ring,* T *a free* R*-module of rank* $r \geq 2$, α_j *and* β_j *two sets of basis for* R, $c = \sum_j \alpha_j \otimes \beta_j \in \mathrm{T}^{\otimes 2}$. *For each m consider the complex* $\mathrm{M}_*^{(m)}$ *given by*

$$\mathrm{M}_m \to \mathrm{M}_{m-1} \to \cdots \to \mathrm{M}_0$$

where $\mathrm{M}_l \subseteq \mathrm{T}^{\otimes m}$ *is the sum over all* $l+1$*-tuples* $(a_0,\ldots,a_l)$ *with* $\sum_i a_i = m-2l$, *of*

$$\mathrm{T}^{\otimes a_0} \otimes c \otimes \mathrm{T}^{\otimes a_1} \otimes c \ldots c \otimes \mathrm{T}^{\otimes a_l}\,.$$

The differentials in the complex are the sums of the inclusions $(0 \leq i < l)$ *with signs*

$$\mathrm{T}^{\otimes a_i} \otimes c \otimes \mathrm{T}^{\otimes a_{i+1}} \subset \mathrm{T}^{\otimes a_i + 2 + a_{i+1}}\,.$$

The sign for the i-th inclusion is

$$(-1)^{a_0 + \ldots a_i + i}\,.$$

Then this complex is exact in degrees $\neq 0$.

Proof. We denote by $\mathbb{T}_m$ the quotient of M_0 under the image of M_1. The direct sum

$$\mathbb{T} = \bigoplus_m \mathbb{T}_m$$

is the quotient of the tensoralgebra of T under the two sided ideal generated by c. We claim first that multiplication by any basiselement α is injective on $\mathbb{T}$.

We may assume that $\alpha = \alpha_1$, as c can be written using any basis. The assertion clearly holds for multiplication on $\mathbb{T}_0$. If the assertion holds on $\mathbb{T}_l$ suppose $z \in \mathbb{T}_{l+1}$ is annihilated by α_1. As $\mathbb{T}_{l+2}$ is the quotient of $\mathrm{T} \otimes \mathbb{T}_{l+1}$ under $c \otimes \mathbb{T}_l$ we have a relation

$$\alpha_1 \otimes z = c \otimes y = \sum \alpha_j \otimes \beta_j y\,.$$

Thus for $j \neq 1$, which is possible as $r \geq 2$, we have

$$\beta_j y = 0$$

and by induction, since β_j is also part of a basis,

$$y = 0,\ z = 0\,.$$

Now for the assertion of the lemma use induction over m. The case $m = 1$, or even $m = 2$, is trivial. In general the subcomplex consisting of direct summands with $a_0 > 0$ is the tensorproduct of T and the complex for $m-1$, thus exact in positive degrees. The quotient is the tensorproduct of c and the complex for $m-2$ and has non-trivial homology $\mathbb{T}_{m-2}$ only in degree 0. The connecting map

$$\mathbb{T}_{m-2} \to \mathrm{T} \otimes \mathbb{T}_{m-1}$$

is given by multiplication by c and injective by the proceeding arguments: decompose the T on the right according to the basis α_i and consider components. □

We derive from this the following proposition.

Proposition 5. *The spectral sequence for the étale homology of* $\underline{\mathrm{P}}_n$ *degenerates after the first differential. The étale homology is free over* $\mathbb{Z}_\ell$ *and concentrated in degrees between* 0 *and* n. *The homology in degree zero is equal to the truncated étale logarithm. The map*

$$\underline{\mathrm{P}}_{n+1} \to \underline{\mathrm{P}}_n$$

induces zero on homology in strictly positive degrees. The pro-object of étale realisations of $\underline{\mathrm{P}}_n$ *is isomorphic in the derived category to the pro-object of étale logarithms.*

Proof. The first step in the spectral sequence is the sum of the complexes $\mathrm{M}_*^{(m)}$ as above, truncated at level n. We derive that after the first differential of the spectral sequence the surviving terms are locally free and either correspond to homology in degree 0, or to higher homology and then are subspaces of the homology of $\mathrm{M(X)}^{\circ,\otimes n+1}$. Furthermore we use a weight argument to show that all higher differentials vanish, so that the homology is locally free (it is the last homology of a truncation of the complex in the lemma) and mixed of certain weights, and this also applies to the homology of $\underline{\mathrm{P}}_n$.

To introduce weights we may assume by base change that S is of finite type over Spec($\mathbb{Z}$), and consider the eigenvalues of Frobenius at closed points of S. Then all terms in the complex $\mathrm{M}_*^{(m)}$ are pure of weight $-m$. Thus for degree 0-homology the weights lie between $-n$ and 0, while for higher homology in degree $i > 0$ the weight is $-(n+i)$. As the higher differentials in the spectral sequence respect weights they must vanish. Also because of weights under the projection $\underline{\mathrm{P}}_{m+1} \to \underline{\mathrm{P}}_m$ the induced maps in strictly positive homological degree vanish. It follows that the projection $\underline{\mathrm{P}}_{2n} \to \underline{\mathrm{P}}_n$ induces in homology a map of complexes which factors canonically over the projection to H_0.

This H_0 has a filtration with subquotients the same as for the étale log, see for example [Fal07], the discussion on page 178. Furthermore for $n = 1$ the extension

$$0 \to \mathrm{T}_\ell \to \mathrm{H}_0 \to \mathbb{Z}_\ell \to 0$$

is induced from the diagonal and thus coincides with the first step of the étale logarithm. Thus we get a homomorphism from the universal étale logarithm to our H_0 which is compatible with the multiplication by the fibre at x, and induces an isomorphism on the first two graded subquotients of the filtration. It follows easily that it induces isomorphisms on all graded subquotients, and that the étale H_0 coincides with the étale log. □

As usual the products and coproducts on the $\underline{\mathrm{P}}(y)$ make the homology of $\underline{\mathrm{P}}$ the affine algebra of a prounipotent group-scheme G_{et} over $\mathbb{Z}_\ell$, and the homology of $\underline{\mathrm{P}}(y)$ that of a torsor over G_{et}.

The pro-ℓ fundamental groups of the fibers of X/S form a *local system*

$$G_{et}$$

on S, or more precisely there exists a profinite group which is an extension of the fundamental group of S by this pro-ℓ group. This extension has a splitting defined by x and thus becomes a semidirect product. The projective system of ℓ-adic homologies of $\underline{P}_n$ is identified with the local system given by the completed group-algebra $\mathcal{A}$, that is it is as a pro-object isomorphic to the projective system of quotients $\mathcal{A}/\mathcal{I}^n$ where $\mathcal{I} \subset \mathcal{A}$ is the augmentation ideal. More precisely the 0-th homology of $\underline{P}_n$ corresponds to the universal unipotent sheaf of length n, that is to $\mathcal{A}/\mathcal{I}^{n+1}$.

From the structure of the fundamental group we know that it is a free pro-ℓ group, divided by one relation if X/S is projective. This relation is the commutator of generators. The completed group-algebra of the free group is a completed free tensoralgebra. The one relation corresponds modulo the cube of the augmentation-ideal to the element c from above, and becomes equal to it after applying a suitable automorphism. It follows that in any case $\mathcal{A}$ is non canonically isomorphic respecting augmentations to the completed tensor-algebra

$$\mathbb{T} = \prod \mathbb{T}_n \,.$$

If we pass to $\mathbb{Q}_\ell$-coefficients the Lie algebra of G_{et}/Z^{n+1} is obtained from $\mathcal{A}/\mathcal{I}^{n+1}$ by applying the operators f_m, for $m \leq n$, to $\underline{P}_n$ or the operators e_n to the subquotients of its homology. The group G_{et}/Z^{n+1} is isomorphic to its Lie algebra via the exponential map, and the multiplication on it is defined by the Hausdorff-series.

5.5 The de Rham and Crystalline Realisation

The arguments are essentially the same as in the previous section. The relative de Rham homology of $\underline{P}_n$ is defined by dualising the double complex derived from the de Rham complexes on $\bar{X}^n$ with logarithmic poles along D. It admits a Hodge filtration. Also, if $S_0 \subset S$ is a closed subscheme defined by an ideal with divided power-structure, the de Rham cohomology (without the Hodge filtration) depends only on the restriction of $(\bar{X}, D)$ to S_0. In fact it is defined for a relative curve over X_0. The remaining arguments carry over verbatim, for example, the weight argument uses that locally the pair $(\bar{X}, D)$ comes by pullback from a smooth $\mathbb{Z}$-scheme. We obtain unipotent flat group-schemes G_{cr} and G_{DR}. The latter admits a Hodge filtration by flat closed subschemes $F^i(G_{DR})$, and the G_{DR}-torsors defined by $\underline{P}(y)$ reduce to $F^0(G_{DR})$-torsors. A possible reference is again [Fal07]. However we should note that while our category $\mathcal{M}$ admits crystalline realisations there is no obvious extension to the motivic category as the action of correspondences on crystalline cohomology still poses some problems.

5.6 The Motivic Realisation

Voevodsky defines a derived category of motives as a quotient of the derived category of Nisnevich sheaves with transfers. The $\underline{P}_n$ obviously correspond to complexes of such sheaves and we obtain objects in this derived category. Again passing to $\mathbb{Q}$-coefficients we may apply the operators f_n to get (super) Lie algebras. All this can be done before taking the quotient under Nisnevich coverings and $\mathbb{A}^1$-homotopy, but passing to it we get objects in the motivic category. Also the $\underline{P}(y)$ become objects in the derived category which obviously should be torsors. However the definition of a *torsor* in this context is not so obvious.

5.7 Torsors in Triangulated Categories

We have to study mixed extensions in derived categories. One of the fundamental difficulties in the theory is that mapping cones are only defined up to non canonical isomorphism. To get around this we have to make vanishing assumptions under which all objects become sufficiently welldefined. Recall the definition of extensions in a triangulated category. An extension of B by A is an exact triangle

$$A \to E \to B \to A[1] .$$

An isomorphism of extensions is an isomorphism of E's commuting with the maps from A and those into B. Isomorphism classes are classified by the map $B \to A[1]$ as follows. If E and E′ induce the same map there is a map of triangles, by axiom TR3, which is necessarily an isomorphism.

The difference of two such maps is induced by a map of E's which induces 0 on A and B. It is induced by a

$$\alpha : E \to A$$

such that its composition with $A \to E$ (from both sides, that is applied twice) vanishes. If

$$\mathrm{Hom}(A, B[-1]) = (0)$$

it induces a trivial endomorphism of A and thus is induced from a map $B \to A$. On the other hand such maps operate obviously as automorphisms of any extension. If in addition

$$\mathrm{Hom}(A, A[-1]) = \mathrm{Hom}(B, B[-1]) = (0)$$

this operation is free.

A similar problem is the classification of mixed extensions. Given A, B, C and extensions, i.e., exact triangles,

$$A \to D \to B \to A[1]$$

and

$$B \to E \to C \to B[1]\,,$$

we consider F's which lie in triangles

$$D \to F \to C \to D[1]$$

as well as

$$A \to F \to E \to A[1]$$

such that the three possible compositions (from A to F, from F to C, and from D to E) coincide. We are interested in automorphisms, isomorphism classes, and existence conditions. The last is easy: the composition of the two classifying maps

$$C \to B[1] \to A[2]$$

must vanish. Indeed, this is necessary as the first map goes to zero if composed with

$$B[1] \to E[1]$$

and the second extends to E[1]. Conversely if the composition vanishes we get a lift

$$C \to D[1]\,,$$

thus an extension F of C by D which induces the extension E.

For the classification of isomorphism classes we show that under suitable vanishing conditions they form a principal homogeneous space under

$$\mathrm{Hom}(C, A[1])\,.$$

Assume right away that

$$\mathrm{Hom}(A, B[-1]) = \mathrm{Hom}(A, C[-1]) = \mathrm{Hom}(B, B[-1]) = \mathrm{Hom}(B, C[-1]) = (0)\,.$$

Then we obtain an operation of $\mathrm{Hom}(C, A[1])$ on the isomorphism classes of mixed extensions. Namely assume we have a mixed extension F and an

$$A \to G \to C \to A[1]\,.$$

The direct sum $F \oplus G$ maps to $C \oplus C$, and we denote by H the preimage of the diagonal, that is H lies in an exact triangle

$$H \to F \oplus G \to C \to H[1]$$

and is unique up to non-unique isomorphism. The inclusions of A and D into the direct summands factor over H, uniquely up to maps into $C[-1]$, so they are unique. Especially the antidiagonal in $A \oplus A$ comes from a unique $A \to H$. Finally define F' is a mapping cokernel of this map, that is it lies in an exact triangle

$$A \to H \to F' \to A[1]\,.$$

The map from H to C factors uniquely over F′, and so does the projection to E. The maps define exact triangles

$$D \to F' \to C \to D[1]$$

and

$$A \to F' \to E \to A[1]\,.$$

Furthermore the composition $D \to E$ coincides with the original composition, that is the factorisation over B, so that F′ is a mixed extension.

Conversely given mixed extensions F and F′ modify $F \oplus F'$ by taking a preimage of the diagonal $E \subset E \oplus E$ and dividing by the diagonal $D \subset D \oplus D$, which amounts to forming mapping cones. Call the result G. For the second we need to lift the inclusion of the diagonal D which is unique up to an element of $\mathrm{Hom}(D, E[-1])$ which by our vanishing conditions is zero. The inclusions of A into F, F' as well as the projections to C extend/factor uniquely and define an exact triangle

$$A \to G \to C \to A[1]\,,$$

and one easily sees that this gives an inverse.

Finally an automorphism of F is given by the sum of the identity and compatible endomorphisms of

$$A \to F \to E \to A[1]$$

and

$$D \to F \to C \to D[1]$$

which vanish on the extremes, that is by compatible morphisms

$$E \to A, C \to D\,.$$

The obstruction that the first factors over C lies in $\mathrm{Hom}(B, A)$ and induces zero in $\mathrm{Hom}(B, D)$, thus comes from an element in

$$\mathrm{Hom}(B, B[-1]) = (0)\,.$$

Thus the factorisation exists and we get a map

$$C \to A\,.$$

By a dual argument also the map from C to D factors over A. The difference of the two maps from C to A vanishes if we compose to get a map

$$E \to D\,.$$

The obstruction for it to vanish is first get an element in $\mathrm{Hom}(E, B[-1])$ and then one in $\mathrm{Hom}(B, A[-1])$. Both groups vanish, so finally the automorphisms of F are the sum of the identity and a unique element of $\mathrm{Hom}(C, A)$.

We subsume the above in the following lemma.

Lemma 6. *Assume the vanishing of*

$$\mathrm{Hom}(\mathrm{A},\mathrm{B}[-1]) = \mathrm{Hom}(\mathrm{A},\mathrm{C}[-1]) = \mathrm{Hom}(\mathrm{B},\mathrm{B}[-1]) = \mathrm{Hom}(\mathrm{B},\mathrm{C}[-1]) = (0)\,.$$

Then the obstruction to extend given extensions of B *by* A *and* C *by* B *to a mixed extension lies in* Hom(C, A[2]). *If it vanishes the isomorphism classes of solutions form a torsor under* Hom(C, A[1]), *and the automorphisms of any solutions are* Hom(C, A).

We apply these generalities to the motivic log, to define **motivic** $\underline{\mathrm{L}}_n$**-torsors**. We work in the derived category of finitely filtered Nisnevich sheaves with transfer, that is Nisnevich sheaves $\mathcal{F}$ with a finite decreasing filtration of length n, for a fixed n,

$$\mathcal{F} = \mathrm{G}^0(\mathcal{F}) \supset \mathrm{G}^1(\mathcal{F}) \supset \cdots \supset \mathrm{G}^{n+1}(\mathcal{F}) = (0)\,.$$

The transfer should respect filtrations. For example $\underline{\mathrm{P}}_n$ with its stupid filtration defines a complex in this category, with graded

$$gr^i_{\mathrm{G}}(\underline{\mathrm{P}}_n) = \mathrm{M}(\mathrm{X})^{\circ,\otimes i}[-i]\,.$$

In the derived category we invert as usual filtered quasi-isomorphisms.

We also assume that our complexes are bounded above, and for the derived product

$$\mathrm{A} \otimes^{\mathbb{L}} \mathrm{B}$$

we take the usual derived product divided by elements of filtration degree $> n+1$. We know that $\underline{\mathrm{P}}_n$ is an algebra and coalgebra in this category. Any such complex can be resolved by complexes whose associated gradeds are injective. If we further assume that the associated gradeds lie in the triangulated category $\mathrm{DM}^{\mathrm{eff}}_{gm}$, see [Voe00], generated by motives of smooth schemes we get by truncation right bounded resolutions which are sufficiently Ext-acyclic as to compute cohomology (defined as maps in the derived category but usually difficult to compute unless one has injective resolutions).

In the following we consider filtered right $\underline{\mathrm{P}}_n$-modules, that is right bounded filtered complexes $\mathrm{K}^\bullet$ together with a map of complexes

$$\mathrm{K}^\bullet \otimes \underline{\mathrm{P}}_n \to \mathrm{K}^\bullet$$

which is strictly associative (no homotopies involved). The coproduct on $\underline{\mathrm{P}}_n$ defines a $\underline{\mathrm{P}}_n$-module structure on the derived tensorproduct of two $\underline{\mathrm{P}}_n$-modules. The exact forgetful functor to filtered complexes has a right-adjoint which maps $\mathrm{K}^\bullet$ to the internal Hom of filtration preserving maps

$$\mathrm{Hom}(\underline{\mathrm{P}}_n, \mathrm{K}^\bullet)\,,$$

defined in the obvious way from the rule

$$\mathrm{Hom}(\mathrm{M(X)},\mathcal{F})(\mathrm{T}) = \mathcal{F}(\mathrm{X}\times\mathrm{T})\,.$$

Applied to filtered injective resolutions without $\underline{\mathrm{P}}_n$-module structure we obtain acyclic resolutions of $\underline{\mathrm{P}}_n$-modules which allow us to compute cohomology. It then coincides with the usual Nisnevich cohomology.

Also as usual we can pass to $\mathbb{A}^1$-homotopy invariant objects by applying the functor C_* with

$$\mathrm{C}_*(\mathcal{F})(\mathrm{T}) = \mathcal{F}(\mathrm{T}\times\Delta^\bullet)\,,$$

and define motivic cohomology by

$$\mathrm{H}^i_{\mathcal{M}}(\mathrm{S},\mathcal{F}) = \mathrm{H}^i(\mathrm{S},\mathrm{C}_*(\mathcal{F}))\,.$$

So from now on we compute in the triangulated category of filtered (with filtration degrees between 0 and n) bounded above complexes of $\mathbb{Q}$-Nisnevich sheaves with transfer which are $\underline{\mathrm{P}}_n$-modules, modulo filtered quasi-isomorphisms and modulo $\mathbb{A}^1$-homotopy. All our objects will lie in the category generated by geometric effective motives, so have finite Ext-dimension. Finally, to apply our previous theory of mixed extensions we make the general assumption that for $i \le j$ the cohomology

$$\mathrm{H}^{-1}_{\mathcal{M}}(\mathrm{S},\mathrm{Hom}(gr^i_{\mathrm{G}}(\underline{\mathrm{P}}_n),gr^j_{\mathrm{G}}(\underline{\mathrm{P}}_n)) = \mathrm{H}^{j-i-1}_{\mathcal{M}}(\mathrm{S},\mathrm{Hom}(\mathrm{M(X)}^{\circ,\otimes i},\mathrm{M(X)}^{\circ,\otimes j})) = (0)$$

vanishes.

We now define a $\underline{\mathrm{L}}_n$-torsor as a filtered $\underline{\mathrm{P}}_n$-module $\underline{\mathrm{Q}}_n$ together with cocommutative and coassociative coproduct

$$c : \underline{\mathrm{Q}}_n \to \underline{\mathrm{Q}}_n \otimes^{\mathbb{L}} \underline{\mathrm{Q}}_n\,.$$

Also we assume that we have isomorphisms compatible with the multiplication

$$gr^i_{\mathrm{G}}(\underline{\mathrm{Q}}_n) \cong \mathrm{C}_*(gr^i_{\mathrm{G}}(\underline{\mathrm{P}}_n))\,.$$

Examples of such objects are the previous $\underline{\mathrm{P}}_n(y)$.

Obviously, for $n = 0$, there is only one isomorphism class of such objects. Furthermore, a $\underline{\mathrm{Q}}_n$ induces naturally a $\underline{\mathrm{Q}}_{n-1}$. Conversely given a $\underline{\mathrm{Q}}_{n-1}$ let us analyse the possible lifts to $\underline{\mathrm{Q}}_n$. Necessarily we have

$$\mathrm{G}^1(\underline{\mathrm{Q}}_n) = \underline{\mathrm{Q}}_{n-1} \otimes^{\mathbb{L}}_{\underline{\mathrm{P}}_{n-1}} \mathrm{G}^1(\underline{\mathrm{P}}_{n-1})\,.$$

Here the tensor product is defined as follows. The complex $\mathrm{G}^1(\underline{\mathrm{P}}_n)$ has the same terms as

$$\underline{\mathrm{P}}_{n-1} \otimes \mathrm{M(X)}^\circ[-1]\,,$$

but with the differential modified by adding multiplication by the diagonal in $\underline{\mathrm{P}}^1_n(\mathrm{X})$. This diagonal satisfies the Maurer-Cartan equation, so we may define the tensor-product as

$$\underline{\mathrm{Q}}_{n-1} \otimes \mathrm{M(X)}^\circ[-1]$$

with multiplication by the same element added to the differential.

Thus $\underline{Q}_n$ becomes a mixed extension with the following subquotients $C_*(\mathbb{Z})$, $\underline{Q}_{n-2} \otimes^{\mathbb{L}}_{\underline{P}_{n-2}} G^1(\underline{P}_{n-1})$, and $C_*(gr^n_G(\underline{P}_n))$. The obstruction z to get such an extensions lies in

$$H^2_{\mathcal{M}}(S, gr^n_G(\underline{P}_n) = H^{2-n}_{\mathcal{M}}(S, L(M(X)^{\circ,\otimes n})) .$$

By the existence of the coproduct on $\underline{Q}_{n-1}$ it satisfies the equation

$$c(z) = z \otimes 1 + 1 \otimes z ,$$

thus lies in the cohomology of the direct summand which is the n-th graded of the Lie algebra.

If z vanishes two mixed extensions differ up to isomorphisms by a class in $H^1_{\mathcal{M}}(S, gr^n_G(\underline{P}_n))$. The obstruction w to extend the coproduct lies in

$$H^1_{\mathcal{M}}(S, gr^n_G(\underline{P}_n \otimes \underline{P}_n))$$

(we have a mixed extension for the tensorproduct), is annihilated by the counits in each factor, and satisfies

$$(\mathrm{id} \otimes c)(w) + 1 \otimes w = (c \otimes \mathrm{id})(w) + w \otimes 1 .$$

If we modify c by the rule

$$\tilde{c}(z) = c(z) - z \otimes 1 - 1 \otimes z$$

this can be written as

$$(\mathrm{id} \otimes \tilde{c})(w) = (\tilde{c} \otimes \mathrm{id})(w) .$$

The obstruction is symmetric and by the general results on Lie algebras, see Lemma 1 and the discussion following it, w lies in the image of $\tilde{c}$:

We know that a certain complex formed from free Lie algebras is acyclic and thus null homotopic. The homotopies are given by elements of group rings $\mathbb{Q}[S_n]$, namely the elements u, v, w from the end of Sect. 5.2. Applying the same elements to the $\underline{P}_n$ gives null-homotopies for the motivic cohomology.

Thus changing the mixed extension $\underline{Q}_n$ by a class in $H^1_{\mathcal{M}}(S, gr^n_G(\underline{P}_n))$ we can make it zero, so the comultiplication extends.

For coassociativity we similarly obtain an obstruction in $H^0_{\mathcal{M}}(S, gr^n_G(\underline{P}^{\otimes 3}))$ which satisfies the cocycle condition and whose (super-)antisymmetric projection vanishes. So it is a coboundary of an element in $H^0_{\mathcal{M}}(S, gr^n_G(\underline{P}^{\otimes 2}))$ which can be chosen symmetric. Thus finally our torsor extends if the obstruction vanishes.

By similar but simpler arguments the isomorphism classes of extensions form a torsor under the first motivic cohomology of the grade n-part of the Lie algebra, and the automorphisms are $H^0_{\mathcal{M}}$.

Remark 7. For complete curves the vanishing assumptions refer to the reduced subspace of the motivic cohomology (where $i \leq j$)

$$H^{2i-1}_{\mathcal{M}}(X^{i+j}, \mathbb{Q}(j)).$$

For an affine curve we need partially compact support. For rational curves (where we deal with Tate-motives) we need the vanishing for $l \geq 0$ of

$$H^{-1}_{\mathcal{M}}(S, \mathbb{Q}(l)).$$

5.8 Representability

Spaces of torsors with an obstruction theory as in the last chapter admit a versal representative as follows. We assume that

$$H^1_{\mathcal{M}}(S, gr^n_G(\underline{L}))$$

is a finitely generated $\mathbb{Q}$-vectorspace. For any $\mathbb{Q}$-algebra R we can by the same procedure as before define torsors with coefficients in R, such that the obstruction to liftings are classified by H^2, the liftings by H^1 and the automorphisms of liftings by H^0, all with coefficients in R. Then there exist a finitely generated commutative $\mathbb{Q}$-algebra R_n and a versal $\underline{L}_n$-torsor over R_n, such that any other such torsor over an R is obtained from it via pushout $R_n \to R$.

Namely the assertion holds for $n = 0$. Assume we have constructed R_{n-1} the obstruction to lift the versal torsor to $\underline{L}_n$ lies in

$$H^2_{\mathcal{M}}(S, gr^n_G(\underline{L}_n)) \otimes_{\mathbb{Q}} R_{n-1}.$$

Writing it as an R_{n-1}-linear combination of basis elements of $H^2_{\mathcal{M}}(S, gr^n_G(\underline{L}_n))$ the coefficients generate an ideal

$$I_{n-1} \subset R_{n-1}.$$

The torsor then lifts over R_{n-1}/I_{n-1} and we chose one such lift. If $H^1_{\mathcal{M}}(S, gr^n_G(\underline{L}_n))$ has dimension l we then obtain a versal torsor over

$$R_n = R_{n-1}/I_{n-1}[T_1, \ldots, T_l]$$

by modifying the lift with the element of H^1 corresponding to the linear combination of basis elements with coefficients T_i.

If in addition the $H^0_{\mathcal{M}}(S, gr^j_G(\underline{L}_n))$ vanish this torsor is universal and R_n represents the motivic torsors. In any case its dimension is bounded by the sum of the dimensions of $H^1_{\mathcal{M}}(S, gr^j_G(\underline{L}_n))$.

Finally by comparison to étale cohomology, for example [SV96, Th. 7.6], the homologies of the $\underline{Q}_n$ define torsors under the étale unipotent fundamental group and also algebraic maps of the representation spaces. The torsors given by points $y \in X(S)$ correspond.

References

[Bou75] N. Bourbaki. *Lie groups and Lie algebras, I.* Hermann, Paris, 1975.

[Del89] P. Deligne. Le groupe fondamental de la droite projective moins trois points, *Galois groups over $\mathbb{Q}$. Math. Sci. Res. Inst. Publ.*, 16:79–297, 1989.

[DG05] P. Deligne and A. Goncharov. Groupes fondamenteaux motiviques de Tate mixte. *Ann. Sci. École Norm. Sup.*, 38:1–56, 2005.

[Fal07] G. Faltings. Mathematics around Kim's new proof of Siegel's theorem. *Diophantine geometry*, pages 173–188, 2007. CRM Series, 4, Ed. Norm, Pisa.

[HJ10] M. Hadian-Jazi. *Motivic fundamental groups and integral points.* PhD thesis, 2010.

[Kim05] M. Kim. The motivic fundamental group of $\mathbb{P}^1\backslash\{0,1,\infty\}$ and the theorem of Siegel. *Invent. Math.*, 161:629–656, 2005.

[SV96] A. Suslin and V. Voevodsky. Singular homology of abstract algebraic varieties. *Invent. Math.*, 123:61–94, 1996.

[Voe00] V. Voevodsky. Triangulated categories of motives over a field. In *Cycles, Transfers, and Motivic Homology Theories*. Princeton University Press, Princeton, 2000.

[Wey46] H. Weyl. *The classical groups*. Princeton University Press, Princeton, 1946.

Chapter 6
On a Motivic Method in Diophantine Geometry

Majid Hadian

Abstract Let k be a totally real number field, S be a finite set of finite places of k, and $\mathcal{O}_S$ be the ring of S-integers in k. Let X be a smooth scheme over $\mathcal{O}_S$ which admits a "nice" projectivization whose generic fiber is unirational. By studying the motivic unipotent fundamental group and path torsors associated to X, we show that S-integral points of X can be covered by the zero loci of finitely many nonzero p-adic analytic functions provided the unipotent fundamental group of the complex points of X is sufficiently non-abelian in a sense that is made precise.

6.1 Introduction

First of all, let us make it clear that this article is an exposition on some recent ideas introduced by Kim in [Kim05] and extended by Faltings in [Fal07] and the author in [Had10] toward studying integral points on varieties over number fields. Being so, instead of giving technical proofs, we refer the reader to proper references for complete treatment. The involved chain of ideas is quite wide and sophisticated, but our main emphasis is on motivic aspects which are introduced in [Had10]. The following notations will be fixed in the sequel.

Let k be a number field with $[k : \mathbb{Q}] = d$, S be a finite set of finite places of k, and $\mathcal{O}_S$ be the ring of S-integers in k. v will denote a generic finite place of k outside S, i.e., any finite place of k outside some finite set of places containing S. Let p denote the rational prime which is divisible by v. For any finite place v of k, k_v (resp. $\mathcal{O}_v$, resp. $\mathbb{F}_v$) denotes the completion of k at v (resp. the ring of integers of k_v, resp. the residue field of $\mathcal{O}_v$). We will be considering triples $(\overline{\mathrm{X}}, \mathrm{X}, \mathrm{D})$ over $\mathcal{O}_S$ consisting of a smooth projective connected scheme $\overline{\mathrm{X}}$ over $\mathrm{Spec}(\mathcal{O}_S)$ with geometrically connected fibers, a relative divisor D in $\overline{\mathrm{X}}$, and the complement $\mathrm{X} := \overline{\mathrm{X}} - \mathrm{D}$. Finally, for any

M. Hadian (✉)
Max-Planck Institute for Mathematics, Bonn, Germany
e-mail: hadian@mpim-bonn.mpg.de

J. Stix (ed.), *The Arithmetic of Fundamental Groups*, Contributions in Mathematical and Computational Sciences 2, DOI 10.1007/978-3-642-23905-2_6,

$n \geqslant 1$, let r_n denotes the rank of the abelian group $Z^n(\pi_1)/Z^{n+1}(\pi_1)$, where $Z^\bullet(\pi_1)$ is the descending central series of the fundamental group of the complex manifold $X(\mathbb{C})$. Note that, although the fundamental group of the complex manifold $X(\mathbb{C})$ might depend on the embedding of k in $\mathbb{C}$, the numbers r_n are independent of this embedding. Let us introduce the following notion.

Definition 1 ($\mathcal{V}$-property). For any scheme X over $\mathcal{O}_S$, any finite place v of k, and any given S-integral point $x \in X(\mathcal{O}_S)$, we say that X satisfies $\mathcal{V}_{S,v}$-property at x if there exists a nonzero p-adic analytic function on the p-adic open unit ball $\mathcal{B}_1^\circ(x_{k_v})$ centered at x_{k_v} which vanishes at all S-integral points of X in $\mathcal{B}_1^\circ(x_{k_v})$. We say that X satisfies $\mathcal{V}_{S,v}$-property if it satisfies $\mathcal{V}_{S,v}$-property at every S-integral point.

Now we would like to propose the following:

Conjecture 2. Assume that $(\overline{X}, X, D)$ is a triple over $\mathcal{O}_S$ such that $r_n > 0$ for all $n \geqslant 1$. Then X satisfies $\mathcal{V}_{S,v}$-property.

Our goal in this article is to sketch the ideas which can be used to prove the above conjecture under some extra conditions. Before making it more precise, let us introduce the following definition which simplifies the statement.

Definition 3. A **standard triple** $(\overline{X}, X, D)$ over $\mathcal{O}_S$ is a triple such that:

(i) The irreducible components of the divisor D are smooth and surjective over $\mathrm{Spec}(\mathcal{O}_S)$ and the generic fiber D_k is a normal crossing divisor whose irreducible components are absolutely irreducible.
(ii) The generic fiber X_k of X is unirational.

Theorem 4 (Main Theorem). *Conjecture 2 is valid provided:*

(i) *k is a totally real number field.*
(ii) *The triple $(\overline{X}, X, D)$ is a standard triple over $\mathcal{O}_S$.*
(iii) $h^1(\overline{X}_{\mathbb{C}}) = 0$.
(iv) *For any constant c, there exists a natural number $n \in \mathbb{N}$ such that*

$$c + d(r_3 + r_5 + \cdots + r_{2\lfloor (n-1)/2 \rfloor + 1}) < r_1 + r_2 + \cdots + r_n\,.$$

Remark 5. Note that in the one dimensional case, the $\mathcal{V}_{S,v}$-property even for a single place v, implies finiteness of S-integral points. This is the case because the curve X can be covered by finitely many p-adic open unit disks and a nonzero p-adic analytic function on a p-adic open unit disk vanishes at finitely many k_v-points. So in the case of the standard triple $(\mathbb{P}^1, X, \{p_1, p_2, \ldots, p_{l+1}\})$, where $l \geqslant 2$ is a natural number and $p_i \in \mathbb{P}^1(\mathcal{O}_S)$ for $1 \leqslant i \leqslant l+1$, the Main Theorem gives rise to finiteness of S-integral points of $\mathbb{P}^1 - \{p_1, p_2, \ldots, p_{l+1}\}$ over any totally real field of degree $d \leqslant l$. In particular, it gives us a *motivic* proof of the Siegel's celebrated finiteness theorem over totally real quadratic fields and the field of rational numbers. We would like to mention that Kim in [Kim05] uses different realizations of the unipotent fundamental group and path torsors to prove Siegel's finiteness theorem over $\mathbb{Q}$. Here in this exposition, we try to explain how motivic fundamental groups and path torsors can be used to promote Kim's method to bigger number fields.

Let us finish up the introduction by sketching the methodology which is going to be exposed in this article. Suppose that we are interested in the Diophantine set $X(\mathcal{O}_S)$ for a scheme X defined over $\mathcal{O}_S$. Fix a base point $x \in X(\mathcal{O}_S)$ if $X(\mathcal{O}_S) \neq \emptyset$. Kim's approach in [Kim05] suggests to study the variation of path torsors from the fixed base point x to a varying point y in étale and de Rham settings in order to obtain some period maps and compare these period maps using p-adic Hodge theory. More precisely:

1. One has to first define the étale and the de Rham notions of the unipotent fundamental group and path torsors over it, see Sect. 6.3.
2. Then, in order to get interesting torsors, and also in order to be able to compare these two notions using p-adic Hodge theory, one needs to endow the de Rham objects with Hodge filtration and Frobenius action and to endow the étale objects with Galois action, see Sect. 6.4.
3. Finally, for having proper target spaces for period maps, one has to construct some algebraic spaces which parametrize the path torsors in a natural way, see Sect. 6.5.

This, together with some sophisticated techniques from p-adic Hodge theory, essentially reduces the $\mathcal{V}_{S,v}$-property for the scheme X to showing that "global Galois cohomology groups are smaller than local Galois cohomology groups", see Sect. 6.6. This last statement is obviously too vague to be a mathematical statement, but hopefully the reader makes sense of it by consulting the current exposition and other articles in the literature.

The general difficulty in the above approach appears in estimating the *size* of global Galois cohomology groups. The most important feature of [Had10] and also of this article is to employ the motivic objects in the sense of Voevodsky to lift this obstacle and replace the global Galois cohomology groups by some algebraic K-groups, see Sects. 6.2 and 6.8. This brings new methods and techniques, which of course involve new difficulties, to the subject.

Acknowledgements. This exposition is based on my Ph.D. thesis. So first of all I would like to thank Professor Doctor G. Faltings for supervising this project, and Max-Planck Institute for Mathematics in Bonn, Germany for providing an excellent atmosphere for doing research. I would also like to thank J. Stix for inviting me to the workshop PIA 2010 and giving me the opportunity to speak there.

6.2 Motivic Fundamental Groups and Path Torsors

In this section, following [DG05], we are going to introduce the most important ingredients in our methodology, namely motivic fundamental groups and path torsors. Let us begin by Voevodsky's construction of the triangulated category of mixed geometric motives over the number field k. Note that most of Voevodsky's construction can be done over any base field and all of it can be done over any perfect field, see for example [Voe00]. But being only interested in the case of number fields and to avoid introducing new notations we will stick to the number field k as our base. Furthermore, though it is very important that the following construction can be made with integral coefficients, we are only interested in the rational coefficient case and hence restrict ourselves to this case for the ease of notation.

Let $\mathrm{SmCor}(k)_{\mathbb{Q}}$ be the category whose objects are smooth separated schemes over k and a morphism from X to Y is an element of the $\mathbb{Q}$-vector space generated by closed integral subschemes $Z \subset X \times_k Y$ where $Z \to X$ is finite and dominates a connected component of X. For any smooth separated scheme X over k (resp. any morphism $f : X \to Y$ between smooth separated schemes over k) let [X] (resp. $[f]$) denotes the corresponding object (resp. morphism) in $\mathrm{SmCor}(k)_{\mathbb{Q}}$. Note that $\mathrm{SmCor}(k)_{\mathbb{Q}}$ is a $\mathbb{Q}$-linear category with disjoint union as direct sum. The triangulated category of mixed geometric motives with rational coefficients over k is obtained from $\mathrm{SmCor}(k)_{\mathbb{Q}}$ as follows:

1. Let T be the thick subcategory in the homotopy category $\mathfrak{h}^b(\mathrm{SmCor}(k)_{\mathbb{Q}})$ of bounded complexes over $\mathrm{SmCor}(k)_{\mathbb{Q}}$ which is generated by complexes of the form
$$[X \times_k \mathbb{A}^1] \xrightarrow{[pr_1]} [X]$$
for all X, and complexes of the form
$$[U \cap V] \to [U] \oplus [V] \to [X]$$
for any Zariski open covering $X = U \cup V$. Then $\mathfrak{h}^b(\mathrm{SmCor}(k)_{\mathbb{Q}})_T$ denotes the Verdier localization of the category $\mathfrak{h}^b(\mathrm{SmCor}(k)_{\mathbb{Q}})$ with respect to T.
2. Define $\mathrm{DM}^{\mathrm{eff}}(k)_{\mathbb{Q}}$ to be the Karoubian envelope of the category $\mathfrak{h}^b(\mathrm{SmCor}(k)_{\mathbb{Q}})_T$. Let $M(X) \in \mathrm{DM}^{\mathrm{eff}}(k)_{\mathbb{Q}}$ denote the object associated to a smooth separated scheme X. Then $\mathrm{DM}^{\mathrm{eff}}(k)_{\mathbb{Q}}$ can be endowed with a tensor product in such a way that
$$M(X \times_k Y) \cong M(X) \otimes M(Y)$$
for all smooth separated schemes X and Y over k.
3. Finally, let $\mathrm{DM}(k)_{\mathbb{Q}}$ be the category resulted by inverting the Tate object in $\mathrm{DM}^{\mathrm{eff}}(k)_{\mathbb{Q}}$, where the Tate object is defined as
$$\mathbb{Q}(1) := \widetilde{M}(\mathbb{P}^1)[-2],$$
with $\widetilde{M}(\mathbb{P}^1)$ being the complement of the direct summand of $M(\mathbb{P}^1)$ induced by the idempotent associated to a constant map $\mathbb{P}^1 \to \mathbb{P}^1$. The tensor product of $\mathrm{DM}^{\mathrm{eff}}(k)_{\mathbb{Q}}$ can be extended to a tensor product on $\mathrm{DM}(k)_{\mathbb{Q}}$. This tensor product together with proper notions of internal Hom-object and dual put a rigid tensor triangulated structure on the category $\mathrm{DM}(k)_{\mathbb{Q}}$.

The main importance of the category $\mathrm{DM}(k)_{\mathbb{Q}}$ is that it is a strong candidate for being the derived category of the undiscovered category of mixed motives over k.

The problem is that there is no known t-structure on $\mathrm{DM}(k)_{\mathbb{Q}}$ whose heart gives the category of mixed motives. Recall that a t-structure $(\mathcal{T}^{\leqslant 0}, \mathcal{T}^{\geqslant 0})$ on a triangulated category $\mathcal{T}$ consists of strictly full subcategories $\mathcal{T}^{\leqslant 0}$ and $\mathcal{T}^{\geqslant 0}$ of $\mathcal{T}$ such that:

(1) $\mathcal{T}^{\leqslant 0}[1] \subset \mathcal{T}^{\leqslant 0}$ and $\mathcal{T}^{\geqslant 0}[-1] \subset \mathcal{T}^{\geqslant 0}$.
(2) $\mathrm{Hom}_{\mathcal{T}}(\mathrm{X},\mathrm{Y}) = 0$ for any $\mathrm{X} \in \mathcal{T}^{\leqslant 0}$ and any $\mathrm{Y} \in \mathcal{T}^{\geqslant 0}[-1]$.
(3) For any object $\mathrm{X} \in \mathcal{T}$, there are objects $\mathrm{A} \in \mathcal{T}^{\leqslant 0}$, $\mathrm{B} \in \mathcal{T}^{\geqslant 0}[-1]$, and a distinguished triangle

$$\mathrm{A} \to \mathrm{X} \to \mathrm{B} \to \mathrm{A}[1]\,.$$

Moreover, a t-structure is said to be non-degenerate if

(4) The intersections $\cap_n \mathcal{T}^{\geqslant 0}[-n]$ and $\cap_n \mathcal{T}^{\leqslant 0}[n]$ consist only of the zero object.

The heart of a t-structure is defined to be the full subcategory $\mathcal{T}^{\leqslant 0} \cap \mathcal{T}^{\geqslant 0}$. As we mentioned above, there is no known t-structure on the whole category $\mathrm{DM}(k)_{\mathbb{Q}}$ which gives the abelian category of mixed motives over k, but on a certain subcategory of $\mathrm{DM}(k)_{\mathbb{Q}}$ the situation is much better in the following sense.

The triangulated category of mixed Tate motives $\mathrm{DMT}(k)_{\mathbb{Q}}$ is defined to be the triangulated subcategory of $\mathrm{DM}(k)_{\mathbb{Q}}$ generated by objects $\mathbb{Q}(n) := \mathbb{Q}(1)^{\otimes n}$, for all $n \geqslant 1$. Using the validity of Beilinson–Soulé vanishing conjecture for number fields, we can put a non-degenerate t-structure on the category $\mathrm{DMT}(k)_{\mathbb{Q}}$ whose heart $\mathrm{MT}(k)_{\mathbb{Q}}$ is called the abelian rigid tensor category of mixed Tate motives. Now we have two important facts at our disposal. The first one is that the Hom-groups in $\mathrm{DMT}(k)_{\mathbb{Q}}$ are given by a certain decomposition of the rational algebraic K-groups of the base number field k and the second one is Borel's explicit calculation of algebraic K-groups of number fields. These two facts can be put together to deduce the following important calculation of the Ext-groups in the category $\mathrm{MT}(k)_{\mathbb{Q}}$, which plays a crucial role in the sequel.

$$\mathrm{Ext}^1_{\mathrm{MT}(k)_{\mathbb{Q}}}(\mathbb{Q}(0), \mathbb{Q}(n)) \cong \mathrm{K}_{2n-1}(k) \otimes_{\mathbb{Z}} \mathbb{Q}\,.$$

Different realizations of Tate objects, with comparison isomorphisms between them, can be extended to a realization functor

$$\mathfrak{real} : \mathrm{MT}(k)_{\mathbb{Q}} \to \mathcal{R}_k\,,$$

where $\mathcal{R}_k$ denotes the Tannakian category of mixed realizations over k. This realization functor enjoys the following properties:

Theorem 6 ([DG05, 2.14 and 2.15]). *The above mentioned realization functor is fully faithful, and the image is essentially stable by sub-objects, that is if an object is in the image of* $\mathfrak{real}$, *so are all its sub-objects.*

Now let X be a "nice" variety over the number field k, for example the complement of a normal crossing divisor in a smooth projective variety. By fixing a base point $x \in \mathrm{X}(k)$ one can define the Betti, the de Rham, and the étale unipotent fundamental group of X with the base point x and furnish them with Hodge filtration and

Frobenius action in the de Rham case and with Galois action in the étale case, see Sects. 6.3 and 6.4 for more details in the one dimensional case. These realizations together with the comparison isomorphisms between them lead to a pro-unipotent affine group scheme $G_{\mathcal{R}}$ over the category $\mathcal{R}_k$. The path torsors $G_{\mathcal{R}}(x,y)$ can be defined similarly for any point $y \in X(k)$. An interesting question is whether or not the unipotent fundamental group $G_{\mathcal{R}}$ and path torsors over it are motivic in the sense that they lie in the image of the realization functor $\mathfrak{real}$. The following important theorem gives a partial affirmative answer.

Theorem 7 ([DG05, Proposition 4.15]). *For any standard triple $(\overline{X}, X, D)$ over k and any rational point $x \in X(k)$ (resp. any two rational points $x, y \in X(k)$), the pro-unipotent fundamental group scheme $G_{\mathcal{R}}$ (resp. the path torsor $G_{\mathcal{R}}(x,y)$) is motivic.*

Let us denote the motivic fundamental group (resp. the motivic path torsor) by the symbol G_{mot} (resp. $G_{mot}(x,y)$), and just mention that there is an integral version of Theorem 7, see for example [DG05, Proposition 4.17]. This integral version is crucial for us, simply because we are interested in studying integral points on varieties, and essentially says that if we start with a standard triple $(\overline{X}, X, D)$ over the ring of S-integers $\mathcal{O}_S$ of k and if the rational points $x, y \in X(k)$ are generic fibers of S-integral points $x_S, y_S \in X(\mathcal{O}_S)$ then the motivic fundamental group and path torsors of Theorem 7 are unramified outside S.

6.3 Different Realizations

The goal of this section is to recall the ideas and techniques in constructing different realizations of the unipotent fundamental group and path torsors of curves. In our applications in coming sections, we are only interested in the punctured projective line over a field of characteristic zero, but our treatment here, following [Fal07], works in a much more general setting. Throughout this section, R is a commutative ring with unit, $C \to \mathrm{Spec}(R)$ is a smooth, projective, connected curve with geometrically connected fibers, and $X := C - D$ where $D \subset C$ is a surjective étale divisor over Spec(R). We have the following categories of unipotent objects.

C_{coh} : The category of vector bundles $\mathcal{E}$ over C which are iterated extensions of trivial bundles, where a trivial bundle is a bundle of the form $P \otimes_R \mathcal{O}_C$ for a finitely generated projective R-module P. The coherent Tate module is defined as:

$$T_{coh} := (H^1(C, \mathcal{O}_C))^\vee .$$

C_{dR} : The category of vector bundles $\mathcal{E}$ over C together with a connection ∇ with logarithmic poles along D such that $(\mathcal{E}, \nabla)$ is an iterated extension of trivial vector bundles with connection. A trivial vector bundle with connection is a pair $(P \otimes_R \mathcal{O}_C, \mathrm{Id}_P \otimes d)$, where P is a finitely generated projective R-module and d from $\mathcal{O}_C$ to $\Omega_{C/R}(D)$ is the canonical differential. The de Rham Tate module is defined as:

$$\mathrm{T}_{dR} := (\mathrm{H}^1_{dR}(\mathrm{C}, \mathcal{O}_\mathrm{C}))^\vee \,.$$

$\mathcal{C}_{\text{ét}}$: The category of smooth ℓ-adic étale sheaves $\mathcal{S}$ on X which are iterated extensions of the trivial ones. Here ℓ is any invertible prime in R and a trivial ℓ-adic étale sheaf on X is the pull back of a smooth ℓ-adic étale sheaf from Spec(R). Finally, the étale Tate module is defined as:

$$\mathrm{T}_{\text{ét}} := (\mathrm{H}^1_{\text{ét}}(\mathrm{X}, \mathbb{Q}_\ell))^\vee \,.$$

In the sequel, $\mathcal{C}_\mathrm{M}$ denotes any of the above categories, no matter which one. The same convention will be applied to T_M, and so on. Moreover, we always refer to the constant ring by the symbol $\mathfrak{R}$, and it should be realized as the base ring R in the coherent and de Rham cases and as the field $\mathbb{Q}_\ell$ in the étale case. The symbol C° stands for the projective curve C in the coherent and de Rham cases and the open complement X in the étale case. An object $\mathcal{E}$ in $\mathcal{C}_\mathrm{M}$ is called unipotent of class n, if it admits a filtration of length n with trivial subquotients. We denote by $\mathcal{C}_{\mathrm{M},n}$ the full subcategory of $\mathcal{C}_\mathrm{M}$ which consists of all objects of unipotent class n.

Now any point $x \in \mathrm{X}(\mathrm{R})$ (resp. any geometric point in the étale case) gives us a functor $\mathcal{F}_\mathrm{M} : \mathcal{E} \mapsto \mathcal{E}[x]$ from the category $\mathcal{C}_\mathrm{M}$ to the category of finitely generated projective $\mathfrak{R}$-modules. The first step in constructing different realizations of the unipotent fundamental group and path torsors is the following pro-representability result for the functor $\mathcal{F}_\mathrm{M}$.

Theorem 8 ([Had10, Sect. 2.1]). *There exists a pro-object $\mathcal{P}_\mathrm{M}$ in the category $\mathcal{C}_\mathrm{M}$, and an element $p \in \mathcal{P}_\mathrm{M}[x]$ such that for any object $\mathcal{E}$ in $\mathcal{C}_\mathrm{M}$ and any element $e \in \mathcal{E}[x]$, there exists a unique morphism $\varphi_e : \mathcal{P}_\mathrm{M} \to \mathcal{E}$ such that $\varphi_{e,x}(p) = e$.*

Using the above pro-representability theorem we can put a co-product map Δ_M on the ring:

$$\mathrm{A}_\mathrm{M} := \mathcal{P}_\mathrm{M}[x] = \mathrm{End}(\mathcal{P}_\mathrm{M}) \,,$$

which is uniquely determined by the property $\Delta_\mathrm{M}(p) = p \otimes p$. Moreover, there is a canonical surjection $\mathrm{A}_\mathrm{M} \twoheadrightarrow \mathfrak{R}$, which can be taken as the co-unit. This makes A_M into a co-commutative Hopf algebra over $\mathfrak{R}$. Now we can define

$$\mathrm{G}_\mathrm{M} := \mathrm{Spec}(\mathrm{A}_\mathrm{M}^\vee)$$

which is a flat pro-unipotent group scheme over $\mathrm{Spec}(\mathfrak{R})$. For any integer $n \geqslant 1$, consider

$$\mathrm{G}_{\mathrm{M},n} := \mathrm{G}_\mathrm{M}/\mathrm{Z}^n(\mathrm{G}_\mathrm{M}) \,,$$

where $\mathrm{Z}^\bullet$ is the descending central series. Furthermore, for any other point $y \in \mathrm{X}(\mathrm{R})$ (resp. for any other geometric point in the étale case) there is a co-associative, co-commutative co-product on $\mathcal{P}_\mathrm{M}[y]$ together with the co-unit $\mathcal{P}_\mathrm{M}[y] \twoheadrightarrow \mathfrak{R}$, which endows $\mathcal{P}_\mathrm{M}[y]$ with a co-commutative co-algebra structure (note that $\mathcal{P}_\mathrm{M}[y]$ does not have a ring structure in general). Using Theorem 8 again, we can show that the affine $\mathfrak{R}$-scheme

$$\mathrm{G}_\mathrm{M}(x,y) := \mathrm{Spec}(\mathcal{P}_\mathrm{M}[y]^\vee) \,,$$

admits a torsor structure on $\mathrm{G_M}$. The affine scheme $\mathrm{G_M}(x,y)$ is called the path torsor from y to x, referring to its torsor structure over $\mathrm{G_M}$. For any $n \geqslant 1$, the push forward of $\mathrm{G_M}(x,y)$ along the projection $\mathrm{G_M} \twoheadrightarrow \mathrm{G}_{\mathrm{M},n}$ is denoted by $\mathrm{G}_{\mathrm{M},n}(x,y)$.

The fiber functor $\mathcal{F}_\mathrm{M} : \mathcal{E} \mapsto \mathcal{E}[x]$ is a tensor functor from the category C_M to the category $\mathrm{Rep}_{\mathfrak{R}}(\mathrm{G_M})$ of $\mathrm{G_M}$-representations over finitely generated $\mathfrak{R}$-modules. The following result says that the group scheme $\mathrm{G_M}$ is in fact the unipotent fundamental group we are looking for.

Theorem 9 ([Had10, Theorem 2.1.9]). *The tensor functor $\mathcal{F}_\mathrm{M}$ gives an equivalence between the category $\mathrm{C_M}$ of unipotent bundles over C° and the category $\mathrm{Rep}_{\mathfrak{R}}(\mathrm{G_M})$ of $\mathrm{G_M}$-representations on finitely generated projective $\mathfrak{R}$-modules.*

Let us finish this section by the following crucial remark. This remark is a direct consequence of the explicit construction of the universal pro-unipotent object $\mathcal{P}_\mathrm{M}$ appeared in Theorem 8.

Remark 10. The fiber $\mathcal{E}[x]$ of any unipotent bundle $\mathcal{E}$ in C_M is a finitely generated projective $\mathfrak{R}$-module. In particular, any short exact sequence of these fibers splits, and hence we obtain a (non-canonical) isomorphism

$$\mathrm{A_M} \cong \prod_{i=0}^{\infty} \mathrm{T}_\mathrm{M}^{\otimes i}\,.$$

Now the Baker-Campbell-Hausdorff formula implies that when $\mathfrak{R}$ is a field of characteristic zero, $\mathrm{G_M}$ is a pro-unipotent group scheme over $\mathfrak{R}$ which is isomorphic to its Lie algebra. By the above observation, this Lie algebra is isomorphic to the free Lie algebra in $\mathrm{T_M}$.

6.4 Extra Structures

Having introduced the de Rham and the étale unipotent fundamental group and path torsors of curves in the previous section, we are going to put some extra structures on them in this section. These extra structures, namely the Hodge filtration and the Frobenius action on the de Rham realization and the Galois action on the étale realization, are important in comparing these two different theories and integrating them into the motivic theory. They are also important since they cause path torsors to be nontrivial.

Let us start with the Hodge filtration. For any object $(\mathcal{E},\nabla)$ of $C_{d\mathrm{R}}$, we consider finite decreasing Hodge filtrations $\mathrm{F}^\bullet$ by subobjects which satisfy Griffiths' transversality property, that is for any integer i

$$\nabla(\mathrm{F}^i(\mathcal{E})) \subset \mathrm{F}^{i-1}(\mathcal{E}) \otimes_{\mathcal{O}_\mathrm{C}} \Omega_{\mathrm{C/R}}(\mathrm{D})\,.$$

The trivial object in this context is $(\mathcal{O}_\mathrm{C}, d)$ equipped with the trivial Hodge filtration which is given as $\mathrm{F}^0(\mathcal{O}_\mathrm{C}) = \mathcal{O}_\mathrm{C}$ and $\mathrm{F}^1(\mathcal{O}_\mathrm{C}) = 0$. Moreover, we consider the filtered morphisms between these filtered objects. Recall that a morphism $f : \mathcal{E} \to \mathcal{G}$ between filtered objects is called filtered if it satisfies $f(\mathrm{F}^i(\mathcal{E})) \subset \mathrm{F}^i(\mathcal{G})$ for all i. These Hodge filtrations on objects, induce Hodge filtrations on the de Rham complexes associated to them and hence on the algebraic de Rham cohomologies as well. More precisely, for any filtered object $(\mathcal{E}, \nabla)$ and any integer i, we define

$$\mathrm{F}^i(\mathcal{E} \xrightarrow{\nabla} \mathcal{E} \otimes_{\mathcal{O}_\mathrm{C}} \Omega_{\mathrm{C}/\mathrm{R}}(\mathrm{D})) := \mathrm{F}^i(\mathcal{E}) \xrightarrow{\nabla} \mathrm{F}^{i-1}(\mathcal{E}) \otimes_{\mathcal{O}_\mathrm{C}} \Omega_{\mathrm{C}/\mathrm{R}}(\mathrm{D}) ,$$

and

$$\mathrm{F}^i\mathrm{H}^*_{d\mathrm{R}}(\mathrm{C}, \mathcal{E}) := \mathbb{H}^*(\mathrm{C}, \mathrm{F}^i(\mathcal{E} \xrightarrow{\nabla} \mathcal{E} \otimes_{\mathcal{O}_\mathrm{C}} \Omega_{\mathrm{C}/\mathrm{R}}(\mathrm{D}))) .$$

Note that the images of the obvious maps from $\mathrm{F}^i\mathrm{H}^*_{d\mathrm{R}}$ to $\mathrm{H}^*_{d\mathrm{R}}$ give the Hodge filtration on $\mathrm{H}^*_{d\mathrm{R}}$. Now we can check that, in this refined category, maps and extensions are determined by the zeroth step of the Hodge filtration of appropriate $\mathcal{H}om$-bundles. In fact, for any two filtered objects $\mathcal{E}$ and $\mathcal{G}$, it is evident that filtered maps from $\mathcal{E}$ to $\mathcal{G}$ are given by $\mathrm{F}^0\mathrm{H}^0_{d\mathrm{R}}(\mathrm{C}, \mathcal{H}om(\mathcal{E}, \mathcal{G}))$. But the less obvious and more important fact is the following:

Proposition 11 ([Had10, Proposition 2.2.3]). *For any two filtered objects $\mathcal{E}$ and $\mathcal{G}$, the set of isomorphism classes of filtered extensions of $\mathcal{G}$ by $\mathcal{E}$ is in bijection with*

$$\mathrm{F}^0\mathrm{H}^1_{d\mathrm{R}}(\mathrm{C}, \mathcal{H}om(\mathcal{G}, \mathcal{E})) .$$

By using the above proposition, we can prove that there is a unique Hodge filtration on $\mathcal{P}_{d\mathrm{R}}$ which makes it the universal object in the category of filtered objects with the distinguished element p in $\mathrm{F}^0(\mathcal{P}_{d\mathrm{R}}[x])$. More precisely, we have the following filtered analogue of Theorem 8 in the de Rham setting.

Theorem 12 ([Had10, Proposition 2.2.7]). *For any filtered object $(\mathcal{E}, \nabla, \mathrm{F}^\bullet)$ and any element $e \in \mathrm{F}^0(\mathcal{E}[x])$, there exists a unique horizontal, strictly compatible homomorphism $\varphi_e : \mathcal{P}_{d\mathrm{R}} \to \mathcal{E}$ such that $\varphi_{e,x}(p) = e$.*

This canonical Hodge filtration on $\mathcal{P}_{d\mathrm{R}}$ induces a Hodge filtration on the fiber $\mathcal{P}_{d\mathrm{R}}[x]$ and hence a Hodge filtration on the coordinate ring $\mathcal{O}_{\mathrm{G}_{d\mathrm{R}}}$ of the de Rham unipotent fundamental group. On the other hand, it is well known in Hodge theory that the Hodge filtration on $\mathrm{H}^1_{d\mathrm{R}}(\mathrm{C}, \mathcal{O}_\mathrm{C})$ is concentrated in degrees 0 and 1, and is given by the following exact sequence

$$0 \to \mathrm{H}^0(\mathrm{C}, \Omega_{\mathrm{C}/\mathrm{R}}(\mathrm{D})) \to \mathrm{H}^1_{d\mathrm{R}}(\mathrm{C}, \mathcal{O}_\mathrm{C}) \to \mathrm{H}^1(\mathrm{C}, \mathcal{O}_\mathrm{C}) \to 0 .$$

Using this we can show that the Hodge filtration on the affine coordinate ring $\mathcal{O}_{\mathrm{G}_{d\mathrm{R}}}$ of $\mathrm{G}_{d\mathrm{R}}$ is concentrated in nonnegative degrees and

$$\mathcal{O}_{\mathrm{G}_{d\mathrm{R}}} / \mathrm{F}^1(\mathcal{O}_{\mathrm{G}_{d\mathrm{R}}}) \cong \mathcal{O}_{\mathrm{G}_{coh}} .$$

So we get a closed immersion from the coherent unipotent fundamental group G_{coh} into the de Rham unipotent fundamental group $\mathrm{G}_{d\mathrm{R}}$ and the defining ideal sheaf of the image is the first step of the Hodge filtration.

The second important decoration that we can put on the de Rham unipotent fundamental group is the Frobenius action. The key idea is to use the logarithmic crystalline theory and show that when the base ring R is a discrete complete valuation ring with unequal characteristic $p > 0$, uniformizer π, and absolute ramification index $e < p$, the category $C_{d\mathrm{R}}$ depends only on the reduction modulo π of the curve C and hence admits a Frobenius action. Let us briefly recall these ideas, following [Kat89].

Recall that a logarithmic structure on a scheme X is a sheaf of monoids $\mathcal{M}$ on the étale site $\mathrm{X}_{\text{ét}}$ of X together with a multiplicative homomorphism $\alpha : \mathcal{M} \to \mathcal{O}_{\mathrm{X}}$ where the induced map

$$\alpha^{-1}(\mathcal{O}_{\mathrm{X}}^{*}) \xrightarrow{\alpha} \mathcal{O}_{\mathrm{X}}^{*}$$

is an isomorphism. The following class of logarithmic structures on schemes will be enough for our purposes.

Example 13. Let X be a regular scheme and D be a reduced divisor with normal crossing on X. Put $\mathrm{Y} := \mathrm{X} - \mathrm{D}$ and let $j : \mathrm{Y} \to \mathrm{X}$ be the open immersion. Then the monoid

$$\mathrm{M} := j_{*}\mathcal{O}_{\mathrm{Y}}^{*} \cap \mathcal{O}_{\mathrm{X}}$$

and the canonical inclusion $\mathrm{M} \hookrightarrow \mathcal{O}_{\mathrm{X}}$ puts a logarithmic structure on X.

Now fix a quadruple $(\mathrm{S}, \mathrm{L}, \mathcal{I}, \gamma)$ as the base, where S is a scheme such that $\mathcal{O}_{\mathrm{S}}$ is killed by a power of a prime number p, L is a fine logarithmic structure on S (all logarithmic structures of Example 13 are fine), $\mathcal{I}$ is a quasi coherent sheaf of ideals in $\mathcal{O}_{\mathrm{S}}$, and finally γ is a DP-structure on $\mathcal{I}$. Assume moreover that $(\mathrm{X}, \mathcal{M})$ is a scheme with a fine logarithmic structure over (S, L) and that γ extends to X. Then we can define the logarithmic crystalline site $\mathrm{Cr}(\mathrm{X}/\mathrm{S})^{\log}$ as follows:

An object of $\mathrm{Cr}(\mathrm{X}/\mathrm{S})^{\log}$ consists of the data $(\mathrm{U} \xrightarrow{i} \mathrm{T}, \mathrm{M}_{\mathrm{T}}, \delta)$ where

$$\mathrm{U} \xrightarrow{f} \mathrm{X}$$

is an étale morphism, $(\mathrm{T}, \mathrm{M}_{\mathrm{T}})$ is a scheme with fine logarithmic structure over (S, L),

$$i : (\mathrm{U}, f^{*}(\mathrm{M})) \to (\mathrm{T}, \mathrm{M}_{\mathrm{T}})$$

is an exact closed immersion over (S, L), and δ is a DP-structure on the defining ideal sheaf of U in T which is compatible with γ. Morphisms in $\mathrm{Cr}(\mathrm{X}/\mathrm{S})^{\log}$ are commutative diagrams

$$\begin{array}{ccc} \mathrm{T} & \longrightarrow & \mathrm{T}' \\ {\scriptstyle i}\uparrow & & \uparrow{\scriptstyle i'} \\ \mathrm{U} & \longrightarrow & \mathrm{U}' \end{array}$$

where $\mathrm{U} \to \mathrm{U}'$ is a morphism in the étale site of X and $\mathrm{T} \to \mathrm{T}'$ is a DP-morphism between fine logarithmic schemes over (S, L). Finally, coverings are defined to be the usual coverings in the étale site of X, regardless of the involved logarithmic structures. The topos of sheaves of sets over $\mathrm{Cr}(\mathrm{X}/\mathrm{S})^{\log}$ will be denoted by $(\mathrm{X}/\mathrm{S})^{\log}_{Cr}$. The structure sheaf $\mathcal{O}_{\mathrm{X}/\mathrm{S}}$ on $\mathrm{Cr}(\mathrm{X}/\mathrm{S})^{\log}$ is defined to be the sheaf which assigns to an object $(\mathrm{U} \to \mathrm{T}, \mathrm{M}_{\mathrm{T}}, \delta)$ the ring of global sections $\Gamma(\mathrm{T}, \mathcal{O}_{\mathrm{T}})$. Now we can define logarithmic crystals as follows:

Definition 14. A logarithmic crystal on $\mathrm{Cr}(\mathrm{X}/\mathrm{S})^{\log}$ is a sheaf of $\mathcal{O}_{\mathrm{X}/\mathrm{S}}$-modules $\mathcal{F}$ in $(\mathrm{X}/\mathrm{S})^{\log}_{Cr}$ such that for any morphism $g : \mathrm{T}' \to \mathrm{T}$ in $\mathrm{Cr}(\mathrm{X}/\mathrm{S})^{\log}$, the induced map $g^*(\mathcal{F}_{\mathrm{T}}) \to \mathcal{F}_{\mathrm{T}'}$ is an isomorphism.

Over a fixed base $(\mathrm{S}, \mathrm{L}, \mathcal{I}, \gamma)$, forgetting the DP-structure gives rise to the forgetful functor from the category of fine logarithmic DP-schemes over S to the category of fine logarithmic schemes over S with a distinguished quasi coherent sheaf of ideals. Very important for us is that this forgetful functor has a right adjoint, see [Kat89, Proposition 5.3], which sends a fine logarithmic scheme over S with a distinguished quasi coherent sheaf of ideals to its divided power envelope, DP-envelope for short. Finally we can state the following important result.

Theorem 15 ([Kat89, Theorem 6.2]). *Let* (Y, N) *be a scheme with fine logarithmic structure which is smooth over* (S, L)*, and let* $(\mathrm{X}, \mathrm{M}) \to (\mathrm{Y}, \mathrm{N})$ *be a closed immersion. Denote by* $(\mathrm{D}, \mathrm{M}_{\mathrm{D}})$ *the DP-envelope of* (X, M) *in* (Y, N)*. Then the following categories are equivalent.*

(a) *The category of crystals on* $\mathrm{Cr}(\mathrm{X}/\mathrm{S})^{\log}$.
(b) *The category of* $\mathcal{O}_{\mathrm{D}}$*-modules with an integrable, quasi nilpotent connection.*

Now let us go back to the case of interest to us. Namely, let the base ring $\mathrm{R} = \mathrm{V}$ be the ring of integers in a complete non-archimedean field K of characteristic zero, and assume that the residue field $\mathfrak{k}$ of V is a perfect field of characteristic $p > 0$. Fix a uniformizing parameter π for V and assume that $v_\pi(p) < p$. Let X_{V} be a smooth curve over V which admits an open immersion into a projective, smooth, connected curve C_{V} over V such that C_{V} has geometrically connected fibers and the complement D_{V} of X_{V} in C_{V} is an étale and surjective relative divisor over V. Then we have the following:

Theorem 16 ([Had10, Theorem 2.3.8]). *The category* $\mathcal{C}_{cr}$ *of unipotent crystals on* $\mathrm{Cr}(\mathrm{C}_{\mathfrak{k}}/\,\mathrm{Spec}(\mathrm{V}))^{\log}$ *is equivalent to the category* $\mathcal{C}_{d\mathrm{R}}$ *of unipotent vector bundles on* C_{V} *with logarithmic connection along* D_{V}.

Now suppose that $\mathfrak{k}$ contains the finite field with $q = p^s$ elements $\mathbb{F}_q$, and assume that $\mathrm{C}_{\mathfrak{k}}$ is defined over this subfield $\mathbb{F}_q$ of $\mathfrak{k}$. Then $\mathrm{C}_{\mathfrak{k}}$ admits a Frobenius action induced by $\mathrm{F} := \mathrm{Fr}^s$, where Fr is the absolute Frobenius. So by the above theorem, the category $\mathcal{C}_{d\mathrm{R}}$ admits also a Frobenius action, which is compatible with tensor products and dual. Moreover, if the reduction modulo π of the base point x is invariant under F, the Frobenius action on $\mathcal{C}_{d\mathrm{R}}$ respects the fiber functor $\mathcal{F}_{d\mathrm{R}}$ associated to x. This means that the de Rham unipotent fundamental group $\mathrm{G}_{d\mathrm{R}}$ can be also furnished with a Frobenius action.

Let us finish this section by shortly mentioning that if the base ring $\mathrm{R} = \mathrm{K}$ is a field, then the étale unipotent fundamental group and path torsors associated to the curve $X_{\overline{K}}$ can be endowed with compatible Galois actions by the absolute Galois group $Gal(\overline{K}/K)$. This Galois action on étale path torsors, like Hodge filtration and Frobenius action on de Rham path torsors, play an important role in the sequel.

6.5 Torsor Spaces and Crystalline Torsors

So far we have introduced different notions of the unipotent fundamental group and path torsors associated to a curve. But in order to define period maps on the curve under study, we need to have some algebraic spaces parametrizing path torsors. These spaces are going to be the target spaces of the period maps, which will be introduced in the next section.

In this section, we are mainly interested in étale torsors. To begin with, we consider a more general situation. Assume that K is a complete Hausdorff topological field of characteristic zero, Γ is a profinite group, and fix finite dimensional continuous representations $\{\rho_i : \Gamma \to GL(\mathbb{L}_i)\}_{i=1}^{r}$ of Γ over K. For any finitely generated K-algebra A and any finitely generated module $\mathcal{M}$ over A, we put the inductive limit topology on $\mathcal{M}$, as an infinite dimensional K-vector space, induced from the natural canonical topology of finite dimensional subspaces. Then one can show that the set of isomorphism classes of continuous representations $\mathbb{M}$ of Γ over A which admit a Γ-stable filtration

$$(0) = W_0(\mathbb{M}) \subset W_1(\mathbb{M}) \subset \cdots \subset W_r(\mathbb{M}) = \mathbb{M}\,,$$

with $W_i(\mathbb{M})/W_{i-1}(\mathbb{M}) \cong_\Gamma \mathbb{L}_i \otimes_K A$ is in a natural bijection with the first non-abelian cohomology set $H^1(\Gamma, \mathbb{G}(A))$, where $\mathbb{G}$ is the following unipotent group scheme over K:

$$\mathbb{G} := \begin{pmatrix} I & Hom_K(\mathbb{L}_2,\mathbb{L}_1) & \dots & Hom_K(\mathbb{L}_r,\mathbb{L}_1) \\ 0 & I & \dots & Hom_K(\mathbb{L}_r,\mathbb{L}_2) \\ \vdots & \vdots & \ddots & \vdots \\ 0 & 0 & \dots & I \end{pmatrix}$$

and Γ acts on $\mathbb{G}$ via

$$g^\gamma = Diag(\rho_i(\gamma)).g.Diag(\rho_i(\gamma))^{-1}\,.$$

The following representability result leads to algebraic spaces which naturally parametrize étale path torsors.

Theorem 17 ([Had10, Corollary 3.1.3 and Theorem 3.1.6]). *Suppose that for all $i < j$ we have*

$$H^0(\Gamma, Hom_K(\mathbb{L}_j, \mathbb{L}_i)) = (0)\,,$$

and

$$dim_K(H^1(\Gamma, Hom_K(\mathbb{L}_j, \mathbb{L}_i))) < \infty\,.$$

Then for any Γ*-stable closed subgroup* $\mathbb{H}$ *of* $\mathbb{G}$*, the functor* $\mathrm{A} \mapsto \mathrm{H}^1(\Gamma, \mathbb{H}(\mathrm{A}))$ *from the category of finitely generated* K*-algebras to sets is representable by an affine scheme over* K.

Now let us restrict to the following situation. Let Γ be the absolute Galois group of a finite extension K of $\mathbb{Q}_p$, and suppose that we are interested in finite dimensional continuous Γ-representations over $\mathbb{Q}_p$. Theorem 17 says that, under the assumed hypotheses, the space of $\mathbb{H}$-torsors is representable by an affine scheme over $\mathbb{Q}_p$. But in this situation there are some interesting torsors, the so called crystalline torsors. We are going to study the subset in the representing affine scheme of torsors which consists of crystalline ones. Our aim in the remaining of this section is to show that in this setting and under the assumptions of Theorem 17, the subfunctor of crystalline torsors is also representable by an affine scheme over $\mathbb{Q}_p$.

Let us first fix some notations. K is a finite extension of $\mathbb{Q}_p$, V is the ring of integers in K, and $\mathfrak{k}$ is the residue field of V. Let K_0 be the maximal unramified subextension of K and V_0 be the ring of integers of K_0. The Frobenius automorphism of K_0 over $\mathbb{Q}_p$ will be denoted by Φ_0. Finally let $\Gamma := \mathrm{Gal}(\overline{\mathrm{K}}/\mathrm{K})$ be the absolute Galois group of K. In order to study finite dimensional representations of Γ over $\mathbb{Q}_p$, Fontaine has introduced the period rings B_{cr} and $\mathrm{B}_{d\mathrm{R}}$. B_{cr} is a K_0-algebra which admits Frobenius action, Galois action by Γ, and a decreasing filtration $\mathrm{F}^\bullet$. $\mathrm{B}_{d\mathrm{R}}$ is defined to be the filtered completion of B_{cr} with respect to $\mathrm{F}^\bullet$. Recall that $\mathrm{B}_{d\mathrm{R}}$ inherits a filtration $\mathrm{F}^\bullet$ and a Galois action from B_{cr}, but it does not admit a Frobenius action.

The importance of these period rings is that they establish connection between the category $\mathrm{Rep}_{\mathbb{Q}_p}(\Gamma)$ of finite dimensional continuous representations of Γ over $\mathbb{Q}_p$ and the category $\mathcal{FM}$ of filtered Frobenius modules. Recall that an object in $\mathcal{FM}$ is a triple $(\mathrm{E}, \Phi, \mathrm{F}^\bullet)$ consisting of a finite dimensional vector space E over K_0, a Frobenius semi-linear automorphism Φ on E, and a decreasing filtration $\mathrm{F}^\bullet$ on $\mathrm{E}_\mathrm{K} := \mathrm{E} \otimes_{\mathrm{K}_0} \mathrm{K}$. Let $\mathbb{L}$ be an object in $\mathrm{Rep}_{\mathbb{Q}_p}(\Gamma)$ and $(\mathrm{E}, \Phi, \mathrm{F}^\bullet)$ be an object in $\mathcal{FM}$ such that there exists a B_{cr}-linear isomorphism

$$\mathbb{L} \otimes_{\mathbb{Q}_p} \mathrm{B}_{cr} \cong_{\mathrm{B}_{cr}} \mathrm{E} \otimes_{\mathrm{K}_0} \mathrm{B}_{cr}$$

which respects Frobenius and Galois actions, and after extension of scalars to K, the filtrations. Then we say that $\mathbb{L}$ is a crystalline representation, $(\mathrm{E}, \Phi, \mathrm{F}^\bullet)$ is a B_{cr}-admissible filtered Frobenius module, and $\mathbb{L}$ and $(\mathrm{E}, \Phi, \mathrm{F}^\bullet)$ are associated to each other. An essential outcome of Fontaine's theory is that the functor of base extension to B_{cr} induces an equivalence between the category of crystalline representations and the category of admissible filtered Frobenius modules.

Now let $\mathbb{L}_1, \ldots, \mathbb{L}_r$ be some fixed crystalline representations in $\mathrm{Rep}_{\mathbb{Q}_p}(\Gamma)$. For any $1 \leqslant i \leqslant r$, let $(\mathrm{E}_i, \Phi_i, \mathrm{F}_i^\bullet)$ be the object in $\mathcal{FM}$ associated to $\mathbb{L}_i$. Moreover, let E be a (K_0, Φ)-module with an increasing filtration

$$(0) = \mathrm{W}_0(\mathrm{E}) \subset \mathrm{W}_1(\mathrm{E}) \subset \cdots \subset \mathrm{W}_r(\mathrm{E}) = \mathrm{E}$$

with $W_i(E)/W_{i-1}(E) \cong E_i$ as (K_0, Φ)-modules. Note that by a (K_0, Φ)-module we mean a finite dimensional K_0 vector space endowed with a Frobenius semi-linear automorphism. Now the isomorphism classes of crystalline representations $\mathbb{L}$ which are iterated extensions of $\mathbb{L}_i$'s is in bijection with the set of possible filtrations on $E_K := E \otimes_{K_0} K$ which induce $F_i^\bullet$ on E_{iK} for all $1 \leqslant i \leqslant r$. To understand the space of such filtrations, consider the following unipotent group scheme G over K_0:

$$G := \begin{pmatrix} I & \mathrm{Hom}_{K_0}(E_2, E_1) & \dots & \mathrm{Hom}_{K_0}(E_r, E_1) \\ 0 & I & \dots & \mathrm{Hom}_{K_0}(E_r, E_2) \\ \vdots & \vdots & \ddots & \vdots \\ 0 & 0 & \dots & I \end{pmatrix}$$

Obviously G has a Frobenius action Φ, induced from E_i's, and $G_K := G \otimes_{K_0} K$ admits a filtration $F^\bullet$ by subgroups, induced by the filtrations on E_{iK}'s. Now we can check that the set of possible filtrations on E_K which induce $F_i^\bullet$ on E_{iK}, for all $1 \leqslant i \leqslant r$, is in bijection with K-points of $G_K/F^0(G_K)$. In other words, crystalline representations $\mathbb{L}$ which are iterated extensions of $\mathbb{L}_i$'s are parametrized by the quotient space $G_K/F^0(G_K)$. By a similar argument over finitely generated $\mathbb{Q}_p$-algebras, we see that for any such algebra R, any element of $(G_K/F^0(G_K))(R \otimes_{\mathbb{Q}_p} K)$ gives a crystalline element in $H^1(\Gamma, \mathbb{G}(R))$. So we obtain an algebraic map

$$c : W_{K/\mathbb{Q}_p}(G_K/F^0(G_K)) \to H^1(\Gamma, \mathbb{G}),$$

where $W_{K/\mathbb{Q}_p}$ stands for the Weil restriction functor. We call this map the comparison map and the points in the image of this map are called crystalline points.

Now since we assumed that $\mathbb{L}_i$'s and $(E_i, \Phi_i, F_i^\bullet)$'s are associated to each other, the group G is B_{cr}-admissible, and is associated to the crystalline group $\mathbb{G}$, which was introduced above. Furthermore, the closed crystalline subgroups of $\mathbb{G}$ correspond to the closed B_{cr}-admissible subgroups of G. By using the above theories and the equivalence between the category of crystalline representations and the category of B_{cr}-admissible filtered Frobenius modules, we can now show the following:

Theorem 18 ([Had10, Proposition 3.2.5]). *Under the hypotheses of Theorem 17 and the ones made above, assume* $\mathbb{H}$ *is a closed* Γ*-stable crystalline subgroup of* $\mathbb{G}$ *and that* H *is the associated subgroup in* G. *Then the comparison map*

$$c : W_{K/\mathbb{Q}_p}(H_K/F^0(H_K)) \to H^1(\Gamma, \mathbb{H})$$

is injective and identifies $W_{K/\mathbb{Q}_p}(H_K/F^0(H_K))$ *with a closed sub-scheme of* $H^1(\Gamma, \mathbb{H})$.

6.6 Period Maps

Using the above theories, we are finally able to define period maps. These period maps are going to be maps from the variety under study to the spaces which parametrize the path torsors of that variety. By definition, period maps encode the variations of path torsors. In this section, we use the notations which are introduced in the introduction, but for the moment we stick to the one dimensional case and leave the higher dimensional case to later sections. Recall that in the one dimensional case, our Main Theorem is equivalent to the finiteness of the Diophantine set $\mathrm{X}(\mathcal{O}_{\mathrm{S}})$. If $\mathrm{X}(\mathcal{O}_{\mathrm{S}}) = \emptyset$ then we are done, otherwise fix an S-integral point $x \in \mathrm{X}(\mathcal{O}_{\mathrm{S}})$ as the base point and define period maps as follows.

The general idea is to pull back the triple $(\mathrm{C}, \mathrm{X}, \mathrm{D})$, which is a priori defined over $\mathcal{O}_{\mathrm{S}}$, to the local field k_v or the global field k and use the base point $\tilde{x}$ induced by x to get unipotent group schemes $\mathrm{G}_{\mathrm{M},n}$. Then for any point y we get a path torsor $\mathrm{G}_{\mathrm{M},n}(\tilde{x}, y)$ over $\mathrm{G}_{\mathrm{M},n}$. Note that, by results of Sect. 6.4, all these unipotent fundamental groups and path torsors admit extra structures. These extra structures are critical in the sense that without them all path torsors, being torsors over unipotent group schemes over a field of characteristic zero, are trivial. Then, as we saw in Sect. 6.5, under some assumptions these path torsors are parametrized by some algebraic spaces. By varying the point y, the isomorphism class of the path torsor $\mathrm{G}_{\mathrm{M},n}(\tilde{x}, y)$, hence the corresponding point in the parametrizing space of path torsors, varies and this variation gives rise to period maps $p_{\mathrm{M}}^{(n)}$. Let us have a closer look at different versions of these period maps.

Let v be a finite place of k and assume that the point $x_{k_v} \in \mathrm{X}(k_v)$ lies in $\mathrm{X}(\mathrm{W}(\mathbb{F}_v))$. For any $n \geqslant 1$ there is a unipotent fundamental group $\mathrm{G}_{d\mathrm{R},n}$ associated to this base point. Moreover, for any point $y \in \mathrm{X}(k_v)$ and any $n \geqslant 1$, we have the path torsor $\mathrm{G}_{d\mathrm{R},n}(x_{k_v}, y)$. These fundamental groups and path torsors admit Hodge filtration and Frobenius action, see Sect. 6.4, and as we mentioned above these extra structures are important in making the path torsors nontrivial. But an important fact here is that if we restrict to the p-aidc open unit disc $\mathcal{D}_1^{\circ}(x_{k_v})$ around the base point x_{k_v}, the Frobenius action alone is not enough to make the path torsors nontrivial. More precisely, the system of path torsors over $\mathcal{D}_1^{\circ}(x_{k_v})$ admits a system of Frobenius invariant elements compatible with concatenation. This is the case because by results of [Fal90] the pull back of any Frobenius isocrystal on X_{k_v} to the p-adic open unit disc centered at a $\mathrm{W}(\mathbb{F}_v)$-point is constant. Hence by forgetting the Hodge filtration, we have

$$\mathcal{O}_{\mathrm{G}_{d\mathrm{R},n}(x_{k_v}, y)} \cong \mathcal{O}_{\mathrm{G}_{d\mathrm{R},n}}, \ \forall n \geqslant 1 ,$$

as Frobenius modules over $\mathrm{W}(\mathbb{F}_v)[1/p] \subset k_v$. So, by varying the point $y \in \mathrm{X}(k_v)$, we obtain a varying family of Hodge filtrations on the coordinate rings $\mathcal{O}_{\mathrm{G}_{d\mathrm{R},n}}$ for any $n \geqslant 1$, which leads to the following de Rham period maps

$$p_{d\mathrm{R}}^{(n)} : \mathrm{X}(k_v) \cap \mathcal{D}_1^{\circ}(x_{k_v}) \rightarrow (\mathrm{G}_{d\mathrm{R},n}/\mathrm{F}^0(\mathrm{G}_{d\mathrm{R},n}))(k_v) .$$

The following result contains two crucial properties enjoyed by the de Rham period maps.

Theorem 19 ([Had10, Theorem 3.3.1]). *For any $n \geqslant 1$, the restriction of the period map $p_{\mathrm{dR}}^{(n)}$ to the p-adic integral points in the p-adic open unit disk around x_{k_v} gives a rigid analytic map with Zariski dense image in* $\mathrm{G}_{\mathrm{dR},n}/\mathrm{F}^0(\mathrm{G}_{\mathrm{dR},n})$.

There are two other versions of period maps which are important for us, namely the local and the global étale period maps. To define the local étale period map, we work with $\mathrm{X}_{\bar{k}_v}$ with base point $x_{\bar{k}_v}$. Now consider the étale version of the theory developed in Sect. 6.3 to get unipotent group schemes $\mathrm{G}_{\text{ét},n}$, on which the absolute Galois group

$$\mathrm{G}_v := \mathrm{Gal}(\bar{k}_v/k_v)$$

acts. Moreover, for any point $y \in \mathrm{X}(k_v)$, if we denote by $\bar{y}$ the induced point in $\mathrm{X}_{\bar{k}_v}$, we obtain path torsors $\mathrm{G}_{\text{ét},n}(x_{\bar{k}_v}, \bar{y})$ equipped with compatible G_v-actions.

Now suppose that we had started with a triple $(\mathbb{P}^1_{\mathcal{O}_\mathrm{S}}, \mathrm{X}, \{p_1, \ldots, p_{d+1}\})$ over $\mathcal{O}_\mathrm{S}$. Then these unipotent groups $\mathrm{G}_{\text{ét},n}$ admit finite increasing filtrations with subquotients being isomorphic to the tensor powers of the étale realization of the Tate object

$$\mathrm{H}^1_{\text{ét}}(\mathrm{X}_{\bar{k}_v}, \mathbb{Q}_\ell)^\vee \cong \mathbb{Q}_\ell(1)^d .$$

Hence, by putting $\Gamma = \mathrm{G}_v$, we can apply Theorem 17 to obtain the following local étale period maps

$$p_{\text{ét}}^{\mathrm{loc},(n)} : \mathrm{X}(k_v) \to \mathrm{H}^1(\mathrm{G}_v, \mathrm{G}_{\text{ét},n})(\mathbb{Q}_p) .$$

The construction of the global étale period map is completely parallel to the construction of the local one. The only difference is that we consider the variety $\mathrm{X}_{\bar{k}}$. Note that in the global case the resulting unipotent group schemes and path torsors are equipped with an action of the global Galois group

$$\mathrm{G}_\mathrm{T} := \mathrm{Gal}(k_\mathrm{T}/k) ,$$

where k_T is the maximal extension of k unramified outside the Galois closure T of $\mathrm{S} \cup \{v\}$. In exactly the same way, for the triple $(\mathbb{P}^1_{\mathcal{O}_\mathrm{S}}, \mathrm{X}, \{p_1, \ldots, p_{d+1}\})$ over $\mathcal{O}_\mathrm{S}$ we get the following global étale period maps

$$p_{\text{ét}}^{\mathrm{gl},(n)} : \mathrm{X}(\mathcal{O}_\mathrm{T}) \to \mathrm{H}^1(\mathrm{G}_\mathrm{T}, \mathrm{G}_{\text{ét},n})(\mathbb{Q}_p) .$$

All these maps and the comparison map of Sect. 6.5 can be put together in the following important commutative diagram.

Remark 20. For the triple $(\mathbb{P}^1_{\mathcal{O}_\mathrm{S}}, \mathrm{X}, \{p_1, \ldots, p_{d+1}\})$ over $\mathcal{O}_\mathrm{S}$ and any integer $n \geqslant 1$, we have the following commutative diagram in which $\mathrm{X}(\mathcal{O}_\mathrm{T})^\circ$ is the set of T-integral points with the same reduction modulo p as the base point x.

$$
\begin{array}{ccccc}
 & & (\mathrm{G}_{d\mathrm{R},n}/\mathrm{F}^0)(k_v) & \xrightarrow{=} & \\
 & & \uparrow p_{d\mathrm{R}}^{(n)} & & \\
\mathrm{X}(\mathcal{O}_\mathrm{T})^\circ & \xrightarrow{i} & \mathrm{X}_{k_v}(\mathcal{O}_v)\cap\mathcal{D}_1^\circ(x_{k_v}) & \dashrightarrow & \mathrm{W}_{k_v/\mathbb{Q}_p}(\mathrm{G}_{d\mathrm{R},n}/\mathrm{F}^0)(\mathbb{Q}_p) \\
\downarrow p_{\text{ét}}^{\mathrm{gl},(n)} & & \downarrow p_{\text{ét}}^{\mathrm{loc},(n)} & \swarrow c & \\
\mathrm{H}^1(\mathrm{G}_\mathrm{T},\mathrm{G}_{\text{ét},n})(\mathbb{Q}_p) & \xrightarrow{\mathrm{res}} & \mathrm{H}^1(\mathrm{G}_v,\mathrm{G}_{\text{ét},n})(\mathbb{Q}_p) & &
\end{array}
$$

Note that "res" is the usual restriction map between group cohomologies induces by the inclusion

$$\mathrm{G}_v \subset \mathrm{G}_\mathrm{T}\,.$$

The left square in the diagram is evidently commutative, but the commutativity of the lower right triangle is a much deeper claim which is a consequence of non-abelian p-adic Hodge theory. We will not address this theory in this article and refer the interested readers to the original source [Fal02] or to a brief review given in [Had10, Sect. 4.1]. Furthermore, note that the non-abelian cohomology sets of the bottom row, by general results of Sect. 6.5, are affine algebraic spaces over $\mathbb{Q}_p$, and "res" is a $\mathbb{Q}_p$-algebraic map with respect to these $\mathbb{Q}_p$-algebraic structures.

Now we get to the key part of this line of ideas. Namely, the very hard problem concerning global Galois cohomology which is the main obstruction in obtaining our main result from the above commutative diagram. The point is that if we could manage to prove that for some $n \geqslant 1$

$$\dim_{\mathbb{Q}_p}(\mathrm{H}^1(\mathrm{G}_\mathrm{T},\mathrm{G}_{\text{ét},n})) < \dim_{k_v}(\mathrm{G}_{d\mathrm{R},n}/\mathrm{F}^0(\mathrm{G}_{d\mathrm{R},n}))$$

then we would be able to prove the finiteness of S-integral points of X from the above commutative diagram. This inequality, in the case $k = \mathbb{Q}$, could be proven using a vanishing result of Soulé which is used in [Kim05] in order to prove Siegel's theorem over $\mathbb{Q}$ along the above line of ideas. Our way to get rid of this difficulty is to use motivic unipotent fundamental groups, in the sense of Voevodsky, in order to replace the global Galois cohomology groups by the algebraic K-groups of the number field k. Before summing up things in Sect. 6.8, let us briefly sketch in the following section how one can apply the above methodology to higher dimensions as well.

6.7 Descent to Lower Dimensions

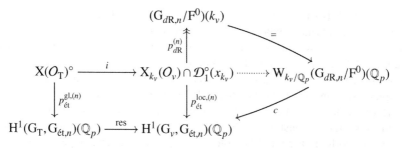

In order to study the structure of different realizations of the unipotent fundamental group of higher dimensional unirational varieties, we can use different versions of the Lefschetz hyperplane section theorem. As a byproduct, we obtain some useful information concerning the motivic unipotent fundamental group of unirational varieties, which is essential in proving our main result. Let us begin with some generalities on Lefschetz hyperplane section theorem.

It is very well known in algebraic topology that the (co)homology and homotopy groups of a CW-complex in low degrees depend only on the low dimensional skeleton of that CW-complex. On the other hand, as Lefschetz has proved, a generic hyperplane section of a high dimensional CW-complex, which is embedded into a projective space, becomes homotopy equivalent to the original CW-complex after attaching high dimensional cells. Hence it is expected that the maps between low degree (co)homology and homotopy groups of a generic hyperplane section of a high dimensional CW-complex and the CW-complex itself are isomorphisms. There are different ways of making this a precise statement, which are known in the literature as Lefschetz hyperplane section theorem or weak Lefschetz theorem. By putting different versions of Lefschetz hyperplane section theorem together and using the fullness assertion in Theorem 6, we obtain the following motivic version of Lefschetz hyperplane section theorem.

Theorem 21 ([Had10, Theorem 4.3.4]). *Let* $(\overline{\mathrm{X}},\mathrm{X},\mathrm{D})$ *be a standard triple over a number field k. Fix a closed immersion* $\overline{\mathrm{X}} \hookrightarrow \mathbb{P}^{\mathrm{N}}_k$ *and consider a generic linear subspace* H *of the ambient projective space through x. If* $\mathrm{X}\cap\mathrm{H}$ *has positive dimension (resp. dimension at least* 2*) then the map*

$$\mathrm{G}_{\mathrm{mot}}(\mathrm{X}\cap\mathrm{H},x)\to\mathrm{G}_{\mathrm{mot}}(\mathrm{X},x)$$

induced by inclusion is surjective (resp. isomorphism).

Note that the same result as above is obviously valid for the algebraic quotients $\mathrm{G}_{\mathrm{mot,n}}$, for any $n\geqslant 1$. Hence, if we denote the kernel of the projection

$$\mathrm{G}_{\mathrm{mot,n+1}}\twoheadrightarrow\mathrm{G}_{\mathrm{mot,n}}$$

by K_n, and we continue with all the assumptions under which the above theorem holds, we get the following commutative diagram.

$$\begin{array}{ccccccccc}
0 & \longrightarrow & \mathrm{K}_n(\mathrm{X}\cap\mathrm{H}) & \longrightarrow & \mathrm{G}_{\mathrm{mot,n+1}}(\mathrm{X}\cap\mathrm{H},x) & \longrightarrow & \mathrm{G}_{\mathrm{mot,n}}(\mathrm{X}\cap\mathrm{H},x) & \longrightarrow & 0\\
 & & \vdots\,\psi & & \downarrow & & \downarrow & & \\
0 & \longrightarrow & \mathrm{K}_n(\mathrm{X}) & \longrightarrow & \mathrm{G}_{\mathrm{mot,n+1}}(\mathrm{X},x) & \longrightarrow & \mathrm{G}_{\mathrm{mot,n}}(\mathrm{X},x) & \longrightarrow & 0
\end{array}$$

But when $\mathrm{X}\cap\mathrm{H}$ is a punctured projective line, $\mathrm{K}_n(\mathrm{X}\cap\mathrm{H})$ is isomorphic to some power of the Tate object $\mathbb{Q}(n)$, and $\mathbb{Q}(n)$ is a simple object in $\mathrm{MT}(k)$. On the other hand, the map ψ in the above diagram can be shown to be a surjection and hence we deduce that $\mathrm{K}_n(\mathrm{X})$ is also isomorphic to some power of $\mathbb{Q}(n)$. This gives us the following important corollary.

Corollary 22 ([Had10, Corollary 4.3.5]). *Under the hypotheses of Theorem 21, for any* $n\geqslant 1$ *we have the following exact sequence*

$$0 \to \mathbb{Q}(n)^{r_n} \to \mathrm{G}_{\mathrm{mot,n+1}}(\mathrm{X},x) \to \mathrm{G}_{\mathrm{mot,n}}(\mathrm{X},x) \to 0\,,$$

where r_n is the dimension of the vector group

$$\mathrm{Z}^n(\mathrm{G}_{\mathrm{mot}}(\mathrm{X},x))/\mathrm{Z}^{n+1}(\mathrm{G}_{\mathrm{mot}}(\mathrm{X},x))\,.$$

Now using the above results, for any standard triple $(\overline{\mathrm{X}},\mathrm{X},\mathrm{D})$ over the ring of S-integers $\mathcal{O}_{\mathrm{S}}$, we can show that the local and global étale torsor spaces are representable by affine varieties. Also we can define different versions of period maps and obtain the analogue for X of the main diagram which appeared in Remark 20. In fact, by the above mentioned observations, all this facts can be reduced to the one dimensional case by using Bertini's theorem over k and enlarging S to a bigger finite set S′ of finite places of k. Note that this enlargement of S to a bigger finite set S′ is harmless for us in the sense that the validity of our Main Theorem for a smaller finite set S of finite places of k is a consequence of its validity for a larger finite set S′ of finite places of k.

6.8 Main Result

Finally, we can put the above techniques together and prove our main result. Following the notations we fixed in Sect. 6.1, for a standard triple $(\overline{\mathrm{X}},\mathrm{X},\mathrm{D})$ over the ring of S-integers $\mathcal{O}_{\mathrm{S}}$ of the number field k and a fixed base point $x \in \mathrm{X}(\mathcal{O}_{\mathrm{S}})$ let

$$\{r_n := \dim(\mathrm{Z}^n(\pi_1^{\mathrm{mot}}(\mathrm{X},x))/\mathrm{Z}^{n+1}(\pi_1^{\mathrm{mot}}(\mathrm{X},x)))\ :\ n \geqslant 1\}$$

be the set of natural numbers appearing in Corollary 22. As we saw in previous sections, the unipotent fundamental group and path torsors of X are motivic. This, in particular, implies that the global étale period maps factor through the motivic period maps

$$p^{(n)}_{\mathrm{mot}}\ :\ \mathrm{X}(\mathcal{O}_{\mathrm{T}})^{\circ} \to \mathrm{H}^1(\mathrm{MT}(\mathcal{O}_{\mathrm{T}}),\mathrm{G}_{\mathrm{mot},n})(\mathbb{Q}_p)\,,$$

where $\mathrm{MT}(\mathcal{O}_{\mathrm{T}})$ is the full subcategory of $\mathrm{MT}(k)$ consisting of the objects for which the étale realization is unramified outside T. Now by using exact sequences of Corollary 22 and using the fact that for any $n \geqslant 2$

$$\mathrm{H}^1(\mathrm{MT}(\mathcal{O}_{\mathrm{T}}),\mathbb{Q}(n)) \cong \mathrm{K}_{2n-1}(k)\otimes_{\mathbb{Z}}\mathbb{Q}\,,$$

we can find good enough upper bounds for the dimension of the Zariski closure of the image of $p^{gl,(n)}_{\text{ét}}$, which shows that under the assumptions of our Main Theorem, for some n this dimension is strictly less than $\dim_{k_v}(\mathrm{G}_{d\mathrm{R}}/\mathrm{F}^0(\mathrm{G}_{d\mathrm{R}}))$. So we get:

Theorem 23 (Main Theorem). *Let $k/\mathbb{Q}$ be a totally real number field of degree d, and assume that $h^{1,0}(\overline{\mathrm{X}}) = 0$. Moreover, assume that for any constant $c \in \mathbb{N}$ there exists a natural number $n \in \mathbb{N}$ such that*

$$c + d(r_3 + r_5 + \cdots + r_{2\lfloor (n-1)/2 \rfloor + 1}) < r_1 + r_2 + \cdots + r_n \,.$$

Then X satisfies $\mathcal{V}_{S,v}$-property for almost all finite places v of k.

Remark 24. For any standard triple of the form $(\mathbb{P}^1, \mathbb{P}^1 - D, D)$ over $\mathcal{O}_S$, where $D = \{p_1, p_2, \ldots, p_{d+1}\}$ consists of at least three S-integral point in $\mathbb{P}^1$, we can compute the numbers r_n as

$$r_n = \frac{1}{n} \sum_{m|n} \mu(m) d^{n/m} \,,$$

where μ is the Möbius function. Hence the numerical assumption of the above theorem is satisfied for any totally real number field of degree less than or equal to d. Now Theorem 23 can be applied to show the $\mathcal{V}_{S,v}$-property for almost all v. But in the one dimensional case, the $\mathcal{V}_{S,v}$-property even for a single place v, implies finiteness of integral points.

Corollary 25. *Combining the above remark and Theorem 23 gives a motivic proof of finiteness of integral points in $\mathbb{P}^1 - \{p_1, p_2, \ldots, p_{d+1}\}$ over totally real fields of degree at most d, where $d \geqslant 2$ is any natural number. In particular, we obtain a motivic proof of Siegel's finiteness theorem of integral points over any totally real quadratic number field and the field of rational numbers.*

References

[Bor74] A. Borel. Stable real cohomology of arithmetic groups. *Ann. Sci. École Norm. Sup. (4) 7*, 235–272, 1974.

[Del89] P. Deligne. Le groupe fondamental de la droite projective moins trois points. *Galois groups over $\mathbb{Q}$, Math. Sci. Res. Inst. Publ., 16*, Springer, New York, 79–297, 1989.

[DG05] P. Deligne, A. Goncharov. Groupes fondamentaux motiviques de Tate mixte. *Ann. Sci. École Norm. Sup. (4) 38*, 1–56, 2005.

[Fal90] G. Faltings. F-isocrystals on open varieties, results and conjectures. *Grothendieck's 60'th birthday festschrift, Vol. II*, Birkhauser, Boston, 219–248, 1990.

[Fal02] G. Faltings. Almost étale coverings. *Astérisque No. 279*, 185–270, 2002.

[Fal07] G. Faltings. Mathematics around Kim's new proof of Siegel's theorem. *Diophantine geometry, CRM Series, 4*, Ed. Norm., Pisa, 173–188, 2007.

[Had10] M. Hadian-Jazi. Motivic fundamental groups and integral points. *Ph.D. Thesis, Max-Planck Institute for Mathematics/Bonn University*, 2010.

[Kat89] K. Kato. Logarithmic structures of Fontaine-Illusie. *Algebraic Analysis, Geometry, and number theory (Baltimore, MD, 1988)*, 191-224, Johns Hopkins Univ. Press, Baltimore, MD, 1989.

[Kim05] M. Kim. The motivic fundamental group of $\mathbb{P}^1 \setminus \{0, 1, \infty\}$ and the theorem of Siegel. *Invent. Math. 161*, 629–656, 2005.

[Voe00] V. Voevodsky. Triangulated categories of motives over a field. *Cycles, transfers, and motivic homology theories, Ann. of Math. Stud. 143*, 188-238, Princeton Univ. Press, Princeton, NJ, 2000.

Chapter 7
Descent Obstruction and Fundamental Exact Sequence

David Harari and Jakob Stix

Abstract We establish a link between the descent obstruction against rational points and sections of the fundamental group extension that has applications to the Brauer–Manin obstruction and to the birational case of the section conjecture in anabelian geometry.

7.1 Introduction

Let k be a field of characteristic zero with algebraic closure $\bar{k}$ and absolute Galois group $\Gamma_k = \mathrm{Gal}(\bar{k}/k)$. Let X be a geometrically connected variety over k. Fix a geometric point $\bar{x} \in \mathrm{X}(\bar{k})$ and let $\pi_1(\mathrm{X}) = \pi_1(\mathrm{X}, \bar{x})$ be the étale fundamental group of X. Set $\overline{\mathrm{X}} = \mathrm{X} \times_k \bar{k}$ and denote by $\pi_1(\overline{\mathrm{X}}) = \pi_1(\overline{\mathrm{X}}, \bar{x})$ the étale fundamental group of $\overline{\mathrm{X}}$. Recall Grothendieck's fundamental exact sequence of profinite groups, cf. [SGA1, IX Thm 6.1],

$$1 \to \pi_1(\overline{\mathrm{X}}) \to \pi_1(\mathrm{X}) \to \Gamma_k \to 1 \ . \tag{7.1}$$

By covariant functoriality of π_1, the existence of a k-point on X implies that the exact sequence (7.1) has a section. Grothendieck's section conjecture predicts that the converse statement is true whenever X is a proper hyperbolic curve over a number field, see [Gro83]. There is also a version of the section conjecture for affine hyperbolic curves when k-rational cusps need to be considered as well, see [Gro83, page 8/9]. For a p-adic version of this conjecture, see for example Conjecture 2 of [PS11]

D. Harari (✉)
Université de Paris-Sud (Orsay), Bâtiment 425, 91405 Orsay Cedex, France
e-mail: david.harari@math.u-psud.fr

J. Stix
MATCH,Universität Heidelberg, Im Neuenheimer Feld 288, 69120 Heidelberg, Germany
e-mail: stix@mathi.uni-heidelberg.de

J. Stix (ed.), *The Arithmetic of Fundamental Groups*, Contributions in Mathematical and Computational Sciences 2, DOI 10.1007/978-3-642-23905-2_7,

or earlier [Koe05], where the p-adic section conjecture can be read inbetween the lines.

The goal of this note is to relate the existence of a section for (7.1) when k is a number field to the fact that X has an adelic point for which there is no descent obstruction, in the sense of [Sko01, II §5.3] as introduced in [HS02], associated to torsors under finite group schemes. The first result of this flavour is Theorem 11. Its main applications are Theorem 15, Theorem 17, and Theorem 21. The latter seems to be (up to date) the most general known statement relating the descent obstruction to the birational fundamental exact sequence.

In Sect. 7.3 we will also prove related statements over arbitrary fields, so that the reader can distinguish between purely formal results and results related to arithmetic properties.

Acknowledgements. We thank U. Görtz for a stimulating question and M. Çiperiani, J-L. Colliot-Thélène, M. Stoll, T. Szamuely, and O. Wittenberg for helpful comments. This work started when both authors visited the Isaac Newton Institute for Mathematical Sciences in Cambridge, whose excellent working conditions and hospitality are gratefully aknowledged.

7.2 Preliminaries

Since we want to deal with variants of the exact sequence (7.1), for example the abelianized fundamental exact sequence, we need to introduce a general setting as follows. Let $\overline{\mathrm{U}} < \pi_1(\overline{\mathrm{X}})$ be a closed subgroup which is a normal subgroup in $\pi_1(\mathrm{X})$. For example $\overline{\mathrm{U}}$ could be a characteristic subgroup in $\pi_1(\overline{\mathrm{X}})$ of which there are plenty, because $\pi_1(\overline{\mathrm{X}})$ is finitely generated as a profinite group. We set $\overline{\mathrm{A}} = \pi_1(\overline{\mathrm{X}})/\overline{\mathrm{U}}$. The pushout of (7.1) by the canonical surjection $\pi_1(\overline{\mathrm{X}}) \to \overline{\mathrm{A}}$ is the exact sequence

$$1 \to \overline{\mathrm{A}} \to \pi_1^{\mathrm{U}}(\mathrm{X}) \to \Gamma_k \to 1 , \tag{7.2}$$

that can be defined because $\overline{\mathrm{U}}$ is also normal in $\pi_1(\mathrm{X})$ and in particular by definition the kernel of the induced quotient map $\pi_1(\mathrm{X}) \to \pi_1^{\mathrm{U}}(\mathrm{X})$. This construction contains as special cases for $\overline{\mathrm{A}}$ the profinite abelianized group $\pi_1(\overline{\mathrm{X}})^{\mathrm{ab}}$ if we take for $\overline{\mathrm{U}}$ the closure of the derived subgroup of $\pi_1(\overline{\mathrm{X}})$, and $\overline{\mathrm{A}} = \pi_1(\overline{\mathrm{X}})$ if we take for $\overline{\mathrm{U}}$ the trivial group.

Let $(k_i)_{i \in \mathrm{I}}$ be a family of field extensions of k. The guiding example is the case of a number field k together with the family of all completions k_v of k, or of all corresponding henselizations[1] k_v^h of k. Fix an algebraic closure $\bar{k}_i$ of k_i and embeddings $\bar{k} \to \bar{k}_i$, so that with $\Gamma_i = \mathrm{Gal}(\bar{k}_i/k_i)$ we have canonical restriction maps $\theta_i : \Gamma_i \to \Gamma_k$. Moreover, the embedding $\bar{k} \to \bar{k}_i$ yields canonically a

[1] By convention if k_v is an archimedean completion of k, the henselization k_v^h means the algebraic closure of k into k_v.

geometric point $\bar{x}_i$ of $X_i = X \times_k k_i$, that projects onto $\bar{x}$, and thus a canonical map $\pi_1(X_i, \bar{x}_i) \to \pi_1(X, \bar{x})$. The reason for assuming characteristic zero in the first place is that by the comparison theorem, cf. [Sza09, p. 186, Remark 5.7.8.], the natural maps induce an isomorphism

$$\pi_1(X_i, \bar{x}_i) \xrightarrow{\sim} \pi_1(X, \bar{x}) \times_{\Gamma_k} \Gamma_i \,.$$

Thus a section s of (7.1) induces canonically a section s_i of the analogue of (7.1) for the k_i-variety X_i. In this formal setting the main goal of this note is to establish a criterion inspired by the descent obstruction for when a collection of sections $(s_i)_{i \in I}$ comes from a section s up to conjugation from $\pi_1(\overline{X})$.

We continue with a reminder on nonabelian H^1. Recall that the étale cohomology set $H^1(X, G)$ for a finite k-group scheme G is the same as the cohomology set

$$H^1(\pi_1(X), G(\bar{k})) \,,$$

where the action of $\pi_1(X)$ on $G(\bar{k})$ is induced by the projection map $\pi_1(X) \to \Gamma_k$ that occurs in (7.1), cf. [SGA1, XI §5].

The identification is natural in both X and G, although for a map $Y \to X$ the induced map $\pi_1(Y) \to \pi_1(X)$ is only well defined up to inner automorphism by an element of $\pi_1(\overline{X})$. In fact, such an inner automorphism acts as the identity on $H^1(\pi_1(X), G(\bar{k}))$ by the following reasoning. Recall that in general if $\varphi = \gamma(-)\gamma^{-1}$ is an inner automorphism of a profinite group π by an element $\gamma \in \pi$ and M is a discrete π-group, then the map (φ^*, γ^{-1}) which is the composite

$$H^1(\pi, M) \xrightarrow{\varphi^*} H^1(\pi, \varphi^* M) \xrightarrow{\gamma^{-1}\cdot} H^1(\pi, M) \,,$$

that exploits the π-map $\varphi^* M \to M$ *multiplication by* γ^{-1} and which on cocycles is given by

$$(\sigma \mapsto a_\sigma) \mapsto (\sigma \mapsto \gamma^{-1}(a_{\gamma\sigma\gamma^{-1}})) \,,$$

is the identity map. This is classical when M is abelian, see [Ser68, VII.5. Proposition 3], and easy to check in the general case by the same direct computation:

$$\gamma^{-1}(a_{\gamma\sigma\gamma^{-1}}) = \gamma^{-1}(a_\gamma) a_{\sigma\gamma^{-1}} = \gamma^{-1}(a_\gamma) a_\sigma \sigma(\gamma^{-1}(a_\gamma)) \,, \tag{7.3}$$

which shows that $(\sigma \mapsto a_\sigma)$ is indeed cohomologous to $\sigma \mapsto \gamma^{-1}(a_{\gamma\sigma\gamma^{-1}})$. In our geometric example the element $\gamma \in \pi_1(\overline{X})$ acts trivially on the coefficients $G(\bar{k})$ such that (φ^*, γ^{-1}) becomes simply the pullback by conjugation with γ, which therefore acts as identity on $H^1(\pi_1(X), G(\bar{k}))$.

The étale cohomology set $H^1(\overline{X}, \overline{G})$ for $\overline{G} = G \times_k \bar{k}$ is naturally the set

$$\mathrm{Hom}^{\mathrm{out}}(\pi_1(\overline{X}), G(\bar{k}))$$

of continuous homomorphisms $\pi_1(\overline{X}) \to G(\bar{k})$ up to conjugation by an element of $G(\bar{k})$, see [Ser94, I.5].

The following interpretation of $H^1(\Gamma, G)$ will become useful later. Let

$$1 \to G \to E \to \bar{\Gamma} \to 1$$

be a short exact sequence of profinite groups, and let $\overline{\varphi} : \Gamma \to \bar{\Gamma}$ be a continuous homomorphism. The set of lifts $\varphi : \Gamma \to E$ of $\overline{\varphi}$ up to conjugation by an element of G is either empty or, with the group G equipped with the conjugation action of Γ via a choice of lift φ_0, in bijection with the corresponding $H^1(\Gamma, G)$. Indeed for a cocycle $a : \Gamma \to G$ the twist of φ_0 by $a = (\gamma \mapsto a_\gamma)$, i.e., the map $\gamma \mapsto a_\gamma \cdot \varphi_0(\gamma)$, is another lift. Any other lift of $\overline{\varphi}$ can be described by such a twist. Two cocycles are cohomologous if and only if they lead to conjugate lifts. The description of lifts via $H^1(\Gamma, G)$ is natural with respect to both Γ and G.

7.3 Results over Arbitrary Fields

The notation and assumptions in this whole section are as above. In particular we consider the exact sequence (7.2) associated to a quotient $\overline{A}$ of $\pi_1(\overline{X})$ by a subgroup $\overline{U}$ that remains normal in $\pi_1(X)$.

For each $i \in I$, let $\sigma_i : \Gamma_i \to \pi_1(X_i, \bar{x}_i)$ be a section of the fundamental sequence associated to X_i. For example σ_i could be the section associated to a k_i-rational point $P_i \in X(k_i)$. By composition we obtain a **section map**

$$s_i : \Gamma_i \xrightarrow{\sigma_i} \pi_1(X_i) \to \pi_1(X) \to \pi_1^U(X) .$$

Let G be a finite k-group scheme, hence $G(\bar{k})$ is a finite discrete Γ_k-group. Via the map $\theta_i : \Gamma_i \to \Gamma_k$ we may view $G(\bar{k}) = \theta_i^* G(\bar{k})$ also as a discrete Γ_i-group that describes the base change $G \times_k k_i$.

A cohomology class $\alpha \in H^1(X, G)$ such that the corresponding geometric element $\bar{\alpha} \in H^1(\overline{X}, \overline{G})$ has trivial restriction to $\overline{U}$, has an evaluation $\alpha(s_i) \in H^1(k_i, G)$ as follows. By the restriction–inflation sequence the class α uniquely comes from $H^1(\pi_1^U(X), G(\bar{k}))$ and so the pullback class

$$\alpha(s_i) := s_i^*(\alpha) \in H^1(k_i, G)$$

is defined. Note that the coefficients G here are indeed the group $G(\bar{k})$ with Γ_i action induced by θ_i because s_i comes from a section σ_i. By formula (7.3) the evaluation does only depend on s_i up to conjugation by an element of $\pi_1^U(X)$ with trivial action on $G(\bar{k})$.

By analogy with [Sto07, Definition 5.2], we say that the tuple of section maps $(s_i)_{i \in I}$ **survives every finite descent obstruction** if the following holds.

(a) For every finite k-group scheme G and every $\alpha \in \mathrm{H}^1(\mathrm{X},\mathrm{G})$ such that the corresponding element $\bar{\alpha} \in \mathrm{H}^1(\overline{\mathrm{X}},\overline{\mathrm{G}})$ has trivial restriction to $\overline{\mathrm{U}}$, the family

$$(\alpha(s_i)) \in \prod_{i \in \mathrm{I}} \mathrm{H}^1(k_i,\mathrm{G})$$

belongs to the diagonal image of $\mathrm{H}^1(k,\mathrm{G})$.

Clearly, if the sections s_i are the sections associated to k_i-rational points, then (s_i) survives every finite descent obstruction if and only if the collection (P_i) of rational points survives every finite descent obstruction in the sense of [Sto07]. We furthermore say that the tuple of section maps $(s_i)_{i \in \mathrm{I}}$ **survives every finite constant descent obstruction** if the following holds.

(a') For every finite constant k-group scheme G and every $\alpha \in \mathrm{H}^1(\mathrm{X},\mathrm{G})$ such that the corresponding element $\bar{\alpha} \in \mathrm{H}^1(\overline{\mathrm{X}},\overline{\mathrm{G}})$ has trivial restriction to $\overline{\mathrm{U}}$, the family

$$(\alpha(s_i)) \in \prod_{i \in \mathrm{I}} \mathrm{H}^1(k_i,\mathrm{G})$$

belongs to the diagonal image of $\mathrm{H}^1(k,\mathrm{G})$.

We first establish a link to continuous homomorphisms $\Gamma_k \to \pi_1^{\mathrm{U}}(\mathrm{X})$. Let us define a **continuous quotient** of a profinite group as a quotient by a normal and closed subgroup.

Proposition 1. *Consider the following assertion:*

(b) There exists a continuous homomorphism $s : \Gamma_k \to \pi_1^{\mathrm{U}}(\mathrm{X})$ such that for each $i \in \mathrm{I}$, we have $s_i = s \circ \theta_i$ up to conjugation in $\pi_1^{\mathrm{U}}(\mathrm{X})$.

Then (b) implies property (a'). If we assume further that the following hypothesis holds:

($*$) For every finite and constant k-group scheme G, the fibres of the diagonal restriction map $\mathrm{H}^1(k,\mathrm{G}) \to \prod_{i \in \mathrm{I}} \mathrm{H}^1(k_i,\mathrm{G})$ are finite.

Then (b) is equivalent to property (a').

Proof. Assume (b). Let G and $\alpha \in \mathrm{H}^1(\mathrm{X},\mathrm{G}) = \mathrm{Hom}^{\mathrm{out}}(\pi_1(\mathrm{X}),\mathrm{G}(\bar{k}))$ be as in property (a'). Since the restriction of α to $\overline{\mathrm{U}}$ is trivial, the class α corresponds to a map denoted again $\alpha : \pi_1^{\mathrm{U}}(\mathrm{X}) \to \mathrm{G}(\bar{k})$ up to conjugation in $\mathrm{G}(\bar{k})$. We get

$$\alpha(s_i) = \alpha \circ s_i = \alpha \circ s \circ \theta_i = \theta_i^*(\alpha(s))$$

up to conjugation in $\mathrm{G}(\bar{k})$, so that $(\alpha(s_i))$ is the image of $\alpha(s) \in \mathrm{H}^1(k,\mathrm{G})$ under the diagonal map, whence property (a').

Suppose now that assertion (a') and the additional hypothesis ($*$) hold. We are going to show that (b) holds as well. For a finite continuous quotient $p : \pi_1^{\mathrm{U}}(\mathrm{X}) \to \mathrm{G}$ we consider the set

$$\mathrm{S_G} := \{s' \in \mathrm{Hom}(\Gamma_k, \mathrm{G})\,; \quad \forall i \in \mathrm{I}, \quad \theta_i^*(s') = p \circ s_i \in \mathrm{H}^1(k_i, \mathrm{G})\}\,,$$

where we view G as a constant k-group scheme. The set $\mathrm{S_G}$ is non empty by assumption (a') and finite thanks to $(*)$ and to the finiteness of G. Therefore $\varprojlim_{\mathrm{G}} \mathrm{S_G}$ where G ranges over all finite continuous quotients of $\pi_1^{\mathrm{U}}(\mathrm{X})$ is not empty, see [Bou98, Chap. I §9.6 Proposition 8]. An element $s \in \varprojlim_{\mathrm{G}} \mathrm{S_G}$ is nothing but a continuous homomorphism

$$s : \Gamma_k \to \varprojlim_{\mathrm{G}} \mathrm{G} = \pi_1^{\mathrm{U}}(\mathrm{X})$$

such that for every $i \in \mathrm{I}$, and every finite continuous quotient $p : \pi_1^{\mathrm{U}}(\mathrm{X}) \to \mathrm{G}$ the equality $p \circ s \circ \theta_i = p \circ s_i$ holds up to conjugation by elements from the finite set

$$\mathrm{C}_{i,\mathrm{G}} = \{c \in \mathrm{G}\,; \quad p \circ s \circ \theta_i = c(p \circ s_i)c^{-1}\} \subseteq \mathrm{G}$$

The set $\varprojlim_{\mathrm{G}} \mathrm{C}_{i,\mathrm{G}}$ is not empty by the same argument, which implies that for each $i \in \mathrm{I}$, we have $s \circ \theta_i = s_i$ up to conjugation in $\pi_1^{\mathrm{U}}(\mathrm{X})$. □

Remark 2. (1) Without additional assumptions, we cannot force the supplementary property that s is a section. Indeed, take $k_i = \bar{k}$ for every $i \in \mathrm{I}$. Then all sets $\mathrm{H}^1(k_i, \mathrm{G})$ are trivial, hence the condition (a) and thus condition (a') is automatically satisfied. Although condition (b) also holds trivially by the choice of the trivial homomorphism, because there is no interpolation property to be satisfied, nevertheless (7.2) does not always admit a section, see for example [Sti10] or [HS09] for counterexamples over local and global fields.

(2) For an example with a nontrivial homomorphism $s : \Gamma_k \to \pi_1(\mathrm{X})$ but no section we consider the case $k = \mathbb{R}$, $k_i = \mathbb{C}$ and a real Godeaux–Serre variety. Computations with SAGE, see [S⁺08], show that the homogenous equations

$$\begin{aligned} z_0^2 + z_1^2 + z_2^2 + z_3^2 + z_4^2 + z_5^2 + z_6^2 &= 0 \\ z_0 z_2 + z_1 z_3 + z_4^2 + z_5 z_6 &= 0 \\ i(z_0^2 - z_1^2) + 3i(z_2^2 - z_3^2) - 2z_6^2 &= 0 \\ i(z_0 z_1 + z_2 z_3) + z_4 z_5 + z_5 z_6 + z_6 z_4 &= 0 \end{aligned}$$

define a smooth surface Y of general type in $\mathbb{P}^6_{\mathbb{C}}$ with ample canonical bundle $\omega_{\mathrm{Y}} = \mathcal{O}(1)|_{\mathrm{Y}}$ as computed by the adjunction formula. By the Lefschetz theorem on hyperplane sections Y is simply connected. The surface Y is preserved by the $\Gamma_{\mathbb{R}}$-semilinear action of $\mathrm{G} = \mathbb{Z}/4\mathbb{Z}$ on $\mathbb{P}^6_{\mathbb{C}}$ generated by

$$[z_0 : z_1 : z_2 : z_3 : z_4 : z_5 : z_6] \mapsto [-\bar{z}_1 : \bar{z}_0 : -\bar{z}_3 : \bar{z}_2 : \bar{z}_4 : \bar{z}_5 : \bar{z}_6]\,,$$

and avoids the fixed point set of the G-action. Hence the quotient map $\mathrm{Y} \to \mathrm{X}$ with $\mathrm{X} = \mathrm{Y}/\mathrm{G}$ is the universal cover of the geometrically connected $\mathbb{R}$-variety X with $\pi_1(\mathrm{X}) = \mathbb{Z}/4\mathbb{Z}$. The analogue of (7.1) for X is given by

$$1 \to \pi_1(\mathrm{X} \times_{\mathbb{R}} \mathbb{C}) \to \mathbb{Z}/4\mathbb{Z} \to \Gamma_{\mathbb{R}} \to 1$$

which clearly does not admit sections. Nevertheless, there is a nontrivial morphism $\Gamma_{\mathbb{R}} \to \pi_1(X)$.

However, if we suppose that (7.2) has a section, then we can prove the following stronger approximation result.

Proposition 3. *Consider the following assertion:*

(c) There exists a section $s : \Gamma_k \to \pi_1^U(X)$ *of (7.2) such that for each* $i \in I$, *we have* $s_i = s \circ \theta_i$ *up to conjugation in* $\overline{A}$.

Then (c) implies (a) which implies the following (a").

(a") For every finite k-group scheme G *and every* $\alpha \in H^1(X,G)$ *such that the corresponding element* $\bar{\alpha} \in H^1(\overline{X},\overline{G})$ *has trivial restriction to* $\overline{U}$ *and is surjective (or equivalently, the* G*-torsor* $Y \to X$ *corresponding to* α *is assumed to be geometrically connected), the family*

$$(\alpha(s_i)) \in \prod_{i \in I} H^1(k_i, G)$$

belongs to the diagonal image of $H^1(k,G)$.

If we moreover assume that $(*)$ *holds and that the exact sequence (7.2) admits a section* s_0, *then the properties (a), (a") and (c) are all equivalent.*

Proof. The implications (c) $\Rightarrow$ (a) $\Rightarrow$ (a") are obvious because by formula (7.3) a section s as in (c) implies for every $\alpha \in H^1(X,G)$ as in (a) that

$$\alpha(s_i) = s_i^*(\alpha) = (s \circ \theta_i)^*(\alpha) = \theta_i^*(s^*(\alpha)) .$$

It remains to show that (a") $\Rightarrow$ (c) under the additional assumption of $(*)$ and the existence of a section s_0 of $\pi_1^U(X) \to \Gamma_k$. The method is similar to [Sto06], Lemma 9.13, which deals with the case when k is a number field and (k_i) is the family of its completions for miscellaneous $\overline{A}$, like $\overline{A} = \pi_1(\overline{X})$ or $\overline{A} = \pi_1^{ab}(\overline{X})$. For the convenience of the reader we give a grouptheoretic version of the argument.

Let us assume (a"). Let $\overline{A}_V$ be the quotient $\pi_1(\overline{X})/\overline{V}$ for an open subgroup $\overline{V} < \pi_1(\overline{X})$ containing $\overline{U}$ and normal in $\pi_1(X)$. Let $p_V : \pi_1^U(X) \twoheadrightarrow \pi_1^V(X)$ be the corresponding quotient map. The composition $s_{0,V} = p_V \circ s_0$ splits the exact sequence (7.2) for $\overline{V}$

$$1 \to \overline{A}_V \to \pi_1^V(X) \to \Gamma_k \to 1 , \tag{7.4}$$

so that $\pi_1^V(X)$ is isomorphic to a semi-direct product. The map p_V and

$$p_{0,V} : \pi_1^U(X) \to \Gamma_k \xrightarrow{s_{0,V}} \pi_1^V(X)$$

lift the natural projection $\pi_1^U(X) \to \Gamma_k$. Their difference $\gamma \mapsto p_V(\gamma)p_{0,V}(\gamma)^{-1}$ is a cohomology class $\alpha_V \in H^1(\pi_1^U(X), \overline{A}_V)$, with $\pi_1^U(X)$ acting via $p_{0,V}$ and conjugation, that corresponds to a class in $\mathrm{H}^1(\mathrm{X},\overline{\mathrm{A}}_\mathrm{V})$ which becomes trivial when restricted to $\overline{\mathrm{U}}$. The restriction of α_V to $\pi_1(\overline{\mathrm{X}})$ equals the surjective map $p_\mathrm{V}|_{\pi_1(\overline{\mathrm{X}})} : \pi_1(\overline{\mathrm{X}}) \twoheadrightarrow \overline{\mathrm{A}}_\mathrm{V}$, hence is geometrically connected.

We now apply (a") to the class α_V. The class $\alpha_\mathrm{V}(s_i) = s_i^*(\alpha_\mathrm{V})$ measures the difference between $p_\mathrm{V} \circ s_i$ and $p_{0,\mathrm{V}} \circ s_i = s_{0,\mathrm{V}} \circ \theta_i$. Twisting $s_{0,\mathrm{V}}$ by a class in $\mathrm{H}^1(k,\overline{\mathrm{A}}_\mathrm{V})$ that diagonally maps to $(\alpha_\mathrm{V}(s_i))$ we obtain a section $s_\mathrm{V} : \Gamma_k \to \pi_1^\mathrm{V}(\mathrm{X})$ such that $s_\mathrm{V} \circ \theta_i$ equals $p_\mathrm{V} \circ s_i$ up to conjugation in $\overline{\mathrm{A}}_\mathrm{V}$.

Assumption $(*)$ now implies that the set of such sections s_V is finite. Again by [Bou98, Chap. I §9.6 Proposition 8], there is a compatible family of sections (s_V) in the projective limit over all possible $\overline{\mathrm{V}}$, which defines a section

$$s : \Gamma_k \to \varprojlim_{\overline{\mathrm{V}}} \pi_1^\mathrm{V}(\mathrm{X}) = \pi_1^\mathrm{U}(\mathrm{X})$$

such that $s \circ \theta_i = s_i$ up to conjugation in $\overline{\mathrm{A}}$ by the projective limit argument as in the proof of Proposition 1. This completes the proof of (c). □

Remark 4. It is worth noting that the additional assumption that (7.2) has a section allows us to find a genuine section s that *interpolates* the s_i up to conjugation even in $\overline{\mathrm{A}}$, and not merely a homomorphism or interpolation up to conjugation in $\pi_1^\mathrm{U}(\mathrm{X})$.

We can prove more under the additional assumption of the collection of fields (k_i) being arithmetically sufficiently rich in a sense to be made precise as follows. Consider the following property.

$(**)$ The union of the conjugates of all the images $\Gamma_i \to \Gamma_k$ is dense in Γ_k.

Lemma 5. *Property $(**)$ is inherited by finite extensions k'/k with respect to the set of all composita $k_i \cdot k'$.*

Proof. For any $\sigma \in \Gamma_k$ let $k'_{i,\sigma}$ be the field extension of k_i associated to the preimage $\Gamma'_{i,\sigma} = \theta_i^{-1}(\sigma^{-1}\Gamma_{k'}\sigma)$ in Γ_i, namely the compositum $k_i\sigma^{-1}(k')$ in $\bar{k}_i$ using the fixed embedding $\sigma^{-1}(k') \subset \bar{k} \subset \bar{k}_i$ that yields θ_i. The inclusion $k' \subset k'_{i,\sigma}$ induces the map

$$\theta'_{i,\sigma} = \sigma(-)\sigma^{-1} \circ \theta_i : \Gamma'_{i,\sigma} \to \Gamma_{k'} .$$

The union of the conjugates of the images of all $\theta'_{i,\sigma}$ is dense in $\Gamma_{k'}$, saying that property $(**)$ is inherited for finite field extensions k'/k for the new family of fields $(k'_{i,\sigma})$. Indeed, we have to show that

$$\bigcup_{i,\sigma} \theta'_{i,\sigma}(\Gamma'_{i,\sigma}) = (\bigcup_{i,\sigma} \sigma\Gamma_i\sigma^{-1}) \cap \Gamma_{k'}$$

surjects onto cofinally any finite continuous quotient of $\Gamma_{k'}$. It is enough to treat quotients $p_0 : \Gamma_{k'} \to \mathrm{G}_0$ with $\ker(p_0)$ normal in Γ_k, i.e., the map p_0 extends to a finite continuous quotient $p : \Gamma_k \to \mathrm{G}$. Then

$$p_0\big((\bigcup_{i,\sigma} \sigma\Gamma_i\sigma^{-1}\big)\cap\Gamma_{k'}\big) = p\big(\bigcup_{i,\sigma} \sigma\Gamma_i\sigma^{-1}\big)\cap \mathrm{G}_0 = \mathrm{G}_0$$

by property (∗∗). □

Lemma 6. *Property (∗∗) implies property (∗).*

Proof. To prove (∗) we consider a finite k-group G and $\alpha \in \mathrm{H}^1(k,\mathrm{G})$. We need to show that the following set is finite:

$$\text{Ш}_\alpha = \{\beta \in \mathrm{H}^1(k,\mathrm{G})\ ;\ \theta_i^*(\beta) = \theta_i^*(\alpha) \in \mathrm{H}^1(k_i,\mathrm{G}) \text{ for all } i \in \mathrm{I}\}\,.$$

By the technique of twisting, see [Ser94, I §5.4], we may assume that α is the trivial class in $\mathrm{H}^1(k,\mathrm{G})$. Let k'/k be a finite Galois extension that trivialises G. With the notation as in Lemma 5, the commutative diagram

$$\begin{array}{ccc} \mathrm{H}^1(k,\mathrm{G}) & \xrightarrow{\text{res}} & \mathrm{H}^1(k',\mathrm{G}) \\ \downarrow{\scriptstyle \theta_i^*} & & \downarrow{\scriptstyle \theta'^*_{i,\sigma}} \\ \mathrm{H}^1(k_i,\mathrm{G}) & \xrightarrow{\text{res}} & \mathrm{H}^1(k'_{i,\sigma},\mathrm{G}) \end{array}$$

shows that under restriction Ш_α maps into

$$\text{Ш}_{\text{trivial}} = \{\chi \in \mathrm{Hom}^{\mathrm{out}}(\Gamma_{k'},\mathrm{G})\ ;\ \chi\circ\theta'_{i,\sigma} = 1 \text{ for all } i,\sigma\}$$

which contains only the trivial class due to property (∗∗) and Lemma 5. Hence, due to the nonabelian inflation–restriction sequence, the group Ш_α is contained in $\mathrm{H}^1(\mathrm{Gal}(k'/k),\mathrm{G}(k'))$ which is a finite set. □

Proposition 7. *Under the assumption of (∗∗) the properties (a) and (c) are equivalent.*

Proof. By Lemma 6 we also have assumption (∗). By Proposition 3 it suffices to show that under assumption (a) the map $\pi_1^{\mathrm{U}}(\mathrm{X}) \to \Gamma_k$ admits a section. As (a) trivially implies (a') we may use Proposition 1 to deduce (b), so that we have found at least a continuous homomorphism $u : \Gamma_k \to \pi_1^{\mathrm{U}}(\mathrm{X})$ such that for all $i \in \mathrm{I}$ we have $s_i = u\circ\theta_i$ up to conjugation in $\pi_1^{\mathrm{U}}(\mathrm{X})$. Let $\varphi : \Gamma_k \to \Gamma_k$ be the composition of u with the projection $p : \pi_1^{\mathrm{U}}(\mathrm{X}) \to \Gamma_k$. To find a section s_0 of (7.2) and thus to complete the proof of Proposition 7, it suffices to prove that φ is bijective because we can then take $s_0 = u\circ\varphi^{-1}$. We have

$$\varphi\circ\theta_i = p\circ(u\circ\theta_i) = p\circ s_i = \theta_i$$

up to conjugation in Γ_k. Thus for every $\gamma \in \bigcup_i \bigcup_{g\in\Gamma_k} g\theta_i(\Gamma_i)g^{-1}$ the image $\varphi(\gamma)$ is conjugate to γ in Γ_k. By assumption (∗∗) the set $\bigcup_i \bigcup_{g\in\Gamma_k} g\theta_i(\Gamma_i)g^{-1}$ is dense in Γ_k so that φ preserves every conjugacy class of Γ_k by continuity and compactness of Γ_k. In particular φ is injective.

In every finite quotient $\Gamma_k \to \mathrm{G}$ the image of $\varphi(\Gamma_k)$ is a subgroup $\mathrm{H} < \mathrm{G}$ such that the union of the conjugates of H covers G. An old argument that goes back to at least Jordan, namely the estimate

$$|\mathrm{G}| = |\bigcup_{g \in \mathrm{G}/\mathrm{H}} g\mathrm{H}g^{-1}| \leqslant (\mathrm{G}:\mathrm{H}) \cdot (|\mathrm{H}| - 1) + 1 = |\mathrm{G}| - (\mathrm{G}:\mathrm{H}) + 1 \leqslant |\mathrm{G}|,$$

shows that necessarily $\mathrm{H} = \mathrm{G}$. Thus φ is also surjective. □

Remark 8. (1) The isomorphism φ that occurs in the proof of Proposition 7 preserves conjugacy classes of elements, hence is of a very special type which is much studied by group theorists.

(2) In the case of a number field, every automorphism of Γ_k is induced by an automorphism of k by a theorem of Neukirch, Uchida and Iwasawa, see [Neu77], and there are also famous extensions of this result by Pop to function fields. In particular every automorphism of $\Gamma_{\mathbb{Q}}$ is an inner automorphism.

Proposition 9. *Under the assumption of (∗∗) the properties (b) and (c) are equivalent.*

Proof. Clearly (c) implies (b). For the converse let $u : \Gamma_k \to \pi_1^{\mathrm{U}}(\mathrm{X})$ be a homomorphism as in (b), so that there are $\gamma_i \in \pi_1^{\mathrm{U}}(\mathrm{X})$ with $u \circ \theta_i = \gamma_i(-)\gamma_i^{-1} \circ s_i$ for all $i \in \mathrm{I}$. With the natural projection $p : \pi_1^{\mathrm{U}}(\mathrm{X}) \to \Gamma_k$ the proof of Proposition 7 says that the homomorphism $\varphi = p \circ u$ is an isomorphism, so that $s = u \circ \varphi^{-1}$ is a section. With $p(\gamma_i) := \sigma_i$ we compute

$$\varphi \circ \theta_i = p \circ u \circ \theta_i = p \circ (\gamma_i(-)\gamma_i^{-1}) \circ s_i = \sigma_i(-)\sigma_i^{-1} \circ \theta_i,$$

since s_i is a section and thus $p \circ s_i = \theta_i$. Applying φ^{-1} to both sides yields with $\tau_i = \varphi^{-1}(\sigma_i^{-1})$ the equation

$$\tau_i(-)\tau_i^{-1} \circ \theta_i = \varphi^{-1} \circ \theta_i.$$

Now the section s interpolates the following

$$s \circ \theta_i = u \circ \varphi^{-1} \circ \theta_i = u \circ (\tau_i(-)\tau_i^{-1}) \circ \theta_i = (u(\tau_i)(-)u(\tau_i)^{-1}) \circ u \circ \theta_i$$

$$= (u(\tau_i)(-)u(\tau_i)^{-1}) \circ (\gamma_i(-)\gamma_i^{-1}) \circ s_i = ((u(\tau_i)\gamma_i)(-)(u(\tau_i)\gamma_i)^{-1}) \circ s_i,$$

and because of

$$p(u(\tau_i)\gamma_i) = \varphi(\tau_i)p(\gamma_i) = \sigma_i^{-1}\sigma_i = 1$$

we find that s actually satisfies the stronger interpolation property of (c). □

Corollary 10. *Under the assumption of (∗∗) the properties (a), (a'), (b) and (c) are equivalent to each other and to (a") together with the existence of section.*

Proof. This follows immediately by Lemma 6, Proposition 1, Proposition 3, Proposition 7, and Proposition 9. □

7.4 Results over Number Fields

From now on we assume that k is a number field. We consider the exact sequence

$$1 \to \overline{\mathrm{A}} \to \pi_1^{\mathrm{U}}(\mathrm{X}) \to \Gamma_k \to 1$$

as above in (7.2). Let k_v be the completion of k at a place v of k. A choice of embeddings $\bar{k} \to \bar{k}_v$ of the respective algebraic closures identifies the absolute Galois group $\Gamma_v = \mathrm{Gal}(\bar{k}_v/k_v)$ of k_v with the decomposition subgroup of v, or more precisely the place of $\bar{k}$ above v corresponding to the embeding $\bar{k} \to \bar{k}_v$. Hence the restriction map $\theta_v : \Gamma_v \to \Gamma_k$ as defined in the introduction is injective.

Theorem 11. *Let* S *be a set of places of k of Dirichlet density* 0*, for example a finite set of places. Assume that* $\mathrm{X}(k_v) \neq \emptyset$ *for* $v \notin \mathrm{S}$*. For each* $v \notin \mathrm{S}$*, let* $s_v : \Gamma_v \to \pi_1^{\mathrm{U}}(\mathrm{X})$ *be the section map associated to a* k_v*-rational point* $\mathrm{P}_v \in \mathrm{X}(k_v)$*.*

Then the following assertions are equivalent:

(i) *For every finite k-group scheme* G *and every* $\alpha \in \mathrm{H}^1(\mathrm{X},\mathrm{G})$ *such that* $\bar{\alpha}$ *has trivial restriction to* $\overline{\mathrm{U}}$*, the family* $(\alpha(\mathrm{P}_v))$ *belongs to the diagonal image of* $\mathrm{H}^1(k,\mathrm{G})$ *in* $\prod_{v\notin \mathrm{S}} \mathrm{H}^1(k_v,\mathrm{G})$*.*

(i') *For every finite constant k-group scheme* G *and every* $\alpha \in \mathrm{H}^1(\mathrm{X},\mathrm{G})$ *such that* $\bar{\alpha}$ *has trivial restriction to* $\overline{\mathrm{U}}$*, the family* $(\alpha(\mathrm{P}_v))$ *belongs to the diagonal image of* $\mathrm{H}^1(k,\mathrm{G})$ *in* $\prod_{v\notin \mathrm{S}} \mathrm{H}^1(k_v,\mathrm{G})$*.*

(i") *There is a section* $s_0 : \Gamma_k \to \pi_1^{\mathrm{U}}(\mathrm{X})$ *and for every finite k-group scheme* G *and every* $\alpha \in \mathrm{H}^1(\mathrm{X},\mathrm{G})$ *such that* $\bar{\alpha}$ *restricts trivially to* $\overline{\mathrm{U}}$ *and the associated* G*-torsor on* X *is geometrically connected, the family* $(\alpha(\mathrm{P}_v))$ *belongs to the diagonal image of* $\mathrm{H}^1(k,\mathrm{G})$ *in* $\prod_{v\notin \mathrm{S}} \mathrm{H}^1(k_v,\mathrm{G})$*.*

(ii) *There exists a homomorphism* $s : \Gamma_k \to \pi_1^{\mathrm{U}}(\mathrm{X})$ *of (7.2) such that for each* $v \notin \mathrm{S}$*, we have* $s_v = s \circ \theta_v$ *up to conjugation in* $\pi_1^{\mathrm{U}}(\mathrm{X})$*.*

(iii) *There exists a section* $s : \Gamma_k \to \pi_1^{\mathrm{U}}(\mathrm{X})$ *of (7.2) such that for each* $v \notin \mathrm{S}$*, we have* $s_v = s \circ \theta_v$ *up to conjugation in* $\overline{\mathrm{A}}$*, i.e, the sections* s_v *come from a global section* s*.*

Proof. This is merely a translation of Corollary 10 into the number field setting, once we notice that assertion (**) follows immediately from Chebotarev's density theorem. □

Remark 12. (1) For $\overline{\mathrm{U}}$ trivial, we have $\pi_1^{\mathrm{U}}(\mathrm{X}) = \pi_1(\mathrm{X})$ and assertion (i) means in the language of [Sto07, Definition 5.2], that the family (P_v) survives every X-torsor under a finite group scheme G/k, while assertion (i') says that (P_v) survives every X-torsor under a finite *constant* group scheme. By (i") this is equivalent to the existence of a section[2] together with (P_v) surviving every *geometrically connected* X-torsor under a finite group scheme.

[2] Thanks to M. Stoll for pointing out the importance of this condition.

(2) Even when $X(k) \neq \emptyset$, it is not sufficient to demand in (i") that (P_v) survives every geometrically connected torsor under a finite and constant group scheme to deduce that (P_v) satisfies the equivalent properties of Theorem 11. Take for example $k = \mathbb{Q}$ and X such that $\pi_1(\overline{X}) = \mu_3$ with the corresponding Galois action. Such examples arise among varieties of general type. Then the only G-torsor over X with G finite constant and Y geometrically connected is X with trivial group G. Nevertheless, there is a torsor $Y \to X$ under μ_3 with Y geometrically connected, and certain families (P_v) do not survive Y, see [Har00, Remark after Corollary 2.4].

(3) An interesting case is when $\overline{U}$ is the closure of the derived subgroup of $\pi_1(\overline{X})$, so that $\overline{A} = \pi_1(\overline{X})/\overline{U}$ is just the abelianized profinite group $\pi_1^{ab}(\overline{X})$. Then the section s in assertion (iii) corresponds to a section of the geometrically abelianized fundamental exact sequence

$$1 \to \pi_1^{ab}(\overline{X}) \to \pi_1^{U}(X) \to \Gamma_k \to 1\,.$$

Then assertion (i) means that the family (P_v) survives every X-torsor Y under a finite group scheme G such that $Y \to X$ has **abelian geometric monodromy**, that is: such that the image of the homomorphism $\pi_1(\overline{X}) \to \overline{G}$ associated to $\overline{Y} \to \overline{X}$ is an abelian group. Similar statements hold for abelian replaced by solvable or nilpotent taking for $\overline{A}$ the maximal prosolvable or pronilpotent quotient of $\pi_1(\overline{X})$.

(4) The analogue of Theorem 11 holds with the same proof if we replace the family of completions (k_v) by the corresponding henselizations (k_v^h), simply because the assertions only depend on the associated sections and the map $\Gamma_{k_v} \to \Gamma_{k_v^h}$ induced by $k_v^h \subset k_v$ is an isomorphism.

7.5 Abelian Applications

Let k be a number field. Denote by Ω_k the set of all places of k. For a smooth and projective k-variety X its **Brauer–Manin set** is the subset $X(\mathbf{A}_k)^{Br}$ of $\prod_{v\in\Omega_k} X(k_v)$ consisting of those adelic points that are orthogonal to the Brauer group for the Brauer–Manin pairing, cf. [Sko01, II. Chap. 5].

The following corollary is a consequence of the implication (iii) $\Rightarrow$ (i) in Theorem 11. Similar results had already been observed independently (at least) by J-L. Colliot-Thélène, O. Wittenberg and the second author.

Corollary 13. *Let* X *be a smooth, projective, geometrically connected curve over a number field k. Assume that the abelianized fundamental exact sequence*

$$1 \to \pi_1^{ab}(\overline{X}) \to \Pi \to \Gamma_k \to 1$$

has a section s, such that for each $v \in \Omega_k$ *the corresponding section* s_v *is induced by a* k_v*-point* P_v *of* X. *Then* $(P_v) \in X(\mathbf{A}_k)^{Br}$.

Proof. We take for $\overline{U}$ the closure of the derived subgroup of $\pi_1(\overline{X})$ in Theorem 11. Then (P_v) satisfies condition (iii) of Theorem 11, hence by (i) it survives every X-torsor with abelian geometric monodromy under a finite k-group G. In particular, the adelic point (P_v) survives any X-torsor under a finite abelian group scheme. It remains to apply [Sto07, Corollary 7.3]. □

Remark 14. (1) Let X be a smooth projective curve of genus 0. Then the assumption of Corollary 13 seems vacuous as $\pi_1(\overline{X}) = 1$ and there is a section with no arithmetic content. But we also assume the existence of an adelic point, whence the curve X has k-rational points by the classical Hasse local–global principle for quadratic forms. Moreover, any adelic point on $X \cong \mathbb{P}^1_k$ satisfies the Brauer–Manin obstruction because $\mathrm{Br}(k) = \mathrm{Br}(\mathbb{P}^1_k)$.

(2) Let X be a smooth projective curve of genus 1 as in Corollary 13 with Jacobian E. Then X corresponds to an element [X] in the Tate–Shafarevich group $\text{Ш}(E/k)$. The existence of an adelic point which survives the Brauer–Manin obstruction then implies by [Sko01, Theorem 6.2.3] that [X] belongs to the maximal divisible subgroup of $\text{Ш}(E/k)$, which also follows from [HS09, Proposition 2.1]. When $\text{Ш}(E/k)$ is finite as is conjecturally always the case, then the curve X has a k-rational point and is actually an elliptic curve E.

(3) It is conjectured that a p-adic version of Grothendieck's section conjecture holds, which would imply that for a smooth, projective, geometrically integral curve of genus at least 2, each local section s_v as in Corollary 13 is automatically induced by a k_v point. See also Remark (2) after Theorem 19.

(4) If we assume further that the Jacobian variety of X has finitely many rational points and finite Tate-Shafarevich group, then the conclusion of Corollary 13 implies $X(k) \neq \emptyset$ by a result due to Scharaschkin and Skorobogatov, see [Sko01, Corollary 6.2.6.] or [Sto07, Corollary 8.1].

The following result and its proof are inspired by Koenigsmann's theorem [Koe05], namely the fact that for a smooth, geometrically connected curve over a p-adic field, the existence of a section for the birational fundamental exact sequence implies the existence of a rational point.

Theorem 15. *Let X be a smooth, projective and geometrically connected curve over a number field k. Assume that the birational fundamental exact sequence*

$$1 \to \Gamma_{\bar{k}(X)} \to \Gamma_{k(X)} \to \Gamma_k \to 1 \tag{7.5}$$

has a section. Then $X(\mathbf{A}_k)^{\mathrm{Br}} \neq \emptyset$*. If we assume further that the Jacobian variety of X has finitely many rational points and finite Tate–Shafarevich group, then* $X(k) \neq \emptyset$*.*

A *non-abelian* version of this theorem will be given in the next section in Theorem 21.

Proof. A section $s : \Gamma_k \to \Gamma_{k(X)}$ of (7.5) induces[3] for every place v of k a section s_v^h for the analogous sequence for k replaced by k_v^h. We follow the argument used by Koenigsmann [Koe05, Proposition 2.4 (1)].

The image of s_v^h defines a field extension $L_v^h/k_v^h(X)$ as the fixed field in the algebraic closure of $k_v^h(X)$. Because the natural maps between absolute Galois groups

$$\Gamma_{L_v^h} \to \Gamma_{k_v^h} \leftarrow \Gamma_{k_v}$$

are isomorphisms, the fields L_v^h, k_v^h and k_v are p-adically closed fields, see [Koe95, Theorem 4.1], and thus L_v^h is an elementary extension of k_v^h, see [Koe05, Fact 2.2]. In particular, the tautological L_v^h point of X given by

$$\operatorname{Spec}(L_v^h) \to \operatorname{Spec} k_v^h(X) \to X$$

implies the existence of a k_v^h-point and thus a k_v-point on X.

The core of the following well-known limit argument goes back at least to Neukirch, and was introduced in anabelian geometry by Nakamura, while Tamagawa emphasized its significance to the section conjecture. We perform the limit argument by applying the above existence result to every connected branched cover $X' \to X$, necessarily geometrically connected over k, with

$$s(\Gamma_k) \subset \Gamma_{k(X')} \subset \Gamma_{k(X)} .$$

Thus the projective system $\varprojlim_{X'} X'(k_v)$ over all such X' is a projective system of nonempty compact spaces, and is therefore itself nonempty by [Bou98, Chap. I §9.6 Proposition 8].

Let (P_v') with $P_v' \in X'(k_v)$ be an element in the projective limit with lowest stage $P_v \in X(k_v)$. It follows that the section $s_{P_v} : \Gamma_{k_v^h} \to \Gamma_{k_v^h(X)}$ composed with the natural projection $\Gamma_{k_v^h(X)} \to \Gamma_{k(X)}$ agrees with the v-local component $s \circ \theta_v$ for the original section s. We may now apply Corollary 13 to the composition

$$\Gamma_k \xrightarrow{s} \Gamma_{k(X)} \to \pi_1(X) ,$$

which shows that the adelic point (P_v) of X is orthogonal to Br X for the Brauer-Manin pairing.

Under the further assumptions that the Jacobian of X has finite Mordell–Weil group and finite Tate–Shafarevich group we now apply the result by Scharaschkin–Skorobogatov, see Remark (4) above, to complete the proof of the theorem. □

Remark 16. In [EW10, Theorem 2.1] H. Esnault and O. Wittenberg discuss a geometrically abelian version of Theorem 15 with the result that an abelian birational section yields a divisor of degree 1 on X under the assumption of the Tate-Shafarevich group of the Jacobian of X being finite.

---

[3] This would not be clear if we had replaced k by k_v instead of k_v^h. Indeed the existence of a birational section is not a condition that is stable by extension of scalars; see [EW10, Remark 3.12(iii)].

We next describe an application towards the birational version of the section conjecture of Grothendieck's. Recall that for a geometrically connected k-variety X a k-rational point $a \in X(k)$ describes by functoriality a $\pi_1(\overline{X})$-conjugacy class of sections s_a of (7.1). In the birational setting the k-rational point leads to the following. Define $\hat{\mathbb{Z}}(1)$ as the inverse limit (over n) of the Γ_k-modules $\mu_n(\bar{k})$. Due to the characteristic zero assumption the decomposition group D_a of $a \in X(k)$ in $\Gamma_{k(X)}$ is an extension

$$1 \to \hat{\mathbb{Z}}(1) \to D_a \to \Gamma_k \to 1 \tag{7.6}$$

that splits for example by the choice of a uniformizer t at a and a compatible choice of nth roots $t^{1/n}$ of t. It follows that up to conjugacy by $\hat{\mathbb{Z}}(1)$, the inertia group at a, we have a **packet of sections** of (7.6) with a free transitive action by the huge uncountable group

$$H^1(k, \hat{\mathbb{Z}}(1)) = \varprojlim_n k^*/(k^*)^n \,.$$

It can be proven in at least two different ways that the map $D_a \to \Gamma_{k(X)}$ maps the $\hat{\mathbb{Z}}(1)$-conjugacy classes of sections of (7.6) injectively into the set of $\Gamma_{\bar{k}(X)}$-conjugacy classes of sections of (7.5), see for example [Koe05, Sect. 1.4], or [Sti08, Sect. 1.3 and Theorem 14+17], or [EH08].

The **birational form of the section conjecture** speculates that for a smooth, projective geometrically connected curve the map from k-rational points to packets of sections of (7.5) is bijective and that there are no other sections of (7.5), see [Koe05, Sect. 1.4+5].

The following theorem is a corollary[4] of Stoll's results [Sto06, Corollary 8.6 and Theorem 9.18].

Theorem 17. *Let X be a smooth, projective and geometrically connected curve over k. If we assume that there is a nonconstant map $X \to A$ to an abelian variety A/k with finitely many k-rational points and finite Tate–Shafarevich group, then every section s of the birational fundamental exact sequence*

$$1 \to \Gamma_{\bar{k}(X)} \to \Gamma_{k(X)} \to \Gamma_k \to 1 \tag{7.7}$$

is the section s_a associated to a k-rational point $a \in X(k)$. In other words, the birational section conjecture is true for such curves X/k.

Proof. Let s be a section of (7.7), and let $X' \to X$ be a finite branched cover, such that upon suitable choices of base points the image of s is contained in $\Gamma_{k(X')} \subset \Gamma_{k(X)}$. Then Theorem 15 shows that $X'(\mathbf{A}_k)^{\mathrm{Br}} \neq \emptyset$. Exploiting the finite map $X' \to X \to A$ we may use Stoll's result [Sto07] Theorem 8.6, to deduce $X'(k) \neq \emptyset$. Cofinally all such X' will have genus at least 2 so that $X'(k)$ then is nonempty and finite by Faltings–Mordell [Fal83, Satz 7]. It follows that

$$\varprojlim_{X'} X'(k) \,,$$

4 Note however that in the proof of [Sto06, Theorem 9.18], it is not explained why the existence of a birational section over k implies the same property over k_v.

where X' ranges over the system of all X' as above, is nonempty by [Bou98, Chap. I §9.6 Proposition 8]. Let $a \in X(k)$ be the projection to $X(k)$ of an element of $\varprojlim_{X'} X'(k)$, then the image of s is contained in the decomposition subgroup

$$D_a \subset \Gamma_{k(X)}$$

and s belongs to the packet of sections associated to the k-rational point a.

It remains to refer to the literature for the injectivity of the (birational) section conjecture, which was already known to Grothendieck [Gro83], see for example [Sti08, Appendix B]. □

Remark 18. (1) The conjecture of Birch and Swinnerton-Dyer predicts that an abelian variety A over a number field k has both finite $A(k)$ and finite $Ш(A/k)$ if and only if its complex L-function $L(s, A/k)$ does not vanish at the critical point $s = 1$. This is known in the case of elliptic curves $E/\mathbb{Q}$ due to work of Coates-Wiles, Rubin, and Kolyvagin. For abelian varieties $A/\mathbb{Q}$ with $L(1, A/\mathbb{Q}) \neq 0$ the work of Kolyvagin-Logachev [KL91] allows to conclude finiteness of $A(\mathbb{Q})$ and $Ш(A/\mathbb{Q})$ subject to an additional technical condition.

Following Mazur [Maz78], every Jacobian $J_0(p)$ of the modular curve $X_0(p)$ for $p = 11$ or a prime $p \geqslant 17$ has a nontrivial Eisenstein quotient, see $\tilde{J}$ [Maz78, II (10.4)], with finite Mordell–Weil group $\tilde{J}(\mathbb{Q})$. But only for an Eisenstein ideal $\mathfrak{p}$ the $\mathfrak{p}$-component of the Tate–Shafarevich group $Ш(\tilde{J}/\mathbb{Q})$ is known to be finite. Building on the work of Mazur, modular quotients $J_0(p) \to A$, which satisfy $L(1, A/\mathbb{Q}) \neq 0$, have been determined in abundance, see for example Duke [Duk95].

Consequently, every smooth, projective geometrically connected curve X over $\mathbb{Q}$ with a nonconstant map $X \to A$ for one of the good abelian varieties A above will (subject to the validity of the technical assumption necessary in [KL91] or unconditionally if $\dim(A) = 1$) satisfy Theorem 17 and thus the birational section conjecture will hold for such X with $k = \mathbb{Q}$.

(2) A recent result of Mazur and Rubin, [MR10, Theorem 1.1], guarantees for any algebraic number field k the existence of infinitely many elliptic curves E/k with $E(k) = 0$. As conjecturally $Ш(E/k)$ is always finite, these elliptic curves and moreover their branched covers $X \to E$ can be used in Theorem 17 to at least conjecturally produce examples of the birational section conjecture over any algebraic number field.

7.6 Non-abelian Applications

We turn our attention to geometrically *non-abelian* applications of Theorem 11.

Theorem 19. *Let* X *be a smooth, projective, geometrically connected curve over a number field k. Let* $(P_v)_{v \in \Omega_k}$ *be an adelic point of* X *that survives every* X*-torsor under a finite group scheme. Then* (P_v) *survives every* X*-torsor under a linear group scheme.*

Proof. We apply Theorem 11 in the case $\overline{\mathrm{U}} = 0$, hence we have $\overline{\mathrm{A}} = \pi_1(\overline{\mathrm{X}})$. The hypothesis means that (P_v) satisfies condition (i) of this theorem, hence there is a section $s : \Gamma_k \to \pi_1(\mathrm{X})$ as in condition (iii).

Let $\mathrm{Y} \to \mathrm{X}$ be a geometrically connected torsor under a finite group scheme. Using the section s, we can lift (P_v) to an adelic point (Q_v) on some twisted torsor Y^σ such that s takes values in the subgroup $\pi_1(\mathrm{Y}^\sigma)$ of $\pi_1(\mathrm{X})$. This means that (Q_v) again satisfies condition (iii) of Theorem 11. In particular Corollary 13 implies that $(\mathrm{Q}_v) \in \mathrm{Y}^\sigma(\mathbf{A}_k)^{\mathrm{Br}}$. So we have proved that for every geometrically connected torsor Y under a finite group scheme G, the adelic point (P_v) can be lifted to an adelic point $(\mathrm{Q}_v) \in \mathrm{Y}^\sigma(\mathbf{A}_k)^{\mathrm{Br}}$ for some twisted torsor Y^σ.

This still holds if Y is not assumed to be geometrically connected: indeed the assumption that there exists an adelic point of X surviving every X-torsor under a finite group scheme implies (by a result of Stoll, see also Demarche's paper [Dem09], beginning of the proof of Lemma 3) that there exists a geometrically connected torsor $\mathrm{Z} \to \mathrm{X}$ under a finite k-group scheme F, a cocycle $\sigma \in \mathrm{Z}^1(k, \mathrm{G})$, and a morphism $\mathrm{F} \to \mathrm{G}^\sigma$ such that Y^σ is obtained by pushout of the torsor Z. We conclude with the functoriality of the Brauer–Manin pairing.

It remains to apply the main result of [Dem09] to finish the proof, namely that the étale Brauer–Manin obstruction is a priori stronger than the descent obstruction imposed by linear algebraic groups. □

Remark 20. (1) The previous result does not hold in higher dimension. For example there are smooth, projective, geometrically integral and geometrically rational surfaces X, in particular we have $\pi_1(\overline{\mathrm{X}}) = 1$, with $\mathrm{X}(k) \neq \emptyset$, but such that some adelic points (P_v) do not belong to $\mathrm{X}(\mathbf{A}_k)^{\mathrm{Br}}$. For an example with an intersection of two quadrics in $\mathbf{P}^4$ see [CTS77, p. 3, Example a]. By [Sko01, Theorem 6.1.2 (a)], such adelic points do not survive the *universal* torsors, which are those torsors under the Néron-Severi torus of X whose type in the sense of Colliot-Thélène and Sansuc's descent theory is an isomorphism, see [Sko01, Definition 2.3.3].

(2) Let X be a curve of genus at least 2 such that the fundamental exact sequence (7.1) has a section. If we knew the p-adic analogue of Grothendieck's section conjecture, Theorem 11 and Theorem 19 would yield the existence of an adelic point (P_v) that survives every torsor under a linear k-group scheme, which is a priori stronger than $(\mathrm{P}_v) \in \mathrm{X}(\mathbf{A}_k)^{\mathrm{Br}}$. Recall that as we have seen before (a result by Scharaschkin/Skorobogatov), the condition $\mathrm{X}(\mathbf{A}_k)^{\mathrm{Br}} \neq \emptyset$ already implies $\mathrm{X}(k) \neq \emptyset$ if the Jacobian variety of X has finitely many rational points and finite Tate-Shafarevich group.

The following result is the *non-abelian* version of Theorem 15.

Theorem 21. *Let* X *be a smooth, projective and geometrically connected curve over a number field k. Assume that the birational fundamental exact sequence*

$$1 \to \Gamma_{\bar{k}(\mathrm{X})} \to \Gamma_{k(\mathrm{X})} \to \Gamma_k \to 1 \tag{7.8}$$

has a section. Then X *contains an adelic point* (P_v) *that survives every torsor under a linear k-group scheme.*

Proof. We proceed exactly as in the proof of Theorem 15, except that at the end we apply Theorem 11 instead of Corollary 13, so that we obtain that the adelic point (P_v) of X survives every torsor under a finite k-group scheme, hence every torsor under a linear k-group scheme by Theorem 19. □

Remark 22. Over a number field or a p-adic field, no example of a smooth and geometrically integral variety X such that the exact sequence (7.5) has a section, but $\mathrm{X}(k) = \emptyset$, is known. According to Grothendieck, a sufficiently small non empty open subset U of X should be *anabelian*, which would imply, if one believes a general form of his section conjecture, see [Gro83], that X has a rational point as soon as the sequence (7.5) is split. We don't know whether Theorem 15 and Theorem 21 still hold in arbitrary dimension.

Let $\Gamma_{\bar{k}(\mathrm{X})} \to \Gamma^{\mathrm{solv}}_{\bar{k}(\mathrm{X})}$ be the maximal pro-solvable quotient of $\Gamma_{\bar{k}(\mathrm{X})}$ and

$$1 \to \Gamma^{\mathrm{solv}}_{\bar{k}(\mathrm{X})} \to \Gamma^{(\mathrm{solv})}_{k(\mathrm{X})} \to \Gamma_k \to 1 \tag{7.9}$$

the pushout of (7.5) by $\Gamma_{\bar{k}(\mathrm{X})} \to \Gamma^{\mathrm{solv}}_{\bar{k}(\mathrm{X})}$. With this exact sequence we can prove the following geometrically pro-solvable version of Theorem 21.

Theorem 23. *Let k be a number field, and let X be a smooth, projective and geometrically connected curve over k. Assume that the geometrically pro-solvable birational fundamental exact sequence*

$$1 \to \Gamma^{\mathrm{solv}}_{\bar{k}(\mathrm{X})} \to \Gamma^{(\mathrm{solv})}_{k(\mathrm{X})} \to \Gamma_k \to 1$$

has a section. Then X contains an adelic point (P_v) that survives every torsor under a finite k-group scheme with geometric monodromy a finite solvable group.

Proof. We start as in the proof of Theorem 15. Let $v \mid p$ be a place of k above p. The local section

$$s_v^h : \Gamma_{k_v^h} \to \Gamma^{(\mathrm{solv})}_{k_v^h(\mathrm{X})} \subset \Gamma^{(\mathrm{solv})}_{k_v^h(\mathrm{X})}$$

restricts, i.e., after adjoining the pth roots of unity $\langle \zeta_p \rangle$, to a liftable section

$$s_v^h|_{\dots} : \Gamma_{k_v^h(\zeta_p)} \to \Gamma^{(p)}_{k_v^h(\zeta_p)(\mathrm{X})}$$

in the sense of [Pop10] for the geometrically pro-p birational fundamental exact sequence of the scalar extension $\mathrm{X} \times_k k_v^h(\zeta_p)$. Now [Pop10, Theorem B 2)] shows that, modulo the geometric commutator, the section $s_v^h|_{\dots}$ belongs to a unique bouquet of sections associated to a point $\mathrm{P}_v \in \mathrm{X}(k_v(\zeta_p))$ with coefficients in the completion $k_v(\zeta_p)$ of $k_v^h(\zeta_p)$. Since $s_v^h|_{\dots}$ is invariant under

$$\mathrm{Gal}(k_v^h(\zeta_p)/k_v^h) = \mathrm{Gal}(k_v(\zeta_p)/k_v)\,,$$

the uniqueness of P_v, structure transport and Galois descent show that $\mathrm{P}_v \in \mathrm{X}(k_v)$. The same limit argument as in the proof of Theorem 15 applies and shows that in fact the local section s_v^h agrees with the composite

$$s_{\mathrm{P}_v} : \Gamma_{k_v^h} = \Gamma_{k_v} \to \Gamma^{(\mathrm{solv})}_{k_v(\mathrm{X})} \twoheadrightarrow \Gamma^{(\mathrm{solv})}_{k_v^h(\mathrm{X})}$$

The rest of the proof follows as in the proof of Theorem 15. □

References

[Bou98] N. Bourbaki. *General topology. Chapters 1–4*. Elements of Mathematics. Springer, 1998. Translated from the French, reprint of the 1989 English translation, vii+437 pp.

[CTS77] J.-L. Colliot-Thélène and J.-J. Sansuc. La descente sur une variété rationnelle définie sur un corps de nombres. *C. R. Acad. Sci. Paris*, 284:1215–1218, 1977.

[Dem09] C. Demarche. Obstruction de descente et obstruction de Brauer-Manin étale. *Algebra and Number Theory*, 3(2):237–254, 2009.

[Duk95] W. Duke. The critical order of vanishing of automorphic L-functions with large level. *Invent. Math.*, 119(1):165–174, 1995.

[EH08] H. Esnault and Ph. H. Hai. Packets in Grothendieck's Section Conjecture. *Adv. Math.*, 218(2):395–416, 2008.

[EW10] H. Esnault and O. Wittenberg. On abelian birational sections. *Journal of the American Mathematical Society*, 23:713–724, 2010.

[Fal83] G. Faltings. Endlichkeitssätze für abelsche Varietäten über Zahlkörpern. *Invent. Math.*, 73(3):349–366, 1983.

[Gro83] A. Grothendieck. Brief an Faltings (27/06/1983). In L. Schneps and P. Lochak, editors, *Geometric Galois Actions 1*, volume 242 of *LMS Lecture Notes*, pages 49–58. Cambridge, 1997.

[Har00] D. Harari. Weak approximation and non-abelian fundamental groups. *Ann. Sci. école Norm. Sup. (4)*, 33(4):467–484, 2000.

[HS02] D. Harari and A. N. Skorobogatov. Non-abelian cohomology and rational points. *Compositio Math.*, 130(3):241–273, 2002.

[HS09] D. Harari and T. Szamuely. Galois sections for abelianized fundamental groups. *Math. Ann.*, 344(4):779–800, 2009. With an appendix by E. V. Flynn.

[KL91] V. A. Kolyvagin and D. Yu. Logachëv. Finiteness of Ш over totally real fields. *Izv. Akad. Nauk SSSR Ser. Mat.*, 55(4):851–876, 1991. Russian, translation in Math. USSR-Izv. 39(1):829–853, 1992.

[Koe95] J. Koenigsmann. From p-rigid elements to valuations (with a Galois-characterization of p-adic fields). *J. Reine Angew. Math.*, 465:165–182, 1995. With an appendix by Florian Pop.

[Koe05] J. Koenigsmann. On the "section conjecture" in anabelian geometry. *J. Reine Angew. Math.*, 588:221–235, 2005.

[Maz78] B. Mazur. Modular curves and the Eisenstein ideal. *IHES Publ. Math.*, 47:33–186, 1978.

[MR10] B. Mazur and K. Rubin. Ranks of twists of elliptic curves and Hilbert's Tenth Problem. *Invent. Math.*, 181(3):541–575, 2010.

[Neu77] J. Neukirch. Über die absoluten Galoisgruppen algebraischer Zahlkörper. *Journées Arithmétiques de Caen 1976, Astérisque*, 41-42:67–79, 1977.

[Pop10] F. Pop. On the birational p-adic section conjecture. *Compositio Math.*, 146(3):621–637, 2010.

[PS11] F. Pop and J. Stix. Arithmetic in the fundamental group of a p-adic curve: on the p-adic section conjecture for curves. Preprint, arXiv:math.AG/1111.1354, 2011.

[S⁺08] W. A. Stein et al. Sage Mathematics Software (Version 3.1.4), The Sage Development Team, 2008, `www.sagemath.org`.

[SGA1] A. Grothendieck. *Revêtements étale et groupe fondamental (SGA 1)*. Séminaire de géométrie algébrique du Bois Marie 1960-61, directed by A. Grothendieck, augmented by two papers by Mme M. Raynaud, *Lecture Notes in Math.* 224, Springer-Verlag, Berlin-New York, 1971. Updated and annotated new edition: *Documents Mathématiques* 3, Société Mathématique de France, Paris, 2003.

[Ser68] J.-P. Serre. *Corps locaux*. Publications de l'Université de Nancago, No. VIII., Hermann, Paris, 1968. (deuxième édition).

[Ser94] J.-P. Serre. *Cohomologie Galoisienne*. Springer Verlag, 1994. (cinquième édition, révisée et complétée).

[Sko01] A. N. Skorobogatov. *Torsors and rational points*. Volume 144 of *Cambridge Tracts in Mathematics*. Cambridge University Press, Cambridge, 2001.

[Sti08] J. Stix. On cuspidal sections of algebraic fundamental groups. Preprint, arXiv:math.AG/0809.0017v1. Philadelphia–Bonn, 2008.

[Sti10] J. Stix. On the period-index problem in light of the section conjecture. *Amer. J. of Math.*, 132(1):157–180, 2010.

[Sto06] M. Stoll. Finite descent obstructions and rational points on curves. Preprint, arXiv:math.NT/0606465v2. Draft version no. 8.

[Sto07] M. Stoll. Finite descent obstructions and rational points on curves. *Algebra and Number Theory*, 1(4):349–391, 2007.

[Sza09] T. Szamuely. *Galois groups and fundamental groups*, volume 117 of *Cambridge Studies in Advanced Mathematics*. Cambridge University Press, Cambridge, 2009.

Chapter 8
On Monodromically Full Points of Configuration Spaces of Hyperbolic Curves

Yuichiro Hoshi*

Abstract We introduce and discuss the notion of monodromically full points of configuration spaces of hyperbolic curves. This notion leads to complements to M. Matsumoto's result concerning the difference between the kernels of the natural homomorphisms associated to a hyperbolic curve and its point from the Galois group to the automorphism and outer automorphism groups of the geometric fundamental group of the hyperbolic curve. More concretely, we prove that any hyperbolic curve over a number field has many nonexceptional closed points, i.e., closed points which do not satisfy a condition considered by Matsumoto, but that there exist infinitely many hyperbolic curves which admit many exceptional closed points, i.e., closed points which do satisfy the condition considered by Matsumoto. Moreover, we prove a Galois-theoretic characterization of equivalence classes of monodromically full points of configuration spaces, as well as a Galois-theoretic characterization of equivalence classes of quasi-monodromically full points of cores. In a similar vein, we also prove a necessary and sufficient condition for quasi-monodromically full Galois sections of hyperbolic curves to be geometric.

8.1 Introduction

In this paper, we discuss **monodromically full** points of configuration spaces of hyperbolic curves. The term monodromically full was introduced in [Hos11], but the corresponding notion was studied by M. Matsumoto and A. Tamagawa in [MT00].

Y. Hoshi (✉)
Research Institute for Mathematical Sciences, Kyoto University, Kyoto 606-8502, Japan
e-mail: yuichiro@kurims.kyoto-u.ac.jp

* This research was supported by Grant-in-Aid for Young Scientists (B) No. 22740012. Various portions of this paper were rewritten by the editorial board without explicit permission of the author. This rewrite might give rise to some errors.

J. Stix (ed.), *The Arithmetic of Fundamental Groups*, Contributions in Mathematical and Computational Sciences 2, DOI 10.1007/978-3-642-23905-2_8,

Let ℓ be a prime number, k a field of characteristic 0 with algebraic closure $\overline{k}$, and X a hyperbolic curve of type (g,r) over k. For an extension $k' \subseteq \overline{k}$ of k, write $\mathrm{G}_{k'} := \mathrm{Gal}(\overline{k}/k')$ for the absolute Galois group of k' determined by the given algebraic closure $\overline{k}$. For a positive integer n, we write X_n for the n-th configuration space of the hyperbolic curve X/k. The natural projection $\mathrm{X}_{n+1} \to \mathrm{X}_n$ to the first n factors may be regarded as a family of hyperbolic curves of type $(g, r+n)$. We shall say that a closed point $x \in \mathrm{X}_n$ of the n-th configuration space is **ℓ-monodromically full** if the $k(x)$-rational point, where $k(x)$ is the residue field at x, of $\mathrm{X}_n \otimes_k k(x)$ determined by x is an ℓ-monodromically full point with respect to the family of hyperbolic curves $\mathrm{X}_{n+1} \otimes_k k(x)$ over $\mathrm{X}_n \otimes_k k(x)$ in the sense of [Hos11, Definition 2.1], i.e., roughly speaking, the image of the pro-ℓ outer monodromy representation of $\pi_1(\mathrm{X}_n \otimes_k \overline{k})$ with respect to the family of hyperbolic curves $\mathrm{X}_{n+1} \to \mathrm{X}_n$ is contained in the image of the pro-ℓ outer Galois representation of $\mathrm{G}_{k(x)}$ with respect to the hyperbolic $k(x)$-curve $\mathrm{X}_{n+1} \times_{\mathrm{X}_n} \mathrm{Spec}(k(x))$, see Definition 8 and Remark 11(i).

We write

$$\Delta^{\{\ell\}}_{\mathrm{X}/k}$$

for the geometric pro-ℓ fundamental group of X, i.e., the maximal pro-ℓ quotient of the étale fundamental group $\pi_1(\mathrm{X} \otimes_k \overline{k})$ of $\mathrm{X} \otimes_k \overline{k}$, and

$$\Pi^{\{\ell\}}_{\mathrm{X}/k} := \pi_1(\mathrm{X}) / \ker(\pi_1(\mathrm{X} \otimes_k \overline{k}) \twoheadrightarrow \Delta^{\{\ell\}}_{\mathrm{X}/k})$$

for the geometrically pro-ℓ fundamental group of X. Then conjugation by elements of $\Pi^{\{\ell\}}_{\mathrm{X}/k}$ determines a commutative diagram of profinite groups

$$\begin{array}{ccccccccc}
1 & \longrightarrow & \Delta^{\{\ell\}}_{\mathrm{X}/k} & \longrightarrow & \Pi^{\{\ell\}}_{\mathrm{X}/k} & \longrightarrow & \mathrm{G}_k & \longrightarrow & 1 \\
 & & \downarrow & & \downarrow {\scriptstyle \overline{\rho}^{\{\ell\}}_{\mathrm{X}/k}} & & \downarrow {\scriptstyle \rho^{\{\ell\}}_{\mathrm{X}/k}} & & \\
1 & \longrightarrow & \mathrm{Inn}(\Delta^{\{\ell\}}_{\mathrm{X}/k}) & \longrightarrow & \mathrm{Aut}(\Delta^{\{\ell\}}_{\mathrm{X}/k}) & \longrightarrow & \mathrm{Out}(\Delta^{\{\ell\}}_{\mathrm{X}/k}) & \longrightarrow & 1
\end{array}$$

with exact rows. The left-hand vertical arrow is, in fact, an isomorphism. On the other hand, for a closed point $x \in \mathrm{X}$, we have a homomorphism

$$\pi_1(x) \colon \mathrm{G}_{k(x)} \to \Pi^{\{\ell\}}_{\mathrm{X}/k}$$

induced by $x \in \mathrm{X}$ and well-defined up to $\Pi^{\{\ell\}}_{\mathrm{X}/k}$-conjugation. In [Mat11], Matsumoto studied the difference between the kernels of the following two homomorphisms

$$\rho^{\{\ell\}}_{\mathrm{X}/k}|_{\mathrm{G}_{k(x)}} \colon \mathrm{G}_{k(x)} \longrightarrow \mathrm{Out}(\Delta^{\{\ell\}}_{\mathrm{X}/k}) \,,$$

$$\mathrm{G}_{k(x)} \stackrel{\pi_1(x)}{\longrightarrow} \Pi^{\{\ell\}}_{\mathrm{X}/k} \stackrel{\overline{\rho}^{\{\ell\}}_{\mathrm{X}/k}}{\longrightarrow} \mathrm{Aut}(\Delta^{\{\ell\}}_{\mathrm{X}/k}) \,.$$

The notion of monodromically full points allows to give some complements to Matsumoto's result [Mat11]. To state these complements, let us review the result given in [Mat11].

We shall say that $\mathrm{E}(\mathrm{X}, x, \ell)$ holds if the kernels of the above two homomorphisms coincide and write

$$\mathrm{X}^{\mathrm{E}_\ell} \subseteq \mathrm{X}^{\mathrm{cl}}$$

for the subset of the set X^{cl} of closed points of X consisting of **exceptional** $x \in X^{cl}$ i.e., $E(X, x, \ell)$ holds, cf. [Mat11, §1, §3], as well as Definition 32 in the present paper. Then the main result of [Mat11] may be stated as follows:

> *Let $g \geqslant 3$ be an integer. Suppose that ℓ divides $2g-2$ and write ℓ^ν for the highest power of ℓ that divides $2g-2$. Then there are infinitely many isomorphism classes of pairs (K, C) of number fields K and proper hyperbolic curves C of genus g over K which satisfy the following condition: For any closed point $x \in C$ of C with residue field $k(x)$, if ℓ^ν does not divide $[k(x) : k]$, then $E(C, x, \ell)$ does not hold.*

In the present paper, we prove that if a closed point $x \in X$ of the hyperbolic curve X is ℓ-monodromically full, then $E(X, x, \ell)$ does *not hold*, cf. Proposition 33 (ii). On the other hand, as a consequence of Hilbert's irreducibility theorem, any hyperbolic curve over a number field has many ℓ-monodromically full points, cf. Proposition 13, as well as, [MT00, Theorem 1.2] or [Hos11, Theorem 2.3]. By applying these observations, one can prove the following result, which may be regarded as a partial generalization of the above theorem due to Matsumoto, cf. Theorem 36.

Theorem A (Existence of many nonexceptional closed points). *Let ℓ be a prime number, k a finite extension of $\mathbb{Q}$, and X a hyperbolic curve over k. Then, regarding the set of closed points X^{cl} of X as a subset of $X(\mathbb{C})$, the complement*

$$X^{cl} \setminus X^{E_\ell} \subseteq X(\mathbb{C})$$

is dense with respect to the complex topology of $X(\mathbb{C})$. Moreover $X(k) \cap X^{E_\ell}$ is finite.

On the other hand, in [Mat11, §2], Matsumoto proved that for any prime number ℓ, the triple

$$(\mathbb{P}^1_{\mathbb{Q}} \setminus \{0, 1, \infty\}, \overrightarrow{01}, \ell),$$

where $\overrightarrow{01}$ is a $\mathbb{Q}$-rational tangential base point, is a triple for which a version of $E(X, x, \ell)$ for rational tangential base points holds. As mentioned in [Mat11, §2], the fact that $E(X, x, \ell)$ holds for this triple was observed by P. Deligne and Y. Ihara.

However, a *tangential base point is not a point.* In this sense, no example of a triple (X, x, ℓ) for which $E(X, x, \ell)$ holds appears in [Mat11]. The following result concerns the existence of triples (X, x, ℓ) for which $E(X, x, \ell)$ holds, cf. Theorem 41.

Theorem B (Existence of many exceptional closed points for certain hyperbolic curves). *Let ℓ be a prime number, k a field of characteristic 0, X a hyperbolic curve which is either of type $(0,3)$ or of type $(1,1)$, and $Y \to X$ a connected finite étale cover over k which arises from an open subgroup of the geometrically pro-ℓ fundamental group $\Pi^{\{\ell\}}_{X/k}$ of X and is geometrically connected over k. Then the subset $Y^{E_\ell} \subseteq Y^{cl}$ is infinite. In particular, the subset $X^{E_\ell} \subseteq X^{cl}$ is infinite.*

Note that in Remark 43, we also give an example of a triple (X, x, ℓ) such that X is a proper hyperbolic curve, and $E(X, x, \ell)$ holds.

A k-rational point $x \in X_n(k)$ of the n-th configuration space X_n of the hyperbolic curve X/k determines *n distinct* k-rational points $x_1, \ldots, x_n$ of X. Write

$$X[x] \subseteq X$$

for the hyperbolic curve of type $(g, r+n)$ over k obtained by $X \setminus \{x_1, \ldots, x_n\}$, i.e., $X[x]$ may be regarded as the fiber product

$$X_{n+1} \times_{X_n, x} \mathrm{Spec}(k) .$$

We shall say that two k-rational points x and y of X_n are **equivalent** if $X[x] \simeq X[y]$ over k. In [Hos11], the author proved that the isomorphism class of a certain (e.g., *split*, cf. [Hos11, Definition 1.5 (i)]) ℓ-monodromically full hyperbolic curve of genus 0 over a finitely generated extension of $\mathbb{Q}$ is completely determined by the kernel of the natural pro-ℓ outer Galois representation associated to the hyperbolic curve, cf. [Hos11, Theorem A]. By a similar argument to the argument used in the proof of [Hos11, Theorem A], one can prove the following Galois-theoretic characterization of equivalence classes of ℓ-monodromically full points of configuration spaces, cf. Theorem 49.

Theorem C (Galois-theoretic characterization of equivalence classes of monodromically full points of configuration spaces). *Let ℓ be a prime number, n a positive integer, k a finitely generated extension of $\mathbb{Q}$, and X a hyperbolic curve over k. Then for two k-rational points x and y of X_n which are ℓ-monodromically full, cf. Definition 8, the following three conditions are equivalent:*

(i) x *is equivalent to* y, *i.e.,* $X[x]$ *is isomorphic to* $X[y]$ *over* k.
(ii) $\mathrm{Ker}(\rho^{\{\ell\}}_{X[x]/k}) = \mathrm{Ker}(\rho^{\{\ell\}}_{X[y]/k})$.
(iii) *We have* $\mathrm{Ker}(\phi_x) = \mathrm{Ker}(\phi_y)$ *for the composites*

$$\phi_x : G_k \xrightarrow{\pi_1(x)} \pi_1(X_n) \xrightarrow{\widetilde{\rho}^{\{\ell\}}_{X_n/k}} \mathrm{Aut}(\Delta^{\{\ell\}}_{X_n/k}) ,$$

$$\phi_y : G_k \xrightarrow{\pi_1(y)} \pi_1(X_n) \xrightarrow{\widetilde{\rho}^{\{\ell\}}_{X_n/k}} \mathrm{Aut}(\Delta^{\{\ell\}}_{X_n/k}) .$$

In [Moc03], S. Mochizuki introduced and studied the notion of a k-**core**, cf. [Moc03, Definition 2.1], as well as [Moc03, Remark 2.1.1]. It follows from [Moc98, Theorem 5.3], together with [Moc03, Proposition 2.3], that if $2g-2+r>2$, then a *general hyperbolic curve of type (g,r) over k is a k-core*, cf. also [Moc03, Remark 2.5.1]. For a hyperbolic curve over k which is a *k-core*, the following stronger Galois-theoretic characterization can be proven, cf. Theorem 51.

Theorem D (Galois-theoretic characterization of equivalence classes of quasi-monodromically full points of cores). *Let ℓ be a prime number, k a finitely generated extension of $\mathbb{Q}$, and X a hyperbolic curve over k which is a k-core, cf. [Moc03, Remark 2.1.1]. Then for two k-rational points x and y of X which are quasi-ℓ-monodromically full, cf. Definition 8, the following four conditions are equivalent:*

(i) $x = y$.
(ii) x *is equivalent to* y.

(*iii*) *With* $\mathrm{U}_x = \mathrm{X} \setminus \{x\}$ *and* $\mathrm{U}_y = \mathrm{X} \setminus \{y\}$, *the intersection* $\mathrm{Ker}(\rho^{\{\ell\}}_{\mathrm{U}_x/k}) \cap \mathrm{Ker}(\rho^{\{\ell\}}_{\mathrm{U}_y/k})$ *is open in* $\mathrm{Ker}(\rho^{\{\ell\}}_{\mathrm{U}_x/k})$ *and* $\mathrm{Ker}(\rho^{\{\ell\}}_{\mathrm{U}_y/k})$.

(*iv*) *For the composites*

$$\phi_x : \mathrm{G}_k \stackrel{\pi_1(x)}{\longrightarrow} \pi_1(\mathrm{X}) \stackrel{\vec{\rho}^{\{\ell\}}_{\mathrm{X}/k}}{\longrightarrow} \mathrm{Aut}(\Delta^{\{\ell\}}_{\mathrm{X}/k}),$$

$$\phi_y : \mathrm{G}_k \stackrel{\pi_1(y)}{\longrightarrow} \pi_1(\mathrm{X}) \stackrel{\vec{\rho}^{\{\ell\}}_{\mathrm{X}/k}}{\longrightarrow} \mathrm{Aut}(\Delta^{\{\ell\}}_{\mathrm{X}/k}),$$

the intersection $\mathrm{Ker}(\phi_x) \cap \mathrm{Ker}(\phi_y)$ *is open in* $\mathrm{Ker}(\phi_x)$ *and* $\mathrm{Ker}(\phi_y)$.

Finally, in a similar vein, we prove a necessary and sufficient condition for a quasi-ℓ-monodromically full Galois section, cf. Definition 46, of a hyperbolic curve to be geometric, cf. Theorem 53.

Theorem E (A necessary and sufficient condition for a quasi-monodromically full Galois section of a hyperbolic curve to be geometric). *Let* ℓ *be a prime number,* k *a finitely generated extension of* $\mathbb{Q}$, *and* X *a hyperbolic curve over* k. *Let* $s\colon \mathrm{G}_k \to \Pi^{\{\ell\}}_{\mathrm{X}/k}$ *be a pro-*ℓ *Galois section of* X, *i.e., a continuous section of the natural surjection* $\Pi^{\{\ell\}}_{\mathrm{X}/k} \twoheadrightarrow \mathrm{G}_k$, *cf. [Hos10, Definition 1.1 (i)], which is quasi-*ℓ*-monodromically full, cf. Definition 46. Write* ϕ_s *for the composite*

$$\mathrm{G}_k \stackrel{s}{\longrightarrow} \Pi^{\{\ell\}}_{\mathrm{X}/k} \stackrel{\vec{\rho}^{\{\ell\}}_{\mathrm{X}/k}}{\longrightarrow} \mathrm{Aut}(\Delta^{\{\ell\}}_{\mathrm{X}/k}) .$$

Then the following four conditions are equivalent:

(*i*) *The pro-*ℓ *Galois section* s *is* geometric, *cf. [Hos10, Definition 1.1 (iii)].*

(*ii*) *The pro-*ℓ *Galois section* s *arises from a* k*-rational point of* X, *cf. [Hos10, Definition 1.1 (ii)].*

(*iii*) *There exists a quasi-*ℓ*-monodromically full* k*-rational point* $x \in \mathrm{X}(k)$, *cf. Definition 8, such that if we write* ϕ_x *for the composite*

$$\mathrm{G}_k \stackrel{\pi_1(x)}{\longrightarrow} \Pi^{\{\ell\}}_{\mathrm{X}/k} \stackrel{\vec{\rho}^{\{\ell\}}_{\mathrm{X}/k}}{\longrightarrow} \mathrm{Aut}(\Delta^{\{\ell\}}_{\mathrm{X}/k}) ,$$

then the intersection $\mathrm{Ker}(\phi_s) \cap \mathrm{Ker}(\phi_x)$ *is open in* $\mathrm{Ker}(\phi_s)$ *and* $\mathrm{Ker}(\phi_x)$.

(*iv*) *There exists a quasi-*ℓ*-monodromically full* k*-rational point* $x \in \mathrm{X}(k)$, *cf. Definition 8, such that with* $\mathrm{U} := \mathrm{X} \setminus \mathrm{Im}(x)$, *the intersection* $\mathrm{Ker}(\phi_s) \cap \mathrm{Ker}(\rho^{\{\ell\}}_{\mathrm{U}/k})$ *is open in* $\mathrm{Ker}(\phi_s)$ *and* $\mathrm{Ker}(\rho^{\{\ell\}}_{\mathrm{U}/k})$.

The present paper is organized as follows: In Sect. 8.3 we introduce and discuss the notion of monodromically full points of configuration spaces of hyperbolic curves. In Sect. 8.4 we consider the fundamental groups of configuration spaces of hyperbolic curves. In Sect. 8.5 we consider the kernels of the outer representations associated to configuration spaces of hyperbolic curves. In Sect. 8.6 we prove Theorems A and B. In Sect. 8.7 we prove Theorems C, D, and E.

Acknowledgements. Part of this research was carried out while the author visited at the University of Heidelberg during the second week of February 2010. The author would like to thank Jakob Stix and the MAThematics Center Heidelberg in the University of Heidelberg for inviting me and for their hospitality during my stay. The author would like to thank Makoto Matsumoto for inspiring me by means of his result given in [Mat11]. The author also would like to thank Shinichi Mochizuki and the referee for helpful comments.

8.2 Notations, Conventions, and Terminology

Numbers. The notation $\mathfrak{Primes}$ will be used to denote the set of all prime numbers. We shall refer to a finite extension of $\mathbb{Q}$ as a **number field**.

Profinite Groups. Let G be a profinite group and $H \subseteq G$ a closed subgroup.

We shall write $N_G(H)$ for the **normalizer** of H in G, $Z_G(H)$ for the **centralizer** of H in G, $Z(G) := Z_G(G)$ for the **center** of G,

$$Z_G^{loc}(H) := \varinjlim_{H' \subseteq H} Z_G(H') \subseteq G,$$

where $H' \subseteq H$ ranges over the open subgroups of H, for the **local centralizer** of H in G, and $Z^{loc}(G) := Z_G^{loc}(G)$ for the **local center** of G. It is immediate from the definitions involved that

$$H \subseteq N_G(H) \supseteq Z_G(H) \subseteq Z_G^{loc}(H),$$

and that if $H_1, H_2 \subseteq G$ are closed subgroups of G such that $H_1 \subseteq H_2$ (respectively, $H_1 \subseteq H_2$; $H_1 \cap H_2$ is open in H_1 and H_2), then $Z_G(H_2) \subseteq Z_G(H_1)$ (respectively, $Z_G^{loc}(H_2) \subseteq Z_G^{loc}(H_1)$; $Z_G^{loc}(H_1) = Z_G^{loc}(H_2)$).

We shall say that G is **center-free** if $Z(G) = \{1\}$. We shall say that G is **slim** if $Z^{loc}(G) = \{1\}$ or equivalently every open subgroup of G is center-free.

We shall denote by Aut(G) the group of continuous automorphisms of the profinite group G, by Inn(G) the group of inner automorphisms of G, and by Out(G) the quotient Aut(G)/Inn(G). If G is topologically finitely generated, then the topology of G admits a basis of characteristic open subgroups, which induces a profinite topology on Aut(G), hence also on Out(G).

Curves. Let S be a scheme and C a scheme over S. For a pair (g,r) of nonnegative integers, we shall say that $C \to S$ is a **smooth curve of type** (g,r) over S if there exist an S-scheme C^{cpt} which is smooth, proper, of relative dimension 1 with geometrically connected fibres of genus g, and a closed subscheme $D \subseteq C^{cpt}$ which is finite étale of degree r over S such that the complement of D in C^{cpt} is isomorphic to C over S.

We shall say that C is a **hyperbolic curve** over S if there exists a pair (g,r) of nonnegative integers with $2g-2+r>0$ such that C is a smooth curve of type (g,r) over S. A **tripod** is a smooth curve of type $(0,3)$.

For a pair (g,r) of nonnegative integers such that $2g-2+r>0$, write $\mathcal{M}_{g,r}$ for the moduli stack over Spec($\mathbb{Z}$) of smooth proper curves of genus g with r ordered marked points, cf. [DM69, Knu83], and $\mathcal{M}_{g,[r]}$ for the moduli stack over Spec($\mathbb{Z}$) of hyperbolic curves of type (g,r). We have a natural finite étale $\mathfrak{S}_r$-Galois cover

$$\mathcal{M}_{g,r} \to \mathcal{M}_{g,[r]} ,$$

where $\mathfrak{S}_r$ is the symmetric group on r letters.

8.3 Monodromically Full Points

We introduce and discuss the notion of monodromically full points of configuration spaces of hyperbolic curves. Let $\Sigma \subseteq \mathfrak{Primes}$ be a nonempty subset of $\mathfrak{Primes}$ and let S be a regular and connected scheme.

Definition 1. Let X be a regular and connected scheme over S.

(i) Let $1 \to \Delta \to \Pi \to \mathrm{G} \to 1$ be an exact sequence of profinite groups. Suppose that Δ is topologically finitely generated. Then conjugation by elements of Π determines a commutative diagram of profinite groups

$$\begin{array}{ccccccccc} 1 & \longrightarrow & \Delta & \longrightarrow & \Pi & \longrightarrow & \mathrm{G} & \longrightarrow & 1 \\ & & \downarrow & & \downarrow & & \downarrow & & \\ 1 & \longrightarrow & \mathrm{Inn}(\Delta) & \longrightarrow & \mathrm{Aut}(\Delta) & \longrightarrow & \mathrm{Out}(\Delta) & \longrightarrow & 1 \end{array}$$

with exact rows. We shall refer to the continuous homomorphism

$$\Pi \longrightarrow \mathrm{Aut}(\Delta) \quad \text{(respectively,} \quad \mathrm{G} \longrightarrow \mathrm{Out}(\Delta))$$

obtained as the middle (respectively, right-hand) vertical arrow in the above diagram as the (respectively, **outer**) **representation associated to** $1 \to \Delta \to \Pi \to \mathrm{G} \to 1$.

(ii) We shall write

$$\Delta^{\Sigma}_{\mathrm{X/S}}$$

for the maximal pro-Σ quotient of $\ker(\pi_1(\mathrm{X}) \to \pi_1(\mathrm{S}))$ and set

$$\Pi^{\Sigma}_{\mathrm{X/S}} := \pi_1(\mathrm{X})/\Big(\ker\big(\ker(\pi_1(\mathrm{X}) \to \pi_1(\mathrm{S})) \twoheadrightarrow \Delta^{\Sigma}_{\mathrm{X/S}}\big)\Big) .$$

Thus, we have a commutative diagram of profinite groups

$$\begin{array}{ccccccc} 1 & \longrightarrow & \mathrm{Ker}(\pi_1(\mathrm{X}) \to \pi_1(\mathrm{S})) & \longrightarrow & \pi_1(\mathrm{X}) & \longrightarrow & \pi_1(\mathrm{S}) \\ & & \downarrow & & \downarrow & & \| \\ 1 & \longrightarrow & \Delta^{\Sigma}_{\mathrm{X/S}} & \longrightarrow & \Pi^{\Sigma}_{\mathrm{X/S}} & \longrightarrow & \pi_1(\mathrm{S}) \end{array}$$

with exact rows and surjective vertical arrows. If S is the spectrum of a ring R, then we shall write $\Delta^{\Sigma}_{\mathrm{X/R}} := \Delta^{\Sigma}_{\mathrm{X/S}}$ and $\Pi^{\Sigma}_{\mathrm{X/R}} := \Pi^{\Sigma}_{\mathrm{X/S}}$.

(iii) Suppose that the natural homomorphism $\pi_1(X) \to \pi_1(S)$ is surjective, or, equivalently, $\Pi^{\Sigma}_{X/S} \to \pi_1(S)$ is surjective. Then we have an exact sequence of profinite groups

$$1 \longrightarrow \Delta^{\Sigma}_{X/S} \longrightarrow \Pi^{\Sigma}_{X/S} \longrightarrow \pi_1(S) \longrightarrow 1\,.$$

We shall write

$$\widetilde{\rho}^{\Sigma}_{X/S} : \Pi^{\Sigma}_{X/S} \longrightarrow \mathrm{Aut}(\Delta^{\Sigma}_{X/S})$$

for the representation associated to the above exact sequence as in (i), and refer to $\widetilde{\rho}^{\Sigma}_{X/S}$ as the **pro-Σ representation associated to** X/S. Moreover, we shall write

$$\rho^{\Sigma}_{X/S} : \pi_1(S) \longrightarrow \mathrm{Out}(\Delta^{\Sigma}_{X/S})$$

for the outer representation associated to the above exact sequence as in (i), and refer to $\rho^{\Sigma}_{X/S}$ as the **pro-Σ outer representation associated to** X/S. If S is the spectrum of a ring R, then we shall write $\widetilde{\rho}^{\Sigma}_{X/R} := \widetilde{\rho}^{\Sigma}_{X/S}$ and $\rho^{\Sigma}_{X/R} := \rho^{\Sigma}_{X/S}$. Moreover, if ℓ is a prime number, then for simplicity, we write **pro-ℓ** instead of **pro-$\{\ell\}$**.

(iv) Suppose that the natural homomorphism $\pi_1(X) \to \pi_1(S)$ is surjective, or, equivalently, $\Pi^{\Sigma}_{X/S} \to \pi_1(S)$ is surjective, and that the profinite group $\Delta^{\Sigma}_{X/S}$ is topologically finitely generated. Then we shall write

$$\Pi^{\Sigma}_{X/S} \twoheadrightarrow \Phi^{\Sigma}_{X/S} := \mathrm{Im}(\widetilde{\rho}^{\Sigma}_{X/S})$$

for the quotient of $\Pi^{\Sigma}_{X/S}$ determined by the pro-Σ representation $\widetilde{\rho}^{\Sigma}_{X/S}$ associated to X/S. Moreover, we shall write

$$\pi_1(S) \twoheadrightarrow \Gamma^{\Sigma}_{X/S} := \mathrm{Im}(\rho^{\Sigma}_{X/S})$$

for the quotient of $\pi_1(S)$ determined by the pro-Σ outer representation $\rho^{\Sigma}_{X/S}$ associated to X/S. If S is the spectrum of a ring R, then we shall write $\Phi^{\Sigma}_{X/R} := \Phi^{\Sigma}_{X/S}$ and $\Gamma^{\Sigma}_{X/R} := \Gamma^{\Sigma}_{X/S}$.

(v) Let $\pi_1(X) \twoheadrightarrow Q$ be a quotient of $\pi_1(X)$. Then we shall say that a finite étale cover $Y \to X$ is a **finite étale** Q**-cover** if Y is connected, and $Y \to X$ arises from an open subgroup of Q, i.e., the open subgroup of $\pi_1(X)$ corresponding to the connected finite étale covering $Y \to X$ contains the kernel of the surjection $\pi_1(X) \twoheadrightarrow Q$.

Remark 2. In the notation of Definition 1, if S is the spectrum of a field k, then it follows from [SGA1, Exposé V, Proposition 6.9] that $\pi_1(X) \to \pi_1(S)$ is surjective if and only if X is geometrically connected over k. Suppose, moreover, that X is geometrically connected and of finite type over S. Then it follows from [SGA1, Exposé IX, Théorème 6.1] that the natural sequence of profinite groups

$$1 \longrightarrow \pi_1(X \otimes_k k^{\mathrm{sep}}) \longrightarrow \pi_1(X) \longrightarrow \pi_1(S) \longrightarrow 1$$

is exact, where k^{sep} is a separable closure of k. Thus, in this case, it follows that $\Delta^{\Sigma}_{X/k}$ is naturally isomorphic to the maximal pro-Σ quotient of the étale fundamental group $\pi_1(X \otimes_k k^{\mathrm{sep}})$ of $X \otimes_k k^{\mathrm{sep}}$. In particular, if k is of characteristic 0, then $\Delta^{\Sigma}_{X/k}$ is topologically finitely generated by [SGA7-I, Exposé II, Théorème 2.3.1].

Remark 3. In the notation of Definition 1, suppose that X is a hyperbolic curve over S. Since S is regular, it follows from [SGA1, Exposé X, Théorème 3.1] that the natural homomorphism $\pi_1(\eta_S) \to \pi_1(S)$, where we write η_S for the generic point of S, is surjective. In light of the surjectivity of $\pi_1(X \times_S \eta_S) \to \pi_1(\eta_S)$, see Remark 2, we conclude that the natural homomorphism $\pi_1(X) \to \pi_1(S)$ is surjective. In particular, we have an exact sequence of profinite groups

$$1 \longrightarrow \Delta^{\Sigma}_{X/S} \longrightarrow \Pi^{\Sigma}_{X/S} \longrightarrow \pi_1(S) \longrightarrow 1 .$$

If, moreover, every element of Σ is invertible on S, then it follows as in the proof of [Hos09, Lemma 1.1] that $\Delta^{\Sigma}_{X/S}$ is naturally isomorphic to the maximal pro-Σ quotient of the étale fundamental group $\pi_1(X \times_S \overline{s})$, where $\overline{s} \to S$ is a geometric point of S. In particular, it follows immediately from the well-known structure of the maximal pro-Σ quotient of the fundamental group of a smooth curve over an algebraically closed field of characteristic $\notin \Sigma$ that $\Delta^{\Sigma}_{X/S}$ is topologically finitely generated and slim. Thus, we have continuous homomorphisms

$$\widetilde{\rho}^{\Sigma}_{X/S} \colon \Pi^{\Sigma}_{X/S} \longrightarrow \mathrm{Aut}(\Delta^{\Sigma}_{X/S}) ,$$
$$\rho^{\Sigma}_{X/S} \colon \pi_1(S) \longrightarrow \mathrm{Out}(\Delta^{\Sigma}_{X/S}) .$$

Moreover, there exists a natural bijection between the set of the cusps of X/S and the set of the conjugacy classes of the cuspidal inertia subgroups of $\Delta^{\Sigma}_{X/S}$.

Lemma 4 (Outer representations arising from certain extensions). *Let*

$$1 \longrightarrow \Delta \longrightarrow \Pi \longrightarrow G \longrightarrow 1 \qquad (8.1)$$

be an exact sequence of profinite groups. Suppose that Δ is topologically finitely generated and center-free. Write $\widetilde{\rho} \colon \Pi \longrightarrow \mathrm{Aut}(\Delta)$ and $\rho \colon G \longrightarrow \mathrm{Out}(\Delta)$ for the continuous homomorphisms arising (8.1)*, see Definition 1 (i). Then we have:*

(i) The natural surjection $\Pi \twoheadrightarrow G$ induces an isomorphism

$$\mathrm{Ker}(\widetilde{\rho}) = Z_{\Pi}(\Delta) \xrightarrow{\sim} \mathrm{Ker}(\rho) .$$

In particular, $\Delta \cap \mathrm{Ker}(\widetilde{\rho}) = \{1\}$.

(ii) The normal closed subgroup $\mathrm{Ker}(\widetilde{\rho}) \subseteq \Pi$ is the maximal normal closed subgroup N of Π such that $N \cap \Delta = \{1\}$.

(iii) Write $\mathrm{Aut}(\Delta \subseteq \Pi) \subseteq \mathrm{Aut}(\Pi)$ for the subgroup of $\mathrm{Aut}(\Pi)$ consisting of automorphisms which preserve the closed subgroup $\Delta \subseteq \Pi$. Suppose that $Z_{\Pi}(\Delta) = \{1\}$. Then the natural homomorphism $\mathrm{Aut}(\Delta \subseteq \Pi) \to \mathrm{Aut}(\Delta)$ is injective, and its image coincides with $N_{\mathrm{Aut}(\Delta)}(\mathrm{Im}(\widetilde{\rho})) \subseteq \mathrm{Aut}(\Delta)$, i.e.,

$$\mathrm{Aut}(\Delta \subseteq \Pi) \xrightarrow{\sim} N_{\mathrm{Aut}(\Delta)}(\mathrm{Im}(\widetilde{\rho})) \subseteq \mathrm{Aut}(\Delta) .$$

Proof. Assertion (i) follows immediately from the definitions involved. Next, we verify assertion (ii). Let $N \subseteq \Pi$ be a normal closed subgroup such that $\Delta \cap N = \{1\}$.

Then since Δ and N are both normal in Π, for any $x \in \Delta$, $y \in \mathrm{N}$, it holds that $xyx^{-1}y^{-1} \in \Delta \cap \mathrm{N} = \{1\}$. In particular, we obtain $\mathrm{N} \subseteq \mathrm{Z}_{\Pi}(\Delta) = \mathrm{Ker}(\widetilde{\rho})$ by (i). This completes the proof of assertion (ii).

Finally, we verify assertion (iii). It follows from the definitions that the natural homomorphism $\mathrm{Aut}(\Delta \subseteq \Pi) \to \mathrm{Aut}(\Delta)$ factors through $\mathrm{N}_{\mathrm{Aut}(\Delta)}(\mathrm{Im}(\widetilde{\rho})) \subseteq \mathrm{Aut}(\Delta)$. On the other hand, since the natural surjection $\Pi \twoheadrightarrow \mathrm{Im}(\widetilde{\rho})$ is an isomorphism, cf. assertion (i), conjugation by elements of $\mathrm{N}_{\mathrm{Aut}(\Delta)}(\mathrm{Im}(\widetilde{\rho}))$ determines a homomorphism

$$\mathrm{N}_{\mathrm{Aut}(\Delta)}(\mathrm{Im}(\widetilde{\rho})) \to \mathrm{Aut}(\mathrm{Im}(\widetilde{\rho})) \xleftarrow{\sim} \mathrm{Aut}(\Pi)\,,$$

which factors through $\mathrm{Aut}(\Delta \subseteq \Pi) \subseteq \mathrm{Aut}(\Pi)$. This homomorphism is the inverse of the homomorphism in question. This completes the proof of assertion (iii). □

Lemma 5 (Certain automorphisms of slim profinite groups). *Let* G *be a topologically finitely generated and slim profinite group and* $\alpha \in \mathrm{Aut}(\mathrm{G})$*. If* α *induces the identity automorphism on an open subgroup, then* α *is the identity.*

Proof. Let $\mathrm{H} \subseteq \mathrm{G}$ be an open subgroup of G such that α induces the identity automorphism of H. By replacing H by the intersection of all G-conjugates of H, we may assume that H is normal in G. Then since $\mathrm{Z}_{\mathrm{G}}(\mathrm{H}) = \{1\}$, it follows from Lemma 4 (iii), that α is the identity automorphism of G. □

Proposition 6 (Fundamental exact sequences associated to certain schemes). *Let* X *be a regular and connected scheme over* S *such that* $\pi_1(\mathrm{X}) \to \pi_1(\mathrm{S})$ *is surjective, and such that the profinite group* $\Delta^{\Sigma}_{\mathrm{X/S}}$ *is topologically finitely generated and center-free. Then the following hold:*

(i) We have a commutative diagram of profinite groups

$$\begin{array}{ccccccccc} 1 & \longrightarrow & \Delta^{\Sigma}_{\mathrm{X/S}} & \longrightarrow & \Pi^{\Sigma}_{\mathrm{X/S}} & \longrightarrow & \pi_1(\mathrm{S}) & \longrightarrow & 1 \\ & & \| & & \downarrow{\scriptstyle \widetilde{\rho}^{\Sigma}_{\mathrm{X/S}}} & & \downarrow{\scriptstyle \rho^{\Sigma}_{\mathrm{X/S}}} & & \\ 1 & \longrightarrow & \Delta^{\Sigma}_{\mathrm{X/S}} & \longrightarrow & \Phi^{\Sigma}_{\mathrm{X/S}} & \longrightarrow & \Gamma^{\Sigma}_{\mathrm{X/S}} & \longrightarrow & 1 \end{array} \qquad (8.2)$$

with exact rows and surjective vertical arrows.

(ii) The quotient $\Pi^{\Sigma}_{\mathrm{X/S}} \twoheadrightarrow \Phi^{\Sigma}_{\mathrm{X/S}}$ *determined by* $\widetilde{\rho}^{\Sigma}_{\mathrm{X/S}}$ *is the minimal quotient* $\Pi^{\Sigma}_{\mathrm{X/S}} \twoheadrightarrow \mathrm{Q}$ *of* $\Pi^{\Sigma}_{\mathrm{X/S}}$ *such that* $\mathrm{Ker}(\Pi^{\Sigma}_{\mathrm{X/S}} \twoheadrightarrow \mathrm{Q}) \cap \Delta^{\Sigma}_{\mathrm{X/S}} = \{1\}$.

Proof. Assertions (i) and (ii) follow from Lemma 4 (i) and (ii) respectively. □

Definition 7. Let n be a nonnegative integer, (g,r) a pair of nonnegative integers such that $2g-2+r>0$, S a regular and connected scheme, and X a hyperbolic curve of type (g,r) over S.

(i) We shall write X_n for the ***n*-th configuration space of** X/S, i.e., the open subscheme of the fiber product of n copies of X over S which represents the functor from the category of schemes over S to the category of sets given by

$$\mathrm{T} \rightsquigarrow \{(x_1,\cdots,x_n) \in \mathrm{X}(\mathrm{T})^{\times n} \mid x_i \neq x_j \text{ if } i \neq j\},$$

in particular, we have $\mathrm{X}_0 = \mathrm{S}$. For a nonnegative integer $m \leqslant n$, we always regard X_n as a scheme over X_m by the natural projection $\mathrm{X}_n \to \mathrm{X}_m$ to the first m factors. Then X_{n+1} is a hyperbolic curve of type $(g, r+n)$ over X_n. If every element of Σ is invertible on S, then we have continuous homomorphisms

$$\widetilde{\rho}^{\Sigma}_{\mathrm{X}_{n+1}/\mathrm{X}_n} : \Pi^{\Sigma}_{\mathrm{X}_{n+1}/\mathrm{X}_n} \longrightarrow \mathrm{Aut}(\Delta^{\Sigma}_{\mathrm{X}_{n+1}/\mathrm{X}_n}),$$

$$\rho^{\Sigma}_{\mathrm{X}_{n+1}/\mathrm{X}_n} : \pi_1(\mathrm{X}_n) \longrightarrow \mathrm{Out}(\Delta^{\Sigma}_{\mathrm{X}_{n+1}/\mathrm{X}_n}),$$

cf. Remark 3. Moreover, it follows immediately that X_n is naturally isomorphic to the $(n-m)$-th configuration space of the hyperbolic curve $\mathrm{X}_{m+1}/\mathrm{X}_m$.

(ii) Let $m \leqslant n$ be a nonnegative integer, T a regular and connected scheme over S, and $x \in \mathrm{X}_m(\mathrm{T})$ a T-valued point of X_m. Then we shall write

$$\mathrm{X}[x] \subseteq \mathrm{X} \times_{\mathrm{S}} \mathrm{T}$$

for the open subscheme of $\mathrm{X} \times_{\mathrm{S}} \mathrm{T}$ obtained as the complement of the images of the m *distinct* T-valued points of $\mathrm{X} \times_{\mathrm{S}} \mathrm{T}$ determined by the T-valued point x. Then $\mathrm{X}[x]$ is naturally a hyperbolic curve of type $(g, r+m)$ over T, and the base change of $\mathrm{X}_n \to \mathrm{X}_m$ via x is naturally isomorphic to the $(n-m)$-th configuration space $\mathrm{X}[x]_{n-m}$ of the hyperbolic curve $\mathrm{X}[x]/\mathrm{T}$, i.e., the following diagram

$$\begin{array}{ccc} \mathrm{X}[x]_{n-m} & \longrightarrow & \mathrm{X}_n \\ \downarrow & & \downarrow \\ \mathrm{T} & \xrightarrow[x]{} & \mathrm{X}_m \end{array}$$

is a cartesian diagram of schemes.

(iii) Let T be a regular and connected scheme over S and $x, y \in \mathrm{X}_n(\mathrm{T})$ two T-valued points of X_n. Then we shall say that x is **equivalent** to y if there exists an isomorphism $\mathrm{X}[x] \xrightarrow{\sim} \mathrm{X}[y]$ over T.

Definition 8. Let n be a nonnegative integer, S a regular and connected scheme, T a regular and connected scheme over S, and X a hyperbolic curve over S, and let $x \in \mathrm{X}_n(\mathrm{T})$ be a T-valued point of the n-th configuration space. For any prime ℓ invertible on S, we write $\Gamma_{\mathrm{T},x,\ell} \subseteq \Gamma^{\{\ell\}}_{\mathrm{X}_{n+1}/\mathrm{X}_n}$ for the image of the composite

$$\pi_1(\mathrm{T}) \stackrel{\pi_1(x)}{\to} \pi_1(\mathrm{X}_n) \stackrel{\rho^{\{\ell\}}_{\mathrm{X}_{n+1}/\mathrm{X}_n}}{\twoheadrightarrow} \Gamma^{\{\ell\}}_{\mathrm{X}_{n+1}/\mathrm{X}_n}$$

and $\Gamma_{\mathrm{geom},\ell} \subseteq \Gamma^{\{\ell\}}_{\mathrm{X}_{n+1}/\mathrm{X}_n}$ for the image of the composite

$$\mathrm{Ker}\big(\pi_1(\mathrm{X}_n) \to \pi_1(\mathrm{S})\big) \hookrightarrow \pi_1(\mathrm{X}_n) \stackrel{\rho^{\{\ell\}}_{\mathrm{X}_{n+1}/\mathrm{X}_n}}{\twoheadrightarrow} \Gamma^{\{\ell\}}_{\mathrm{X}_{n+1}/\mathrm{X}_n},$$

cf. Definitions 1 (iii), (iv), 7 (i).

(i) We shall say that a T-valued point $x \in \mathrm{X}_n(\mathrm{T})$ is **ℓ-monodromically full** if $\Gamma_{\mathrm{T},x,\ell}$ contains $\Gamma_{\mathrm{geom},\ell}$. We shall say that x is **quasi-ℓ-monodromically full** if the intersection $\Gamma_{\mathrm{T},x,\ell} \cap \Gamma_{\mathrm{geom},\ell}$ is an open subgroup of $\Gamma_{\mathrm{geom},\ell}$.

(ii) Suppose that every element of Σ is invertible on S. Then we shall say that a T-valued point $x \in \mathrm{X}_n(\mathrm{T})$ is **Σ-monodromically full** (respectively, **quasi-Σ-monodromically full**) if it is ℓ-monodromically full (respectively, quasi-ℓ-monodromically full) for every $\ell \in \Sigma$ in the sense of (i).

(iii) Moreover, we shall say that a point $x \in \mathrm{X}_n$ of X_n is **Σ-monodromically full** (respectively, **quasi-Σ-monodromically full**) if, for any $\ell \in \Sigma$, the corresponding $k(x)$-valued point of X_n, where we write $k(x)$ for the residue field at x, is Σ-monodromically full (respectively, quasi-Σ-monodromically full).

Remark 9. (i) Note that in Definition 8, since the closed subgroup $\Gamma_{\mathrm{geom},\ell} \subseteq \Gamma^{\{\ell\}}_{\mathrm{X}_{n+1}/\mathrm{X}_n}$ is normal, whether or not $\Gamma_{\mathrm{T},x,\ell}$ contains $\Gamma_{\mathrm{geom},\ell}$ (respectively, $\Gamma_{\mathrm{T},x,\ell} \cap \Gamma_{\mathrm{geom},\ell}$ is an open subgroup of $\Gamma_{\mathrm{geom},\ell}$) does *not* depend on the choice of the homomorphism $\pi_1(x) : \pi_1(\mathrm{T}) \to \pi_1(\mathrm{X}_n)$ induced by $x \in \mathrm{X}_n(\mathrm{T})$ among the various $\pi_1(\mathrm{X}_n)$-conjugates.

(ii) As the terminologies suggest, it follows immediately from the definitions that Σ-monodromic fullness implies quasi-Σ-monodromic fullness.

Remark 10. The notion of (quasi-)monodromic fullness defined in Definition 8, as well as the notion of (quasi-)monodromic fullness defined in [Hos11, Definitions 2.1 and 2.2], is motivated by the study by Matsumoto and Tamagawa of the difference between the profinite and pro-ℓ outer Galois representaions associated to a hyperbolic curve, cf. [MT00]. A consequence obtained from the main result of [MT00] is the following: Suppose that S is the spectrum of a finite extension $k/\mathbb{Q}$. Let $\overline{k}$ be an algebraic closure of k. Then for *any* closed point x of X_n, the image of the *profinite* outer Galois representation associated to the hyperbolic curve $\mathrm{X}[x]$ has trivial intersection with the image of the profinite outer monodromy representation associated to $\mathrm{X}_{n+1} \otimes_k \overline{k}/\mathrm{X}_n \otimes_k \overline{k}$. On the other hand, for any prime number ℓ, there exist *many* ℓ-monodromically full closed points of X_n, i.e., a closed point x such that the image of the *pro-ℓ* outer Galois representation associated to the hyperbolic curve $\mathrm{X}[x]$ contains the image of the *pro-ℓ* outer monodromy representation associated to $\mathrm{X}_{n+1} \otimes_k \overline{k}/\mathrm{X}_n \otimes_k \overline{k}$, cf. Remark 14 below.

Remark 11. (i) In the notation of Definition 8, if S is the spectrum of a field k of characteristic 0, then it follows that for a closed point $x \in \mathrm{X}_n$ of X_n with residue field $k(x)$, the following two conditions are equivalent:

- The closed point $x \in \mathrm{X}_n$ is a Σ-monodromically full (respectively, quasi-Σ-monodromically full) point in the sense of Definition 8.
- The $k(x)$-rational point of $\mathrm{X}_n \otimes_k k(x)$ determined by x is a Σ-monodromically full (respectively, quasi-Σ-monodromically full) point with respect to the hyperbolic curves $\mathrm{X}_{n+1} \otimes_k k(x)/\mathrm{X}_n \otimes_k k(x)$ in the sense of [Hos11, Definition 2.1].

(ii) If $\mathrm{X} = \mathbb{P}^1_k \setminus \{0, 1, \infty\}$, then $\mathrm{X}_n = \mathcal{M}_{0,n+3} \otimes_{\mathbb{Z}} k$, the moduli stack of smooth proper curves of genus 0 with $n+3$ ordered marked points, and for a closed point $x \in \mathrm{X}_n$ with residue field $k(x)$, the following two conditions are equivalent:

- The closed point $x \in \mathrm{X}_n$ is a Σ-monodromically full (respectively, quasi-Σ-monodromically full) point in the sense of Definition 8.
- The hyperbolic curve $\mathrm{X}[x]$ over $k(x)$, cf. Definition 7 (ii), is a Σ-monodromically full (respectively, quasi-Σ-monodromically full) hyperbolic curve over $k(x)$ in the sense of [Hos11, Definition 2.2].

Remark 12. In the notation of Definition 8, suppose that $\mathrm{S} = \mathrm{T}$. Then it follows from the various definitions involved that the following two conditions are equivalent:

(i) The S-valued point $x \in \mathrm{X}_n(\mathrm{S})$ is a Σ-monodromically full (respectively, quasi-Σ-monodromically full) point.
(ii) For any $\ell \in \Sigma$, the following composite is surjective (resp. has open image)

$$\pi_1(\mathrm{S}) \overset{\pi_1(x)}{\to} \pi_1(\mathrm{X}_n) \overset{\rho^{\{\ell\}}_{\mathrm{X}_{n+1}/\mathrm{X}_n}}{\twoheadrightarrow} \Gamma^{\{\ell\}}_{\mathrm{X}_{n+1}/\mathrm{X}_n} .$$

Proposition 13 (Existence of many monodromically full points). *Let Σ be a nonempty finite set of prime numbers, k a finitely generated extension of $\mathbb{Q}$, and X a hyperbolic curve over k. Let n be a positive integer, and X_n the n-th configuration space of X/k. Let $\mathrm{X}_n^{\mathrm{cl}}$ be the set of closed points of X_n, and $\mathrm{X}_n^{\Sigma\text{-MF}} \subseteq \mathrm{X}_n^{\mathrm{cl}}$ the subset of $\mathrm{X}_n^{\mathrm{cl}}$ consisting of closed points of X_n which are Σ-monodromically full. If we regard $\mathrm{X}_n^{\mathrm{cl}}$ as a subset of $\mathrm{X}_n(\mathbb{C})$, then the subset*

$$\mathrm{X}_n^{\Sigma\text{-MF}} \subseteq \mathrm{X}_n(\mathbb{C})$$

is dense with respect to the complex topology of $\mathrm{X}_n(\mathbb{C})$. Moreover, if X has genus 0, then the complement in $\mathrm{X}_n(k)$ of $\mathrm{X}_n(k) \cap \mathrm{X}_n^{\Sigma\text{-MF}}$ forms a thin set in $\mathrm{X}_n(k)$ in the sense of Serre.

Proof. This follows from [Hos11, Theorem 2.3], together with Remark 11(i). □

Remark 14. Let $n > 0$ be an integer, Σ a nonemtpy subset of $\mathfrak{Primes}$, k a finite extension of $\mathbb{Q}$, and X a hyperbolic curve over k. For a closed point $x \in \mathrm{X}_n$ with residue field $k(x)$, as in Definition 8, write $\Gamma_{k,x,\Sigma} \subseteq \Gamma^{\Sigma}_{\mathrm{X}_{n+1}/\mathrm{X}_n}$ for the image of the composite

$$\pi_1(\mathrm{Spec}(k(x))) \overset{\pi_1(x)}{\to} \pi_1(\mathrm{X}_n) \overset{\rho^{\Sigma}_{\mathrm{X}_{n+1}/\mathrm{X}_n}}{\twoheadrightarrow} \Gamma^{\Sigma}_{\mathrm{X}_{n+1}/\mathrm{X}_n}$$

and write $\Gamma_{\mathrm{geom},\Sigma} \subseteq \Gamma^{\Sigma}_{\mathrm{X}_{n+1}/\mathrm{X}_n}$ for the image of the composite

$$\mathrm{Ker}\Big(\pi_1(\mathrm{X}_n) \to \pi_1(\mathrm{Spec}(k))\Big) \hookrightarrow \pi_1(\mathrm{X}_n) \overset{\rho^{\Sigma}_{\mathrm{X}_{n+1}/\mathrm{X}_n}}{\twoheadrightarrow} \Gamma^{\Sigma}_{\mathrm{X}_{n+1}/\mathrm{X}_n} .$$

Now, Proposition 13 asserts that *many* closed points of X_n satisfy $\Gamma_{\mathrm{geom},\{\ell\}} \subset \Gamma_{k,x,\{\ell\}}$. On the other hand, it follows immediately from [MT00, Theorem 1.1] and [HM11,

Corollary 6.4] that for *any* closed point $x \in X_n$, we have $\Gamma_{k,x,\mathfrak{Primes}} \cap \Gamma_{\mathrm{geom},\mathfrak{Primes}} = \{1\}$. In particular, since $\Gamma_{\mathrm{geom},\mathfrak{Primes}} \neq \{1\}$, the inclusion $\Gamma_{\mathrm{geom},\mathfrak{Primes}} \subset \Gamma_{k,x,\mathfrak{Primes}}$ never holds.

8.4 Fundamental Groups of Configuration Spaces

We consider the fundamental groups of configuration spaces of hyperbolic curves. Let Σ be a nonempty subset of $\mathfrak{Primes}$, S a regular and connected scheme, and X a hyperbolic curve over S. Suppose that every element of Σ is invertible on S.

Lemma 15 (Fundamental groups of configuration spaces). *Let $0 \leqslant m < n$ be integers. Suppose that $\Sigma = \mathfrak{Primes}$ or that $\Sigma = \{\ell\}$ for one prime ℓ. Then we have:*

(i) *The natural homomorphism $\pi_1(X_n) \to \pi_1(X_m)$ is surjective. Thus, we have an exact sequence of profinite groups*

$$1 \longrightarrow \Delta^{\Sigma}_{X_n/X_m} \longrightarrow \Pi^{\Sigma}_{X_n/X_m} \longrightarrow \pi_1(X_m) \longrightarrow 1\,.$$

(ii) *For a geometric point $\overline{x} \to X_m$, the group $\Delta^{\Sigma}_{X_n/X_m}$ is naturally isomorphic to the maximal pro-Σ quotient of the étale fundamental group $\pi_1(X_n \times_{X_m} \overline{x})$.*

(iii) *Let T be a regular and connected scheme over S and $x \in X_m(T)$ a T-valued point of X_m. Then the homomorphism*

$$\Delta^{\Sigma}_{X[x]_{n-m}/T} \longrightarrow \Delta^{\Sigma}_{X_n/X_m}$$

determined by the cartesian square of schemes of Definition 7(ii)

$$\begin{array}{ccc} X[x]_{n-m} & \longrightarrow & X_n \\ \downarrow & & \downarrow \\ T & \xrightarrow{\;x\;} & X_m \end{array}$$

is an isomorphism. In particular, the right-hand square of the diagram

$$\begin{array}{ccccccccc} 1 & \longrightarrow & \Delta^{\Sigma}_{X[x]_{n-m}/T} & \longrightarrow & \Pi^{\Sigma}_{X[x]_{n-m}/T} & \longrightarrow & \pi_1(T) & \longrightarrow & 1 \\ & & \wr\downarrow & & \downarrow & & \downarrow{\scriptstyle \pi_1(x)} & & \\ 1 & \longrightarrow & \Delta^{\Sigma}_{X_n/X_m} & \longrightarrow & \Pi^{\Sigma}_{X_n/X_m} & \longrightarrow & \pi_1(X_m) & \longrightarrow & 1, \end{array}$$

with exact rows, is cartesian.

(iv) *We have a natural exact sequence of profinite groups*

$$1 \longrightarrow \Delta^{\Sigma}_{X_n/X_m} \longrightarrow \Delta^{\Sigma}_{X_n/S} \longrightarrow \Delta^{\Sigma}_{X_m/S} \longrightarrow 1\,.$$

(v) *The profinite group $\Delta^{\Sigma}_{X_n/X_m}$ is topologically finitely generated and slim. Thus, we have representations*

$$\widetilde{\rho}^{\Sigma}_{X_n/X_m} : \Pi^{\Sigma}_{X_n/X_m} \longrightarrow \mathrm{Aut}(\Delta^{\Sigma}_{X_n/X_m}),$$
$$\rho^{\Sigma}_{X_n/X_m} : \pi_1(X_m) \longrightarrow \mathrm{Out}(\Delta^{\Sigma}_{X_n/X_m}).$$

(vi) *Let* T *be a regular and connected scheme over* S *and* $x \in X_m(T)$ *a* T*-valued point of* X_m. *Then the diagram of profinite groups*

$$\begin{array}{ccc} \pi_1(T) & \xrightarrow{\rho^{\Sigma}_{X[x]_{n-m}/T}} & \mathrm{Out}(\Delta^{\Sigma}_{X[x]_{n-m}/T}) \\ {\scriptstyle \pi_1(x)}\downarrow & & \downarrow\wr \\ \pi_1(X_m) & \xrightarrow[\rho^{\Sigma}_{X_n/X_m}]{} & \mathrm{Out}(\Delta^{\Sigma}_{X_n/X_m}) \end{array}$$

cf. assertion (v), where the right-hand vertical arrow is the isomorphism determined by the isomorphism obtained in assertion (iii), commutes.

(vii) *The centralizer*

$$Z_{\Delta^{\Sigma}_{X_n/S}}(\Delta^{\Sigma}_{X_n/X_m})$$

of $\Delta^{\Sigma}_{X_n/X_m}$ *in* $\Delta^{\Sigma}_{X_n/S}$ *is trivial.*

(viii) *The pro-*Σ *outer representation associated to* X_n/X_m

$$\rho^{\Sigma}_{X_n/X_m} : \pi_1(X_m) \longrightarrow \mathrm{Out}(\Delta^{\Sigma}_{X_n/X_m})$$

factors through the natural surjection $\pi_1(X_m) \twoheadrightarrow \Pi^{\Sigma}_{X_m/S}$, *and, moreover, the composite of the natural inclusion* $\Delta^{\Sigma}_{X_m/S} \hookrightarrow \Pi^{\Sigma}_{X_m/S}$ *and the resulting homomorphism* $\Pi^{\Sigma}_{X_m/S} \to \mathrm{Out}(\Delta^{\Sigma}_{X_n/X_m})$ *is injective.*

Proof. First, we verify assertion (i). By induction on $n-m$, we may assume without loss of generality that $n = m+1$, Then $X_n \to X_m$ is a hyperbolic curve over X_m. The desired surjectivity follows from Remark 3.

Next, we verify assertion (ii). It is immediate that there exists a connected finite étale cover $Y \to X_m$ of X_m which satisfies the condition (c) in the statement of [MT08, Proposition 2.2], hence also (a), (b), and (c) of [MT08, Proposition 2.2], which therefore implies that if $\overline{y} \to Y$ is a geometric point, then $\Delta^{\Sigma}_{X_n \times_{X_m} Y/Y}$ is naturally isomorphic to the maximal pro-Σ quotient of $\pi_1(X_n \times_{X_m} \overline{y})$. On the other hand, it follows from the various definitions involved that $\Delta^{\Sigma}_{X_n \times_{X_m} Y/Y}$ is naturally isomorphic to $\Delta^{\Sigma}_{X_n/X_m}$. Thus, assertion (ii) follows from the fact that any geometric point of X_m arises from a geometric point of Y. This completes the proof of assertion (ii).

Assertion (iii) follows immediately from assertion (ii). Assertion (iv) (respectively, (v)) follows immediately from [MT08, Proposition 2.2 (iii) (respectively, (ii))], together with assertion (ii). Assertion (vi) follows immediately from assertion (iii).

Next, we verify assertion (vii). Since $\Delta^{\Sigma}_{X_n/X_m}$ is center-free, cf. assertion (v), it holds that $Z_{\Delta^{\Sigma}_{X_n/S}}(\Delta^{\Sigma}_{X_n/X_m}) \cap \Delta^{\Sigma}_{X_n/X_m} = \{1\}$. Thus, to verify assertion (vii), by replacing $\Delta^{\Sigma}_{X_n/S}$ by the quotient

$$\Delta^{\Sigma}_{X_{m+1}/S} \xleftarrow{\sim} \Delta^{\Sigma}_{X_n/S}/\Delta^{\Sigma}_{X_n/X_{m+1}} ,$$

cf. assertion (iv), of $\Delta^{\Sigma}_{X_n/S}$ by $\Delta^{\Sigma}_{X_n/X_{m+1}} \subseteq (\Delta^{\Sigma}_{X_n/X_m} \subseteq) \Delta_{X_n/S}$, we may assume without loss of generality that $n = m+1$. Then it follows from Lemma 4 (i) that it suffices to show that the outer representation $\Delta^{\Sigma}_{X_m/S} \to \mathrm{Out}(\Delta^{\Sigma}_{X_{m+1}/X_m})$ associated to the exact sequence of profinite groups

$$1 \longrightarrow \Delta^{\Sigma}_{X_{m+1}/X_m} \longrightarrow \Delta^{\Sigma}_{X_{m+1}/S} \longrightarrow \Delta^{\Sigma}_{X_m/S} \longrightarrow 1 ,$$

cf. assertion (iv), is injective. On the other hand, this injectivity follows immediately from [Asa01, Theorem 1], together with [Asa01, Remark following the proof of Theorem 1]. This completes the proof of assertion (vii).

Finally, we verify assertion (viii). The fact that the pro-Σ outer representation $\rho^{\Sigma}_{X_n/X_m}$ factors through the natural surjection $\pi_1(X_m) \twoheadrightarrow \Pi^{\Sigma}_{X_m/S}$ follows immediately from assertion (iv). The fact that the composite in question is injective follows immediately from assertion (vii), together with Lemma 4 (i). □

Proposition 16 (Monodromic fullness and base changing). *Let n be a positive integer, T a regular and connected scheme over S, and $x \in X_n(T)$ a T-valued point of X_n. Then the following hold:*

(i) The point x is (quasi-)Σ-monodromically full if and only if the T-valued point of $X_n \times_S T$ determined by x is (quasi-)Σ-monodromically full.

(ii) Let T' be a regular and connected scheme over S and $T' \to T$ a morphism over S such that the natural outer homomorphism $\pi_1(T') \to \pi_1(T)$ is surjective (respectively, has open image, e.g., $T' \to T$ is a connected finite étale cover). Then the point x is Σ-monodromically full (respectively, quasi-Σ-monodromically full) if and only if the T'-valued point of X_n determined by x is Σ-monodromically full (respectively, quasi-Σ-monodromically full).

Proof. This follows immediately from Lemma 15 (vi), and Remark 12. □

Lemma 17 (Extensions arising from FC-admissible outer automorphisms). *Let $m < n$ be positive integers, G a profinite group, and*

$$1 \longrightarrow \Delta^{\Sigma}_{X_n/S} \longrightarrow E_n \longrightarrow G \longrightarrow 1$$

an exact sequence of profinite groups with associated outer representation

$$\phi\colon G \longrightarrow \mathrm{Out}(\Delta^{\Sigma}_{X_n/S}).$$

Suppose that Σ is either $\mathfrak{Primes}$ or $\{\ell\}$ for one prime number ℓ, and that ϕ factors through the closed subgroup

$$\mathrm{Out}^{\mathrm{FC}}(\Delta^{\Sigma}_{X_n/S}) \subseteq \mathrm{Out}(\Delta^{\Sigma}_{X_n/S}) ,$$

where we refer to [Moc10, Definition 1.1 (ii)] concerning $\mathrm{Out}^{\mathrm{FC}}$.

With[2] $B_m = E_n/\Delta^{\Sigma}_{X_n/X_m}$ *and the natural isomorphism* $\Delta^{\Sigma}_{X_n/S}/\Delta^{\Sigma}_{X_n/X_m} \xrightarrow{\sim} \Delta^{\Sigma}_{X_m/S}$, *cf. Lemma 15 (iv), we have exact sequences of profinite groups*

$$1 \longrightarrow \Delta^{\Sigma}_{X_n/X_m} \longrightarrow E_n \longrightarrow B_m \longrightarrow 1\,,$$
$$1 \longrightarrow \Delta^{\Sigma}_{X_m/S} \longrightarrow B_m \longrightarrow G \longrightarrow 1\,.$$

and the corresponding continuous representations

$$\rho\colon B_m \longrightarrow \mathrm{Out}(\Delta^{\Sigma}_{X_n/X_m})\,,$$
$$\widetilde{\rho}\colon B_m \longrightarrow \mathrm{Aut}(\Delta^{\Sigma}_{X_m/S})\,.$$

Then the following hold:

(i) *The natural surjection* $B_m \twoheadrightarrow G$ *induces an isomorphism* $\mathrm{Ker}(\widetilde{\rho}) \xrightarrow{\sim} \mathrm{Ker}(\phi)$.
(ii) $\mathrm{Ker}(\rho) = \mathrm{Ker}(\widetilde{\rho}) = Z_{B_m}(\Delta^{\Sigma}_{X_m/S})$.
(iii) $Z_{E_n}(\Delta^{\Sigma}_{X_n/X_m}) = Z_{E_n}(\Delta^{\Sigma}_{X_n/S})$.
(iv) *The natural surjections* $E_n \twoheadrightarrow B_m \twoheadrightarrow G$ *induce isomorphisms*

$$Z_{E_n}(\Delta^{\Sigma}_{X_n/X_m}) = Z_{E_n}(\Delta^{\Sigma}_{X_n/S}) \xrightarrow{\sim} \mathrm{Ker}(\rho) = \mathrm{Ker}(\widetilde{\rho}) = Z_{B_m}(\Delta^{\Sigma}_{X_m/S}) \xrightarrow{\sim} \mathrm{Ker}(\phi)\,.$$

Proof. First, we verify assertion (i). Write $Z \subseteq B_m$ for the image of the centralizer $Z_{E_n}(\Delta^{\Sigma}_{X_n/S})$ of $\Delta^{\Sigma}_{X_n/S}$ in E_n via the natural surjection $E_n \twoheadrightarrow B_m$. Then it follows immediately that

$$Z \subseteq Z_{B_m}(\Delta^{\Sigma}_{X_m/S})\,.$$

Now I claim that

($\star_1$) the surjection $B_m \twoheadrightarrow G$ induces an isomorphism $Z \xrightarrow{\sim} \mathrm{Ker}(\phi)$.

Indeed, since $\Delta^{\Sigma}_{X_n/S}$ is center-free by Lemma 15 (v), we have

$$\Delta^{\Sigma}_{X_n/S} \cap Z_{E_n}(\Delta^{\Sigma}_{X_n/S}) = \{1\}\,.$$

In particular, the natural surjection $Z_{E_n}(\Delta^{\Sigma}_{X_n/S}) \twoheadrightarrow Z$ is an isomorphism. Thus claim ($\star_1$) is equivalent to the fact that the surjection $E_n \twoheadrightarrow G$ induces an isomorphism

$$Z_{E_n}(\Delta^{\Sigma}_{X_n/S}) \xrightarrow{\sim} \mathrm{Ker}(\phi)\,.$$

This follows immediately from $\Delta^{\Sigma}_{X_n/S}$ being center-free, together with Lemma 4 (i). This completes the proof of claim ($\star_1$).

Next, I claim that

($\star_2$) $Z = Z_{B_m}(\Delta^{\Sigma}_{X_m/S})$.

Indeed, by Lemma 4 (i), the image of $Z_{B_m}(\Delta^{\Sigma}_{X_m/S}) \subseteq B_m$ via the natural surjection $B_m \twoheadrightarrow G$ coincides with the kernel of the composite

[2] Note that since ϕ factors through $\mathrm{Out}^{\mathrm{FC}}(\Delta^{\Sigma}_{X_n/S})$, it follows immediately that $\Delta^{\Sigma}_{X_n/X_m}$ is a normal closed subgroup of E_n.

$$G \xrightarrow{\phi} \mathrm{Out}^{\mathrm{FC}}(\Delta^{\Sigma}_{X_n/S}) \longrightarrow \mathrm{Out}^{\mathrm{FC}}(\Delta^{\Sigma}_{X_m/S}),$$

where the second arrow is the homomorphism induced by the natural surjection $\Delta^{\Sigma}_{X_n/S} \twoheadrightarrow \Delta^{\Sigma}_{X_m/S}$, cf. Lemma 15 (iv). Thus, by [HM11, Theorem B], the image of $Z_{B_m}(\Delta^{\Sigma}_{X_m/S}) \subseteq B_m$ via the surjection $B_m \twoheadrightarrow G$ coincides with $\mathrm{Ker}(\phi)$. On the other hand, since $\Delta^{\Sigma}_{X_m/S}$ is center-free, it holds that $\Delta^{\Sigma}_{X_m/S} \cap Z_{B_m}(\Delta^{\Sigma}_{X_m/S}) = \{1\}$. Thus, since $Z \subseteq Z_{B_m}(\Delta^{\Sigma}_{X_m/S})$, by considering the images of Z and $Z_{B_m}(\Delta^{\Sigma}_{X_m/S})$ in G, claim $(\star_1)$ implies that $Z = Z_{B_m}(\Delta^{\Sigma}_{X_m/S})$. This completes the proof of claim $(\star_2)$.

Now it follows from Lemma 4 (i) that $\mathrm{Ker}(\widetilde{\rho}) = Z_{B_m}(\Delta^{\Sigma}_{X_m/S})$. Thus, assertion (i) follows from the claims $(\star_1)$, $(\star_2)$.

Next, we verify assertion (ii). Now I claim that

$(\star_3)$ $\mathrm{Ker}(\widetilde{\rho}) \subseteq \mathrm{Ker}(\rho)$.

Indeed, claim $(\star_2)$ and Lemma 4 (i) imply that $\mathrm{Ker}(\widetilde{\rho}) = Z_{B_m}(\Delta^{\Sigma}_{X_m/S}) = Z$. On the other hand, since $Z_{E_n}(\Delta^{\Sigma}_{X_n/S}) \subseteq Z_{E_n}(\Delta^{\Sigma}_{X_n/X_m})$, cf. Lemma 15 (iv), it follows from Lemma 4 (i), together with the definition of $Z \subseteq B_m$, that $Z \subseteq \mathrm{Ker}(\rho)$. This completes the proof of claim $(\star_3)$.

Now, by Lemma 15 (viii), we have $\mathrm{Ker}(\rho) \cap \Delta^{\Sigma}_{X_m/S} = \{1\}$. Thus, assertion (ii) follows immediately from claim $(\star_3)$, together with Lemma 4 (ii).

We verify assertion (iii). Observe that $Z_{E_n}(\Delta^{\Sigma}_{X_n/S}) \subseteq Z_{E_n}(\Delta^{\Sigma}_{X_n/X_m})$, cf. Lemma 15 (iv). It follows from Lemma 4 (i) (respectively claim $(\star_2)$, together with Lemma 4 (i)) that the image of $Z_{E_n}(\Delta^{\Sigma}_{X_n/X_m})$ (respectively, $Z_{E_n}(\Delta^{\Sigma}_{X_n/S})$) via the natural surjection $E_n \twoheadrightarrow B_m$ coincides with $\mathrm{Ker}(\rho)$ (respectively, $\mathrm{Ker}(\widetilde{\rho})$). On the other hand, since $\Delta^{\Sigma}_{X_n/X_m}$ is center-free, we have $\Delta^{\Sigma}_{X_n/X_m} \cap Z_{E_n}(\Delta^{\Sigma}_{X_n/X_m}) = \{1\}$. Therefore, by considering the images of $Z_{E_n}(\Delta^{\Sigma}_{X_n/S})$ and $Z_{E_n}(\Delta^{\Sigma}_{X_n/X_m})$ in B_m, assertion (iii) follows from assertion (ii).

Assertion (iv) follows immediately from assertions (i), (ii), and (iii), together with Lemma 4 (i). □

Remark 18. A similar result to Lemma 17 (ii) can be found in [Bog09, Thm. 2.5].

Proposition 19 (Two quotients of the fundamental group of a configuration space). *Let $m < n$ be positive integers,* T *a regular and connected scheme over* S*, and $x \in X_m(T)$ a* T*-valued point of X_m. Suppose that Σ is either* $\mathfrak{Primes}$ *or $\{\ell\}$ for one prime ℓ. Then the following hold:*

(i) The kernel of the pro-Σ representation associated to X_m/S

$$\pi_1(X_m) \twoheadrightarrow \Pi^{\Sigma}_{X_m/S} \overset{\widetilde{\rho}^{\Sigma}_{X_m/S}}{\rightarrow} \mathrm{Aut}(\Delta^{\Sigma}_{X_m/S})$$

coincides with the kernel of the pro-Σ outer representation

$$\rho^{\Sigma}_{X_n/X_m} : \pi_1(X_m) \to \mathrm{Out}(\Delta^{\Sigma}_{X_n/X_m})$$

associated to X_n/X_m*, i.e., the two quotients* $\Phi^{\Sigma}_{X_m/S}$ *and* $\Gamma^{\Sigma}_{X_n/X_m}$ *of* $\Pi^{\Sigma}_{X_m/S}$ *coincide. In particular, we obtain a commutative diagram of profinite groups*

$$\begin{array}{ccccccccc} 1 & \longrightarrow & \Delta^{\Sigma}_{X_m/S} & \longrightarrow & \Pi^{\Sigma}_{X_m/S} & \longrightarrow & \pi_1(S) & \longrightarrow & 1 \\ & & \| & & \downarrow & & \downarrow & & \\ 1 & \longrightarrow & \Delta^{\Sigma}_{X_m/S} & \longrightarrow & \Gamma^{\Sigma}_{X_n/X_m} & \longrightarrow & \Gamma^{\Sigma}_{X_m/S} & \longrightarrow & 1 \end{array}$$

with exact rows and surjective vertical arrows.

(ii) *The kernel of the pro-*Σ *outer representation associated to* $X[x]_{n-m}/T$

$$\rho^{\Sigma}_{X[x]_{n-m}/T} : \pi_1(T) \to \mathrm{Out}(\Delta^{\Sigma}_{X[x]_{n-m}/T})$$

and the kernel of the composite

$$\pi_1(T) \stackrel{\pi_1(x)}{\to} \pi_1(X_m) \twoheadrightarrow \Pi^{\Sigma}_{X_m/S} \stackrel{\tilde{\rho}^{\Sigma}_{X_m/S}}{\to} \mathrm{Aut}(\Delta^{\Sigma}_{X_m/S})$$

coincide. In particular, for a point $x \in X(T)$ *and* $U := (X \times_S T) \setminus \mathrm{Im}(x)$*, the kernel of the pro-*Σ *outer representation associated to* U/T

$$\rho^{\Sigma}_{U/T} : \pi_1(T) \to \mathrm{Out}(\Delta^{\Sigma}_{U/T})$$

coincides with the kernel of the composite

$$\pi_1(T) \stackrel{\pi_1(x)}{\to} \pi_1(X) \twoheadrightarrow \Pi^{\Sigma}_{X/S} \stackrel{\tilde{\rho}^{\Sigma}_{X/S}}{\to} \mathrm{Aut}(\Delta^{\Sigma}_{X/S}) .$$

(iii) *The following two conditions are equivalent:*

(iii-1) *The* T*-valued point* $x \in X_n(T)$ *is (resp. quasi-)*Σ*-monodromically full.*
(iii-2) *For any* $\ell \in \Sigma$*, if we write* $\Phi_T \subseteq \Phi^{\{\ell\}}_{X_m/S}$ *for the image of the composite*

$$\pi_1(T) \stackrel{\pi_1(x)}{\to} \pi_1(X_m) \twoheadrightarrow \Pi^{\{\ell\}}_{X_m/S} \stackrel{\tilde{\rho}^{\{\ell\}}_{X_m/S}}{\twoheadrightarrow} \Phi^{\{\ell\}}_{X_m/S}$$

and $\Phi_{\mathrm{geom}} \subseteq \Phi^{\{\ell\}}_{X_m/S}$ *for the image of the composite*

$$\mathrm{Ker}\big(\pi_1(X_m) \to \pi_1(S)\big) \hookrightarrow \pi_1(X_m) \twoheadrightarrow \Pi^{\{\ell\}}_{X_m/S} \stackrel{\tilde{\rho}^{\{\ell\}}_{X_m/S}}{\twoheadrightarrow} \Phi^{\{\ell\}}_{X_m/S} ,$$

then Φ_T *contains* Φ_{geom} *(respectively,* $\Phi_T \cap \Phi_{\mathrm{geom}}$ *is an open subgroup of* Φ_{geom}*).*

(iv) *If* $S = T$*, then the following two conditions are equivalent:*

(iv-1) *The* S*-valued point* $x \in X_n(S)$ *is (resp. quasi-)*Σ*-monodromically full.*
(iv-2) *For any* $\ell \in \Sigma$*, the composite*

$$\pi_1(\mathrm{S}) \overset{\pi_1(x)}{\to} \pi_1(\mathrm{X}_m) \twoheadrightarrow \Pi^{\{\ell\}}_{\mathrm{X}_m/\mathrm{S}} \overset{\widetilde{\rho}^{\{\ell\}}_{\mathrm{X}_m/\mathrm{S}}}{\twoheadrightarrow} \Phi^{\{\ell\}}_{\mathrm{X}_m/\mathrm{S}}$$

is surjective (respectively, has open image).

Proof. Assertion (i) follows from Lemma 17 (ii) and Proposition 6 (i). Assertion (ii) follows from assertion (i) and Lemma 15 (vi). Assertion (iii) follows from assertion (i), and assertion (iv) follows immediately from assertion (iii). □

Lemma 20 (Extension via the outer universal monodromy representations). *Let (g,r) be a pair of nonnegative integers such that $2g-2+r>0$. Let n be a positive integer and k a field of characteristic 0. Let s: $\mathrm{Spec}(k) \to \mathcal{M}_{g,r}$ be a morphism of stacks corresponding to an r-pointed smooth curve C of genus g over k, and let $\mathrm{C}_n \to \mathcal{M}_{g,r+n}$ be the natural morphism from the n-th configuration space C_n of C/k, namely the base change of s by the morphism $\mathcal{M}_{g,r+n} \to \mathcal{M}_{g,r}$ obtained by forgetting the last n sections.*

Let Σ either be $\mathfrak{Primes}$ or $\{\ell\}$ for one prime ℓ. Then $\mathrm{C}_n \to \mathcal{M}_{g,r+n}$ and the pro-Σ outer universal monodromy representations, cf. [Hos11, Definition 1.3(ii)],

$$\rho^{\Sigma}_{g,r}\colon \pi_1(\mathcal{M}_{g,r}\otimes_{\mathbb{Z}} k) \longrightarrow \mathrm{Out}(\Delta^{\Sigma}_{g,r}),$$

$$\rho^{\Sigma}_{g,r+n}\colon \pi_1(\mathcal{M}_{g,r+n}\otimes_{\mathbb{Z}} k) \longrightarrow \mathrm{Out}(\Delta^{\Sigma}_{g,r+n}),$$

determine a commutative diagram of profinite groups

$$\begin{array}{ccccccccc}
1 & \longrightarrow & \pi_1(\mathrm{C}_n\otimes_k \overline{k}) & \longrightarrow & \pi_1(\mathcal{M}_{g,r+n}\otimes_{\mathbb{Z}} k) & \longrightarrow & \pi_1(\mathcal{M}_{g,r}\otimes_{\mathbb{Z}} k) & \longrightarrow & 1 \\
 & & \downarrow & & \rho^{\Sigma}_{g,r+n}\downarrow & & \downarrow\rho^{\Sigma}_{g,r} & & \\
1 & \longrightarrow & \Delta^{\Sigma}_{\mathrm{C}_n/k} & \longrightarrow & \mathrm{Im}(\rho^{\Sigma}_{g,r+n}) & \longrightarrow & \mathrm{Im}(\rho^{\Sigma}_{g,r}) & \longrightarrow & 1
\end{array}$$

with exact rows and surjective vertical arrows.

Proof. The sequence of stacks

$$\mathrm{C}_n\otimes_k \overline{k} \to \mathcal{M}_{g,r+n}\otimes_{\mathbb{Z}} k \to \mathcal{M}_{g,r}\otimes_{\mathbb{Z}} k$$

gives rise to the commutative diagram of the statement of Lemma 20, cf. Lemma 15 (viii). The exactness of the top row follows from [MT00, Lemma 2.1].

To verify the exactness of the bottom row, we write Π for the quotient of $\pi_1(\mathcal{M}_{g,r+n}\otimes_{\mathbb{Z}} k)$ by the kernel of the natural surjection $\pi_1(\mathrm{C}_n\otimes_k \overline{k}) \twoheadrightarrow \Delta^{\Sigma}_{\mathrm{C}_n/k}$, i.e.,

$$\Pi = \Pi^{\Sigma}_{\mathcal{M}_{g,r+n}\otimes_{\mathbb{Z}} k/\mathcal{M}_{g,r}\otimes_{\mathbb{Z}} k}$$

with notation as in Definition 1 (ii), and

$$\widetilde{\rho}\colon \Pi \to \mathrm{Aut}(\Delta^{\Sigma}_{\mathrm{C}_n/k}) \quad \text{and} \quad \rho\colon \pi_1(\mathcal{M}_{g,r}\otimes_{\mathbb{Z}} k) \to \mathrm{Out}(\Delta^{\Sigma}_{\mathrm{C}_n/k})$$

for the (outer) representation associated to the natural exact sequence

$$1 \longrightarrow \Delta^{\Sigma}_{C_n/k} \longrightarrow \Pi \longrightarrow \pi_1(\mathcal{M}_{g,r} \otimes_{\mathbb{Z}} k) \longrightarrow 1 .$$

Then it follows from [HM11, Theorem B] that $\mathrm{Ker}(\rho) = \mathrm{Ker}(\rho^{\Sigma}_{g,r})$. On the other hand, by Lemma 15 (viii), $\rho^{\Sigma}_{g,r+n} \colon \pi_1(\mathcal{M}_{g,r+n} \otimes_{\mathbb{Z}} k) \to \mathrm{Out}(\Delta^{\Sigma}_{g,r+n})$ factors through the natural surjection $\pi_1(\mathcal{M}_{g,r+n} \otimes_{\mathbb{Z}} k) \twoheadrightarrow \Pi$. Moreover, by Lemma 17 (ii), the kernel of the homomorphism $\Pi \to \mathrm{Out}(\Delta^{\Sigma}_{g,r+n})$ determined by $\rho^{\Sigma}_{g,r+n}$ coincides with $\mathrm{Ker}(\widetilde{\rho})$. Therefore, the exactness of the bottom rows follows from Lemma 15 (v), together with Lemma 4 (i). This completes the proof of Lemma 20. □

Proposition 21 (Monodromically full curves and monodromically full points). *Suppose that* S *is the spectrum of a field* k *of characteristic* 0. *Let* n *be a positive integer and* $x \in \mathrm{X}_n(k)$. *Then the hyperbolic curve* $\mathrm{X}[x]$ *over* k *is (quasi-)*Σ*-monodromically full, cf. [Hos11, Definition 2.2], if and only if the following two conditions are satisfied:*

(i) The hyperbolic curve X *over* k *is (quasi-)*Σ*-monodromically full.*
(ii) The k*-rational point* $x \in \mathrm{X}_n(k)$ *is (quasi-)*Σ*-monodromically full.*

Proof. Suppose that X is of type (g,r). By the definition of (quasi-)monodromic fullness, we may replace Σ by $\{\ell\}$ for $\ell \in \Sigma$ and treat one prime at a time.

Moreover, again by the definition of (quasi-)monodromic fullness, we may replace k by the minimal Galois extension over which X is split, see [Hos11, Definition 1.5 (i)] for the term split. Hence, we may assume that X is split over k, i.e., the classifying morphism $\mathrm{Spec}(k) \to \mathcal{M}_{g,[r]}$ factors as $s_{\mathrm{X}} \colon \mathrm{Spec}(k) \to \mathcal{M}_{g,r}$ through the natural $\mathcal{M}_{g,r} \to \mathcal{M}_{g,[r]}$. Then, by Lemma 20, we have a commutative diagram

$$\begin{array}{ccccccccc}
 & & & & \pi_1(\mathrm{Spec}(k)) & & & & \\
 & & & & \downarrow{\scriptstyle \pi_1(x)} & & & & \\
1 & \longrightarrow & \pi_1(\mathrm{X}_n \otimes_k \overline{k}) & \longrightarrow & \pi_1(\mathrm{X}_n) & \longrightarrow & \pi_1(\mathrm{Spec}(k)) & \longrightarrow & 1 \\
 & & \| & & \downarrow & & \downarrow{\scriptstyle \pi_1(s_{\mathrm{X}})} & & \\
1 & \longrightarrow & \pi_1(\mathrm{X}_n \otimes_k \overline{k}) & \longrightarrow & \pi_1(\mathcal{M}_{g,r+n} \otimes_{\mathbb{Z}} k) & \longrightarrow & \pi_1(\mathcal{M}_{g,r} \otimes_{\mathbb{Z}} k) & \longrightarrow & 1 \\
 & & \downarrow & & \downarrow{\scriptstyle \rho^{\Sigma}_{g,r+n}} & & \downarrow{\scriptstyle \rho^{\Sigma}_{g,r}} & & \\
1 & \longrightarrow & \Delta^{\Sigma}_{\mathrm{X}_n/k} & \longrightarrow & \mathrm{Im}(\rho^{\Sigma}_{g,r+n}) & \longrightarrow & \mathrm{Im}(\rho^{\Sigma}_{g,r}) & \longrightarrow & 1
\end{array}$$

with exact rows, where $\pi_1(x)$ and $\pi_1(s_{\mathrm{X}})$ are the outer homomorphisms induced by x, s_{X}, respectively. By definition, the composite of the three middle vertical arrows $\pi_1(\mathrm{Spec}(k)) \to \mathrm{Im}(\rho^{\Sigma}_{g,r+n})$ coincides with the outer pro-Σ representation $\rho^{\Sigma}_{\mathrm{X}[x]/k}$ associated to $\mathrm{X}[x]/k$, and the composite of the two right-hand vertical arrows $\pi_1(\mathrm{Spec}(k)) \to \mathrm{Im}(\rho^{\Sigma}_{g,r})$ coincides with the outer pro-Σ representation $\rho^{\Sigma}_{\mathrm{X}/k}$ associated to X/k. Therefore, again by the various definitions involved, cf. also Remark 12, if we write $\rho \colon \pi_1(\mathrm{X}_n) \to \mathrm{Im}(\rho^{\Sigma}_{g,r+n})$ for the composite of the middle vertical arrow $\pi_1(\mathrm{X}_n) \to \pi_1(\mathcal{M}_{g,r+n} \otimes_{\mathbb{Z}} k)$ and the lower middle vertical arrow $\rho^{\Sigma}_{g,r+n}$, then

- the hyperbolic curve $\mathrm{X}[x]$ is Σ-monodromically full (respectively, quasi-Σ-monodromically full) if and only if the composite of the three middle vertical arrows, i.e., $\rho^{\Sigma}_{\mathrm{X}[x]/k}$, is surjective (respectively, has open image in $\mathrm{Im}(\rho^{\Sigma}_{g,r+n})$),
- the hyperbolic curve X is Σ-monodromically full (respectively, quasi-Σ-monodromically full) if and only if the composite of the two right-hand vertical arrows, i.e., $\rho^{\Sigma}_{\mathrm{X}/k}$, is surjective (respectively, has open image in $\mathrm{Im}(\rho^{\Sigma}_{g,r})$), and
- the k-rational point $x \in \mathrm{X}_n(k)$ of X_n is Σ-monodromically full (respectively, quasi-Σ-monodromically full) if and only if the image of the composite of the three middle vertical arrows, i.e., $\rho^{\Sigma}_{\mathrm{X}[x]/k}$, coincides with the image of ρ (respectively, is an open subgroup of the image of ρ).

One now may easily verify that Proposition 21 holds. □

Remark 22. One verifies easily that $\mathbb{P}^1_k \setminus \{0, 1, \infty\}$ is a $\mathfrak{Primes}$-monodromically full hyperbolic curve, cf. [Hos11, Definition 2.2 (ii)]. Thus, one may regard Proposition 21 as a generalization of Remark 11(ii).

8.5 Kernels of the Outer Representations Associated to Configuration Spaces

In the present section, we consider the kernels of the outer representations associated to configuration spaces of hyperbolic curves. Let $\Sigma \subseteq \mathfrak{Primes}$ be a nonempty subset of $\mathfrak{Primes}$, S a regular and connected scheme such that every element of Σ is invertible on S, and X a hyperbolic curve over S.

Lemma 23 (Difference between kernels of outer representations arising from extensions). *In the commutative diagram of profinite groups with exact rows*

$$\begin{array}{ccccccccc} 1 & \longrightarrow & \Delta' & \longrightarrow & \Pi' & \longrightarrow & \mathrm{G} & \longrightarrow & 1 \\ & & {\scriptstyle\alpha}\downarrow & & \downarrow & & \| & & \\ 1 & \longrightarrow & \Delta & \longrightarrow & \Pi & \longrightarrow & \mathrm{G} & \longrightarrow & 1 \end{array}$$

where the right-hand vertical arrow is the identity of G, *we suppose that* Δ *and* Δ' *are topologically finitely generated. Write*

$$\rho\colon \mathrm{G} \longrightarrow \mathrm{Out}(\Delta) \quad \textit{and} \quad \rho'\colon \mathrm{G} \longrightarrow \mathrm{Out}(\Delta')$$

for the outer representations associated to the lower and top rows. Then the following hold:

(i) *If* α *is injective, then we have a natural exact sequence of profinite groups*

$$1 \longrightarrow \mathrm{Ker}(\rho) \cap \mathrm{Ker}(\rho') \longrightarrow \mathrm{Ker}(\rho) \stackrel{\rho'}{\longrightarrow} \mathrm{Im}(\phi)\,,$$

where we write ϕ *for the outer representation induced by conjugation*

$$\mathrm{N}_{\Delta}(\Delta')/\Delta' \longrightarrow \mathrm{Out}(\Delta').$$

(ii) *If α is surjective, then we have an inclusion* $\mathrm{Ker}(\rho') \subseteq \mathrm{Ker}(\rho)$.
(iii) *If α is an open injection, and Δ is slim, then*

$$\mathrm{Z}_{\Pi'}(\Delta') = \mathrm{Z}_{\Pi}(\Delta) \cap \Pi' \quad \textit{and} \quad \mathrm{Ker}(\rho') \subseteq \mathrm{Ker}(\rho).$$

Moreover, $\mathrm{Z}_{\Pi}(\Delta) \subseteq \Pi'$ *if and only if* $\mathrm{Ker}(\rho') = \mathrm{Ker}(\rho)$.

Proof. Assertions (i) and (ii) follow immediately from the definitions. For assertion (iii), since Δ and Δ' are center-free, it follows from Lemma 4 (i) that it suffices to prove $\mathrm{Z}_{\Pi'}(\Delta') = \mathrm{Z}_{\Pi}(\Delta) \cap \Pi'$. As Δ is topologically finitely generated and slim, this follows from Lemma 5. This completes the proof of assertion (iii). □

Proposition 24 (Monodromic fullness and partial compactifications). *Let n be a positive integer,* T *a regular and connected scheme over* S, *and* $x \in \mathrm{X}_n(\mathrm{T})$ *a* T*-valued point of the n-th configuration space* X_n *of* X/S. *Suppose that the* T*-valued point* $x \in \mathrm{X}_n(\mathrm{T})$ *is Σ-monodromically full (respectively, quasi-Σ-monodromically full). Then the following hold:*

(i) *Let* Z *be a* ***hyperbolic partial compactification*** *of* X/S, *i.e., a hyperbolic curve* Z *over* S *which contains* X *as an open subscheme over* S. *Then the* T*-valued point of* Z_n *determined by x and the natural open immersion* $\mathrm{X}_n \hookrightarrow \mathrm{Z}_n$ *is Σ-monodromically full (respectively, quasi-Σ-monodromically full).*
(ii) *Let $m < n$ be a positive integer. Then the* T*-valued point of* X_m *determined by x is Σ-monodromically full (respectively, quasi-Σ-monodromically full).*

Proof. By Proposition 16 (i), we may replace X by $\mathrm{X} \times_{\mathrm{S}} \mathrm{T}$, and so we may assume without loss of generality that S = T. Let ℓ be a prime number in Σ. First, we verify assertion (i). The natural open immersion $\mathrm{X} \hookrightarrow \mathrm{Z}$ induces a commutative diagram of schemes

$$\begin{array}{ccc} \mathrm{X}_{n+1} & \longrightarrow & \mathrm{X}_n \\ \downarrow & & \downarrow \\ \mathrm{Z}_{n+1} & \longrightarrow & \mathrm{Z}_n; \end{array}$$

thus, we obtain a commutative diagram of profinite groups

$$\begin{array}{ccccccccc} 1 & \longrightarrow & \Delta^{\{\ell\}}_{\mathrm{X}_{n+1}/\mathrm{X}_n} & \longrightarrow & \Pi^{\{\ell\}}_{\mathrm{X}_{n+1}/\mathrm{S}} & \longrightarrow & \Pi^{\{\ell\}}_{\mathrm{X}_n/\mathrm{S}} & \longrightarrow & 1 \\ & & \downarrow & & \downarrow & & \downarrow & & \\ 1 & \longrightarrow & \Delta^{\{\ell\}}_{\mathrm{Z}_{n+1}/\mathrm{Z}_n} & \longrightarrow & \Pi^{\{\ell\}}_{\mathrm{Z}_{n+1}/\mathrm{S}} & \longrightarrow & \Pi^{\{\ell\}}_{\mathrm{Z}_n/\mathrm{S}} & \longrightarrow & 1 \end{array}$$

with exact rows, cf. Lemma 15 (iv), and surjective vertical arrows, cf. Lemma 15 (ii). Then the right vertical arrow induces a surjection $\Gamma^{\{\ell\}}_{\mathrm{X}_{n+1}/\mathrm{X}_n} \twoheadrightarrow \Gamma^{\{\ell\}}_{\mathrm{Z}_{n+1}/\mathrm{Z}_n}$, and assertion (i) follows from Remark 12.

Next, we verify assertion (ii). We have a commutative diagram of schemes

$$\begin{array}{ccc} X_{n+1} & \longrightarrow & X_n \\ \downarrow & & \downarrow \\ X_{m+1} & \longrightarrow & X_m \end{array}$$

where the left-hand vertical arrow is the projection obtained as

$$(x_1,\cdots,x_{n+1}) \mapsto (x_1,\cdots,x_m,x_{n+1})$$

and the other arrows are the natural projections. We obtain a commutative diagram

$$\begin{array}{ccccccccc} 1 & \longrightarrow & \Delta^{\{\ell\}}_{X_{n+1}/X_n} & \longrightarrow & \Pi^{\{\ell\}}_{X_{n+1}/S} & \longrightarrow & \Pi^{\{\ell\}}_{X_n/S} & \longrightarrow & 1 \\ & & \downarrow & & \downarrow & & \downarrow & & \\ 1 & \longrightarrow & \Delta^{\{\ell\}}_{X_{m+1}/X_m} & \longrightarrow & \Pi^{\{\ell\}}_{X_{m+1}/S} & \longrightarrow & \Pi^{\{\ell\}}_{X_m/S} & \longrightarrow & 1 \end{array}$$

with exact rows, cf. Lemma 15 (iv), and surjective vertical arrows, cf. Lemma 15 (i) (ii). Now the right-hand vertical arrow induces a surjective map $\Gamma^{\{\ell\}}_{X_{n+1}/X_n} \twoheadrightarrow \Gamma^{\{\ell\}}_{X_{m+1}/X_m}$. Thus, assertion (ii) follows from Remark 12. □

Proposition 25 (Kernels of the outer representations associated to the fundamental groups of configuration spaces). *Let n be a positive integer, T a regular and connected scheme over S, $Y \to X_n$ a finite étale $\Pi^{\Sigma}_{X_n/S}$-cover, cf. Definition 1(v), $y \in Y(T)$ a T-valued point of Y. Write $x \in X_n(T)$ for the T-valued point of X_n determined by y. Suppose that $\Sigma = \mathfrak{Primes}$ or that $\Sigma = \{\ell\}$ for one prime number ℓ. Suppose, moreover, that the natural outer homomorphism $\pi_1(Y) \to \pi_1(S)$ is surjective. Thus, we have a commutative diagram of profinite groups*

$$\begin{array}{ccccccccc} 1 & \longrightarrow & \Delta^{\Sigma}_{Y/S} & \longrightarrow & \Pi^{\Sigma}_{Y/S} & \longrightarrow & \pi_1(S) & \longrightarrow & 1 \\ & & \downarrow & & \downarrow & & \| & & \\ 1 & \longrightarrow & \Delta^{\Sigma}_{X_n/S} & \longrightarrow & \Pi^{\Sigma}_{X_n/S} & \longrightarrow & \pi_1(S) & \longrightarrow & 1 \end{array}$$

with exact rows, where the vertical arrows are open injections, and $\Delta^{\Sigma}_{Y/S}$ is topologically finitely generated and slim, cf. Lemma 15 (v). Then the following hold:

(i) $\mathrm{Ker}(\rho^{\Sigma}_{Y/S})$ is an open subgroup of $\mathrm{Ker}(\rho^{\Sigma}_{X_n/S})$, and $\mathrm{Ker}(\rho^{\Sigma}_{X_n/S}) = \mathrm{Ker}(\rho^{\Sigma}_{Y/S})$ if and only if the cover $Y \to X_n$ is a finite étale $\Phi^{\Sigma}_{X_n/S}$-cover.

(ii) The natural inclusion $\Pi^{\Sigma}_{Y/S} \hookrightarrow \Pi^{\Sigma}_{X_n/S}$ induces a commutative diagram

$$\begin{array}{ccccccccc} 1 & \longrightarrow & \Delta^{\Sigma}_{Y/S} & \longrightarrow & \Phi^{\Sigma}_{Y/S} & \longrightarrow & \Gamma^{\Sigma}_{Y/S} & \longrightarrow & 1 \\ & & \downarrow & & \downarrow & & \downarrow & & \\ 1 & \longrightarrow & \Delta^{\Sigma}_{X_n/S} & \longrightarrow & \Phi^{\Sigma}_{X_n/S} & \longrightarrow & \Gamma^{\Sigma}_{X_n/S} & \longrightarrow & 1 \end{array}$$

with exact rows, where the left-hand and middle vertical arrows are open injection, and the right-hand vertical arrow is a surjection with finite kernel.

(iii) The kernels of the following two composites coincide:

$$\pi_1(\mathrm{T}) \stackrel{\pi_1(y)}{\to} \pi_1(\mathrm{Y}) \stackrel{\widetilde{\rho}^{\Sigma}_{\mathrm{Y/S}}}{\to} \mathrm{Aut}(\Delta^{\Sigma}_{\mathrm{Y/S}}),$$

$$\pi_1(\mathrm{T}) \stackrel{\pi_1(x)}{\to} \pi_1(\mathrm{X}_n) \stackrel{\widetilde{\rho}^{\Sigma}_{\mathrm{X}_n/\mathrm{S}}}{\to} \mathrm{Aut}(\Delta^{\Sigma}_{\mathrm{X}_n/\mathrm{S}}).$$

(iv) If $n = 1$, and Y *is a hyperbolic curve over* S*, then, for* $\mathrm{U_Y} := (\mathrm{Y} \times_{\mathrm{S}} \mathrm{T}) \setminus \mathrm{Im}(y)$ *and* $\mathrm{U_X} := (\mathrm{X} \times_{\mathrm{S}} \mathrm{T}) \setminus \mathrm{Im}(x)$*, we have*

$$\mathrm{Ker}(\rho^{\Sigma}_{\mathrm{U_Y/T}}) = \mathrm{Ker}(\rho^{\Sigma}_{\mathrm{U_X/T}}).$$

(v) Suppose that $\mathrm{S} = \mathrm{T}$*, that* $\Sigma = \{\ell\}$*, and that* $x \in \mathrm{X}_n(\mathrm{T})$ *is* ℓ*-monodromically full. Then the cover* $\mathrm{Y} \to \mathrm{X}_n$ *is an isomorphism if and only if* $\mathrm{Ker}(\rho^{\Sigma}_{\mathrm{Y/S}}) = \mathrm{Ker}(\rho^{\Sigma}_{\mathrm{X}_n/\mathrm{S}})$*, i.e., if the cover* $\mathrm{Y} \to \mathrm{X}_n$ *is not an isomorphism, then the cokernel of the natural inclusion* $\mathrm{Ker}(\rho^{\Sigma}_{\mathrm{Y/S}}) \hookrightarrow \mathrm{Ker}(\rho^{\Sigma}_{\mathrm{X}_n/\mathrm{S}})$ *is nontrivial. If, moreover,* $\Delta^{\Sigma}_{\mathrm{Y/S}} \subseteq \Delta^{\Sigma}_{\mathrm{X}_n/\mathrm{S}}$ *is normal, then we have an exact sequence of profinite groups*

$$1 \longrightarrow \mathrm{Ker}(\rho^{\Sigma}_{\mathrm{Y/S}}) \longrightarrow \mathrm{Ker}(\rho^{\Sigma}_{\mathrm{X}_n/\mathrm{S}}) \longrightarrow \Delta^{\Sigma}_{\mathrm{X}_n/\mathrm{S}}/\Delta^{\Sigma}_{\mathrm{Y/S}} \longrightarrow 1.$$

Proof. Assertion (i) follows from Lemma 4 (i) and Lemma 23 (i) (iii). Assertion (ii) follows from Lemma 4 (i) and Lemma 23 (i) (iii), together with Proposition 6 (i).

Assertion (iii) follows from assertion (ii). Assertion (iv) follows from assertion (iii) together with Proposition 19 (ii).

Finally, we verify assertion (v). Since $\mathrm{S} = \mathrm{T}$, and $x \in \mathrm{X}_n(\mathrm{T})$ is ℓ-monodromically full it follows from Proposition 19 (iv) that the composite

$$\pi_1(\mathrm{S}) \stackrel{\pi_1(x)}{\to} \pi_1(\mathrm{X}_n) \stackrel{\widetilde{\rho}^{\Sigma}_{\mathrm{X}_n/\mathrm{S}}}{\twoheadrightarrow} \Phi^{\Sigma}_{\mathrm{X}_n/\mathrm{S}}$$

is surjective. Thus, by assertion (ii), we have a commutative diagram

$$\begin{array}{ccccccccc}
1 & \longrightarrow & \Delta^{\Sigma}_{\mathrm{Y/S}} & \longrightarrow & \Phi^{\Sigma}_{\mathrm{Y/S}} & \longrightarrow & \Gamma^{\Sigma}_{\mathrm{Y/S}} & \longrightarrow & 1 \\
 & & \downarrow & & \wr\downarrow & & \downarrow & & \\
1 & \longrightarrow & \Delta^{\Sigma}_{\mathrm{X}_n/\mathrm{S}} & \longrightarrow & \Phi^{\Sigma}_{\mathrm{X}_n/\mathrm{S}} & \longrightarrow & \Gamma^{\Sigma}_{\mathrm{X}_n/\mathrm{S}} & \longrightarrow & 1
\end{array}$$

with exact rows, where the middle vertical arrow is an isomorphism. Therefore, if the cover $\mathrm{Y} \to \mathrm{X}_n$ is *not* an isomorphism, then the kernel of the natural surjection $\Gamma^{\Sigma}_{\mathrm{Y/S}} \twoheadrightarrow \Gamma^{\Sigma}_{\mathrm{X}_n/\mathrm{S}}$ is nontrivial. If, moreover, $\Delta^{\Sigma}_{\mathrm{Y/S}} \subseteq \Delta^{\Sigma}_{\mathrm{X}_n/\mathrm{S}}$ is normal, then the exactness of the sequence as in assertion (v) follows from Lemma 23 (i). This completes the proof of assertion (v). □

Remark 26. Let $k/\mathbb{Q}$ be a finite extension, (g_0, r_0) a pair of nonnegative integers such that $2g_0 - 2 + r_0 > 0$, and $\mathrm{N} \subseteq \pi_1(\mathrm{Spec}(k))$ a normal closed subgroup. Write

$$\mathcal{I}^{\mathrm{Gal}}(\ell,k,g_0,r_0,\mathrm{N})$$

for the set of the isomorphism classes over k of hyperbolic curves C of type (g_0,r_0) over k such that

$$\mathrm{N} = \ker(\rho^{\{\ell\}}_{\mathrm{C}/k} : \pi_1(\mathrm{Spec}(k)) \longrightarrow \mathrm{Out}(\Delta^{\{\ell\}}_{\mathrm{C}/k})) .$$

Then, by [Hos11, Theorem C], the set $\mathcal{I}^{\mathrm{Gal}}(\ell,k,g_0,r_0,\mathrm{N})$ is finite. On the other hand, by Proposition 25 (i), in general,

$$\mathcal{I}^{\mathrm{Gal}}(\ell,k,\mathrm{N}) := \bigcup_{2g-2+r>0} \mathcal{I}^{\mathrm{Gal}}(\ell,k,g,r,\mathrm{N})$$

is *not finite*. Indeed, let C be a hyperbolic curve over k such that there exists a k-rational point $x \in \mathrm{C}(k)$ which is *not* quasi-ℓ-monodromically full, e.g., C is a tripod, cf. Proposition 33 (ii) and Theorem 41 below. Then, by Remark 12, the image of the composite

$$\phi : \pi_1(\mathrm{Spec}(k)) \overset{\pi_1(x)}{\to} \pi_1(\mathrm{C}) \twoheadrightarrow \Phi^{\{\ell\}}_{\mathrm{C}/k}$$

is not open. Thus, there exists an infinite sequence of open subgroups of $\Phi^{\{\ell\}}_{\mathrm{C}/k}$ which contain $\mathrm{Im}(\phi)$

$$\mathrm{Im}(\phi) \subseteq \cdots \subsetneq \Phi_n \subsetneq \cdots \subsetneq \Phi_2 \subsetneq \Phi_1 \subsetneq \Phi_0 = \Phi^{\{\ell\}}_{\mathrm{C}/k} ,$$

hence also an infinite sequence of nontrivial connected finite étale covers of C

$$\cdots \longrightarrow \mathrm{C}^n \longrightarrow \cdots \longrightarrow \mathrm{C}^2 \longrightarrow \mathrm{C}^1 \longrightarrow \mathrm{C}^0 = \mathrm{C},$$

where we write C^n for the finite étale cover of C corresponding to the open subgroup $\Phi_n \subseteq \Phi^{\{\ell\}}_{\mathrm{C}/k}$. Then since $\mathrm{Im}(\phi) \subset \Phi_n$, C^n is a hyperbolic curve over k. Moreover, since C^n is a finite étale $\Phi^{\{\ell\}}_{\mathrm{C}/k}$-cover, it follows from Proposition 25 (i) that $\mathrm{Ker}(\rho^{\{\ell\}}_{\mathrm{C}/k}) = \mathrm{Ker}(\rho^{\{\ell\}}_{\mathrm{C}^n/k})$. In particular, the set

$$\mathcal{I}^{\mathrm{Gal}}(\ell,k,\mathrm{Ker}(\rho^{\{\ell\}}_{\mathrm{C}/k})) = \bigcup_{2g-2+r>0} \mathcal{I}^{\mathrm{Gal}}(\ell,k,g,r,\mathrm{Ker}(\rho^{\{\ell\}}_{\mathrm{C}/k}))$$

is not finite.

Proposition 27 (Monodromic fullness and finite étale covers). *Let* $\mathrm{Y} \to \mathrm{X}$ *be a finite étale* $\Pi^{\Sigma}_{\mathrm{X}/\mathrm{S}}$*-cover over* S, T *a regular and connected scheme over* S, *and* $y \in \mathrm{Y}(\mathrm{T})$ *a* T*-valued point of* Y. *Write* $x \in \mathrm{X}(\mathrm{T})$ *for the* T*-valued point of* X *determined by* y. *Suppose that* Y *is a hyperbolic curve over* S, *and that* $\Sigma = \{\ell\}$ *for some prime number* ℓ. *Then the following hold:*

(i) *If* $x \in \mathrm{X}(\mathrm{T})$ *is* ℓ*-monodromically full, then* $y \in \mathrm{Y}(\mathrm{T})$ *is* ℓ*-monodromically full.*
(ii) $x \in \mathrm{X}(\mathrm{T})$ *is quasi-*ℓ*-monodromically full if and only if* $y \in \mathrm{Y}(\mathrm{T})$ *is quasi-*ℓ*-monodromically full.*

Proof. By Proposition 16 (i), we may replace X by $X \times_S T$, and thus we may assume without loss of generality that $S = T$. Then Proposition 27 follows immediately from Proposition 19 (iv) and Proposition 25 (ii). □

Lemma 28 (Kernels of outer representations associated to certain finite étale covers). *Let* $Y \to X$ *be a finite étale* $\Pi^{\Sigma}_{X/S}$*-cover over* S. *Suppose that the following five conditions are satisfied:*

(i) Y *is a hyperbolic curve over* S. *In particular,* $n = 1$.
(ii) $\Sigma = \{\ell\}$ *for some prime number* ℓ.
(iii) The subgroup $\Delta^{\Sigma}_{Y/S} \subseteq \Delta^{\Sigma}_{X/S}$ *is normal, i.e., there exists a geometric point* $\overline{s} \to S$ *such that the connected finite étale cover* $Y \times_S \overline{s} \to X \times_S \overline{s}$ *is Galois.*
(iv) The action of the Galois group $\mathrm{Gal}(Y \times_S \overline{s}/X \times_S \overline{s})$*, cf. condition (iii), on the set of the cusps of* $Y \times_S \overline{s}$ *is faithful. In particular, if the cover* $Y \to X$ *is* not *an isomorphism, then* Y*, hence also* X*, is not proper over* S.
(v) Every cusp of Y/S *is defined over the connected (possibly infinite) étale cover of* S *corresponding to* $\mathrm{Ker}(\rho^{\Sigma}_{X/S}) \subseteq \pi_1(S)$.

Then $\mathrm{Ker}(\rho^{\Sigma}_{X/S}) = \mathrm{Ker}(\rho^{\Sigma}_{Y/S})$.

Proof. It follows from Proposition 25 (i) that $\mathrm{Ker}(\rho^{\Sigma}_{Y/S}) \subseteq \mathrm{Ker}(\rho^{\Sigma}_{X/S})$. Thus it suffices to show the other inclusion. Let

$$\phi\colon \Delta^{\Sigma}_{X/S}/\Delta^{\Sigma}_{Y/S} \longrightarrow \mathrm{Out}(\Delta^{\Sigma}_{Y/S})$$

be the outer representation induced by conjugation. By condition (iii), together with Lemma 23 (i), the image of $\mathrm{Ker}(\rho^{\Sigma}_{X/S}) \subseteq \pi_1(S)$ via $\rho^{\Sigma}_{Y/S}$ is contained in $\mathrm{Im}(\phi)$. Now, by condition (iv), the action of $\mathrm{Im}(\phi)$ on the set of the $\Delta^{\Sigma}_{Y/S}$-conjugacy classes of cuspidal inertia subgroups of $\Delta^{\Sigma}_{Y/S}$ is faithful, cf. Remark 3. On the other hand, by condition (v), the action of $\rho^{\Sigma}_{Y/S}(\mathrm{Ker}(\rho^{\Sigma}_{X/S}))$ on the set of the $\Delta^{\Sigma}_{Y/S}$-conjugacy classes of cuspidal inertia subgroups of $\Delta^{\Sigma}_{Y/S}$ is trivial. Therefore, it follows that $\rho^{\Sigma}_{Y/S}(\mathrm{Ker}(\rho^{\Sigma}_{X/S})) = \{1\}$. This completes the proof of Lemma 28. □

We abbreviate $\mathcal{M}_{0,4} = \mathrm{Spec}\left(\mathbb{Z}[t^{\pm 1}, 1/(t-1)]\right)$ by P and denote, for any scheme S, the base change $P \times_{\mathbb{Z}} S$ by P_S.

Proposition 29 (Kernels of outer representations associated to certain covers of tripods). *Let* ℓ *be a prime number which is invertible on* S. *Let* $U \to P_S$ *be a finite étale* $\Pi^{\{\ell\}}_{P_S/S}$*-cover over* S *such that* U *is a hyperbolic curve over* S. *Suppose that the following three conditions are satisfied:*

(i) The subgroup $\Delta^{\{\ell\}}_{U/S} \subseteq \Delta^{\{\ell\}}_{P_S/S}$ *is normal, i.e., there exists a geometric point* $\overline{s} \to S$ *such that the connected finite étale cover* $U \times_S \overline{s} \to P_S \times_S \overline{s}$ *is Galois.*
(ii) The action of the Galois group $\mathrm{Gal}(U \times_S \overline{s}/P_S \times_S \overline{s})$*, cf. condition (i), on the set of the cusps of* $U \times_S \overline{s}$ *is faithful.*
(iii) Every cusp of U/S *is defined over the connected (possibly infinite) étale Galois cover of* S *corresponding to the kernel* $\mathrm{Ker}(\rho^{\{\ell\}}_{P_S/S}) \subseteq \pi_1(S)$ *of* $\rho^{\{\ell\}}_{P_S/S}$.

Let V *be a hyperbolic partial compactification of* U *over* S*, i.e., a hyperbolic curve over* S *which contains* U *as an open subscheme over* S*. Then,* $\mathrm{Ker}(\rho^{\{\ell\}}_{\mathrm{V/S}}) = \mathrm{Ker}(\rho^{\{\ell\}}_{\mathrm{P_S/S}})$. *In particular,* $\mathrm{Ker}(\rho^{\{\ell\}}_{\mathrm{U/S}}) = \mathrm{Ker}(\rho^{\{\ell\}}_{\mathrm{P_S/S}})$.

Proof. It follows from [HM11, Theorem C] and Lemma 23 (ii) that we have inclusions

$$\mathrm{Ker}(\rho^{\{\ell\}}_{\mathrm{U/S}}) \subseteq \mathrm{Ker}(\rho^{\{\ell\}}_{\mathrm{V/S}}) \subseteq \mathrm{Ker}(\rho^{\{\ell\}}_{\mathrm{P_S/S}}) .$$

Thus it suffices to show $\mathrm{Ker}(\rho^{\{\ell\}}_{\mathrm{P_S/S}}) \subseteq \mathrm{Ker}(\rho^{\{\ell\}}_{\mathrm{U/S}})$, that follows from Lemma 28. □

Remark 30. A similar result to Proposition 29 can be found in [AI88, Corollary 3.8.1].

Proposition 31 (Kernels of outer representations associated to certain hyperbolic curves arising from elliptic curves). *Let* ℓ *be a prime number which is invertible on* S*,* N *a positive integer, and* E *an elliptic curve over* S*. Write* $o \in \mathrm{E(S)}$ *for the identity section of* E*, and* $\mathrm{E}[l^{\mathrm{N}}] \subseteq \mathrm{E}$ *for the kernel of multiplication by* ℓ^{N}*. We set*

$$\mathrm{Z} := \mathrm{E} \setminus \mathrm{Im}(o) \quad \textit{and} \quad \mathrm{U} := \mathrm{E} \setminus \mathrm{E}[\ell^{\mathrm{N}}] .$$

Let V *be an open subscheme of* Z *such that* $\mathrm{U} \subseteq \mathrm{V} \subseteq \mathrm{Z}$*. Then* $\mathrm{Ker}(\rho^{\{\ell\}}_{\mathrm{V/S}}) = \mathrm{Ker}(\rho^{\{\ell\}}_{\mathrm{Z/S}})$. *In particular,* $\mathrm{Ker}(\rho^{\{\ell\}}_{\mathrm{U/S}}) = \mathrm{Ker}(\rho^{\{\ell\}}_{\mathrm{Z/S}})$.

Proof. Note that Z (respectively, U) is a hyperbolic curve of type $(1,1)$ (respectively, $(1,\ell^{2\mathrm{N}})$) over S. Observe that multiplication by l^{N} determines an open injection $\Pi^{\{\ell\}}_{\mathrm{U/S}} \hookrightarrow \Pi^{\{\ell\}}_{\mathrm{Z/S}}$ over $\pi_1(\mathrm{S})$.

First, the natural action of $\mathrm{Ker}(\rho^{\{\ell\}}_{\mathrm{Z/S}})$ on $\mathrm{E}[\ell^{\mathrm{N}}] \times_{\mathrm{S}} \overline{s} \cong \Delta^{\{\ell\}}_{\mathrm{Z/S}} / \Delta^{\{\ell\}}_{\mathrm{U/S}}$ is trivial, where $\overline{s} \to \mathrm{S}$ is a geometric point of S. Now, by Lemma 23 (ii), the natural open immersions $\mathrm{U} \hookrightarrow \mathrm{V} \hookrightarrow \mathrm{Z}$ induce inclusions

$$\mathrm{Ker}(\rho^{\{\ell\}}_{\mathrm{U/S}}) \subseteq \mathrm{Ker}(\rho^{\{\ell\}}_{\mathrm{V/S}}) \subseteq \mathrm{Ker}(\rho^{\{\ell\}}_{\mathrm{Z/S}}) .$$

Thus, it suffices to show $\mathrm{Ker}(\rho^{\{\ell\}}_{\mathrm{Z/S}}) \subseteq \mathrm{Ker}(\rho^{\{\ell\}}_{\mathrm{U/S}})$. By the above, the finite étale cover $\mathrm{U} \to \mathrm{Z}$ arising from multiplication by l^{N} satisfies condition (v) of Lemma 28. Thus, by applying Lemma 28 to $\mathrm{U} \to \mathrm{Z}$, we conclude that $\mathrm{Ker}(\rho^{\{\ell\}}_{\mathrm{Z/S}}) = \mathrm{Ker}(\rho^{\{\ell\}}_{\mathrm{U/S}})$. □

8.6 Some Complements to Matsumoto's Result Concerning the Representations Arising from Hyperbolic Curves

We give some complements to Matsumoto's result obtained in [Mat11] concerning the difference between the kernels of the natural homomorphisms associated to a hyperbolic curve and its point from the Galois group to the automorphism and outer automorphism groups of the geometric fundamental group of the hyperbolic curve. Let ℓ be a prime number, k a field of characteristic $\neq \ell$, and X a hyperbolic curve over k. Write

$$\mathrm{X}^{\mathrm{cl}}$$

for the set of closed points of X.

Definition 32. Let $x \in \mathrm{X}^{\mathrm{cl}}$ be a closed point of X. Then we shall say that $\mathrm{E}(\mathrm{X}, x, \ell)$ **holds** if the kernel of the composite

$$\pi_1(\mathrm{Spec}(k(x))) \stackrel{\pi_1(x)}{\longrightarrow} \pi_1(\mathrm{X}) \stackrel{\widetilde{\rho}^{\{\ell\}}_{\mathrm{X}/k}}{\longrightarrow} \mathrm{Aut}(\Delta^{\{\ell\}}_{\mathrm{X}/k}) \tag{8.3}$$

coincides with the kernel of the composite

$$\pi_1(\mathrm{Spec}(k(x))) \longrightarrow \pi_1(\mathrm{Spec}(k)) \stackrel{\rho^{\{\ell\}}_{\mathrm{X}/k}}{\longrightarrow} \mathrm{Out}(\Delta^{\{\ell\}}_{\mathrm{X}/k}),$$

i.e., the intersection of the closed subgroup $\mathrm{Inn}(\Delta^{\{\ell\}}_{\mathrm{X}/k}) \subseteq \mathrm{Aut}(\Delta^{\{\ell\}}_{\mathrm{X}/k})$ and the image of (8.3) is trivial, cf. [Mat11, §1 and §3]. Since $\mathrm{Inn}(\Delta^{\{\ell\}}_{\mathrm{X}/k})$ is normal in $\mathrm{Aut}(\Delta^{\{\ell\}}_{\mathrm{X}/k})$, whether or not the intersection in question is trivial does *not depend* on the choice of the homomorphism $\pi_1(x) : \pi_1(\mathrm{Spec}(k(x))) \to \pi_1(\mathrm{X})$ induced by $x \in \mathrm{X}$ among the $\pi_1(\mathrm{X})$-conjugates. Moreover, we shall write

$$\mathrm{X}^{\mathrm{E}_\ell} \subseteq \mathrm{X}^{\mathrm{cl}}$$

for the set of closed points x of X such that $\mathrm{E}(\mathrm{X}, x, \ell)$ holds.

Proposition 33 (Properties of exceptional points). *Let $x \in \mathrm{X}(k)$ be a k-rational point of* X. *Then the following hold:*

(i) *We set* $\mathrm{U} := \mathrm{X} \setminus \mathrm{Im}(x)$. *Then the following four conditions are equivalent:*

  (1) $\mathrm{E}(\mathrm{X}, x, \ell)$ *holds.*
  (2) *The section of the natural surjection* $\pi_1(\mathrm{X}) \twoheadrightarrow \pi_1(\mathrm{Spec}(k))$ *induced by x determines a section of the natural surjection* $\Phi^{\{\ell\}}_{\mathrm{X}/k} \twoheadrightarrow \Gamma^{\{\ell\}}_{\mathrm{X}/k}$, *cf. Proposition 6 (i).*
  (3) $\mathrm{Ker}(\rho^{\{\ell\}}_{\mathrm{X}/k}) = \mathrm{Ker}(\rho^{\{\ell\}}_{\mathrm{U}/k})$.
  (4) *The cokernel of the inclusion* $\mathrm{Ker}(\rho^{\{\ell\}}_{\mathrm{U}/k}) \subseteq \mathrm{Ker}(\rho^{\{\ell\}}_{\mathrm{X}/k})$, *cf. Lemma 23 (ii), is finite.*

(ii) *The following holds: x is an ℓ-monodromically full point* $\Longrightarrow$ *x is a quasi-ℓ-monodromically full point* $\Longrightarrow$ $\mathrm{E}(\mathrm{X}, x, \ell)$ *does not hold.*

Proof. First, we verify assertion (i). The equivalence (1) ⇔ (2) follows from the definitions. The equivalence (1) ⇔ (3) follows from Proposition 19 (ii), and (3) ⇒ (4) is immediate. Finally, we verify (4) ⇒ (3). By Proposition 19 (ii), the natural surjection $\Gamma^{\{\ell\}}_{\mathrm{U}/k} \twoheadrightarrow \Gamma^{\{\ell\}}_{\mathrm{X}/k}$ factors through the natural injection $\Gamma^{\{\ell\}}_{\mathrm{U}/k} \hookrightarrow \Phi^{\{\ell\}}_{\mathrm{X}/k}$. In particular, the cokernel of the natural inclusion $\mathrm{Ker}(\rho^{\{\ell\}}_{U/k}) \subseteq \mathrm{Ker}(\rho^{\{\ell\}}_{X/k})$ may be regarded as a closed subgroup of $\Delta^{\{\ell\}}_{X/k}$. Therefore, since $\Delta^{\{\ell\}}_{X/k}$ is torsion-free, if the cokernel of $\mathrm{Ker}(\rho^{\{\ell\}}_{U/k}) \subseteq \mathrm{Ker}(\rho^{\{\ell\}}_{X/k})$ is finite, then it is trivial. This completes the proof of the implication (4) $\Rightarrow$ (3), hence also of assertion (i).

Assertion (ii) follows immediately from Proposition 19 (iv). □

Remark 34. In the notation of Proposition 33 (ii), in general, the implication

$$\mathrm{E}(\mathrm{X}, x, \ell) \text{ does not hold} \Longrightarrow x \text{ is a quasi-}\ell\text{-monodromically full point}$$

does *not hold*. Indeed, for $\mathrm{X} := \mathbb{P}^1_{\mathbb{Q}} \setminus \{0,1,\infty\}$, the point $x = 2 \in \mathrm{X}(\mathbb{Q}) = \mathbb{Q} \setminus \{0,1\}$ is *not* quasi-ℓ-monodromically full, i.e., the hyperbolic curve $\mathrm{U} := \mathbb{P}^1_{\mathbb{Q}} \setminus \{0,1,2,\infty\}$ is *not* quasi-ℓ-monodromically full, cf. Remark 11(ii), together with [Hos11, Corollary 7.12].

On the other hand, since $\mathrm{U} = \mathbb{P}^1_{\mathbb{Q}} \setminus \{0,1,2,\infty\}$ has *bad reduction* at the prime 2, if $\ell \neq 2$, then it follows from [Tam97, Theorem 0.8] that the extension of $\mathbb{Q}$ corresponding to $\mathrm{Ker}(\rho^{\{\ell\}}_{U/\mathbb{Q}})$ is ramified at 2. In particular, the natural surjection $\Gamma^{\{\ell\}}_{U/\mathbb{Q}} \twoheadrightarrow \Gamma^{\{\ell\}}_{X/\mathbb{Q}}$ is not an isomorphism. Thus, by the equivalence (1) $\Leftrightarrow$ (3) in Proposition 33 (i), we have that $\mathrm{E}(\mathbb{P}^1_{\mathbb{Q}} \setminus \{0,1,\infty\}, 2, \ell)$ does not hold.

Proposition 35 (Exceptional points and finite étale covers). *Let* $\mathrm{Y} \to \mathrm{X}$ *be a finite étale* $\Pi^{\{\ell\}}_{X/k}$*-cover over* k *and* $y \in \mathrm{Y}(k)$*. In particular,* Y *is a hyperbolic curve over* k*. Write* $x \in \mathrm{X}(k)$ *for the* k*-rational point of* X *determined by* y*. Then* $\mathrm{E}(\mathrm{X},x,\ell)$ *holds if and only if* $\mathrm{E}(\mathrm{Y},y,\ell)$ *holds, and if both hold, then* $\mathrm{Ker}(\rho^{\{\ell\}}_{X/k}) = \mathrm{Ker}(\rho^{\{\ell\}}_{Y/k})$.

Proof. We write $\mathrm{U_X} := \mathrm{X} \setminus \mathrm{Im}(x)$ and $\mathrm{U_Y} := \mathrm{Y} \setminus \mathrm{Im}(y)$. Then the maps $\mathrm{U_Y} \subset \mathrm{Y} \to \mathrm{X}$ and $\mathrm{U_X} \subset \mathrm{X}$ induce a commutative diagram of profinite groups

$$\begin{array}{ccc} \Gamma^{\{\ell\}}_{U_Y/k} & \longrightarrow & \Gamma^{\{\ell\}}_{Y/k} \\ \wr\downarrow & & \downarrow \\ \Gamma^{\{\ell\}}_{U_X/k} & \longrightarrow & \Gamma^{\{\ell\}}_{X/k} \end{array}$$

where the left vertical arrow is the isomorphism obtained by Proposition 25 (iv). Now observe that

- the horizontal arrows are surjective, cf. Lemma 23 (ii), and
- the right vertical arrow is surjective and has finite kernel, cf. Proposition 25 (i).

If $\mathrm{E}(\mathrm{X},x,\ell)$ holds, then by the equivalence (1) $\Leftrightarrow$ (3) in Proposition 33 (i), the lower horizontal arrow is an isomorphism. Thus, also the top horizontal arrow and the right-hand vertical arrow are isomorphisms. In particular, again by the equivalence (1) $\Leftrightarrow$ (3) in Proposition 33 (i), $\mathrm{E}(\mathrm{Y},y,\ell)$ holds, and $\mathrm{Ker}(\rho^{\{\ell\}}_{X/k}) = \mathrm{Ker}(\rho^{\{\ell\}}_{Y/k})$.

On the other hand, if $\mathrm{E}(\mathrm{X},x,\ell)$ does *not hold*, then it follows from the equivalence (1) $\Leftrightarrow$ (4) in Proposition 33 (i) that the kernel of the lower horizontal arrow is infinite. Since the kernel of the right-hand vertical arrow in the above diagram is finite, the kernel of the top horizontal arrow is infinite. In particular, again by the equivalence

(1) $\Leftrightarrow$ (4) in Proposition 33 (i), $\mathrm{E}(\mathrm{Y}, y, \ell)$ does not hold. This completes the proof of Proposition 35. □

Theorem 36 (Existence of many nonexceptional points). *Let ℓ be a prime number, $k/\mathbb{Q}$ a finite extension, and X a hyperbolic curve over k. We regard X^{cl} as a subset of $\mathrm{X}(\mathbb{C})$. Then the complement*

$$\mathrm{X}^{\mathrm{cl}} \setminus \mathrm{X}^{\mathrm{E}_\ell} \subseteq \mathrm{X}(\mathbb{C})$$

is dense in the complex topology of $\mathrm{X}(\mathbb{C})$. Moreover, $\mathrm{X}(k) \cap \mathrm{X}^{\mathrm{E}_\ell}$ is finite.

Proof. That $\mathrm{X}^{\mathrm{cl}} \setminus \mathrm{X}^{\mathrm{E}_\ell}$ is *dense* in the complex topology of $\mathrm{X}(\mathbb{C})$ follows from Proposition 13 and Proposition 33 (ii). Finiteness of $\mathrm{X}(k) \cap \mathrm{X}^{\mathrm{E}_\ell}$ follows from the equivalence (1) $\Leftrightarrow$ (3) in Proposition 33 (i), [Hos11, Thm. C], and Lemma 37 below. □

Lemma 37 (Finiteness of the set consisting of equivalent points). *Let n be a positive integer and $x \in \mathrm{X}_n(k)$ a k-rational point of X_n. Then the set of k-rational points of X_n which are equivalent to x is finite.*

Proof. Write $\mathrm{X}^{\mathrm{cpt}}$ for the smooth compactification of X over k and $x_1, \ldots, x_n \in \mathrm{X}(k)$ for the the n distinct k-rational points determined by $x \in \mathrm{X}_n(k)$. Now by replacing k by a suitable finite separable extension of k, we may assume without loss of generality that every cusp of X is defined over k. Write g for the genus of X and set

$$n_g := \begin{cases} 3 \text{ if } g = 0, \\ 1 \text{ if } g = 1, \\ 0 \text{ if } g \geqslant 2, \end{cases}$$

and fix n_g distinct cusps

$$a_1, \ldots, a_{n_g} \in \mathrm{S} := \mathrm{X}^{\mathrm{cpt}}(k) \setminus \mathrm{X}(k)$$

of X. Then finiteness of the set of k-rational points of X_n which are equivalent to $x \in \mathrm{X}_n(k)$ follows from the finiteness of the set

$$\{f \in \mathrm{Aut}_k(\mathrm{X}^{\mathrm{cpt}}) \mid \{a_i\}_{i=1}^{n_g} \subseteq f(\mathrm{S}) \cup \{f(x_1), \ldots, f(x_n)\}\}\,.$$

This finiteness follows immediately from the well-known finiteness of the automorphism group of a hyperbolic curve. This completes the proof of Lemma 37. □

Remark 38. Matsumoto proved the following theorem, cf. [Mat11, Theorem 1]:

> *Let ℓ be a prime number and $g \geqslant 3$ an integer. Suppose that ℓ divides $2g-2$ and write ℓ^ν for the highest power of ℓ that divides $2g-2$. Then there are infinitely many isomorphism classes of pairs (k, X) of number fields k and hyperbolic curves X of type $(g, 0)$ over k which satisfy the following condition: For any closed point $x \in \mathrm{X}$ with residue field $k(x)$, if ℓ^ν does not divide $[k(x) : k]$, then $\mathrm{E}(\mathrm{X}, x, \ell)$ does not hold.*

Theorem 36 may be regarded as a partial generalization of this theorem.

Recall that we set $\mathrm{P}_k := \mathrm{Spec}(k[t^{\pm 1}, 1/(t-1)])$.

Proposition 39 (Exceptional points via tripods). *The following hold:*

(i) *Let* $\mathrm{U} \to \mathrm{P}_k$ *be a finite étale* $\Pi^{\{\ell\}}_{\mathrm{P}_k/k}$*-cover such that the following holds:*

(1) For a separable closure k^{sep} *of* k*, the cover* $\mathrm{U} \otimes_k k^{\mathrm{sep}} \to \mathrm{P}_k \otimes_k k^{\mathrm{sep}}$ *is Galois. In particular,* U *is geometrically connected over* k*.*

(2) The action of the Galois group $\mathrm{Gal}(\mathrm{U} \otimes_k k^{\mathrm{sep}}/\mathrm{P}_k \otimes_k k^{\mathrm{sep}})$*, cf. condition (1), on the set of cusps of* $\mathrm{U} \otimes_k k^{\mathrm{sep}}$ *is faithful.*

(3) Every cusp of U *is defined over the (possibly infinite) Galois extension of* k *corresponding to the kernel* $\mathrm{Ker}(\rho^{\{\ell\}}_{\mathrm{P}_k/k}) \subseteq \pi_1(\mathrm{Spec}(k))$ *of* $\rho^{\{\ell\}}_{\mathrm{P}_k/k}$*.*

Let V *be a hyperbolic partial compactification of* U *over* k*, and let* $x \in \mathrm{V}^{\mathrm{cl}}$ *be a closed point of* V *such that the complement* $\mathrm{V} \setminus \{x\}$ *contains* U*. Then* $\mathrm{E}(\mathrm{V}, x, \ell)$ *holds.*

(ii) *Let* N *be a positive integer. If a closed point* $x \in \mathrm{P}^{\mathrm{cl}}_k$ *of* P_k *is contained in the closed subscheme of* P_k *determined by the principal ideal*

$$(t^{\ell^{\mathrm{N}}} - 1) \subseteq k[t^{\pm 1}, 1/(t-1)] ,$$

then $\mathrm{E}(\mathrm{P}_k, x, \ell)$ *holds.*

Proof. First, we verify assertion (i). After replacing k by the residue field $k(x)$ at x, we may assume without loss of generality that $x \in \mathrm{V}(k)$. Then it follows immediately from Proposition 29 that

$$\mathrm{Ker}(\rho^{\{\ell\}}_{\mathrm{V}/k}) = \mathrm{Ker}(\rho^{\{\ell\}}_{(\mathrm{V}\setminus\{x\})/k}) .$$

Thus, assertion (i) follows from the equivalence (1) $\Leftrightarrow$ (3) in Proposition 33 (i). Next, we verify assertion (ii). By replacing k by the residue field $k(x)$ at x, we may assume without loss of generality that $x \in \mathrm{P}_k(k)$. Write

$$\mathrm{U} := \mathrm{Spec}\,(k[s^{\pm 1}, 1/(s^{\ell^{\mathrm{N}}} - 1)]) \longrightarrow \mathrm{P}_k$$

for the finite étale cover given by $t \mapsto s^{\ell^{\mathrm{N}}}$. Then $\mathrm{U} \to \mathrm{P}_k$ satisfies the three conditions in the statement of assertion (i). Now by assumption, we have an open immersion $\mathrm{U} \hookrightarrow \mathrm{P}_k \setminus \mathrm{Im}(x)$, so that assertion (ii) follows from assertion (i). □

Proposition 40 (Exceptional points via elliptic curves). *Let* N *be a positive integer and* E *an elliptic curve over* k*. Write* $o \in \mathrm{E}(k)$ *for the identity section of* E*, and* $\mathrm{E}[\ell^{\mathrm{N}}]$ *for the kernel of multiplication by* ℓ^{N}*. We set* $\mathrm{Z} := \mathrm{E} \setminus \mathrm{Im}(o)$ *and* $\mathrm{U} := \mathrm{E} \setminus \mathrm{E}[\ell^{\mathrm{N}}]$*.*

Let V *be an open subscheme of* Z *such that* $\mathrm{U} \subseteq \mathrm{V} \subseteq \mathrm{Z}$ *and* $x \in \mathrm{V}^{\mathrm{cl}}$ *a closed point such that the complement* $\mathrm{V} \setminus \{x\}$ *contains* U*. Then* $\mathrm{E}(\mathrm{V}, x, \ell)$ *holds. In particular, for any closed point* $z \in \mathrm{Z}^{\mathrm{cl}}$ *contained in* $\mathrm{E}[\ell^{\mathrm{N}}]$*, property* $\mathrm{E}(\mathrm{Z}, z, \ell)$ *holds.*

Proof. By replacing k by the residue field $k(x)$ at x, we may assume without loss of generality that $x \in \mathrm{X}(k)$. Then it follows immediately from Proposition 31 that

$$\mathrm{Ker}(\rho^{\{\ell\}}_{\mathrm{V}/k}) = \mathrm{Ker}(\rho^{\{\ell\}}_{(\mathrm{V}\setminus\{x\})/k}) ,$$

and we conclude by the equivalence (1) ⇔ (3) in Proposition 33 (i). □

Theorem 41 (Many exceptional points for certain hyperbolic curves). *Let ℓ be a prime number, k a field of characteristic 0, X a hyperbolic curve over k which is either of type $(0,3)$ or type $(1,1)$, and $Y \to X$ a finite étale $\Pi^{\{\ell\}}_{X/k}$-cover which is geometrically connected over k. Then the subset $Y^{E_\ell} \subseteq Y^{cl}$ is infinite. In particular, the subset $X^{E_\ell} \subseteq X^{cl}$ is infinite.*

Proof. By definition of the set $(-)^{E_\ell}$, we may replace k by a finite Galois extension over which the cusps of X are defined. We may thus assume without loss of generality that every cusp of X is defined over k. Then it follows from Proposition 39 (ii) and Proposition 40 that the set X^{E_ℓ} is infinite. Therefore, it follows from Proposition 35 that the set Y^{E_ℓ} is infinite. □

Remark 42. By a similar argument to the argument used in the proof of Theorem 41, one may also prove the following assertion:

> *Let ℓ be a prime number, k a field of characteristic 0, and $r \geqslant 3$ (respectively, $r \geqslant 1$) an integer. Then there exist a finite extension k' of k and a hyperbolic curve X over k' of type $(0,r)$ (respectively, $(1,r)$) such that the subset $X^{E_\ell} \subseteq X^{cl}$ is infinite.*

Indeed, write $C_0 := \mathrm{Spec}(k[t^{\pm 1}, 1/(t-1)])$ and C_1 for the complement in an elliptic curve E over k of the origin. Let N be a positive integer such that $r \leqslant \ell^N$. Moreover, write $F_0^N \subseteq C_0$ for the closed subscheme of C_0 defined by the principal ideal $(t^{\ell^N} - 1)$ and $F_1^N \subseteq C_1$ for the closed subscheme obtained as the kernel of the multiplication by ℓ^N of E. After replacing k by a finite extension, we may assume without loss generality that every geometric point of F_0^N and of F_1^N can be defined over k. Then it is immediate that for $g = 0$ or 1, there exists an open subscheme $X_g \subseteq C_g$ of C_g such that X_g is of *type* (g,r), and, moreover, X_g contains the complement of F_g^N in C_g. Now by Proposition 29 and Proposition 31, we have $\mathrm{Ker}(\rho^{\{\ell\}}_{C_g/k}) = \mathrm{Ker}(\rho^{\{\ell\}}_{X_g/k})$. Therefore, by Proposition 39 (i) and Proposition 40, together with the equivalence (1) ⇔ (3) in Proposition 33 (i), one may verify easily that the set $X_g^{E_\ell}$ is *infinite*.

Remark 43. An example of a triple (X, x, ℓ) such that X is a proper hyperbolic curve over a number field k, and, moreover, $E(X, x, \ell)$ holds is as follows: Suppose that $\ell > 3$. Let k be a number field and set

$$U := \mathrm{Spec}\left(k[t_1^{\pm 1}, t_2^{\pm 1}]/(t_1^\ell + t_2^\ell - 1)\right).$$

Then the connected finite étale cover

$$U \longrightarrow \mathrm{Spec}(k[t^{\pm 1}, 1/(t-1)])$$

given by $t \mapsto t_1^\ell$ satisfies the three conditions of Proposition 39 (i). In particular, for

$$X := \mathrm{Proj}(k[t_1, t_2, t_3]/(t_1^\ell + t_2^\ell - t_3^\ell))$$

and $x := [1, -1, 0] \in X(k)$, it follows from Proposition 39 (i) that $E(X, x, \ell)$ holds.

Remark 44. In [Mat11, §2], Matsumoto proved that for any prime number ℓ, the triple

$$(\mathbb{P}^1_{\mathbb{Q}} \setminus \{0,1,\infty\}, \overrightarrow{01}, \ell),$$

where $\overrightarrow{01}$ is a $\mathbb{Q}$-rational tangential base point, is a triple for which $\mathrm{E}(\mathrm{X},x,\ell)$ holds, as observed by P. Deligne and Y. Ihara. However, a tangential base point is not a point. In this sense, [Mat11] contains no example (X,x,ℓ) for which $\mathrm{E}(\mathrm{X},x,\ell)$ holds.

8.7 Galois-Theoretic Characterization of Equivalence Classes of Monodromically Full Points

We prove that the equivalence class of a monodromically full point of a configuration space of a hyperbolic curve is completely determined by the kernel of the representation associated to the point, cf. Theorem 49 and Theorem 51 below. Moreover, we also give a necessary and sufficient condition for a quasi-monodromically full Galois section, cf. Definition 46 below, of a hyperbolic curve to be geometric, cf. Theorem 53 below.

In this section, let n be a positive integer, k a field of characteristic 0, $\overline{k}$ an algebraic closure of k, and X a hyperbolic curve over k. As before, we write $\mathrm{G}_k := \mathrm{Gal}(\overline{k}/k)$ for the absolute Galois group of k determined by $\overline{k}$.

Lemma 45 (Equivalence and automorphisms). *For $x, y \in \mathrm{X}_n(k)$, if there is an automorphism $\alpha \in \mathrm{Aut}_k(\mathrm{X}_n)$ with $y = \alpha(x)$, then x is equivalent to y, cf. Definition 7(iii).*

Proof. If the hyperbolic curve X is of type $(0,3)$ (respectively, neither of type $(0,3)$ nor of type $(1,1)$), then Lemma 45 follows immediately from [Hos11, Lemma 4.1 (i) (ii)], (respectively, [NT98, Theorem A, Corollary B] and [Moc99, Theorem A]). Thus, we may assume that X is of type $(1,1)$.

Write E for the smooth compactification of X over k and $o \in \mathrm{E}(k)$ for the k-rational point $\mathrm{E} \setminus \mathrm{X}$. The pair (E, o) is canonically an elliptic curve with origin $o \in \mathrm{E}$. By [NT98, Theorem A, Corollary B] and [Moc99, Theorem A], the group $\mathrm{Aut}_k(\mathrm{X}_n)$ is generated by the images of the natural inclusions

$$\mathrm{Aut}_k(\mathrm{X}) \hookrightarrow \mathrm{Aut}_k(\mathrm{X}_n) \quad \text{and} \quad \mathfrak{S}_n \hookrightarrow \mathrm{Aut}_k(\mathrm{X}_n),$$

where $\mathfrak{S}_n$ is the symmetric group on n letters, together with the automorphism of X_n induced by

$$\begin{array}{ccc} \mathrm{E} \times_k \ldots \times_k \mathrm{E} & \longrightarrow & \mathrm{E} \times_k \ldots \times_k \mathrm{E} \\ (x_1, \ldots, x_n) & \mapsto & (x_1, x_1 - x_2, x_1 - x_3, \ldots, x_1 - x_n). \end{array}$$

Therefore it suffices to verify that for any n distinct $x_1, \ldots, x_n \in \mathrm{X}(k)$, the hyperbolic curve of type $(1, n+1)$

$$\mathrm{E} \setminus \{o, x_1, \ldots, x_n\}$$

is isomorphic over k to the hyperbolic curve of type $(1, n+1)$

$$\mathrm{E} \setminus \{o, x_1, x_1 - x_2, x_1 - x_3, \ldots, x_1 - x_n\}.$$

This is the composite of multiplication by -1 and translation $a \mapsto a + x_1$. □

Definition 46. Let $\Sigma' \subseteq \Sigma$ be nonempty subsets of $\mathfrak{Primes}$ and let

$$s : \mathrm{G}_k \to \Pi^{\Sigma}_{\mathrm{X}_n/k}$$

be a pro-Σ Galois section of X_n/k, i.e., a continuous section of $\Pi^{\Sigma}_{\mathrm{X}_n/k} \twoheadrightarrow \mathrm{G}_k$. Then we shall say that s is **Σ'-monodromically full** (respectively, **quasi-Σ'-monodromically full**) if, for any $\ell \in \Sigma'$, the composite

$$\mathrm{G}_k \xrightarrow{s} \Pi^{\Sigma}_{\mathrm{X}_n/k} \twoheadrightarrow \Pi^{\{\ell\}}_{\mathrm{X}_n/k} \overset{\rho^{\{\ell\}}_{\mathrm{X}_{n+1}/\mathrm{X}_n}}{\twoheadrightarrow} \Gamma^{\{\ell\}}_{\mathrm{X}_{n+1}/\mathrm{X}_n},$$

cf. Lemma 15 (viii), is surjective (respectively, has open image).

Remark 47. Let $x \in \mathrm{X}_n(k)$ be a k-rational point of X_n. Then by Remark 12, the following two conditions are equivalent:

(i) The k-rational point $x \in \mathrm{X}_n(k)$ is Σ-monodromically full (respectively, quasi-Σ-monodromically full).
(ii) The pro-$\mathfrak{Primes}$ Galois section of X_n arising from $x \in \mathrm{X}_n(k)$ is Σ-monodromically full (respectively, quasi-Σ-monodromically full).

Lemma 48 (Certain two monodromically full Galois sections). *Let ℓ be a prime number and s, t pro-ℓ Galois sections of X_n/k. We write $\phi_{\mathfrak{a}}$ for the composite*

$$\mathrm{G}_k \xrightarrow{\mathfrak{a}} \Pi^{\{\ell\}}_{\mathrm{X}_n/k} \xrightarrow{\overline{\rho}^{\{\ell\}}_{\mathrm{X}_n/k}} \mathrm{Aut}(\Delta^{\{\ell\}}_{\mathrm{X}_n/k})$$

for $\mathfrak{a} \in \{s, t\}$. Then the following hold:

(i) If s and t are ℓ-monodromically full, and $\mathrm{Ker}(\phi_s) = \mathrm{Ker}(\phi_t)$, then there exists an automorphism α of $\Pi^{\{\ell\}}_{\mathrm{X}_n/k}$ over G_k such that $\alpha \circ s = t$.
(ii) If s and t are quasi-ℓ-monodromically full, and $\mathrm{Ker}(\phi_s) \cap \mathrm{Ker}(\phi_t)$ is open in $\mathrm{Ker}(\phi_s)$ and $\mathrm{Ker}(\phi_t)$, then, after replacing G_k by a suitable open subgroup, we have for $\mathfrak{a} \in \{s, t\}$ a finite étale $\Phi^{\{\ell\}}_{\mathrm{X}_n/k}$-cover $\mathrm{C}_{\mathfrak{a}} \to \mathrm{X}_n$ over k which is geometrically connected over k, and an isomorphism $\alpha : \Pi^{\{\ell\}}_{\mathrm{C}_s/k} \xrightarrow{\sim} \Pi^{\{\ell\}}_{\mathrm{C}_t/k}$ over G_k, such that

- *for $\mathfrak{a} \in \{s, t\}$, the pro-$\ell$ Galois section $\mathfrak{a} : \mathrm{G}_k \to \Pi^{\{\ell\}}_{\mathrm{X}_n/k}$ factors through $\Pi^{\{\ell\}}_{\mathrm{C}_{\mathfrak{a}}/k} \subseteq \Pi^{\{\ell\}}_{\mathrm{X}_n/k}$, and*
- *the composite $\alpha \circ s$ coincides with t.*

Proof. First, we verify assertion (i). Since s and t are ℓ-monodromically full, it follows from Proposition 19 (i), that $\mathrm{Im}(\phi_s) = \mathrm{Im}(\phi_t) = \Phi^{\{\ell\}}_{\mathrm{X}_n/k}$. Write β for the automorphism of $\Phi^{\{\ell\}}_{\mathrm{X}_n/k}$ obtained as the composite

$$\Phi^{\{\ell\}}_{X_n/k} = \mathrm{Im}(\phi_s) \overset{\sim}{\longleftarrow} G_k/\mathrm{Ker}(\phi_s) = G_k/\mathrm{Ker}(\phi_t) \overset{\sim}{\longrightarrow} \mathrm{Im}(\phi_t) = \Phi^{\{\ell\}}_{X_n/k}.$$

Then β is an automorphism over $\Gamma^{\{\ell\}}_{X_n/k}$. Thus, since the right-hand square in (8.2) from Proposition 6 is cartesian, by base-changing β via the natural surjection $G_k \twoheadrightarrow \Gamma^{\{\ell\}}_{X_n/k}$, we obtain an automorphism α of $\Pi^{\{\ell\}}_{X_n/k}$ over G_k. It follows from the definitions that α satisfies the condition of assertion (i).

Next, we verify assertion (ii). By replacing G_k by an open subgroup, we may assume that $\mathrm{Ker}(\phi_s) = \mathrm{Ker}(\phi_t)$. Now since s and t are quasi-ℓ-monodromically full, the images $\mathrm{Im}(\phi_s)$ and $\mathrm{Im}(\phi_t)$ are open in $\Phi^{\{\ell\}}_{X_n/k}$ by Proposition 19 (i). We write $C_s \to X_n$ and $C_t \to X_n$ for the corresponding connected finite étale covers. By Proposition 25 (ii), for $\mathfrak{a} \in \{x, y\}$, the open injection $\Pi^{\{\ell\}}_{C_\mathfrak{a}} \hookrightarrow \Pi^{\{\ell\}}_{X_n/k}$ determines a diagram

$$\Pi^{\{\ell\}}_{C_\mathfrak{a}/k} \twoheadrightarrow \Phi^{\{\ell\}}_{C_\mathfrak{a}/k} = \mathrm{Im}(\phi_\mathfrak{a}) \subseteq \Phi^{\{\ell\}}_{X_n/k}.$$

By Proposition 25 (i), the natural surjection $\Gamma^{\{\ell\}}_{C_\mathfrak{a}/k} \twoheadrightarrow \Gamma^{\{\ell\}}_{X_n/k}$ is an isomorphism. Write β for the isomorphism obtained as the composite

$$\Phi^{\{\ell\}}_{C_s/k} = \mathrm{Im}(\phi_s) \overset{\sim}{\longleftarrow} G_k/\mathrm{Ker}(\phi_s) = G_k/\mathrm{Ker}(\phi_t) \overset{\sim}{\longrightarrow} \mathrm{Im}(\phi_t) = \Phi^{\{\ell\}}_{C_t/k}$$

Then we obtain a commutative diagram of profinite groups

$$\begin{array}{ccccc}
\Phi^{\{\ell\}}_{C_s/k} = \mathrm{Im}(\phi_s) & \longrightarrow & \Phi^{\{\ell\}}_{X_n/k} & \longrightarrow & \Gamma^{\{\ell\}}_{C_s/k} = \Gamma^{\{\ell\}}_{X_n/k} \\
\downarrow \beta & & & & \| \\
\mathrm{Im}(\phi_t) = \Phi^{\{\ell\}}_{C_t/k} & \longrightarrow & \Phi^{\{\ell\}}_{X_n/k} & \longrightarrow & \Gamma^{\{\ell\}}_{X_n/k} = \Gamma^{\{\ell\}}_{C_t/k}
\end{array}$$

Thus, since the right-hand square in (8.2) from Proposition 6 is cartesian, by base-changing β via the natural surjection $G_k \twoheadrightarrow \Gamma^{\{\ell\}}_{X_n/k} = \Gamma^{\{\ell\}}_{C_s/k} = \Gamma^{\{\ell\}}_{C_t/k}$, we obtain an isomorphism $\alpha : \Pi^{\{\ell\}}_{C_s/k} \overset{\sim}{\to} \Pi^{\{\ell\}}_{C_t/k}$ over G_k that satisfies the condition of assertion (ii). □

Theorem 49 (Galois-theoretic characterization of equivalence classes of monodromically full points of configuration spaces). *Let ℓ be a prime number, n a positive integer, k a finitely generated extension of $\mathbb{Q}$, $\overline{k}$ an algebraic closure of k, and X a hyperbolic curve over k. Then for two k-rational points x and y of X_n which are ℓ-monodromically full, the following three conditions are equivalent:*

(i) *x is equivalent to y.*
(ii) $\mathrm{Ker}(\rho^{\{\ell\}}_{X[x]/k}) = \mathrm{Ker}(\rho^{\{\ell\}}_{X[y]/k})$.
(iii) *We have $\mathrm{Ker}(\phi_x) = \mathrm{Ker}(\phi_y)$ for the composites*

$$\phi_x : G_k \overset{\pi_1(x)}{\longrightarrow} \pi_1(X_n) \overset{\tilde{\rho}^{\{\ell\}}_{X_n/k}}{\longrightarrow} \mathrm{Aut}(\Delta^{\{\ell\}}_{X_n/k})$$

$$\phi_y : \mathrm{G}_k \xrightarrow{\pi_1(y)} \pi_1(\mathrm{X}_n) \xrightarrow{\overline{\rho}^{\{\ell\}}_{\mathrm{X}_n/k}} \mathrm{Aut}(\Delta^{\{\ell\}}_{\mathrm{X}_n/k}) .$$

Proof. The implication (i) ⇒ (ii) is immediate, while (ii) ⇔ (iii) follows from Proposition 19 (ii). Thus it suffices to show (iii) ⇒ (i). Suppose that condition (iii) is satisfied. Then it follows from Lemma 48 (i) that there exists an automorphism α of $\Pi^{\{\ell\}}_{\mathrm{X}_n/k}$ over G_k such that the two homomorphisms

$$\mathrm{G}_k \xrightarrow{\pi_1(x)} \Pi^{\{\ell\}}_{\mathrm{X}_n/k} \xrightarrow[\sim]{\alpha} \Pi^{\{\ell\}}_{\mathrm{X}_n/k}$$

$$\mathrm{G}_k \xrightarrow{\pi_1(y)} \Pi^{\{\ell\}}_{\mathrm{X}_n/k}$$

coincide. Now by [NT98, Corollary B], [Moc99, Theorem A], [Hos11, Lemma 4.1 (i) and Lemma 4.3 (iii)], the automorphism α of $\Pi^{\{\ell\}}_{\mathrm{X}_n/k}$ arises from an automorphism f_α of X_n over k. Thus, it follows from [Moc99, Theorem C], that $f_\alpha \circ x = y$. In particular, it follows from Lemma 45 that condition (i) is satisfied. □

Remark 50. If, in Theorem 49, one drops the assumption that x and y are ℓ-monodromically full, then the conclusion no longer holds in general. Such a counter-example is as follows. Suppose that $\ell \neq 2$. Let $\overline{\mathbb{Q}}$ be an algebraic closure of $\mathbb{Q}$ and $\zeta_\ell \in \overline{\mathbb{Q}}$ a primitive ℓ-th root of unity. Write $k = \mathbb{Q}(\zeta_\ell)$ and $\mathrm{X} = \mathbb{P}^1_k \setminus \{0, 1, \infty\}$, and set

$$x = \zeta_\ell, \quad y = \zeta_\ell^2 \in \mathrm{X}(k) = k \setminus \{0, 1\}$$

and moreover $\mathrm{U}_x = \mathrm{U} \setminus \mathrm{Im}(x)$ and $\mathrm{U}_y = \mathrm{U} \setminus \mathrm{Im}(y)$. Then it follows from Proposition 39 (ii) that $\mathrm{E}(\mathrm{X}, x, \ell)$ and $\mathrm{E}(\mathrm{X}, y, \ell)$ hold. Thus, it follows from the equivalence (1) ⇔ (3) in Proposition 33 (i) that

$$\mathrm{Ker}(\rho^{\{\ell\}}_{\mathrm{X}/k}) = \mathrm{Ker}(\rho^{\{\ell\}}_{\mathrm{U}_x/k}) = \mathrm{Ker}(\rho^{\{\ell\}}_{\mathrm{U}_y/k}) .$$

In particular, x and y satisfy condition (ii) in the statement of Theorem 49. On the other hand, if, moreover, $\ell \neq 3$, then one verifies easily that U_x is not isomorphic to U_y over k, i.e., x and y do not satisfy condition (i) in the statement of Theorem 49.

Theorem 51 (Galois-theoretic characterization of equivalence classes of quasi-monodromically full points of cores). *Let ℓ be a prime number, k a finitely generated extension of $\mathbb{Q}$, $\overline{k}$ an algebraic closure of k, and X a hyperbolic curve over k which is a k-core, cf. [Moc03, Remark 2.1.1]. Then for two k-rational points x and y of X which are quasi-ℓ-monodromically full, the following four conditions are equivalent:*

(i) $x = y$.

(ii) *x is equivalent to y.*

(iii) *With $\mathrm{U}_x = \mathrm{X} \setminus \mathrm{Im}(x)$ and $\mathrm{U}_y = \mathrm{X} \setminus \mathrm{Im}(y)$, then $\mathrm{Ker}(\rho^{\{\ell\}}_{\mathrm{U}_x/k}) \cap \mathrm{Ker}(\rho^{\{\ell\}}_{\mathrm{U}_y/k})$ is open in $\mathrm{Ker}(\rho^{\{\ell\}}_{\mathrm{U}_x/k})$ and $\mathrm{Ker}(\rho^{\{\ell\}}_{\mathrm{U}_y/k})$.*

(iv) *For the composite*

$$\phi_x : G_k \stackrel{\pi_1(x)}{\longrightarrow} \pi_1(X) \stackrel{\overline{\rho}^{\{\ell\}}_{X/k}}{\longrightarrow} \mathrm{Aut}(\Delta^{\{\ell\}}_{X/k}) \quad \textit{and} \quad \phi_y : G_k \stackrel{\pi_1(y)}{\longrightarrow} \pi_1(X) \stackrel{\overline{\rho}^{\{\ell\}}_{X/k}}{\longrightarrow} \mathrm{Aut}(\Delta^{\{\ell\}}_{X/k}),$$

the intersection $\mathrm{Ker}(\phi_x) \cap \mathrm{Ker}(\phi_y)$ *is open in* $\mathrm{Ker}(\phi_x)$ *and* $\mathrm{Ker}(\phi_y)$.

Proof. It is immediate that the implications (i) $\Rightarrow$ (ii) $\Rightarrow$ (iii) hold. On the other hand, it follows, by Proposition 19 (ii), that the equivalence (iii) $\Leftrightarrow$ (iv) holds. Thus, it suffices to show (iv) $\Rightarrow$ (i). Suppose that condition (iv) is satisfied. Then it follows from Lemma 48 (ii), after replacing G_k by a suitable open subgroup, that there exist

- for $\mathfrak{a} \in \{x,y\}$, a finite étale $\Phi^{\{\ell\}}_{X/k}$-cover $C_{\mathfrak{a}} \to X$ over k which is geometrically connected over k, and
- an isomorphism $\alpha\colon \Pi^{\{\ell\}}_{C_x/k} \xrightarrow{\sim} \Pi^{\{\ell\}}_{C_y/k}$ over G_k

such that

- for $\mathfrak{a} \in \{x,y\}$, the pro-ℓ Galois section $\pi_1(\mathfrak{a})\colon G_k \to \Pi^{\{\ell\}}_{X/k}$ determined by $\mathfrak{a} \in X(k)$ factors through $\Pi^{\{\ell\}}_{C_{\mathfrak{a}}/k} \subseteq \Pi^{\{\ell\}}_{X/k}$, and
- the composite $\alpha \circ \pi_1(x)$ coincides with $\pi_1(y)$.

Note that if $k' \subseteq \overline{k}$ is a finite extension of k, then by [Moc03, Proposition 2.3 (i)], the curve $X \otimes_k k'$ is a k'-core.

Now by [Moc99, Theorem A], the isomorphism $\alpha\colon \Pi^{\{\ell\}}_{C_x/k} \xrightarrow{\sim} \Pi^{\{\ell\}}_{C_y/k}$ arises from an isomorphism $f_\alpha\colon C_x \xrightarrow{\sim} C_y$ over k. Moreover, since X is a k-core, it follows that the isomorphism f_α is an isomorphism over X. Therefore, by [Moc99, Theorem C], condition (i) is satisfied. □

Remark 52. A general hyperbolic curve of type (g,r) with $2g-2+r>2$ over a field k of characteristic 0 is a k-core, by [Moc98, Theorem 5.3], together with [Moc03, Proposition 2.3], cf. also [Moc03, Remark 2.5.1].

Theorem 53 (A necessary and sufficient condition for a quasi-monodromically full Galois section of a hyperbolic curve to be geometric). *Let ℓ be a prime number, k a finitely generated extension of $\mathbb{Q}$, X a hyperbolic curve over k, and $s\colon G_k \to \Pi^{\{\ell\}}_{X/k}$ a pro-ℓ Galois section of X which is quasi-ℓ-monodromically full. Write ϕ_s for the composite*

$$G_k \stackrel{s}{\longrightarrow} \Pi^{\{\ell\}}_{X/k} \stackrel{\overline{\rho}^{\{\ell\}}_{X/k}}{\longrightarrow} \mathrm{Aut}(\Delta^{\{\ell\}}_{X/k}),$$

Then the following four conditions are equivalent:

(i) s is geometric, cf. [Hos10, Definition 1.1 (iii)].
(ii) s arises from a k-rational point of X, cf. [Hos10, Definition 1.1 (ii)].
(iii) There is a quasi-ℓ-monodromically full point $x \in X(k)$ such that if we write ϕ_x for the composite

$$G_k \stackrel{\pi_1(x)}{\longrightarrow} \Pi^{\{\ell\}}_{X/k} \stackrel{\overline{\rho}^{\{\ell\}}_{X/k}}{\longrightarrow} \mathrm{Aut}(\Delta^{\{\ell\}}_{X/k}),$$

then the intersection $\mathrm{Ker}(\phi_s) \cap \mathrm{Ker}(\phi_x)$ *is open in* $\mathrm{Ker}(\phi_s)$ *and* $\mathrm{Ker}(\phi_x)$.

(iv) There is a quasi-ℓ-monodromically full point $x \in \mathrm{X}(k)$ such that with $\mathrm{U} = \mathrm{X} \setminus \{x\}$ the intersection $\mathrm{Ker}(\phi_s) \cap \mathrm{Ker}(\rho^{\{\ell\}}_{\mathrm{U}/k})$ *is open in* $\mathrm{Ker}(\phi_s)$ *and* $\mathrm{Ker}(\rho^{\{\ell\}}_{\mathrm{U}/k})$.

Proof. By Remark 47, we have (ii) ⇒ (iii), while (iii) ⇔ (iv) follows from Proposition 19 (ii). Thus, it suffices to show (i) ⇒ (ii) and (iii) ⇒ (i).

To verify (i) ⇒ (ii), suppose that condition (i) is satisfied. Since the pro-ℓ Galois section s is geometric, there exists a k-rational point $x \in \mathrm{X}^{\mathrm{cpt}}(k)$ of the smooth compactification $\mathrm{X}^{\mathrm{cpt}}$ of X such that the image of s is contained in a decomposition subgroup $\mathrm{D} \subseteq \Pi^{\{\ell\}}_{\mathrm{X}/k}$ of $\Pi^{\{\ell\}}_{\mathrm{X}/k}$ associated to x. Suppose that x is a cusp of X, i.e., an element of $\mathrm{X}^{\mathrm{cpt}}(k) \setminus \mathrm{X}(k)$. Write $\mathrm{I} \subseteq \mathrm{D}$ for the inertia subgroup of D. Then since x is a cusp of X, it follows immediately from [Moc04, Lemma 1.3.7] that $\mathrm{D} = \mathrm{N}_{\Pi^{\{\ell\}}_{\mathrm{X}/k}}(\mathrm{I})$ and $\mathrm{I} = \mathrm{N}_{\Pi^{\{\ell\}}_{\mathrm{X}/k}}(\mathrm{I}) \cap \Delta_{\mathrm{X}/k}$. Therefore, since the composite

$$\Delta^{\{\ell\}}_{\mathrm{X}/k} \hookrightarrow \Pi^{\{\ell\}}_{\mathrm{X}/k} \twoheadrightarrow \Phi^{\{\ell\}}_{\mathrm{X}/k}$$

is injective, cf. Lemma 15 (v), if we write $\overline{\mathrm{D}}, \overline{\mathrm{I}} \subseteq \Phi^{\{\ell\}}_{\mathrm{X}/k}$ for the images of the composites

$$\mathrm{D} \hookrightarrow \Pi^{\{\ell\}}_{\mathrm{X}/k} \twoheadrightarrow \Phi^{\{\ell\}}_{\mathrm{X}/k} \quad \text{and} \quad \mathrm{I} \hookrightarrow \Pi^{\{\ell\}}_{\mathrm{X}/k} \twoheadrightarrow \Phi^{\{\ell\}}_{\mathrm{X}/k},$$

respectively, then it holds that $\overline{\mathrm{D}} \subseteq \mathrm{N}_{\Phi^{\{\ell\}}_{\mathrm{X}/k}}(\overline{\mathrm{I}})$ and $\overline{\mathrm{I}} = \mathrm{N}_{\Phi^{\{\ell\}}_{\mathrm{X}/k}}(\overline{\mathrm{I}}) \cap \Delta^{\{\ell\}}_{\mathrm{X}/k}$. In particular, $\overline{\mathrm{D}} \subseteq \Phi^{\{\ell\}}_{\mathrm{X}/k}$ is not open in $\Phi^{\{\ell\}}_{\mathrm{X}/k}$. On the other hand, since s is quasi-ℓ-monodromically full, it follows immediately from Proposition 19 (ii) that the image of ϕ_s, hence also $\overline{\mathrm{D}} \subseteq \Phi^{\{\ell\}}_{\mathrm{X}/k}$, is open in $\Phi^{\{\ell\}}_{\mathrm{X}/k}$. Thus, we obtain a contradiction. Therefore, x is not a cusp of X. This completes the proof of the implication (i) ⇒ (ii).

Next, to verify the implication (iii) ⇒ (i), suppose that condition (iii) is satisfied. Then it follows from Lemma 48 (ii), after replacing G_k by a suitable open subgroup of G_k, that there exist:

- for $\mathfrak{a} \in \{s, x\}$, a finite étale $\Phi^{\{\ell\}}_{\mathrm{X}/k}$-cover $\mathrm{C}_{\mathfrak{a}} \to \mathrm{X}$ over k, cf. Definition 1 (v), which is geometrically connected over k, and
- an isomorphism $\alpha\colon \Pi^{\{\ell\}}_{\mathrm{C}_x/k} \xrightarrow{\sim} \Pi^{\{\ell\}}_{\mathrm{C}_s/k}$ over G_k

such that

- the pro-ℓ Galois section $\pi_1(x)\colon \mathrm{G}_k \to \Pi^{\{\ell\}}_{\mathrm{X}/k}$ factors through $\Pi^{\{\ell\}}_{\mathrm{C}_x/k} \subseteq \Pi^{\{\ell\}}_{\mathrm{X}/k}$, and
- the composite $\alpha \circ \pi_1(x)$ coincides with s.

Now it follows from [Moc99, Theorem A] that the isomorphism

$$\alpha\colon \Pi^{\{\ell\}}_{\mathrm{C}_x/k} \xrightarrow{\sim} \Pi^{\{\ell\}}_{\mathrm{C}_s/k}$$

arises from an isomorphism $\mathrm{C}_x \xrightarrow{\sim} \mathrm{C}_s$ over k. Therefore, it follows from Lemma 54 below that condition (i) is satisfied. This completes the proof of Theorem 53. □

Lemma 54 (Geometricity and base-changing). *Suppose that k is a finitely generated extension of $\mathbb{Q}$. Let $\Sigma \subseteq \mathfrak{Primes}$ be a nonempty subset of $\mathfrak{Primes}$,* $s\colon \mathrm{G}_k \to \Pi^{\Sigma}_{\mathrm{X}/k}$

a pro-Σ Galois section of X/k, *and* $k' \subseteq \overline{k}$ *a finite extension of* k. *Then* s *is geometric if and only if the restriction* $s|_{G_{k'}}$ *of* s *to* $G_{k'} \subseteq G_k$ *is geometric.*

Proof. The necessity of the condition is immediate. Thus, it suffices to verify the sufficiency of the condition. Suppose that the restriction $s|_{G_{k'}}$ of s to $G_{k'} \subseteq G_k$ is geometric. Now by replacing k' by a finite Galois extension of k, we may assume without loss of generality that k' is Galois over k. Moreover, by replacing $\Pi^{\Sigma}_{X/k}$ by an open subgroup of $\Pi^{\Sigma}_{X/k}$ which contains the image $\mathrm{Im}(s)$ of s, we may assume without loss of generality that X is of genus $\geqslant 2$. Let

$$\cdots \subseteq \Delta_n \subseteq \cdots \subseteq \Delta_2 \subseteq \Delta_1 \subseteq \Delta_0 = \Delta^{\Sigma}_{X/k}$$

be a sequence of characteristic open subgroups of $\Delta^{\Sigma}_{X/k}$ such that $\bigcap_i \Delta_i = \{1\}$.

Write $\Pi_n := \Delta_n \cdot \mathrm{Im}(s) \subseteq \Pi^{\Sigma}_{X/k}$, $X^n \to X$ for the connected finite étale cover corresponding to Π_n and $s_n \colon G_k \to \Pi_n = \Pi^{\Sigma}_{X^n/k}$ for the pro-Σ Galois section of X^n/k determined by s. Note that it holds that

$$\bigcap_i \Pi_i = \mathrm{Im}(s)\,.$$

Now $(X^n)^{\mathrm{cpt}}(k) \neq \emptyset$, where $(X^n)^{\mathrm{cpt}}$ denotes the smooth compactification of X^n. Indeed, since the restriction $s|_{G_{k'}}$ is geometric, it holds that the image of the composite

$$G_{k'} \overset{s_n|_{G_{k'}}}{\hookrightarrow} \Pi^{\Sigma}_{X^n/k} \twoheadrightarrow \Pi^{\Sigma}_{(X^n)^{\mathrm{cpt}}/k}$$

is a decomposition subgroup D associated to a k'-rational point x_n of $(X^n)^{\mathrm{cpt}}$. On the other hand, since this decomposition subgroup D associated to x_n is contained in the image of the homomorphism s_n from G_k, and k' is Galois over k, by considering the $\mathrm{Im}(s_n)$-conjugates of D, it follows from [Moc99, Theorem C] that the k'-rational point x_n is defined over k. In particular, it holds that $(X^n)^{\mathrm{cpt}}(k) \neq \emptyset$.

Since the set $(X^n)^{\mathrm{cpt}}(k)$ is finite by Mordell-Faltings' theorem, the projective limit $(X^{\infty})^{\mathrm{cpt}}(k)$ of the sequence of sets

$$\cdots \longrightarrow (X^n)^{\mathrm{cpt}}(k) \longrightarrow \cdots \longrightarrow (X^2)^{\mathrm{cpt}}(k) \longrightarrow (X^1)^{\mathrm{cpt}}(k) \longrightarrow (X^0)^{\mathrm{cpt}}(k)$$

is nonempty. Let $x_{\infty} \in (X^{\infty})^{\mathrm{cpt}}(k)$ be an element of $(X^{\infty})^{\mathrm{cpt}}(k)$. Then it follows from $\bigcap_i \Pi_i = \mathrm{Im}(s)$ that the pro-Σ Galois section of X/k arising from x_{∞} coincides with s. In particular, s is geometric. This completes the proof of Lemma 54. □

Remark 55. In [Hos10], the author proved that there exist a prime number ℓ, a number field k, a hyperbolic curve X over k, and a pro-ℓ Galois section s of X/k such that the pro-ℓ Galois section s is not geometric, cf. [Hos10, Theorem A]. On the other hand, it seems to the author that the nongeometric pro-ℓ Galois sections appearing in [Hos10] are not quasi-ℓ-monodromically full. It is not clear to the author at the time of writing whether or not there exists a pro-ℓ Galois section of a hyperbolic curve over a number field which is nongeometric and quasi-ℓ-monodromically full.

References

[AI88] G. Anderson and Y. Ihara. Pro-ℓ branched coverings of $\mathbb{P}^1$ and higher circular ℓ-units. *The Annals of Mathematics*, 128(2):271–293, 1988.

[Asa01] M. Asada. The faithfulness of the monodromy representations associated with certain families of algebraic curves. *J. Pure Appl. Algebra*, 159(2-3):123–147, 2001.

[Bog09] M. Boggi. The congruence subgroup property for the hyperelliptic modular group: the open surface case. *Hiroshima Math. J.*, 39(3):351–362, 2009.

[DM69] P. Deligne and D. Mumford. The irreducibility of the space of curves of given genus. *Inst. Hautes Études Sci. Publ. Math.*, 36:75–109, 1969.

[HM11] Y. Hoshi and S. Mochizuki. On the combinatorial anabelian geometry of nodally nondegenerate outer representations. to appear in Hiroshima Math. J.

[Hos09] Y. Hoshi. Absolute anabelian cuspidalizations of configuration spaces of proper hyperbolic curves over finite fields. *Publ. Res. Inst. Math. Sci.*, 45(3):661–744, 2009.

[Hos10] Y. Hoshi. Existence of nongeometric pro-p Galois sections of hyperbolic curves. *Publ. Res. Inst. Math. Sci.*, 46(4):829–848, 2010.

[Hos11] Y. Hoshi. Galois-theoretic characterization of isomorphism classes of monodromically full hyperbolic curves of genus zero. to appear in Nagoya Math. J.

[Knu83] F. Knudsen. The projectivity of the moduli space of stable curves. II. The stacks $M_{g,n}$. *Math. Scand.*, 52(2):161–199, 1983.

[Mat11] M. Matsumoto. Difference between Galois representations in automorphism and outer-automorphism groups of a fundamental group. *Proc. Amer. Math. Soc.*, 139(4):1215–1220, 2011.

[Moc98] S. Mochizuki. Correspondences on hyperbolic curves. *J. Pure Appl. Algebra*, 131(3):227–244, 1998.

[Moc99] S. Mochizuki. The local pro-p anabelian geometry of curves. *Invent. Math.*, 138(2):319–423, 1999.

[Moc03] S. Mochizuki. The absolute anabelian geometry of canonical curves. *Doc. Math.*, Extra Volume: Kazuya Kato's fiftieth birthday:609–640, 2003.

[Moc04] S. Mochizuki. The absolute anabelian geometry of hyperbolic curves. *Galois theory and modular forms*, pages 77–122, 2004.

[Moc10] S. Mochizuki. On the combinatorial cuspidalization of hyperbolic curves. *Osaka J. Math*, 47(3):651–715, 2010.

[MT00] M. Matsumoto and A. Tamagawa. Mapping class group action versus Galois action on profinite fundamental groups. *Amer. J. Math.*, 122(5):1017–1026, 2000.

[MT08] S. Mochizuki and A. Tamagawa. The algebraic and anabelian geometry of configuration spaces. *Hokkaido Math. J.*, 37(1):75–131, 2008.

[NT98] H. Nakamura and N. Takao. Galois rigidity of pro-ℓ pure braid groups of algebraic curves. *Trans. Amer. Math. Soc*, 350(3):1079–1102, 1998.

[SGA1] A. Grothendieck. *Revêtements étale et groupe fondamental (SGA 1)*. Séminaire de géométrie algébrique du Bois Marie 1960-61, directed by A. Grothendieck, augmented by two papers by Mme M. Raynaud, *Lecture Notes in Math.* 224, Springer-Verlag, Berlin-New York, 1971. Updated and annotated new edition: *Documents Mathématiques* 3, Société Mathématique de France, Paris, 2003.

[SGA7-I] A. Grothendieck, M. Raynaud, and D. S. Rim. *Groupes de monodromie en géométrie algébrique. I.* Lecture Notes in Mathematics, Vol. 288. Springer, 1972. Séminaire de Géométrie Algébrique du Bois-Marie 1967–1969 (SGA 7 I).

[Tam97] A. Tamagawa. The Grothendieck conjecture for affine curves. *Compositio Math*, 109(2):135–194, 1997.

Chapter 9
Tempered Fundamental Group and Graph of the Stable Reduction

Emmanuel Lepage

Abstract The tempered fundamental group of a hyperbolic curve over an algebraically closed nonarchimedean field is an invariant that does not depend only on the genus of the curve. In this paper we review what can be recovered of a hyperbolic curve from its tempered fundamental group. S. Mochizuki proved that, for a curve over $\overline{\mathbb{Q}}_p$, one can recover the graph of the stable reduction of the curve. For Mumford curves, one can also recover a natural metric on this graph.

9.1 Introduction

In characteristic 0, one cannot recover much of a proper curve over an algebraically closed field from the geometric fundamental group: it only depends on the genus. In p-adic analytic geometry, the homotopy type of a curve cannot be described in terms of the genus of the curve. Here we will be interested in what one can recover of a p-adic curve from a category of geometric analytic coverings, including finite étale coverings and infinite coverings from analytic geometry. More precisely, we will be interested in tempered coverings, i.e., coverings that become topological coverings after pullback by some finite étale covering. These coverings are classified by a topological group called the **tempered fundamental group**.

This paper reviews what can be recovered of a hyperbolic curve from its geometric tempered fundamental group. Mochizuki proved in [Moc06] that one can recover the graph of its stable reduction:

Theorem 1 ([Moc06, Cor. 3.11]). *If* X_α *and* X_β *are two hyperbolic* $\overline{\mathbb{Q}}_p$*-curves, every (outer) isomorphism* $\phi : \pi_1^{\mathrm{temp}}(\mathrm{X}_{\alpha,\mathbb{C}_p}) \simeq \pi_1^{\mathrm{temp}}(\mathrm{X}_{\beta,\mathbb{C}_p})$ *determines, functorially in* ϕ*, an isomorphism of graphs of the stable reductions* $\bar{\phi} : \mathbf{G}_{\mathrm{X}_\alpha} \simeq \mathbf{G}_{\mathrm{X}_\beta}$.

E. Lepage (✉)
Université Pierre et Marie Curie, 4 place Jussieu, 75005 Paris, France
e-mail: `lepage@math.jussieu.fr`

J. Stix (ed.), *The Arithmetic of Fundamental Groups*, Contributions in Mathematical and Computational Sciences 2, DOI 10.1007/978-3-642-23905-2_9,

More precisely, one can recover from the tempered fundamental group a (p')-version of the fundamental group, which classifies coverings that become topological after pullback by some finite Galois covering of order prime to p. One can describe the graph of the stable reduction from this (p')-tempered fundamental group in the following way:

- Vertices correspond to conjugacy classes of maximal compact subgroups of the (p')-tempered fundamental group.
- Edges correspond to conjugacy classes of nontrivial intersection of two different maximal compact subgroups.

In [Lep10], we were interested in recovering a natural metric of the graph of the stable reduction from the tempered fundamental group. The metric is defined so that the length of an edge is the width of the annulus which is the generic fiber of the formal completion of the node corresponding to this edge. We proved the following:

Theorem 2 ([Lep10, Thm. 4.13]). *If* X_1 *and* X_2 *are hyperbolic Mumford* $\overline{\mathbb{Q}}_p$*-curves, i.e., with totally degenerate stable reduction, then for an isomorphism*

$$\phi : \pi_1^{\mathrm{temp}}(\mathrm{X}_{1,\mathbb{C}_p}) \simeq \pi_1^{\mathrm{temp}}(\mathrm{X}_{2,\mathbb{C}_p}),$$

the induced $\bar{\phi} : \mathbf{G}_{\mathrm{X}_1} \to \mathbf{G}_{\mathrm{X}_2}$ *is an isomorphism of metric graphs.*

In this paper, we will explain the proof of this result. In contrast to the previous result, one cannot recover this metric from the (p')-tempered fundamental group. We will also have to study the topological behavior of wildly ramified coverings, in particular, wildly ramified abelian torsors on a Mumford curve X and follow the study made in [vdP83b] and [vdP83a].

The abelian torsors on Mumford curves can be described in terms of currents on the graph of the stable reduction. Indeed, the pullback of a μ_{p^n}-torsor of X to the universal covering Ω of X can be obtained by pulling back the canonical μ_{p^n}-torsor on $\mathbb{G}_{\mathrm{m}}$ along some theta function $\Omega \to \mathbb{G}_{\mathrm{m}}^{\mathrm{an}}$.

Theta functions can be described in terms of currents on the graph of the stable reduction: given a theta function $f : \Omega \to \mathbb{G}_{\mathrm{m}}^{\mathrm{an}}$, the potential associated to the corresponding current is the function $x \mapsto |f(x)|$. This gives a surjective map

$$\mathrm{Hom}(\pi_1^{\mathrm{temp}}(\mathrm{X}), \mu_n) \to \mathrm{C}(\mathbf{G}_{\mathrm{X}}, \mathbb{Z}/n\mathbb{Z})$$

from the set of μ_n-torsors on X to the set of currents on X with value in $\mathbb{Z}/n\mathbb{Z}$. We will show in Proposition 14 that these morphisms for X_1 and X_2 are compatible with ϕ and $\bar{\phi}$ up to a scalar. For the canonical μ_{p^h}-torsor on $\mathbb{G}_{\mathrm{m}}$, the splitting of the torsor at a Berkovich point depends on the distance of this point to the skeleton of $\mathbb{G}_{\mathrm{m}}^{\mathrm{an}}$, i.e., the line linking 0 to ∞. For a given theta function $\Omega \to \mathbb{G}_{\mathrm{m}}$, one can then get information about the splitting of the torsor in a point in terms of the distance of the point to the support of the corresponding current.

In Sect. 9.2 we recall the Berkovich space of an algebraic variety and define the tempered fundamental group. In Sect. 9.3 we explain Theorem 1, and in Sect. 9.4 we study abelian coverings of Mumford curves, and prove Theorem 2 in Sect. 9.5.

9.2 Tempered Fundamental Group

Let K be a complete nonarchimedean field. We will mostly be interested later on in the case where $K = \mathbb{C}_p$. The norm will be chosen so that $|p| = p^{-1}$ and the valuation so that $v(p) = 1$. All valued fields will have valuations with values in $\mathbb{R}$.

9.2.1 Berkovich Analytification of Algebraic Varieties and Curves

If X is an algebraic variety over K, one can associate to X a topological set X^{an} with a continuous map $\phi : X^{an} \to X$ defined in the following way. A point of X^{an} is an equivalence class of morphisms $\operatorname{Spec} K' \to X$ over $\operatorname{Spec} K$ where K' is a complete valued extension of K. Two morphisms $\operatorname{Spec} K' \to X$ and $\operatorname{Spec} K'' \to X$ are equivalent if there exists a common valued extension L of K' and K'' such that

$$\begin{array}{ccc} \operatorname{Spec} L & \longrightarrow & \operatorname{Spec} K'' \\ \downarrow & & \downarrow \\ \operatorname{Spec} K' & \longrightarrow & X \end{array}$$

commutes. In fact, for any point $x \in X^{an}$, there is a unique smallest such complete valued field defining x denoted by $\mathcal{H}(x)$ and called the completed residue field of x. Forgetting the valuation, one gets points $\operatorname{Spec}(K) \to X$ from the same equivalence class of points: this defines a point of X, hence the map $X^{an} \to X$. If $U = \operatorname{Spec} A$ is an affine open subset of X, every $x \in \phi^{-1}(U)$ defines a seminorm $| \ |_x$ on A. The topology on $\phi^{-1}(U)$ is defined to be the coarsest such that $x \mapsto |f|_x$ is continuous for every $f \in A$.

The space X^{an} is locally compact, and even compact if X is proper. In fact X^{an} is more than just a topological space: it can be enriched into a K-analytic space, in the sense of and as defined by Berkovich in [Ber90].

Let us assume for simplicity that X is irreducible and reduced. One can describe the sheaf $\mathcal{O}$ of analytic functions on X^{an} as follows. But recall that an analytic space in the sense of Berkovich is not just given by a locally ringed space, thus more data should be given to get a well defined analytic space. If U is an open subset of X^{an}, then $\mathcal{O}(U)$ is the ring of functions $f : U \to \bigsqcup_x \mathcal{H}(x)$ such that $f(x) \in \mathcal{H}(x)$ and f is locally a uniform limit of rational functions: for every $x \in U$, there is an open neighborhood V of x and a sequence (g_n) of rational functions on X with no poles in U such that

$$\sup_{x \in V} |f(x) - g_n(x)| \to 0 .$$

The sheaf $\mathcal{M}$ of meromorphic functions on X is the sheaf associated to the presheaf mapping an open subset U of X^{an} to the total ring of fractions of $\mathcal{O}(U)$.

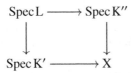

For a hyperbolic curve X over an algebraically closed complete nonarchimedean field K, the homotopy type of X^{an} can be described in terms of the stable model $\mathcal{X}$ of X, see [Knu83, Def. 1.1] for the definition of stable curves in the non-proper case. Indeed, consider the **graph $\mathbf{G}_X$ of the stable reduction** of X: the vertices correspond to the irreducible component of the stable reduction and the edges correspond to the nodes of the stable reduction. There is a canonical embedding $\mathbf{G}_X \hookrightarrow X^{an}$ which admits a canonical strong deformation retraction Φ. In particular, it is a homotopy equivalence. The image of $\mathbf{G}_X$ is called the **skeleton** of X.

Let e be an edge corresponding to a double point of the special fiber $\mathcal{X}_s$ of the stable model of X. Locally for the étale topology, $\mathcal{X}$ is isomorphic to

$$\mathrm{Spec}\, O_K[x,y]/(xy-a)$$

with $a \in K$ and $|a| < 1$. If $\mathring{e}$ is an open edge of $\mathbf{G}_X$, its preimage $\Phi^{-1}(\mathring{e})$ is isomorphic to the open annulus $\{z\ ;\ |a| < |z| < 1\}$. One can define a metric on $\mathbf{G}_X$ by setting the **length** of e to be

$$\lg(e) = v(a)\,.$$

A metric graph $\mathbf{G}_{\mathcal{X}'}$ can also be defined for any semistable model $\mathcal{X}'$ of X, and there is also a natural embedding $\mathbf{G}_{\mathcal{X}'} \hookrightarrow X^{an}$. Those metrics are compatible under blow-up.

Let us assume $K = \mathbb{C}_p$. Points of $\mathbb{A}^{1,an}$ are of four different types and are described in the following way:

- A closed ball $B = B(a,r) \subset \mathbb{C}_p$ of center a and radius r defines a point $b = b_{a,r}$ of $\mathbb{A}^{1,an}$ by

  $$|f|_b = \sup_{x \in B} |f(x)|\,.$$

  The point $b_{a,r}$ is said to be of type 1 if $r = 0$, of type 2 if $r \in p^{\mathbb{Q}}$ and of type 3 otherwise. The pairs (a,r) and (a',r') define the same point if and only if $r = r'$ and $|a-a'| \leqslant r$, that is, if and only if

  $$B(a,r) = B(a',r')\,.$$

- A decreasing family of balls $E = (B_i)$ with empty intersection defines a point by

  $$|f|_E = \inf |f|_{b_i}\,.$$

  Such a point is said to be of type 4.

The analytic projective line $\mathbb{P}^{1,an}$ is obtained from $\mathbb{A}^{1,an}$ by adding a point at infinity. There is a natural metric on the set of points of type 2 and 3 of $\mathbb{P}^{1,an}$ defined by the following formula:

$$d(b_{a,r}, b_{a',r'}) = \begin{cases} \log_p(|a-a'|/r) + \log_p(|a-a'|/r') & \text{if } |a-a'| \geqslant \max(r,r') \\ |\log_p(r'/r)| & \text{if } |a-a'| \leqslant \max(r,r')\,. \end{cases}$$

This metric is compatible with the metrics of the graphs of the semistable reductions. It is invariant under automorphisms of $\mathbb{P}^{1,\mathrm{an}}$. The metric topology defined on the set of points of type 2 and 3 is much finer than the Berkovich topology.

If $x \neq y \in \mathbb{P}^{1,\mathrm{an}}$, there is a unique smallest connected subset of $\mathbb{P}^{1,\mathrm{an}}$ that contains x and y. It is homeomorphic to a closed interval and is denoted $[x,y]$. We also set

$$]x,y[= [x,y] \backslash \{x,y\} ,$$

which consists of points of type 2 and 3. The topology induced by the restriction of the metric d to $]x,y[$ is the topology induced by the topology of $\mathbb{P}^{1,\mathrm{an}}$. For example, if $x,y \in \mathbb{A}^1(\mathbb{C}_p)$,

$$]x,y[= \{b_{x,r}\}_{0<r\leqslant|x-y|} \cup \{b_{y,r}\}_{0<r\leqslant|x-y|} .$$

9.2.2 *Definition of the Tempered Fundamental Group*

Let K be a complete nonarchimedean field, and let X be a connected smooth algebraic variety over K. The usual definition of tempered fundamental groups, as given in [And03, Def. 2.1.1], uses a notion of étale topology on Berkovich spaces defined in [Ber93]. However, we will start with a description of the tempered fundamental group that only uses the topology of the analytifications of the finite étale coverings of X. This description will be enough for our purposes.

Let x : $\operatorname{Spec} \mathrm{K}' \to \mathrm{X}$ be a geometric point of X and assume that K′ is provided with a complete valuation extending the valuation of K so that x also defines a point of X^{an}.

Let $(\mathrm{Y},y) \to (\mathrm{X},x)$ be a pointed Galois finite étale covering of (X,x). Then y also defines a point of the Berkovich space Y^{an}. Let

$$\phi : (\mathrm{Y}^{\infty}, y^{\infty}) \to (\mathrm{Y}^{\mathrm{an}}, y)$$

be the pointed universal covering of $(\mathrm{Y}^{\mathrm{an}}, y)$. Let us consider the following group:

$$\mathrm{H}_{\mathrm{Y}} := \{(g,h) \in \mathrm{Gal}(\mathrm{Y}/\mathrm{X}) \times \mathrm{Aut}_{\mathrm{X}^{\mathrm{an}}}(\mathrm{Y}^{\infty}) \ ; \ \phi h = g^{\mathrm{an}}\phi\} .$$

Heuristically, the group H_{Y} can be thought of as the Galois group of Y^{∞} over X. There is a natural homomorphism $\pi_1^{\mathrm{top}}(\mathrm{Y}^{\mathrm{an}}, y) \to \mathrm{H}_{\mathrm{Y}}$ that maps $h \in \mathrm{Gal}(\mathrm{Y}^{\infty}/\mathrm{Y})$ to $(\mathrm{id}_{\mathrm{Y}}, h)$ and a natural homomorphism $\mathrm{H}_{\mathrm{Y}} \to \mathrm{Gal}(\mathrm{Y}/\mathrm{X})$ mapping (g,h) to g. One thus gets an exact sequence

$$1 \to \pi_1^{\mathrm{top}}(\mathrm{Y}^{\mathrm{an}}, y) \to \mathrm{H}_{\mathrm{Y}} \to \mathrm{Gal}(\mathrm{Y}/\mathrm{X}) \to 1 .$$

The surjectivity of the morphism on the right comes from the extension property of universal topological coverings.

By the strong deformation retraction recalled above, the group $\pi_1^{\mathrm{top}}(\mathrm{Y}^{\mathrm{an}})$ is isomorphic to $\pi_1^{\mathrm{top}}(\mathbf{G}_{\mathrm{Y}})$ and the extension of $\mathrm{Gal}(\mathrm{Y}/\mathrm{X})$ by $\pi_1^{\mathrm{top}}(\mathbf{G}_{\mathrm{Y}})$ can also be directly described in terms of the action of $\mathrm{Gal}(\mathrm{Y}/\mathrm{X})$ on $\mathbf{G}_{\mathrm{Y}}$.

For a morphism $\psi : (\mathrm{Y}_1, y_1) \to (\mathrm{Y}_2, y_2)$ of pointed Galois finite étale coverings, let $\psi^\infty : (\mathrm{Y}_1^\infty, y_1^\infty) \to (\mathrm{Y}_2^\infty, y_2^\infty)$ be the morphism of pointed topological spaces extending ψ. One defines a morphism $\mathrm{H}_{\mathrm{Y}_1} \to \mathrm{H}_{\mathrm{Y}_2}$ by mapping (g,h) to (g',h') such that $h'\psi = \psi h$ and $g'\psi^\infty = \psi^\infty g$.

Definition 3. The **tempered fundamental group** of X, pointed at x, is the topological group

$$\pi_1^{\mathrm{temp}}(\mathrm{X},x) = \varprojlim_{(\mathrm{Y},y)\in C_0} \mathrm{H}_{\mathrm{Y}} ,$$

where C_0 is the filtered category of pointed Galois finite étale coverings.

By a result of J. de Jong, the morphism of groups $\pi_1^{\mathrm{temp}}(\mathrm{X},x) \to \mathrm{H}_{\mathrm{Y}}$ is surjective for any (Y,y) and the group $\pi_1^{\mathrm{temp}}(\mathrm{X},x)$ does not depend on x up to inner automorphism.

The tempered fundamental group is functorial: if $(\mathrm{Y},y) \to (\mathrm{X},x)$ is a morphism of geometrically pointed smooth varieties, one gets a morphism of topological groups $\pi_1^{\mathrm{temp}}(\mathrm{X},x) \to \pi_1^{\mathrm{temp}}(\mathrm{Y},y)$. If we forget base points, one gets a functor π_1^{temp} from smooth K-varieties to topological groups with outer morphisms.

We will be mainly interested in curves X over $\overline{\mathbb{Q}}_p$ in this paper, and abbreviate $\mathrm{X}^{\mathrm{an}} = \mathrm{X}^{\mathrm{an}}_{\mathbb{C}_p}$ and

$$\pi_1^{\mathrm{temp}}(\mathrm{X}) = \pi_1^{\mathrm{temp}}(\mathrm{X}_{\mathbb{C}_p}) .$$

As stated before, the tempered fundamental group classifies a category of analytic coverings. A morphism of K-analytic spaces $f : \mathrm{S} \to \mathrm{X}^{\mathrm{an}}$ is said to be an **étale covering** if X^{an} is covered by open subsets U such that $f^{-1}(\mathrm{U}) = \bigsqcup \mathrm{V}_j$ and $\mathrm{V}_j \to \mathrm{U}$ is finite étale, see [dJ95]. For example, finite étale coverings, also called **algebraic coverings**, and coverings in the usual topological sense for the Berkovich topology, also called **topological coverings**, are tempered coverings. Then, André defines tempered coverings as follows:

Definition 4 ([And03, Def. 2.1.1]). An étale covering $\mathrm{S} \to \mathrm{X}^{\mathrm{an}}$ is **tempered** if it is a quotient of the composition of a topological covering $\mathrm{T}' \to \mathrm{T}$ with a finite étale covering $\mathrm{T} \to \mathrm{X}$.

Here are two properties of the category of tempered coverings.

Proposition 5. *Let* X *be a proper curve over* $\overline{\mathbb{Q}}_p$.

(1) The category of tempered coverings of X *is equivalent to the category of locally constant sheaves for the Berkovich étale topology on* X^{an}.

(2) There is an equivalence between the category of sets endowed with an action of $\pi_1^{\mathrm{temp}}(\mathrm{X},x)$ *that goes through a discrete quotient and the category of tempered coverings of* X^{an}.

9.3 Mochizuki's Results on the pro-(p') Tempered Group of a Curve

Mochizuki proves in [Moc06] the folllowing theorem.

Theorem 6 ([Moc06, Cor. 3.11]). *If* X_α *and* X_β *are two hyperbolic* $\overline{\mathbb{Q}}_p$*-curves, every isomorphism*

$$\gamma : \pi_1^{\mathrm{temp}}(\mathrm{X}_{\alpha,\mathbb{C}_p}) \simeq \pi_1^{\mathrm{temp}}(\mathrm{X}_{\beta,\mathbb{C}_p})$$

determines, functorially in γ*, an isomorphism of graphs* $\bar{\gamma} : \mathbf{G}_{\mathrm{X}_\alpha} \simeq \mathbf{G}_{\mathrm{X}_\beta}$.

Let us explain this result. In fact, the graph of the stable reduction of the curve can even be recovered from a prime-to-p version of the fundamental group. Let

$$\pi_1^{\mathrm{temp}}(\mathrm{X},x)^{(p')} = \varprojlim_{(\mathrm{Y},y)\in C} \mathrm{H}_\mathrm{Y}$$

where C is the category of pointed Galois finite étale coverings (Y,y) of (X,x) such that the order of $\mathrm{Gal}(\mathrm{Y}/\mathrm{X})$ is prime to p. Any morphism $\pi_1^{\mathrm{temp}}(\mathrm{X}_1) \to \pi_1^{\mathrm{temp}}(\mathrm{X}_2)$ induces a morphism

$$\pi_1^{\mathrm{temp}}(\mathrm{X}_1)^{(p')} \to \pi_1^{\mathrm{temp}}(\mathrm{X}_2)^{(p')} .$$

Indeed, the system $(\mathrm{H}_\mathrm{Y})_{p\wedge\sharp\mathrm{Gal}(\mathrm{Y}/\mathrm{X})=1}$ is cofinal among discrete quotients of $\pi_1^{\mathrm{temp}}(\mathrm{X})$ that are extensions of a finite prime-to-p group by a torsionfree group. Hence, if C is the class of discrete groups that have a normal torsionfree subgroup of finite prime-to-p index, then $\pi_1^{\mathrm{temp}}(\mathrm{X})^{(p')}$ is the pro-C completion of $\pi_1^{\mathrm{temp}}(\mathrm{X})$.

A finite prime-to-p covering $\mathrm{Y} \to \mathrm{X}$ extends as a Kummer covering $\mathcal{Y} \to \mathcal{X}$, where $\mathcal{Y}$ and $\mathcal{X}$ are the stable models of Y and X. The map $\mathcal{Y} \to \mathcal{X}$ induces a morphism of graphs $\mathbf{G}_\mathrm{Y} \to \mathbf{G}_\mathrm{X}$ and a commutative diagram

$$\begin{array}{ccc} \mathbf{G}_\mathrm{Y} & \hookrightarrow & \mathrm{Y}^{\mathrm{an}} \\ \downarrow & & \downarrow \\ \mathbf{G}_\mathrm{X} & \hookrightarrow & \mathrm{X}^{\mathrm{an}} . \end{array}$$

Let z be a vertex (resp. an edge) of $\mathbf{G}_\mathrm{X}$. Let us consider a compatible family $(z_\mathrm{Y}^\infty)_{\mathrm{Y}\in C}$ where z_Y^∞ is a vertex (resp. an edge) of $\mathbf{G}_\mathrm{Y}^\infty$ over z. Then $\pi_1^{\mathrm{temp}}(\mathrm{X},x)^{(p')}$ acts on $\mathbf{G}_\mathrm{Y}^\infty$ for every Y. Let D_z be the subgroup of $\pi_1^{\mathrm{temp}}(\mathrm{X},x)^{(p')}$ that stabilizes z_Y for every Y. Changing the family $(z_\mathrm{Y}^\infty)_\mathrm{Y}$ would replace D_z by a conjugate subgroup, so that D_z only depends on z up to conjugacy. The group D_z is called the **decomposition subgroup** of z. It is a profinite subgroup of $\pi_1^{\mathrm{temp}}(\mathrm{X})^{(p')}$, and in fact it can be identified with the decomposition group of z in $\pi_1^{\mathrm{alg}}(\mathrm{X})^{(p')}$, which is the prime-to-$p$ completion of $\pi_1^{\mathrm{temp}}(\mathrm{X})^{(p')}$.

If e is an edge that ends at the vertex v, then D_e is a subgroup of D_v. This gives a natural structure of a graph of profinite groups on $\mathbf{G}_\mathrm{X}$ The important facts for Theorem 6 are that (1) every compact subgroup of $\pi_1^{\mathrm{temp}}(\mathrm{X})^{(p')}$ is contained in some decomposition group of some vertex of $\mathbf{G}_x$, and that (2) if the intersection of two different decomposition subgroups of a vertex is non trivial, then this intersection is the decomposition subgroup of a unique edge, and that (3) the intersection of three different decomposition subgroups of a vertex is trivial. Hence $\mathbf{G}_{\mathrm{X}}$ can be recovered from $\pi_1^{\mathrm{temp}}(\mathrm{X})^{(p')}$ together with the structure of graph of profinite groups on it in the following way.

- The vertices of $\mathbf{G}_{\mathrm{X}}$ correspond to conjugacy classes of maximal compact subgroups of $\pi_1^{\mathrm{temp}}(\mathrm{X})^{(p')}$. Such a maximal compact subgroup is called a **vertical subgroup** of $\pi_1^{\mathrm{temp}}(\mathrm{X})^{(p')}$.
- The edges of $\mathbf{G}_{\mathrm{X}}$ correspond to conjugacy classes of nontrivial intersections of two different maximal compact subgroups. Such an intersection is called an **edge-like** subgroup of $\pi_1^{\mathrm{temp}}(\mathrm{X})^{(p')}$.

A connected finite étale covering $f : \mathrm{Y} \to \mathrm{X}$ induces a morphism of stable models $\mathcal{Y} \to \mathcal{X}$. One can recover $\mathbf{G}_{\mathrm{X}}$ from $\pi_1^{\mathrm{temp}}(\mathrm{X})$ and $\mathbf{G}_{\mathrm{Y}}$ from $\pi_1^{\mathrm{temp}}(\mathrm{Y})$. We are now interested in the combinatorial data of the morphism $\mathcal{Y}_s \to \mathcal{X}_s$ which can be recovered from the embedding $\iota : \pi_1^{\mathrm{temp}}(\mathrm{Y}) \hookrightarrow \pi_1^{\mathrm{temp}}(\mathrm{X})$. If H is a vertical subgroup of $\pi_1^{\mathrm{temp}}(\mathrm{Y})^{(p')}$ corresponding to an irreducible component y of $\mathcal{Y}_s$, then $\iota^{(p')}(\mathrm{H})$ is

- Either a finite index subgroup of a unique vertical subgroup H' of $\pi_1^{\mathrm{temp}}(\mathrm{X})^{(p')}$ if y maps onto an irreducible component x of $\mathcal{X}_s$, and then H' is the vertical subgroup corresponding to x.
- Or a commutative group and hence not a finite index subgroup of any vertical subgroup.

Thus, for a given irreducible component x of $\mathcal{X}_s$, one can recover from ι the set of irreducible components of $\mathcal{Y}_s$ that map onto x. Translated into Berkovich spaces, one gets the following.

Proposition 7. *One can recover from ι the preimage of any vertex of the skeleton of X^{an}. In particular, one can know if the covering is split at this vertex.*

9.4 Abelian Coverings of Mumford Curves

Definition 8. A proper curve X over $\overline{\mathbb{Q}}_p$ is a **Mumford curve** if the following equivalent properties are satisfied:

(a) All normalized irreducible components of its stable reduction are isomorphic to $\mathbb{P}^1$.
(b) X^{an} is locally isomorphic to $\mathbb{P}^{1,\mathrm{an}}$.
(c) Its Jacobian variety J has multiplicative reduction.
(d) The universal topological covering of J^{an} is a torus $\widetilde{\mathrm{J}}$.

The universal topological covering Ω of X^{an} for a Mumford curve X is an open subset of $\mathbb{P}^{1,\mathrm{an}}$. More precisely there is a Shottky subgroup Γ of $\mathrm{PGL}_2(\mathbb{C}_p)$, i.e., a free finitely generated discrete subgroup of $\mathrm{PGL}_2(\mathbb{C}_p)$, such that $\Omega = \mathbb{P}^{1,\mathrm{an}} \backslash \mathcal{L}$, where $\mathcal{L}$ is the closure of the set of $\mathbb{C}_p$-points stabilized by some nontrivial element of Γ, and X is p-adic analytically uniformized as

$$\mathrm{X}^{\mathrm{an}} = \Omega/\Gamma$$

with $\Gamma = \pi_1^{\mathrm{top}}(\mathrm{X})$. The points of $\mathcal{L}$ are of type 1, i.e., are $\mathbb{C}_p$-points.

Let $\mathbf{G}_{\mathrm{X}}$ be the graph of the stable reduction of X and $\mathbf{T}_{\mathrm{X}}$ be its universal topological covering. The graph $\mathbf{T}_{\mathrm{X}}$ embeds in Ω and can be described as the smallest subset of Ω such that $\mathbf{T}_{\mathrm{X}} \cup \mathcal{L}$ is connected, i.e.,

$$\mathbf{T}_{\mathrm{X}} = \bigcup_{(x,y)\in\mathcal{L}^2} \left]x,y\right[.$$

9.4.1 Abelian Torsors and Invertible Functions on Ω

Let X be a Mumford curve of genus $g \geqslant 2$ over $\overline{\mathbb{Q}}_p$, let $\Omega \subset \mathbb{P}^1$ be the universal topological covering of X^{an}, and $\Gamma = \mathrm{Gal}(\Omega/\mathrm{X})$, so that $\mathrm{X}^{\mathrm{an}} = \Omega/\Gamma$. All the cohomology groups will be cohomology groups for étale cohomology in the sense of algebraic geometry or in the sense of Berkovich. One can replace étale cohomology of X^{an} by étale cohomology of X thanks to [Ber95, Thm. 3.1]. Kummer theory gives us the following diagram with exact lower row, see [Ber93, Prop. 4.1.7] for the Kummer exact sequence in Berkovich étale topology.

$$\begin{array}{ccccccc} & & & & \mathrm{H}^1(\mathrm{X},\mu_n) & \longrightarrow & \mathrm{H}^1(\mathrm{X},\mathcal{O}^*) \\ & & & & \downarrow & & \downarrow \\ 1 & \longrightarrow & \mathcal{O}(\Omega)^*/(\mathcal{O}(\Omega)^*)^n & \longrightarrow & \mathrm{H}^1(\Omega,\mu_n) & \longrightarrow & \mathrm{H}^1(\Omega,\mathcal{O}^*) \end{array}$$

The map $\mathrm{H}^1(\mathrm{X},\mathcal{O}^*) \to \mathrm{H}^1(\Omega,\mathcal{O}^*)$ is zero, and thus $\mathrm{H}^1(\mathrm{X},\mu_n) \to \mathrm{H}^1(\Omega,\mu_n)$ goes through $\mathcal{O}(\Omega)^*/(\mathcal{O}(\Omega)^*)^n$. Let us explain why $\mathrm{H}^1(\mathrm{X},\mathcal{O}^*) \to \mathrm{H}^1(\Omega,\mathcal{O}^*)$ is zero. There is a commutative diagram with exact lines:

$$\begin{array}{ccccc} \mathbb{C}_p(\mathrm{X})^* & \longrightarrow & \mathrm{Div}(\mathrm{X}) & \longrightarrow & \mathrm{H}^1(\mathrm{X},\mathcal{O}^*) \\ \downarrow & & \downarrow & & \downarrow \\ \mathcal{M}(\Omega)^* & \longrightarrow & \mathrm{Div}(\Omega) & \longrightarrow & \mathrm{H}^1(\Omega,\mathcal{O}^*) \end{array}$$

where $\mathcal{M}(\Omega)^*$ is the group of nonzero meromorphic functions on Ω and $\mathrm{Div}(\Omega)$ is the group of divisors on $\Omega(\mathbb{C}_\mathrm{P})$ with discrete support, i.e.,

$$\mathrm{Div}(\Omega) = \{f : \Omega(\mathbb{C}_p) \to \mathbb{Z}\ ;\ \text{the support } \{x \in \Omega(\mathbb{C}_p) \mid f(x) \neq 0\} \text{ is discrete}\} .$$

Since $\mathrm{Div}(X) \to H^1(X,\mathcal{O}^*)$ is surjective, the next proposition tells us that the map $H^1(X,\mathcal{O}^*) \to H^1(\Omega,\mathcal{O}^*)$ is zero.

Proposition 9 ([vdP83b, Prop. 1.3]). $\mathcal{M}(\Omega)^* \to \mathrm{Div}(\Omega)$ *is surjective.*

Proof. Assume $\infty \in \Omega$. Let $D = \sum_{i\in I} n_i x_i \in \mathrm{Div}(\Omega)$. Let us choose $y_i \in \mathcal{L}$ such that $|y_i - x_i| = \min_{z\in\mathcal{L}} |z - x_i|$. Then the infinite product $\prod_{i\in I}(\frac{z-x_i}{z-y_i})^{n_i}$ is uniformly convergent on every compact of Ω and thus defines a meromorphic function f such that $\mathrm{div}(f) = D$. □

Combining the above, we find morphisms for every $n \in \mathbb{N}$

$$H^1(X,\mu_n) \to \mathcal{O}(\Omega)^*/(\mathcal{O}(\Omega)^*)^n \hookrightarrow H^1(\Omega,\mu_n)\,.$$

Moreover, $H^1(X,\mu_n)$ is mapped into the set of Γ-equivariant elements of $H^1(\Omega,\mu_n)$. There is a spectral sequence

$$H^p(\Gamma, H^q(\Omega,\mu_n)) \quad\Longrightarrow\quad H^n(X^{an},\mu_n) = H^n(X,\mu_n)$$

associated to the Galois étale covering $\Omega \to X^{an}$. It gives a five-term exact sequence

$$0 \to H^1(\Gamma,\mu_n) \to H^1(X,\mu_n) \to H^1(\Omega,\mu_n)^\Gamma \to H^2(\Gamma,\mu_n) \to H^2(X,\mu_n)\,.$$

We have $H^2(\Gamma,\mu_n) = 0$, since Γ is free. Thus $H^1(X,\mu_n) \to H^1(\Omega,\mu_n)^\Gamma$ is surjective, which in turn implies that

$$(\mathcal{O}(\Omega)^*/(\mathcal{O}(\Omega)^*)^n)^\Gamma \to H^1(\Omega,\mu_n)^\Gamma$$

is an isomorphism. The kernel of $H^1(X,\mu_n) \to H^1(\Omega,\mu_n)$ is denoted by $H^1_{\mathrm{top}}(X,\mu_n)$ and consists of μ_n-torsors that are already locally constant for the topology of X^{an}.

9.4.2 Invertible Functions on Ω and Currents on $\mathbf{T}$

We recall the combinatorial description of $\mathcal{O}(\Omega)^*$ in terms of currents on the graph $\mathbf{T} = \mathbf{T}_X$ as given in [vdP83b]. If $\mathbf{G}_0$ is a locally finite graph and A is an abelian group, a **current** C on $\mathbf{G}_0$ with coefficients in A is a function

$$C : \{\text{oriented edges of } \mathbf{G}_0\} \to A$$

such that

- $C(e) = -C(e')$ if e and e' are the same edge but with reversed orientation.
- If v is a vertex of $\mathbf{G}_0$, then $\sum_{e \text{ ending at } v} C(e) = 0$.

The group of currents on $\mathbf{G}_0$ with coefficients in A will be denoted $C(\mathbf{G}_0, A)$. We will simply write $C(\mathbf{G}_0)$ for $C(\mathbb{G}_0,\mathbb{Z})$.

Proposition 10 ([vdP83b, Prop. 1.1]). *There is an exact sequence*

$$1 \to \mathbb{C}_p^* \to \mathrm{O}(\Omega)^* \to \mathrm{C}(\mathbf{T}) \to 0 .$$

Proof (sketch). The morphism $\mathrm{O}(\Omega)^* \to \mathrm{C}(\mathbf{T})$ assigns to an $f \in \mathrm{O}(\Omega^*)$ a current C_f on $\mathbf{T}$ as follows. For any oriented open edge $\mathring{e}$, the preimage $\Phi^{-1}(\mathring{e})$ is isomorphic to an open annulus

$$\{z \in \mathbb{P}^1 \ ; \ 1 < |z| < r\}$$

where the beginning of the edge tends to 1 and the end tends to r. Let

$$i : \{z \in \mathbb{P}^1 \ ; \ 1 < |z| < r\} \to \Phi^{-1}(\mathring{e})$$

be such an isomorphism. Then fi can be written in a unique way as $z \mapsto z^m g(z)$ with $m \in \mathbb{Z}$ and $|g|$ is constant. Then we set $C_f(e) = m$, which does not depend on the choice of i.

For $x, y \in \mathcal{L}$, let $f_{x,y} : \mathbb{P}^1 \to \mathbb{P}^1$ be an automorphism of $\mathbb{P}^1$ that maps x to 0 and y to ∞. It restricts to a map $\Omega \to \mathbb{G}_{\mathrm{m}}$, i.e., an element $f_{x,y} \in \mathcal{O}(\Omega)^*$ that depends on the chosen homography only up to multiplication by a scalar. Let us choose $x_0 \in \Omega(\mathbb{C}_p)$ and fix $f_{x,y}$ by imposing $f_{x,y}(x_0) = 1$. The corresponding current is denoted by $c_{]x,y[}$. We find $c_{]x,y[}(e) = \pm 1$ for every edge that belongs to $]x,y[$ and $c_{]x,y[}(e) = 0$ for any other edge.

Every current $c \in \mathrm{C}(\mathbf{T})$ can be written as a locally finite sum $c = \sum_{i \in \mathrm{I}} n_i c_{]x_i,y_i[}$. Then $f = \prod f_{x_i,y_i}$ is a locally uniformly convergent product and defines a preimage of c in $\mathcal{O}(\Omega)^*$. The exactness in the middle comes from the fact that every bounded analytic function on Ω is constant. □

The Kummer exact sequence of Ω gives a map

$$\mathrm{C}(\mathbf{T}, \mathbb{Z}/n\mathbb{Z}) = \mathrm{C}(\mathbf{T})/n\mathrm{C}(\mathbf{T}) \simeq \mathcal{O}(\Omega)^*/(\mathcal{O}(\Omega)^*)^n \to \mathrm{H}^1(\Omega, \mu_n) ,$$

that combined with $(\mathcal{O}(\Omega)^*/(\mathcal{O}(\Omega)^*)^n)^\Gamma = \mathrm{C}(\mathbf{T}, \mathbb{Z}/n\mathbb{Z})^\Gamma = \mathrm{C}(\mathbf{G}, \mathbb{Z}/n\mathbb{Z})$ yields an exact sequence

$$0 \to \mathrm{H}^1_{\mathrm{top}}(\mathrm{X}, \mu_n) \to \mathrm{H}^1(\mathrm{X}, \mu_n) \to \mathrm{C}(\mathbf{G}, \mathbb{Z}/n\mathbb{Z}) \to 0. \qquad (9.1)$$

9.4.3 Splitting of Abelian Torsors

We now assume $n = p^h$ for some positive integer h. Let c be a current on $\mathbf{T}$ with coefficients in $\mathbb{Z}/p^h\mathbb{Z}$. The subtree $\cup_{e|c(e)\neq 0}$ of $\mathbf{T}$ is called the **support** of c and is denoted by $\operatorname{supp} c$. It can also be viewed as a subset of Ω. We show that the corresponding μ_{p^h}-torsor of Ω is split at some point if this point is far enough from the support of c.

Let us begin with the simplest case, the torsor corresponding to the current $c_{]x,y[}$.

Lemma 11. *Let $z \in \Omega$ be of type 2 or 3, and let h be a positive integer. The μ_{p^h}-torsor corresponding to the current $c_{]x,y[}$ is split over z if and only if*

$$d(z,]x,y[) > h + 1/(p-1)$$

where d is the metric defined in 9.2.1.

Proof. Up to changing the embedding $\Omega \to \mathbb{P}^1$, one can assume $x = 0$ and $y = \infty$. Then the μ_{p^h}-torsor is just $f : \mathbb{G}_m \to \mathbb{G}_m$, with $f(t) = t^{p^h}$.

We first assume that $h = 1$. Let $b_{a,r}$ be a preimage of z. Then $f(b_{a,r}) = b_{a^p,r'} = z$ with $r' = \sup_{|x-a|\leqslant r} |f(x) - a^p|$. If $a = 0$, then $r' = p^h$. Otherwise we compute

$$f(y+a) - a^p = \sum_{k=1}^{p} \binom{p}{k} a^{p-k} y^k$$

and $r' = \sup_{k=1\dots p} |\binom{p}{k}| \cdot |a|^{p-k} r^k$ with $|\binom{p}{k}| = p^{-1}$ if $1 \leqslant k \leqslant p-1$ and $|\binom{p}{p}| = 1$. Thus

$$r' = \begin{cases} r^p & \text{if } r \leqslant |a| p^{-\frac{1}{p-1}} \\ |a|^{p-1} r & \text{if } r \geqslant |a| p^{-\frac{1}{p-1}} \end{cases}.$$

Let ζ be a generator of μ_p. The torsor is split over z if and only if the orbit of $b_{a,r}$ under the action of μ_p is not reduced to one point, if and only if $b_{\zeta a,r} \neq b_{a,r}$, if and only if $|\zeta a - a| > r$, i.e., $a \neq 0$ and $|\zeta - 1| > r/|a|$. But $|\zeta - 1| = p^{-\frac{1}{p-1}}$. If $a = 0$, $z \in]0,\infty[$ and the torsor is not split. Otherwise, one can assume $r' < |a^p|$. The torsor is split over $z = b_{a^p,r'}$ if and only if $r'/|a^p| < p^{-\frac{p}{p-1}}$. But

$$d(b_{a^p,r'},]0,\infty[) = \inf_{r''} d(b_{a^p,r'}, b_{0,r''}) = \log_p |a^p|/r'.$$

The result follows for any h by induction. □

Every current c on $\mathbf{T}$ can be decomposed as a locally finite sum $c = \sum_{i\in I} n_i c_{]x_i,y_i[}$. For any point z of type 2 or 3, the set

$$I_z = \{i \,;\, d(z,]x_i,y_i[) \leqslant h + 1/(p-1)\}$$

is finite, and locally around z, the μ_{p^h} torsor $Y_c \to \Omega$ defined by c is isomorphic to the μ_{p^h}-torsor defined by $c' = \sum_{i\in I_z} n_i c_{]x_i,y_i[}$. One can moreover choose the decomposition $c = \sum_{i\in I} n_i c_{]x_i,y_i[}$ such that $\operatorname{supp} c = \cup]x_i,y_i[$. This proves the following proposition.

Proposition 12. *If z is a point of type 2 or 3 of Ω such that $d(z, \operatorname{supp} c) > h + \frac{1}{p-1}$, then the μ_{p^h}-torsor of Ω defined by c is split over z.*

9.5 Metric Graph and Tempered Fundamental Group

We consider now two Mumford curves X_1 and X_2 over $\overline{\mathbb{Q}}_p$ of genus $g \geqslant 2$, and an isomorphism

$$\phi : \pi_1^{\mathrm{temp}}(\mathrm{X}_1) \xrightarrow{\sim} \pi_1^{\mathrm{temp}}(\mathrm{X}_2)\,,$$

together with the induced isomorphism of graphs

$$\bar{\phi} : \mathbf{G}_1 \xrightarrow{\sim} \mathbf{G}_2\,,$$

hence an isomorphism $\bar{\phi} : \mathbf{T}(\Omega_1) \xrightarrow{\sim} \mathbf{T}(\Omega_2)$.

Theorem 13. *The isomorphism* $\bar{\phi} : \mathbf{G}_1 \to \mathbf{G}_2$ *is an isomorphism of metric graphs.*

We will sketch the proof of this result. The metric d_i on $\mathbf{T}_i$ obtained by pullback of the metric on $\mathbf{G}_i$ is equal to the one induced by the natural metric of $\mathbb{P}^{1,\mathrm{an}}$. Consider the diagram

$$\begin{array}{ccc} \mathrm{H}^1(\mathrm{X}_2,\mu_n) & \longrightarrow & \mathrm{H}^1(\mathrm{X}_1,\mu_n) \\ \downarrow & & \downarrow \\ \mathrm{C}(\mathbf{G}_2,\mathbb{Z}/n\mathbb{Z}) & \longrightarrow & \mathrm{C}(\mathbf{G}_1,\mathbb{Z}/n\mathbb{Z}) \end{array} \qquad (9.2)$$

where the vertical arrows are given by equation (9.1), the upper arrow is $\mathrm{Hom}(\phi,\mu_n)$ after identifying $\mathrm{H}^1(\mathrm{X}_i,\mu_n) = \mathrm{Hom}(\pi_1^{\mathrm{temp}}(\mathrm{X}_i),\mu_n)$. The lower arrow is $\mathrm{C}(\bar{\phi},\mathbb{Z}/n\mathbb{Z})$.

Proposition 14. *The diagram (9.2) above commutes up to multiplication by a scalar* $\lambda \in (\mathbb{Z}/n\mathbb{Z})^*$.

Proof. For $i = 1,2$, a μ_n-torsor on X_i is in $\mathrm{H}^1_{\mathrm{top}}(\mathrm{X}_i,\mu_n)$ if and only if it is dominated by a Galois tempered covering with torsion free Galois group. Thus the isomorphism $\mathrm{H}^1(\mathrm{X}_2,\mu_n) \to \mathrm{H}^1(\mathrm{X}_1,\mu_n)$ is compatible with a unique isomorphism

$$\tilde{\phi} : \mathrm{C}(\mathbf{G}_2,\mathbb{Z}/n\mathbb{Z}) \to \mathrm{C}(\mathbf{G}_1,\mathbb{Z}/n\mathbb{Z})\,.$$

We have to show that there exists $\lambda \in (\mathbb{Z}/n\mathbb{Z})^*$ such that for every $c \in \mathrm{C}(\mathbf{G}_2,\mathbb{Z}/n\mathbb{Z})$ and every edge e of $\mathbf{G}_1$, we have $\tilde{\phi}(c)(e) = \lambda c(\bar{\phi}(e))$.

Let e_1 be an edge of $\mathbf{G}_1$ such that $\mathbf{G}_1\backslash e_1$ is connected, and let $e_2 = \bar{\phi}(e_1)$. By contracting $\mathbf{G}_i\backslash e_i$ to a point, one gets a map $\mathbf{G}_i \to \mathbf{S}^1$ to the circle $\mathbf{S}^1$. There is a unique connected Galois covering $\mathbf{S}^1 \to \mathbf{S}^1$ of order n and Galois group $\mathrm{G} = \mathbb{Z}/n\mathbb{Z}$, the pullback of which to $\mathbf{G}_i$ we denote by $\psi_i : \mathbf{G}_i^{(n)} \to \mathbf{G}_i$. Let $\mathrm{X}_i^{(n)}$ be the corresponding topological covering of X_i. We will use the following lemma, where for sake of clarity we have omitted the indices $i = 1,2$.

Lemma 15. *For a current* $c \in \mathrm{C}(\mathbf{G},\mathbb{Z}/n\mathbb{Z})$ *we have* $c(e) = 0$ *if and only if there exists* $c' \in \mathrm{C}(\mathbf{G}^{(n)},\mathbb{Z}/n\mathbb{Z})$ *such that* $\psi^*c = \sum_{g\in\mathrm{G}} g^*c'$.

Proof. Assume there is such a current c'. Then, the fact that c' is a current implies that $c'(e')$ is the same for every preimage e' of e. Thus if e' is such a preimage of e, then

$$c(e) = \psi^*c(e') = \sum_{e''\in\psi^{-1}(e)} c'(e'') = n\cdot c'(e') = 0\,.$$

If $c(e) = 0$, then c induces a current on $\mathbf{G}\backslash e$. Let A be a connected component of $\psi^{-1}(\mathbf{G}\backslash e)$. Since ψ is a trivial covering above $\mathbf{G}\backslash e$, the component A is isomorphic to $\mathbf{G}\backslash e$ and c thus induces a current on A. One extends this current by 0 on $\mathbf{G}^{(n)}\backslash \mathrm{A}$ to get a current c' on $\mathbf{G}^{(n)}$ for which $\psi^* c = \sum_{g\in\Gamma} g^* c'$. □

Let us come back to the proof of Proposition 14. The map ϕ induces a commutative diagram

$$\begin{array}{ccc} \pi_1^{\mathrm{temp}}(\mathrm{X}_1^{(n)}) & \xrightarrow{\phi^{(n)}} & \pi_1^{\mathrm{temp}}(\mathrm{X}_2^{(n)}) \\ \big\downarrow & & \big\downarrow \\ \pi_1^{\mathrm{temp}}(\mathrm{X}_1) & \xrightarrow{\phi} & \pi_1^{\mathrm{temp}}(\mathrm{X}_2)\,. \end{array}$$

It induces a commutative diagram compatible with the actions of G

$$\begin{array}{ccc} \mathrm{C}(\mathbf{G}_2,\mathbb{Z}/n\mathbb{Z}) & \xrightarrow{\tilde{\phi}} & \mathrm{C}(\mathbf{G}_1,\mathbb{Z}/n\mathbb{Z}) \\ \big\downarrow \psi_2^* & & \big\downarrow \psi_1^* \\ \mathrm{C}(\mathbf{G}_2^{(n)},\mathbb{Z}/n\mathbb{Z}) & \xrightarrow{\tilde{\phi}^{(n)}} & \mathrm{C}(\mathbf{G}_1^{(n)},\mathbb{Z}/n\mathbb{Z})\,. \end{array}$$

Lemma 15 shows that $\tilde{\phi}(c)(e_1) = 0$ if and only if $\psi_1^*\tilde{\phi}(c)$ is a norm in $\mathrm{C}(\mathbf{G}_1^{(n)},\mathbb{Z}/n\mathbb{Z})$ for the G-action. This holds if and only if $\psi_2^*(c)$ is a norm in $\mathrm{C}(\mathbf{G}_2^{(n)},\mathbb{Z}/n\mathbb{Z})$ for the G-action, or again by Lemma 15 if and only if $c(e_2) = 0$.

An edge e of a graph $\mathbf{G}$ is said to be **unconnecting** if $\pi_0(\mathbf{G}\backslash\{e\}) \to \pi_0(\mathbf{G})$ is injective. If the evaluation map $\mathrm{ev}_e : \mathrm{C}(\mathbf{G}_1,\mathbb{Z}/n\mathbb{Z}) \to \mathbb{Z}/n\mathbb{Z}$ is nonzero, then e is unconnecting and ev_e is surjective.

Let e be an unconnecting edge of $\mathbf{G}_1$. Since $\tilde{\phi}$ maps $\mathrm{Ker}(\mathrm{ev}_{\bar{\phi}(e)})$ to $\mathrm{Ker}(\mathrm{ev}_e)$, one gets an isomorphism

$$\mathbb{Z}/n\mathbb{Z} = \mathrm{C}(\mathbf{G}_2,\mathbb{Z}/n\mathbb{Z})/\,\mathrm{Ker}(\mathrm{ev}_{\bar{\phi}(e)}) \to \mathrm{C}(\mathbf{G}_1,\mathbb{Z}/n\mathbb{Z})/\,\mathrm{Ker}(\mathrm{ev}_e) = \mathbb{Z}/n\mathbb{Z}$$

induced by $\tilde{\phi}$, that is multiplication by a unique $\lambda_e \in (\mathbb{Z}/n\mathbb{Z})^*$. This means that for every $c \in \mathrm{C}(\mathbf{G}_1,\mathbb{Z}/n\mathbb{Z})$, we have $\tilde{\phi}(c)(e) = \lambda_e c(\bar{\phi}(e))$. One has now to prove that λ_e does not depend of e.

Let $\pi_2 : \mathbf{G}_2' \to \mathbf{G}_2$ be a finite topological covering, let $\mathrm{X}_2' \to \mathrm{X}_2$ be the corresponding finite topological covering of X_2, let $\pi_1 : \mathbf{G}_1' = \bar{\phi}^*\mathbf{G}_2' \to \mathbf{G}_1$ and let X_1' be the corresponding finite topological covering of X_1. Let $\phi' : \pi_1^{\mathrm{temp}}(\mathrm{X}_1) \to \pi_1^{\mathrm{temp}}(\mathrm{X}_2)$ be the induced isomorphism. Let e' be a preimage of e which is also unconnecting. The scalar $\lambda_{e'} \in \mathbb{Z}/n\mathbb{Z}$ induced by ϕ' turns out to be equal to λ_e because for every current $c \in \mathrm{C}(\mathbf{G}_2,\mathbb{Z}/n\mathbb{Z})$ we have

$$\begin{aligned} \lambda_e \cdot \tilde{\phi}'(\pi_2^* c)(e') &= \lambda_e \lambda_{e'} \cdot \pi_2^* c(\bar{\phi}'(e')) = \lambda_e \lambda_{e'} \cdot c(\bar{\phi}(e)) \\ &= \lambda_{e'} \cdot \tilde{\phi}(c)(e) = \lambda_{e'} \cdot \pi_1^* \tilde{\phi}(c)(e') = \lambda_{e'} \cdot \tilde{\phi}'(\pi_2^* c)(e')\,. \end{aligned}$$

If e_a and e_b are two unconnecting edges of $\mathbf{G}_1$, there exists a finite topological covering $\mathbf{G}'_1 \to \mathbf{G}_1$, a preimage e'_a (resp. e'_b) of e_a (resp. e_b) and a cycle of $\mathbf{G}'_1$ that goes through e'_a and e'_b. Let c be the $\mathbb{Z}/n\mathbb{Z}$-current on $\mathbf{G}'_2$ which follows this cycle and is zero everywhere else. Since $\tilde{\phi}'(c)$ must also be a current, one gets that λ must be constant on the cycle and thus $\lambda_{e_a} = \lambda_{e'_a} = \lambda_{e'_b} = \lambda_{e_b}$. Therefore the scalar $\lambda := \lambda_e$ does not depend of e and, for every $c \in \mathrm{C}(\mathbf{G}_2, \mathbb{Z}/n\mathbb{Z})$ and every edge e of $\mathbf{G}_1$, we have $\tilde{\phi}(c)(e) = \lambda c(\bar{\phi}(e))$. □

We continue the proof of Theorem 13. Let L_1 be an oriented loop in $\mathbf{G}_1$, and set $\mathrm{L}_2 = \bar{\phi}(\mathrm{L}_1)$. Let $\widetilde{\mathrm{L}}_1$ be an oriented path lifting L_1 in $\mathbf{T}_1$ and set $\widetilde{\mathrm{L}}_2 = \bar{\phi}(\widetilde{\mathrm{L}}_1)$. Then $\widetilde{\mathrm{L}}_i =]x_i, y_i[$ for some points $x_i, y_i \in \Omega_i$. Let z_1 be a vertex of $\mathbf{T}_1$ and set $z_2 = \bar{\phi}(z_1)$. The stabilizer $\mathrm{H} \subset \Gamma_1$ of $\widetilde{\mathrm{L}}_1$ is the image of $\pi_1(\mathrm{L}_1) \to \pi_1(\mathbf{G}_1) = \Gamma_1$.

We fix an integer $h' > 0$. Let Γ' be a finite index subgroup of Γ_1 such that, for every $g \in \Gamma' \backslash \mathrm{H}$,

$$\min\{d_1(\widetilde{\mathrm{L}}_1, g\widetilde{\mathrm{L}}_1), d_2(\widetilde{\mathrm{L}}_2, \bar{\phi}(g\widetilde{\mathrm{L}}_1))\} > h' .$$

The current

$$c_1 = \sum_{g \in \Gamma'/(\mathrm{H} \cap \Gamma')} g^* c_{]x_1, y_1[}$$

is Γ'-equivariant, and we set $c_2 = (\bar{\phi}^{-1})^* c_1$. We consider the finite topological covering $\mathrm{X}'_1 = \Omega_1/\Gamma' \to \mathrm{X}_1$, and set $\mathrm{X}'_2 = \phi^* \mathrm{X}'_1 = \Omega_2/\phi(\Gamma')$.

Let Y_1 be a μ_{p^h}-torsor of X'_1, whose pullback S_1 to Ω_1 is induced by c_1. Let $\mathrm{Y}_2 = (\phi^{-1})^* \mathrm{Y}_1$ and let S_2 be its pullback to Ω_2. According to Proposition 14, the torsor S_2 is the μ_{p^h}-torsor induced by λc_2 for some $\lambda \in \mathbb{Z}/p^h\mathbb{Z}$.

For h' chosen big enough, locally around z_1 (resp. z_2), S_1 (resp. S_2) is isomorphic to the μ_{p^h}-torsor induced by $c_{]x_1,y_1[}$ (resp. $\lambda c_{]x_2,y_2[}$), hence is split at z_1 (resp. z_2) if and only if $d_1(]x_1, y_1[, z_1) > h + \frac{1}{p-1}$ (resp. $d_2(]x_2, y_2[, z_2) > h + \frac{1}{p-1}$). According to Proposition 7, those two conditions must be equivalent for any z_1 and h. In particular,

$$\left| d_1(]x_1, y_1[, z_1) - d_2(]x_2, y_2[, z_2) \right| \leqslant 2 . \tag{9.3}$$

Let L'_1 be a loop in $\mathbf{G}_1$ and $\widetilde{\mathrm{L}}'_1$ be a lifting of the universal covering of L'_1 to $\mathbf{T}_1$ such that $\widetilde{\mathrm{L}}'_1 \neq \widetilde{\mathrm{L}}_1$. This is possible since X_1 is hyperbolic and thus $\widetilde{\mathrm{L}}_1 \neq \mathbf{T}_1$. Let $\mathrm{lg}_1(\mathrm{L}'_1)$ be the length of this loop and $\mathrm{lg}_2(\bar{\phi}(\mathrm{L}'_1))$ be the length of $\bar{\phi}(\mathrm{L}'_1)$.

Let $(z_1^n)_{n \in \mathbb{Z}}$ be the family of preimages in $\widetilde{\mathrm{L}}'_1$ of a vertex of L'_1, numbered compatibly with an orientation of $\widetilde{\mathrm{L}}'_1$. Let z_2^n be the image of z_1^n in $\mathbf{T}_2$. Then there exists constants c_1 and c_2 such that, for $n \gg 0$,

$$d_1(]x_1, y_1[, z_1^n) = n\,\mathrm{lg}_1(\mathrm{L}'_1) + c_1 ,$$

$$d_2(]x_2, y_2[, z_2^n) = n\,\mathrm{lg}_2(\bar{\phi}(\mathrm{L}'_1)) + c_2 .$$

Thus, for $n \gg 0$,

$$n|\mathrm{lg}_1(\mathrm{L}'_1) - \mathrm{lg}_2(\bar{\phi}(\mathrm{L}'_1))| = |d_1(]x_1, y_1[, z_1) - d_2(]x_2, y_2[, z_2) + c_1 - c_2| \leqslant 2 + |c_1| + |c_2| .$$

Since the sequence $(n|\lg_1(\mathrm{L}'_1) - \lg_2(\bar{\phi}(\mathrm{L}'_1))|)_{n\in\mathbb{N}}$ is bounded,

$$\lg_1(\mathrm{L}'_1) = \lg_2(\bar{\phi}(\mathrm{L}'_1)) .$$

Theorem 13 is thus a consequence of the following purely combinatorial statement applied to $\mathbf{G} = \mathbf{G}_1$ and $f = \lg_1 - \lg_2 \circ \bar{\phi}$.

Proposition 16 ([Lep10, Prop. A.1]). *Let* **G** *be finite graph such that the valency of every vertex is at least 3. Let*

$$f : \{\textit{edges of}\,\mathbf{G}\} \to \mathbb{R}$$

be any function. Let us denote also by f the induced function on the set of edges of a topological covering of **G**. *Let us set $f(\mathrm{C}) = \sum_{x\in\{\textit{edges of}\,\mathrm{C}\}} f(x)$ for C a loop of a covering of* **G**.

If $f(\mathrm{C}) = 0$ for every loop C of every covering of **G**, *then $f = 0$.*

Remark 17. Thm. 13 is also true for open Mumford curves, see [Lep09, Cor. 3.4.7].

References

[And03] Y. André. Period mappings and differential equations: From $\mathbb{C}$ to $\mathbb{C}_p$. *MSJ Memoirs*, 12, 2003.

[Ber90] V. G. Berkovich. *Spectral theory and analytic geometry over non-archimedian fields*, volume 33 of *Mathematical Surveys and Monographs*. American Mathematical Society, Providence, 1990.

[Ber93] V. G. Berkovich. Étale cohomology for non-archimedean analytic spaces. *Publication mathématiques de l'Institut des hautes études scientifiques*, 78:5–161, 1993.

[Ber95] V. G. Berkovich. On the comparison theorem for étale cohomology of non-Archimedean analytic spaces. *Israel J. Math.*, 92(1-3):45–59, 1995.

[dJ95] A. J. de Jong. Étale fundamental group of non archimedean analytic spaces. *Compositio mathematica*, 97:89–118, 1995.

[Knu83] F. F. Knudsen. The projectivity of the moduli space of stable curves. II. The stacks $\mathrm{M}_{g,n}$. *Math. Scand.*, 52(2):161–199, 1983.

[Lep09] E. Lepage. *Géométrie anabélienne tempérée*. PhD thesis, Université Paris Diderot, 2009.

[Lep10] E. Lepage. Tempered fundamental group and metric graph of a mumford curve. *Publications of the Research Institute for Mathematical Sciences*, 46(4):849–897, 2010.

[Moc06] S. Mochizuki. Semi-graphs of anabelioids. *Publications of the Research Institute of Mathematical Sciences*, 42(1):221–322, 2006.

[vdP83a] M. van der Put. Étale coverings of a Mumford curve. *Annales de l'Institut Fourier*, 33(1):29–52, 1983.

[vdP83b] M. van der Put. Les fonctions thêta d'une courbe de Mumford. *Groupe de travail d'Analyse Ultramétrique, 9e Année: 1981/82, No.1, Exposé No.10*, 1983.

Chapter 10
$\mathbb{Z}/\ell$ Abelian-by-Central Galois Theory of Prime Divisors

Florian Pop*

Abstract In this manuscript I show how to recover some of the inertia structure of (quasi) divisors of a function field $\mathrm{K}|k$ over an algebraically closed base field k from its maximal mod ℓ *abelian-by-central* Galois theory of K, provided $\mathrm{td}(\mathrm{K}|k) > 1$. This is a first technical step in trying to extend Bogomolov's birational anabelian program beyond the full pro-ℓ situation, which corresponds to the limit case mod ℓ^∞.

10.1 Introduction

At the beginning of the 1990s, Bogomolov [Bog91] initiated a program whose final aim is to recover function fields $\mathrm{K}|k$ over algebraically closed base fields k from their pro-ℓ abelian-by-central Galois theory. That program goes beyond Grothendieck's birational anabelian program as initiated in [Gro83], [Gro84], because k being algebraically closed, there is no arithmetical Galois action in the game. In a few words, the precise context for Bogomolov's birational anabelian program is as follows:

- Let ℓ be a fixed rational prime number.
- Consider function fields $\mathrm{K}|k$ with k algebraically closed of characteristic $\neq \ell$.
- Let $\mathrm{K}' \hookrightarrow \mathrm{K}''$ be maximal pro-ℓ abelian, respectively *abelian-by-central*, extensions of K.
- Let $pr : \Pi^{\mathrm{c}}_{\mathrm{K}} \to \Pi_{\mathrm{K}}$ be the corresponding projection of Galois groups.

Notice that $pr : \Pi^{\mathrm{c}}_{\mathrm{K}} \to \Pi_{\mathrm{K}}$ can be recovered group theoretically from $\Pi^{\mathrm{c}}_{\mathrm{K}}$, as its kernel is exactly the topological closure of the commutator subgroup of $\Pi^{\mathrm{c}}_{\mathrm{K}}$. Actually, we set $\mathrm{G}^{(1)} = \mathrm{G}_{\mathrm{K}}$ for the absolute Galois group of K, and for $i \geqslant 1$ we let

F. Pop (✉)
Department of Mathematics, University of Pennsylvania, DRL, 209 S 33rd Street, Philadelphia, PA 19104, USA
e-mail: pop@math.upenn.edu

* Supported by NSF grant DMS-0801144.

J. Stix (ed.), *The Arithmetic of Fundamental Groups*, Contributions in Mathematical and Computational Sciences 2, DOI 10.1007/978-3-642-23905-2_10,

$$G^{(i+1)} := [G^{(i)}, G^{(1)}](G^{(i)})^{\ell^\infty}$$

be the closed subgroup of $G^{(i)}$ generated by all the commutators $[x,y]$ with $x \in G^{(i)}$, $y \in G^{(1)}$ and the ℓ^∞-powers of all the $z \in G^{(i)}$. Then the $G^{(i)}$, $i \geqslant 1$, are the descending central ℓ^∞ terms of the absolute Galois group G_K, and

$$\Pi_K^c = G^{(1)}/G^{(3)} \to \Pi_K = G^{(1)}/G^{(2)}.$$

Further, denoting $G^{(\infty)} = \cap_i G^{(i)}$, it follows that $G_K(\ell) := G_K/G^{(\infty)}$ is the maximal pro-ℓ quotient of G_K, see e.g. [NSW08, page 220]. The program initiated by Bogomolov[1] mentioned above has as ultimate goal to recover function fields $K|k$ as above from $\mathrm{Gal}(K''|K)$ in a functorial way. If completed, this program would go far beyond Grothendieck's birational anabelian geometry, see [Pop98] for a historical note on birational anabelian geometry, and [Pop11, Introduction], for a historical note on Bogomolov's Program and an outline of a strategy to tackle this program. For an early history beginning even before [Gro83], [Gro84], see [NSW08] Chap. XII, the original sources [Neu69], [Uch79], as well as Szamuely's Séminaire Bourbaki talk [Sza04]. To conclude, I would like to mention that in contrast to Grothendieck's birational anabelian program, which is completed to a large extent, Bogomolov's birational anabelian program is completed only in the case the base field k is an *algebraic closure of a finite field*, see Bogomolov–Tschinkel [BT08] for the case $\mathrm{td}(K|k) = 2$ and Pop [Pop11] in general.

The results of the present manuscript represent a first step and hints at the possibility that a mod ℓ *abelian-by-central* form of birational anabelian geometry might hold, which would then go beyond Bogomolov's birational anabelian program in many ways.

In order to put the results of this paper in the right perspective, let me mention that the present paper shares similarities with [Pop06] and [Pop10], where similar results were obtained, but working with the full pro-ℓ Galois group $G_K(\ell)$, respectively the maximal pro-ℓ abelian-by-central Galois group Π_K^c. Whereas a key technical tool used in [Pop10] is the theory of $\mathbb{Z}_\ell$ *commuting liftable pairs* as developed in Bogomolov–Tschinkel [BT02], which was used as a black box, we use here the theory of $\mathbb{Z}/\ell$ *commuting liftable pairs*. The fact that such a mod ℓ variant might hold was already suggested by previous results from Mahé–Mináč–Smith [MMS04] in the case $\ell = 2$. The details for the theory of $\mathbb{Z}/\ell$ commuting liftable pairs can be found in the manuscript Topaz [Top11].

Before going into the details of the manuscript, let me introduce notations which will be used throughout the manuscript and mention briefly facts used later on.

- Let ℓ be a prime number.
- Consider function fields $K|k$ with k algebraically closed of characteristic $\neq \ell$.
- Let $K' \hookrightarrow K''$ be a maximal $\mathbb{Z}/\ell$ abelian extension, respectively a maximal $\mathbb{Z}/\ell$ abelian-by-central extension of K.
- Let $pr : \overline{\Pi}_K^c \to \overline{\Pi}_K$ be the corresponding quotient map of Galois groups.

[1] Recall that Bogomolov denotes $\mathrm{Gal}(K''|K)$ by PGal_K^c.

Note that $pr : \overline{\Pi}^c_K \to \overline{\Pi}_K$ can be recovered group theoretically from $\overline{\Pi}^c_K$, as its kernel is the topological closure of the commutator group of $\overline{\Pi}^c_K$.

Let v be a valuation of K, and v' some prolongation of v to K'. Let $T_{v'} \subseteq Z_{v'}$ be the inertia group, respectively decomposition group, of v' in $\overline{\Pi}_K$. By Hilbert decomposition theory for valuations, the groups $T_{v'} \subseteq Z_{v'}$ of the several prolongations v' of v to K' are conjugated. Thus, since $\overline{\Pi}_K$ is abelian, the groups $T_{v'} \subseteq Z_{v'}$ depend on v only, and not on its prolongations v' to K'. We will denote these groups by $T_v \subseteq Z_v$, and call them the inertia group, respectively decomposition group, at v. And we denote by $K'^{Z_v} =: K^Z \subseteq K^T := K'^{T_v}$ the corresponding fixed fields. It turns out that the residue field $K'v'$ of v' is actually a maximal $\mathbb{Z}/\ell$ extension of the residue field Kv of v, i.e., $K'v' = (Kv)'[\sqrt[\ell]{Kv}\,]$, which is abelian if $\mathrm{char}(Kv) \neq \ell$. We further set $K^{Z^1} := K^Z[\sqrt[\ell]{1+\mathbf{m}_v}\,]$ and $K^{T^1} := K^Z[\sqrt[\ell]{\mathcal{O}_v^\times}\,]$ and denote $Z^1_v := \mathrm{Gal}(K'|K^{Z^1})$ and $T^1_v := \mathrm{Gal}(K'|K^{T^1})$, and call these groups the **minimized** decomposition, respectively inertia, groups of v. We notice that by Lemma 2, one has $Z^1_v \subseteq Z_v$ and $T^1_v \subseteq T_v$, and $Z^1_v = Z_v$ and $T^1_v = T_v$, provided $\mathrm{char}(Kv) \neq \ell$.

We next recall that for a k-valuation of K, i.e., a valuation of K whose valuation ring $\mathcal{O}_v$ contains k, and thus k canonically embeds in the residue field $Kv := \mathcal{O}_v/\mathbf{m}_v$, the following conditions are equivalent:

(i) The valuation ring $\mathcal{O}_v$ equals the local ring $\mathcal{O}_{X,x_v}$ of the generic point x_v of some Weil prime divisor of some normal model $X \to \mathrm{Spec}(k)$ of $K|k$.

(ii) The transcendence degrees satisfy $\mathrm{td}(Kv\,|\,k) = \mathrm{td}(K|k) - 1$.

A **prime divisor** of $K|k$ is any k-valuation v of K which satisfies the above equivalent conditions. In particular, if v is a prime divisors of $K|k$, then $vK \cong \mathbb{Z}$ and $Kv|k$ is a function field satisfying $\mathrm{td}(Kv\,|\,k) = \mathrm{td}(K|k) - 1$. By Hilbert decomposition theory for valuations, see e.g. [Bou64] Chap. 6, it follows that the following hold:

$$T_v \cong \mathbb{Z}/\ell \quad \text{and} \quad Z_v \cong T_v \times \mathrm{Gal}(K'v'|Kv) \cong \mathbb{Z}/\ell \times \mathrm{Gal}(K'v'|Kv).$$

For a prime divisor v, we will call Z_v endowed with T_v a **divisorial subgroup** of $\overline{\Pi}_K$ or of the function field $K|k$.

As a first step in recovering $K|k$ from its mod ℓ abelian-by-central Galois theory, one would like to recover the divisorial subgroups of $\overline{\Pi}_K$ from $\overline{\Pi}^c_K$. This is indeed possible if k is the algebraic closure of a finite field, see below. Unfortunately, there are serious difficulties when one tries to do the same in the case k is not an algebraic closure of a finite field, as the non-trivial valuations of k interfere. Therefore one is led to considering the following generalization of prime divisors, see e.g. [Pop06] Appendix. The valuation v of K is called **quasi divisorial**, or a **quasi prime divisor** of K, if the valuation ring $\mathcal{O}_v$ of v is maximal among the valuation rings of valuations of K satisfying:

(i) The relative value group vK/vk is isomorphic to $\mathbb{Z}$ as abstract groups.

(ii) The residue extension $Kv\,|\,kv$ is a function field with $\mathrm{td}(Kv\,|\,kv) = \mathrm{td}(K|k) - 1$.

Notice that a quasi prime divisor v of K is a prime divisor if and only if v is trivial on k. In particular, in the case where k is an *algebraic closure of a finite field*, the quasi prime divisors and the prime divisors of $K|k$ coincide, as all valuations of K are trivial on k.

For a Galois extension $\tilde{K}|K$ and its Galois group $\mathrm{Gal}(\tilde{K}|K)$, we will say that a subgroup Z of $\mathrm{Gal}(\tilde{K}|K)$ endowed with a subgroup T of Z is a **quasi-divisorial subgroup** of $\mathrm{Gal}(\tilde{K}|K)$— or of $K|k$ in case $\mathrm{Gal}(\tilde{K}|K)$ is obvious from the context— if $T \subseteq Z$ are the inertia group, respectively the decomposition group, above some quasi-divisor v of $K|k$. One defines the corresponding **minimized** quasi-divisorial subgroups $T^1 \subseteq Z^1$ of the quasi-divisorial subgroups $T \subset Z$ of $\mathrm{Gal}(\tilde{K}|K)$, and notice that if $\mathrm{char}(Kv) \neq \ell$, then $T^1 = T$ and $Z^1 = Z$. We notice that by Proposition 7, it follows that $T \subset Z$ and $T^1 \subset Z^1$ determine each other uniquely, provided $T^1_v \neq 1$.

It was the main result in Pop [Pop10] to show that the quasi-divisorial subgroups of the Galois group Π_K can be recovered by a group theoretical recipe from the canonical projection $\Pi^c_K \to \Pi_K$, which itself can be recovered from Π^c_K. Paralleling that result, the main result of this paper can be summarized as follows:

Theorem 1. *Let $K|k$ be a function field over the algebraically closed field k of characteristic $\mathrm{char}(k) \neq \ell$, and let*

$$\overline{\Pi}^c_K \to \overline{\Pi}_K$$

be the canonical projection. For subgroups T, Z, Δ of $\overline{\Pi}_K$, let T'', Z'', Δ'' denote their preimages in $\overline{\Pi}^c_K$. Then one has:

(1) The transcendence degree $d = \mathrm{td}(K|k)$ is the maximal integer d such that there exist closed subgroups $\Delta \cong (\mathbb{Z}/\ell)^d$ of $\overline{\Pi}_K$ with Δ'' abelian.
(2) Suppose that $d := \mathrm{td}(K|k) > 1$. Let $T \subset Z$ be closed subgroups of $\overline{\Pi}_K$. Then Z endowed with T is a minimized quasi divisorial subgroup of $\overline{\Pi}_K$ if and only Z and T are maximal in the set of closed subgroups of $\overline{\Pi}_K$ which satisfy:

(i) Z contains a closed subgroup $\Delta \cong (\mathbb{Z}/\ell)^d$ such that Δ'' is abelian.
(ii) $T \cong \mathbb{Z}/\ell$, and T'' is the center of Z''.

Actually the above theorem is a special case of the more general assertions Proposition 15, and Theorems 17 and 19, which deal with generalized [almost] (quasi) prime r-divisors. The above Theorem corresponds to the case $r = 1$.

Acknowledgements. I would like to thank all who showed interested in this work, among whom: Jakob Stix, Tamás Szamuely and Adam Topaz for technical discussions and help, and Viktor Abrashkin, Minhyong Kim, Pierre Lochak, Hiroaki Nakamura, Mohamed Saïdi and Akio Tamagawa for discussions at the INI Cambridge during the *NAG Programme* in 2009.

10.2 Basic Facts from Valuation Theory

Let K be a field of characteristic $\neq \ell$ containing all ℓ-th roots of unity μ_ℓ. In particular, we can fix a (non-canonically) isomorphism $\iota_K : \mu_\ell \to \mathbb{Z}/\ell$ as G_K modules. Kummer theory provides a canonical non-degenerate pairing

$$K^{\times}/\ell \times \overline{\Pi}_K \to \mu_\ell,$$

that by Pontrjagin duality gives rise to canonical isomorphisms $\overline{\Pi}_K = \mathrm{Hom}(K^{\times}, \mu_\ell)$ and $K^{\times}/\ell = \mathrm{Hom}_{\mathrm{cont}}(G_K, \mu_\ell)$. Thus we finally get isomorphisms:

$$\overline{\Pi}_K = \mathrm{Hom}(K^{\times}, \mu_\ell) \xrightarrow{\imath_K} \mathrm{Hom}(K^{\times}, \mathbb{Z}/\ell).$$

10.2.1 Hilbert Decomposition in Abelian Extensions

In the above context, let v be a valuation of K, and v' some prolongation of v to K′. Further, let $p = \mathrm{char}(Kv)$ be the residual characteristic. We denote by $T_v \subseteq Z_v$ the inertia, respectively decomposition, groups of $v'|v$ in $\overline{\Pi}_K$. We notice that one has: First, if $\ell \neq p$, then the ramification group V_v of $v'|v$ is trivial, as $p = \mathrm{char}(Kv) \neq \ell$ does not divide the order of $\overline{\Pi}_K$, thus that of Z_v. Second, if $\ell = p$, then $\mathrm{char}(K) = 0$, and $T_v = V_v$, because $p = \ell$ is the only prime dividing the order of $\overline{\Pi}_K$. We also notice that $V_v \subseteq T_v \subseteq Z_v$ do depend on v only, and not on the prolongation $v'|v$ used to define them. Finally, we denote by K^T and K^Z the corresponding fixed fields in K′.

Let $\zeta_\ell \in \mu_p$ be primitive, and set $\theta := \zeta_\ell - 1 \in K^{\times}$. The following are well known facts, and we reproduce them here for the reader's convenience.

Lemma 2. *In the above context one has the following:*

(1) *Let* $U_v^1 = 1 + \mathbf{m}_v \leqslant K^{\times}$ *be the group of principal v-units in* K. *Then, if* $p = \ell$,

$$K[\sqrt[\ell]{1 + \theta^\ell \mathbf{m}_v}\,] \subseteq K^Z \subseteq K[\sqrt[\ell]{U_v^1}\,] = K^{Z^1}$$

and else

$$K^Z = K[\sqrt[\ell]{U_v^1}\,] = K^{Z^1}$$

In particular, $K^{Z^1} = K'$ *if and only if* $K^{\times}/\ell = U_v^1/\ell$.

(2) *Let* U_v *be the group of v-units of* K. *Then, if* $p = \ell$,

$$K[\sqrt[\ell]{1 + \theta^\ell U_v}\,] \subseteq K^T \subseteq K[\sqrt[\ell]{U_v}\,] = K^{T^1}$$

and else

$$K^T = K[\sqrt[\ell]{U_v}\,] = K^{T^1}.$$

(3) *The field extension* $(Kv)' \subseteq K'v'$ *is purely inseparable, and equality holds if* $\mathrm{char}(Kv) \neq \ell$. *Setting* $\delta_v = \dim(vK/\ell)$, *the following holds:*

- *(a)* $T_v^1 = \mathrm{Hom}(vK, \mathbb{Z}/\ell) \cong (\mathbb{Z}/\ell)^{\delta_v}$ *non-canonically.*
- *(b)* $G_v := Z_v^1/T_v^1 = Z_v/T_v = \mathrm{Hom}(Kv/\ell, \mu_\ell)$.
- *(c)* *There are isomorphisms (the latter non-canonical) of ℓ-torsion groups:*

$$Z_v^1 \cong T_v^1 \times G_v^1 \cong (\mathbb{Z}/\ell)^{\delta_v} \times \mathrm{Hom}(Kv/\ell, \mu_\ell).$$

Proof. (1) Let K^h be some Henselization of K containing K^Z. Then by general decomposition theory, $K^Z = K^h \cap K'$. We first prove that

$$\sqrt[\ell]{1+\theta^\ell \mathbf{m}_v} \subseteq K^Z.$$

Equivalently, if $a = 1+\theta^\ell x$ with $x \in \mathbf{m}_v$, then we have to show $\sqrt[\ell]{a} \in K^Z$. First, consider the case $p = \ell$. Then the equation $X^p = 1+\theta^\ell x$ is equivalent to

$$(X_0+1)^p = 1+\theta^\ell x,$$

and setting $Y = X_0/\theta^\ell$, we get the equation of the form $Y^p - Y = y$ with $y \in \mathbf{m}_v$. The latter has a solution in K^Z by Hensel's Lemma, hence in $K' \cap K^h$.

Second, if $p \neq \ell$, one has $a \equiv 1 \pmod{\mathbf{m}_v}$ and it follows that

$$X^\ell - a \equiv X^\ell - 1 \mod \mathbf{m}_v,$$

hence $X^\ell - a$ has ℓ distinct roots $(\mathrm{mod}\, \mathbf{m}_v)$. By Hensel's Lemma, $X^\ell - a$ has a root in K^h, hence in $K' \cap K^h$.

For the second inclusion, let $a \in K^\times$ be such that $\sqrt[\ell]{a} \in K^Z$. Since K and K^Z have equal value groups, and $\sqrt[\ell]{a} \in K^Z$, it follows that there is an element $b \in K$ such that $v\sqrt[\ell]{a} = vb$, hence $va = \ell \cdot vb$. We set $c := a/b^\ell \in U_v$ and find

$$\sqrt[\ell]{c} = \sqrt[\ell]{a}/b \in K^Z.$$

Since K and K^Z have equal residue fields, it follows that there is $d \in U_v$ such that $\sqrt[\ell]{c} \equiv d \,\mathrm{mod}\, \mathbf{m}_v$, hence $c = d^\ell \cdot a_1$ with $a_1 \in U_v^1$. Thus finally as claimed

$$\sqrt[\ell]{a} = bd\sqrt[\ell]{a_1} \in K[\sqrt[\ell]{U_v^1}].$$

Finally, if $p \neq \ell$, then $\theta \in U_v$, hence $1+\theta^\ell \mathbf{m}_v = U_v^1$.

In the case $p \neq \ell$, the last assertion of (1) is just the translation via Kummer theory of the fact that we have equalities $K' = K[\sqrt[\ell]{K^\times}]$ and $K[\sqrt[\ell]{U_v^1}] = K^Z$.

The proof of (2) is similar, and therefore we will omit the details. And finally, (3) is just a translation in Galois terms of the assertions (1) and (2). □

Recall that for given valuation v, one can recover $\mathcal{O}_v$ from $\mathbf{m}_v$, respectively U_v^1, respectively U_v. Indeed, if U_v is given, then

$$U_v^1 = \{x \in U_v \mid x \notin U_v - 1\},$$

by which we deduce $\mathbf{m}_v = U_v^1 - 1$, and finally recover $\mathcal{O}_v$ through its complement

$$K \backslash \mathcal{O}_v = \{x \in K^\times \mid x^{-1} \in \mathbf{m}_v\}.$$

Finally, recall that for given valuations v, w of K, with valuation rings $\mathcal{O}_v$, respectively $\mathcal{O}_w$, we say that $w \leqslant v$, or that w is a coarsening of v, if $\mathcal{O}_v \subseteq \mathcal{O}_w$. From the discussion above we deduce that for given valuations v, w, of K, the following assertions are equivalent:

(i) w is a coarsening of v.
(ii) $\mathbf{m}_w \subseteq \mathbf{m}_v$.
(iii) $\mathrm{U}^1_w \subseteq \mathrm{U}^1_v$.
(iv) $\mathrm{U}_v \subseteq \mathrm{U}_w$.

These facts have the following Galois theoretic translation.

Fact 3. In the above context and notations, the following hold:

(1) $\mathrm{Z}_v \subseteq \mathrm{Z}_w$ if and only if $\mathrm{K}^{\mathrm{Z}}_w \subseteq \mathrm{K}^{\mathrm{Z}}_v$, and $\mathrm{Z}^1_v \subseteq \mathrm{Z}^1_w$ if and only if $\mathrm{U}^1_w/\ell \subseteq \mathrm{U}^1_v/\ell$.
(2) $\mathrm{T}_w \subseteq \mathrm{T}_v$ if and only if $\mathrm{K}^{\mathrm{T}}_v \subseteq \mathrm{K}^{\mathrm{T}}_w$, and $\mathrm{T}^1_v \subseteq \mathrm{T}^1_w$ if and only if $\mathrm{U}_v/\ell \subseteq \mathrm{U}_w/\ell$.
(3) In particular[2], if $w \leqslant v$, then $\mathrm{T}_w \subseteq \mathrm{T}_v \subseteq \mathrm{Z}_v \subseteq \mathrm{Z}_w$, and $\mathrm{T}^1_w \subseteq \mathrm{T}^1_v$ and $\mathrm{Z}^1_v \subseteq \mathrm{Z}^1_w$.

10.2.2 The $\mathbb{Z}/\ell$ Abelian Form of Two Results of F. K. Schmidt

In this subsection we give the abelian pro-ℓ form of two results of F. K. Schmidt and generalizations of these like the ones in Pop [Pop94, The local theory]. See also Endler–Engler [EE77].

Let v be a fixed valuation of K, and $v'|v$ a fixed prolongation of v to K'. Let further $\Lambda|\mathrm{K}$ be a fixed sub-extension of $\mathrm{K}'|\mathrm{K}$ containing K^{Z}. Let $\mathcal{V}'_{\Lambda,v'}$ be the set of all coarsenings w' of v' such that $\Lambda w' \supseteq (\mathrm{K}w)'$, or equivalently, $\Lambda w' \subseteq \mathrm{K}'v'$ is purely inseparable. Let w be the restriction of w' to K, and let $\mathcal{V}_{\Lambda,v}$ be the restriction of $\mathcal{V}'_{\Lambda,v'}$ to K. We set $\mathcal{V}^0_{\Lambda,v} = \mathcal{V}_{\Lambda,v} \cup \{v\}$.

Lemma 4. *(1) The set $\mathcal{V}_{\Lambda,v}$ depends on v and Λ only, and not on the specific prolongation v' of v. In fact, $\mathcal{V}_{\Lambda,v}$ consists of all the coarsenings w of v such that $\Lambda' w \supseteq (\mathrm{K}w)'$ for some prolongation w of w to K' (and equivalently, for every prolongation w' of w to K').*

(2) More precisely, we have $w \in \mathcal{V}_{\Lambda,v} \iff \mathrm{K}^{\mathrm{T}}_w \subseteq \Lambda \iff \mathrm{Gal}(\mathrm{K}'|\Lambda) \subseteq \mathrm{T}_w$, and in particular, $v \in \mathcal{V}_{\Lambda,v} \iff \mathrm{K}^{\mathrm{T}}_v \subseteq \Lambda \iff \mathrm{Gal}(\mathrm{K}'|\Lambda) \subseteq \mathrm{T}_v$.

Proof. (1) If $\tilde{v}$ is another prolongation of v to K', then there exists some σ in $\overline{\Pi}_{\mathrm{K}}$ such that $\tilde{v} = v' \circ \sigma^{-1} := \sigma(v')$, and so σ defines a bijection

$$\mathcal{V}_{v',\Lambda} \to \mathcal{V}_{\tilde{v},\sigma(\Lambda)}$$

by $w \mapsto w \circ \sigma^{-1}$. Note that since $\Lambda|\mathrm{K}$ is abelian, thus in particular Galois, one has $\Lambda = \sigma(\Lambda)$. Thus for $w' \in \mathcal{V}_{\Lambda,v'}$, and $\tilde{w} = \sigma(w') := w' \circ \sigma$ one has: σ gives rise to an $\mathrm{K}w$-isomorphism of the residue fields

$$(\mathrm{K}w)' = \Lambda w' \to \sigma(\Lambda)\sigma(w') = \Lambda\tilde{w}.$$

[2] This is actually true for all Galois extensions $\tilde{\mathrm{K}}|\mathrm{K}$, and not just for $\mathrm{K}'|\mathrm{K}$. But then one has to start with valuations $\tilde{v}$ and coarsenings $\tilde{w}$ of those on $\tilde{\mathrm{K}}$, etc.

The proof of the remaining assertions is clear.

(2) Let w' be a coarsening of v'. By general decomposition theory for valuations it follows that $\mathrm{K}^{\mathrm{Z}}_{w} \subseteq \mathrm{K}^{\mathrm{Z}}_{v}$. Further, by general decomposition theory, $\mathrm{K}^{\mathrm{T}}_{w}|\mathrm{K}^{\mathrm{Z}}_{w}$ is the unique minimal one among all the sub-extensions of $\mathrm{K}'|\mathrm{K}^{\mathrm{Z}}_{w}$ having residue field equal to $(\mathrm{K}w)'$. Now since by hypothesis $\mathrm{K}^{\mathrm{Z}}_{w} \subseteq \mathrm{K}^{\mathrm{Z}}_{v} \subseteq \Lambda$, we have: $\mathrm{K}^{\mathrm{T}}_{w} \subseteq \Lambda$, provided $\Lambda w' \supseteq (\mathrm{K}w)'$. Thus, $w \in \mathcal{V}^0_{\Lambda,v}$ if and only if $\mathrm{K}^{\mathrm{T}}_{w} \subseteq \Lambda$, by the discussion above. □

Definition 5. By general valuation theory, the set $\mathcal{V}^0_{\Lambda,v}$ has an infimum whose valuation ring is the union of all the valuation rings $\mathcal{O}_w$ with $w \in \mathcal{V}^0_{\Lambda,v}$. We denote this valuation by

$$v_\Lambda := \inf \mathcal{V}^0_{\Lambda,v}$$

and call it the **$\mathbb{Z}/\ell$-abelian Λ-core** of v.

Proposition 6. *In the above context and notations, suppose that $\Lambda \neq \mathrm{K}'$ is a proper sub-extension of $\mathrm{K}'|\mathrm{K}$ containing K^{Z}, thus in particular, $\mathrm{K}^{\mathrm{Z}} \neq \mathrm{K}'$. Then the $\mathbb{Z}/\ell$-abelian Λ-core v_Λ of v is non-trivial and lies in $\mathcal{V}^0_{v,\Lambda}$. Consequently:*

(1) If v_1 is a valuation of K satisfying $v_1 < v_\Lambda$, then $\Lambda v_1' \not\supseteq (\mathrm{K}v_1)'$. If $\Lambda v' \supseteq (\mathrm{K}v)'$, then $\Lambda v_\Lambda' \supseteq (\mathrm{K}v)'$, and v_Λ is the minimal coarsening of v with this property.
(2) $(\mathrm{K}v_{\mathrm{Z}})' = \mathrm{K}v_{\mathrm{Z}} \Leftrightarrow (\mathrm{K}v)' = \mathrm{K}v$, where v_{Z} is the $\mathbb{Z}/\ell$-abelian K^{Z}-core of v.
(3) If v has rank one, or if $\mathrm{K}v \neq (\mathrm{K}v)'$, then v equals its $\mathbb{Z}/\ell$-abelian K^{Z}-core v_{Z}.

Proof. If $\mathcal{V}_{\Lambda,v}$ is empty, i.e., $\Lambda v' \not\supseteq (\mathrm{K}v)'$, then $\mathcal{V}^0_{\Lambda,v} = \{v\}$, hence $v_\Lambda = v$, and there is nothing to show. Now suppose that $\mathcal{V}_{\Lambda,v}$ is non-empty. Then $\Lambda v' \supseteq (\mathrm{K}v)'$, hence $v \in \mathcal{V}_{\Lambda,v}$, and we will show that actually $v_\Lambda \in \mathcal{V}_{\Lambda,v}$. Equivalently, by Lemma 4 (2), above, we have to show that $\mathrm{K}^{\mathrm{T}}_{v_\Lambda} \subseteq \Lambda$. Since $\mathrm{K}^{\mathrm{T}}_{v_\Lambda} v_\Lambda'$ is the maximal separable subextension of $(\mathrm{K}v)'v' \mid \mathrm{K}v$, the inclusion $\mathrm{K}^{\mathrm{T}}_{v_\Lambda} \subseteq \Lambda$ is equivalent to showing that $(\mathrm{K}v)'v_\Lambda' \mid \Lambda v_\Lambda'$ is purely inseparable. On the other hand, since by the definition of v_Λ we have $\mathcal{O}_{v_\Lambda} = \cup_w \mathcal{O}_w$, $w \in \mathcal{V}^0_{\Lambda,v}$, and $\mathbf{m}_{v_\Lambda} = \cap_w \mathbf{m}_w$, and since each $(\mathrm{K}w)'w' \mid \Lambda w'$ are purely inseparable, by "taking limits" the same is true for $(\mathrm{K}v)'v_\Lambda' \mid \Lambda v_\Lambda'$.

The assertions (1), (2), (3) are immediate consequences of the main assertion of the proposition proved above, and we omit their proof. □

Proposition 7. *(1) Let v_1, v_2 be valuations of K such that $\mathrm{K}^{\mathrm{Z}}_{v_1}$, $\mathrm{K}^{\mathrm{Z}}_{v_2}$ are contained in some $\Lambda \neq \mathrm{K}'$. Then the $\mathbb{Z}/\ell$-abelian Λ-cores of v_1 and v_2 are comparable.*

(2) Let v_1, v_2 be valuations of K which equal their $\mathbb{Z}/\ell$-abelian K^{Z}-cores, respectively. If $\mathrm{K}^{\mathrm{Z}}_{v_1} = \mathrm{K}^{\mathrm{Z}}_{v_2}$, then v_1 and v_2 are comparable. (Obviously, if $\mathrm{K}^{\mathrm{Z}}_{w} \subset \mathrm{K}^{\mathrm{Z}}_{v}$ strictly, then $w < v$ strictly.)

Proof. (1) Recalling that $\theta := \zeta_\ell - 1$, we first remark for independent valuations v_1 and v_2 of K one has $\mathrm{K}^\times = (1+\theta^\ell \mathbf{m}_{v_1}) \cdot (1+\theta^\ell \mathbf{m}_{v_2})$. Indeed, this follows immediately from the Approximation Theorem for independent valuations. In particular, if v_1 and v_2 are independent, then K' equals the compositum $\mathrm{K}^{\mathrm{Z}}_{v_1}\mathrm{K}^{\mathrm{Z}}_{v_2}$ inside K'. Now since by hypothesis we have $\mathrm{K}^{\mathrm{Z}}_{v_1}, \mathrm{K}^{\mathrm{Z}}_{v_2} \subseteq \Lambda \not\supseteq \mathrm{K}'$, it follows that v_1 and v_2 are not independent. Let v be the maximal common coarsening of v_1 and v_2. By general valuation theory, the valuation ideal $\mathbf{m}_v$ is the maximal common ideal of $\mathcal{O}_{v_1}$ and $\mathcal{O}_{v_2}$. Denote $w_i = v_i/v$ on the residue field $\mathrm{L} := \mathrm{K}v$. Then we have:

- If both w_1 and w_2 are non-trivial, then they are independent.
- For $i = 1,2$ we have
$$\mathrm{L}^{\mathrm{Z}}_{w_i} = (\mathrm{K}^{\mathrm{Z}}_{v_i})v' \subseteq \Lambda v' \subseteq \mathrm{K}'v'.$$

Therefore, by the discussion above, we either have $\Lambda v' \supseteq (\mathrm{K}v)'$, or otherwise at least one of the w_i is the trivial valuation.

First, we consider the case when one of the w_i is trivial. Equivalently, we have $v_i = \min(v_1, v_2)$, hence v_1 and v_2 are comparable. Thus any two coarsenings of v_1 and v_2 are comparable, hence their $\mathbb{Z}/\ell$-abelian Λ-cores are comparable, too.

Second, consider the case $\Lambda v' \supseteq (\mathrm{K}v)'$. Then by the definition of $\mathcal{V}'_{\Lambda,v_i}$, it follows that $v \in \mathcal{V}'_{\Lambda,v_i}$, as v is by definition a coarsening of v_i, $i = 1,2$. Hence finally the $\mathbb{Z}/\ell$-abelian Λ-cores of v_1 and v_2 are both some coarsenings of v, thus comparable.

(2) We apply assertion (1) with $\Lambda = \mathrm{K}^{\mathrm{Z}}_{v_1} = \mathrm{K}^{\mathrm{Z}}_{v_2}$. □

10.3 Hilbert Decomposition in $\mathbb{Z}/\ell$ Abelian-by-Central Extensions

We keep the notations from the introduction and the previous sections concerning field extensions $\mathrm{K}|k$ and the canonical projection $pr : \overline{\Pi}^{\mathrm{c}}_{\mathrm{K}} \to \overline{\Pi}_{\mathrm{K}}$.

Fact/Definition 8. In the above notations we have the following:

(1) For a family $\Sigma = (\sigma_i)_i$ of elements of $\overline{\Pi}_{\mathrm{K}}$, let Δ_Σ be the closed subgroup generated by Σ. Then the following are equivalent:

(i) There are preimages $\sigma'_i \in \overline{\Pi}^{\mathrm{c}}_{\mathrm{K}}$, for all i, which commute with each other.

(ii) The preimage Δ''_Σ in $\overline{\Pi}^{\mathrm{c}}_{\mathrm{K}}$ of Δ_Σ is abelian.

We say that a family of elements $\Sigma = (\sigma_i)_i$ of $\overline{\Pi}_{\mathrm{K}}$ is **commuting liftable**, for short c.l., if Σ satisfies the above equivalent conditions (i), (ii).

(2) For a family $(\Delta_i)_i$ of subgroups of $\overline{\Pi}_{\mathrm{K}}$ the following are equivalent:

(i) All families $(\sigma_i)_i$ with $\sigma_i \in \Delta_i$ are c.l.

(ii) If Δ''_i is the preimage of Δ_i in $\overline{\Pi}^{\mathrm{c}}_{\mathrm{K}}$, then $[\Delta''_i, \Delta''_j] = 1$ for all $i \neq j$.

We say that a family of subgroups $(\Delta_i)_i$ of $\overline{\Pi}_{\mathrm{K}}$, is **commuting liftable**, for short c.l., if it satisfies the equivalent conditions (i), (ii) above.

(3) We will say that a subgroup Δ of $\overline{\Pi}_{\mathrm{K}}$ is **commuting liftable**, for short c.l., if its preimage Δ'' in $\overline{\Pi}^{\mathrm{c}}_{\mathrm{K}}$ is commutative.

(4) We finally notice the following: For subgroups $\mathrm{T} \subseteq \mathrm{Z}$ of $\overline{\Pi}_{\mathrm{K}}$, let $\mathrm{T}'' \subseteq \mathrm{Z}''$ be their preimages in $\overline{\Pi}^{\mathrm{c}}_{\mathrm{K}}$. Then the following are equivalent:

(i) The pair (T, Z) is c.l.

(ii) T'' is contained in the center of Z''.

(5) In particular, given a closed subgroup Z of $\overline{\Pi}_{\mathrm{K}}$, there exists a unique maximal (closed) subgroup T of Z such that T and (T,Z) are c.l. Indeed, denoting by Z'' the preimage of Z in $\overline{\Pi}^{\mathrm{c}}_{\mathrm{K}}$, and denoting by T'' its center, the group T is the image of T'' in $\overline{\Pi}_{\mathrm{K}}$ under the canonical projection $pr : \overline{\Pi}^{\mathrm{c}}_{\mathrm{K}} \to \overline{\Pi}_{\mathrm{K}}$.

We next recall the following fundamental fact concerning $\mathbb{Z}/\ell$ liftable commuting pairs. It might be well possible that one could work out a proof along the technical steps in the proofs from Bogomolov–Tschinkel [BT02]. But there is a much simpler/easier way to get the result by using the theory of **rigid elements**, originating in work by Ware [War81], and further developed by Arason–Jacob–Ware [AEJ87], Koenigsmann [Koe01], and others; see Topaz [Top11] for complete proofs.

Key Fact 9. Let K|k be an extension with k algebraically closed and char(k) $\neq \ell$. Let $\sigma, \tau \in \overline{\Pi}_{\mathrm{K}}$ be such that $\langle \sigma, \tau \rangle$ is isomorphic to $(\mathbb{Z}/\ell)^2$. Then σ, τ is c.l. if and only if there exists a valuation v of K with the following properties:

(i) The group $\langle \sigma, \tau \rangle$ is contained in the minimized decomposition group Z^1_v.
(ii) The intersection $\langle \sigma, \tau \rangle \cap \mathrm{T}^1_v$ is non-trivial.

Proof. We only give a sketch of a proof, see Topaz [Top11] for details. Let $\mathrm{T} \subset \mathrm{K}^\times/\ell$ be the orthogonal complement of $\langle \sigma, \tau \rangle$ under the Kummer pairing. Equivalently, $\mathrm{K}_{\mathrm{T}} := \mathrm{K}[\sqrt[\ell]{\mathrm{T}}]$ is the fixed field of $\langle \sigma, \tau \rangle$ in K', and $(\mathrm{K}^\times/\ell)/\mathrm{T}$ is the Kummer dual of $\langle \sigma, \tau \rangle$ under the Kummer pairing. Let $x, y \in \mathrm{K}^\times$ be such that $\mathrm{K}^\times/\ell$ is generated by T, x, y as an abelian group. Then the fact that σ, τ is a commuting liftable pair implies that for the characters $\chi_x, \chi_y \in \mathrm{Hom}(\overline{\Pi}_{\mathrm{K}}, \mu_\ell)$ defined by x, y, it follows that their cup product

$$\chi_x \cup \chi_y \in \mathrm{H}^2(\mathrm{K}, \mu_\ell^{\otimes 2})$$

is non-trivial. From this fact it follows instantly that

$$\mathrm{T} + \mathrm{T}z \subset \cup_{1=0}^{\ell-1} z^i \mathrm{T}$$

for all $z \in \mathrm{K}^\times/\ell$. Thus all the elements of $\mathrm{K}^\times/\ell$ are quasi rigid with respect to T. One concludes that there T-rigid elements in $\mathrm{K}^\times$, thus there exists a valuation v of K such that $1 + \mathbf{m}_v \subseteq \mathrm{T}$, and $(v\mathrm{K}/\ell)/v\mathrm{T}$ is not trivial, etc. □

10.3.1 Inertia Elements

Recall that in the notations from above, we say that an element $\sigma \in \overline{\Pi}_{\mathrm{K}}$ is an ℓ-**inertia element**, for short **inertia element**, if there exists a valuation v of K such that $\sigma \in \mathrm{T}_v$ and char($\mathrm{K}v$) $\neq \ell$. Clearly, the set of all the inertia elements at v is exactly T_v.

Lemma 10. *Let $\sigma \neq 1$ be an inertia element of $\overline{\Pi}_{\mathrm{K}}$. Then there exists a valuation v_σ of* K*, which we call the canonical valuation for σ such that the following hold:*

(i) $\sigma \in \mathrm{T}_{v_\sigma}$, *i.e.,* σ *is inertia element at* v_σ.
(ii) *If* σ *is inertia element at some valuation* v, *then* $v_\sigma \leqslant v$.

Proof. We construct v_σ as follows. Let Λ be the fixed field of σ in K'. For every valuation v such that $\sigma \in \mathrm{T}_v$, let v_Λ be the $\mathbb{Z}/\ell$-abelian Λ-core of v. We claim that $v_\sigma := v_\Lambda$ satisfies the conditions (i) and (ii). Indeed, since $\sigma \in \mathrm{T}_v$, and $\mathrm{K}^{\mathrm{T}}v' \supseteq (\mathrm{K}v)'$, we have $\mathrm{K}_v^{\mathrm{T}} \subseteq \Lambda$, hence $\Lambda v' \supseteq (\mathrm{K}v)'$. Therefore, by Proposition 6 (1), it follows that $\Lambda v'_\Lambda \supseteq (\mathrm{K}v_\Lambda)'$. But then one must have $\mathrm{K}_{v_\Lambda}^{\mathrm{T}} \subseteq \Lambda$, and therefore $\mathrm{Gal}(\mathrm{K}'|\Lambda) \subseteq \mathrm{T}_{v_\Lambda}$. Hence $\sigma \in \mathrm{T}_{v_\Lambda}$ verifies (i).

In order to prove (ii), let v_1 be another valuation of K such that $\sigma \in \mathrm{T}_{v_1}$. For the $\mathbb{Z}/\ell$-abelian Λ-core $v_{1,\Lambda}$ of v_1, by the discussion above, we have $\Lambda v'_{1,\Lambda} \supseteq (\mathrm{K}v_{1,\Lambda})'$. We claim that actually $v_\Lambda = v_{1,\Lambda}$. Indeed, both v_Λ and $v_{1,\Lambda}$ equal their $\mathbb{Z}/\ell$-abelian Λ-cores. Hence they are comparable by Proposition 7 (1). By contradiction, suppose that $v_{1,\Lambda} \neq v_\Lambda$, say $v_{1,\Lambda} < v_\Lambda$. Since v_Λ equals its $\mathbb{Z}/\ell$-abelian Λ-core, and $v_{1,\Lambda} < v_\Lambda$, it follows by Proposition 6 (1), that $\Lambda v'_{1,\Lambda} \not\supseteq (\mathrm{K}v_{1,\Lambda})'$, contradiction. Thus $v_\Lambda = v_{1,\Lambda}$. Since $v_{1,\Lambda} \leqslant v_1$, we finally get $v_\Lambda \leqslant v_1$. This completes the proof of (ii). □

Proposition 11. *In the context and the notations from above, the following hold:*

(1) Let $\Sigma = (\sigma_i)_i$ *be a c.l. family of inertia elements. Then the canonical valuations* v_{σ_i} *are pairwise comparable. Moreover, denoting by* $v_\Sigma = \sup_i v_{\sigma_i}$ *their supremum, and by* Λ *the fixed field of* Σ *in* K', *one has:*

(a) $\sigma_i \in \mathrm{T}_{v_\Sigma}$ *for all i.*
(b) v_Σ *equals its* $\mathbb{Z}/\ell$*-abelian* Λ*-core.*

(2) Let $\mathrm{Z} \subseteq \overline{\Pi}_{\mathrm{K}}$ *be some subgroup, and* $\Sigma_{\mathrm{Z}} = (\sigma_i)_i$ *be the family of all inertia elements* σ_i *in* Z *such that* (σ_i, Z) *is c.l. for each i. Suppose that* Z *is not cyclic. Then the valuation* $v := v_{\Sigma_{\mathrm{Z}}}$ *as constructed above with respect to* Σ_{Z} *satisfies:*

(a) $\mathrm{Z} \subseteq \mathrm{Z}_v^1$.
(b) $\Sigma_{\mathrm{Z}} = \mathrm{Z} \cap \mathrm{T}_v^1$.

Proof. (1) We may assume that all σ_i are nontrivial. For each σ_i let T_i be the closed subgroup of $\overline{\Pi}_{\mathrm{K}}$ generated by σ_i, and Λ_i the fixed field of σ_i in K'. Thus one has that $\mathrm{T}_i = \mathrm{Gal}(\mathrm{K}'|\Lambda_i)$, and $\sigma_i \neq 1$ implies that $\mathrm{T}_i \cong \mathbb{Z}/\ell$. Setting $\mathrm{T} := \mathrm{T}_i \cap \mathrm{T}_j$, we have the following possibilities:

Case $\mathrm{T} \neq \{1\}$: Since $\mathrm{T}_i \cong \mathbb{Z}/\ell \cong \mathrm{T}_j$, we have $\mathrm{T} = \mathrm{T}_i = \mathrm{T}_j$, and $\Lambda = \Lambda_i = \Lambda_j$ is the fixed field of T in K'. Hence by Proposition 7 (2), the $\mathbb{Z}/\ell$-abelian Λ-core of v_{σ_i} and v_{σ_j} are comparable. But then reasoning as at the end of the proof of (ii) from Lemma 10, it follows that $v_{\Lambda,i} = v_{\Lambda,j}$. Hence finally $v_{\sigma_j} = v_{\sigma_i}$, as claimed.

Case $\mathrm{T} = \{1\}$: Let T_{ij} be the subgroup generated by σ_i, σ_j in $\overline{\Pi}_{\mathrm{K}}$, and let Λ_{ij} be the fixed field of T_{ij} in K'. Since $\mathrm{T}_i \cap \mathrm{T}_j = \{1\}$, we have $\mathrm{T}_{ij} \cong \mathbb{Z}/\ell \times \mathbb{Z}/\ell$, thus T_{ij} is not pro-cyclic. Hence by the Key Fact 9 above, there exists a valuation v such that $\mathrm{T}_{ij} \subseteq \mathrm{Z}_v^1$, and $\mathrm{T} := \mathrm{T}_v^1 \cap \mathrm{T}_{ij}$ is non-trivial. Moreover, by replacing v by its $\mathbb{Z}/\ell$-abelian Λ_{ij}-core, we can suppose that actually v equals its Λ_{ij}-core. Finally let us notice that we have $\mathrm{T}_{ij}/\mathrm{T} = \mathrm{Gal}((\mathrm{K}v)'|\Lambda_{ij}v')$, and by Kummer theory, $\mathrm{Gal}((\mathrm{K}v)'|\Lambda_{ij}v')$ is of the form $(\mathbb{Z}/\ell)^r$ for some r. Now since T is non-trivial, and $\mathrm{T}_{ij} \cong \mathbb{Z}/\ell \times \mathbb{Z}/\ell$, we finally get: $\mathrm{T}_{ij}/\mathrm{T}$ is either trivial, or $\mathrm{T}_{ij}/\mathrm{T} \cong \mathbb{Z}/\ell$ else.

Next let v_i be the $\mathbb{Z}/\ell$-abelian Λ_i-core of v. The we have the following case discussion:

Suppose $v_i > v_{\sigma_i}$: Reasoning as in the proof of (ii) of Lemma 10, by Proposition 6 (1), we get: Since v_i is the $\mathbb{Z}/\ell$-abelian Λ_i-core of v, and $v_{\sigma_i} < v_i$, we have $\Lambda_i v'_{\sigma_i} \not\supseteq (\mathrm{K}v_{\sigma_i})'$, contradiction. The same holds correspondingly for v_{σ_j} and the corresponding v_j. Hence we must have $v_i \leqslant v_{\sigma_i}$, and $v_j \leqslant v_{\sigma_j}$.

Suppose $v_i < v_{\sigma_i}$: Recall that v_{σ_i} equals its $\mathbb{Z}/\ell$-abelian Λ_i-core, and further $\Lambda_i v'_{\sigma_i} \supseteq (\mathrm{K}v_{\sigma_i})'$ by the definition/construction of v_{σ_i}. Since $v_i < v_{\sigma_i}$, it follows by Proposition 6, (1), that $\Lambda_i v' \not\supseteq (\mathrm{K}v_i)'$. But then by Proposition 6, (3), it follows that $v_i = v$. Hence finally we have the following situation: $v = v_i < v_{\sigma_i}$, and $\Lambda_i v' \not\supseteq (\mathrm{K}v)'$. Equivalently, $\Lambda_i \not\subseteq \mathrm{K}^{\mathrm{T}}_v$, and so, $\mathrm{T}_i \not\subseteq \mathrm{T}_v$. On the other hand, since $\overline{\Pi}_{\mathrm{K}}$ is ℓ-torsion, it follows that $\mathrm{T}_i \cap \mathrm{T}_v$ is trivial, hence $\mathrm{T}_i \cap \mathrm{T}$ is trivial. Hence by the remarks above we have: $\mathrm{T}_{ij} = \mathrm{T}_i\mathrm{T}$. On the other hand, $v < v_{\sigma_i}$ implies that $\mathrm{T}_v \subseteq \mathrm{T}_{v_{\sigma_i}}$, and therefore $\mathrm{T} \subseteq \mathrm{T}_{v_{\sigma_i}}$. Since $\mathrm{T}_i \subseteq \mathrm{T}_{v_{\sigma_i}}$ by the definition of v_{σ_i}, we finally get: T_{ij} is contained in $\mathrm{T}_{v_{\sigma_i}}$. Therefore, σ_i, σ_j are both inertia elements at v_{σ_i}. But then reasoning as at the end of the proof of (ii) from Lemma 10, we deduce that the $\mathbb{Z}/\ell$-abelian Λ_j-core of v_{σ_i}, say w_i, equals v_{σ_j}. Thus finally we have $v_{\sigma_j} = w_i \leqslant v_{\sigma_i}$, hence v_{σ_j} and v_{σ_i} are comparable, as claimed.

By symmetry, we come to the same conclusion in the case $v_j < v_{\sigma_j}$, etc. Thus it remains to analyze the case when $v_{\sigma_i} = v_i$ and $v_{\sigma_j} = v_j$. Now since both v_i and v_j are coarsenings of v, it follows that they are comparable. Equivalently, $v_{\sigma_i} = v_i$ and $v_{\sigma_j} = v_j$ are comparable, as claimed.

(2) Let $\sigma \in \Sigma_{\mathrm{Z}}$ be a non-trivial element, and let v_σ be the canonical valuation attached to σ as defined above at Lemma 10.

Claim. $\mathrm{Z} \subseteq \mathrm{Z}^1_{v_\sigma}$.

Let $\tau \in \mathrm{Z}$ be any element such that the subgroup $\mathrm{Z}_{\sigma,\tau}$ generated by σ, τ is not pro-cyclic. We claim that $\tau \in \mathrm{Z}^1_{v_\sigma}$. Indeed, by the Key Fact 9 above, it follows that there exists a valuation v having the following properties: $\mathrm{Z}_{\sigma,\tau} \subseteq \mathrm{Z}^1_v$, $\mathrm{T} := \mathrm{Z}_{\sigma,\tau} \cap \mathrm{T}^1_v$ is non-trivial, etc. Let ρ be a generator of T. Then since (σ, Z) is c.l., and $\rho \in \mathrm{Z}$, it follows that (σ, ρ) is a c.l. pair of inertia elements of $\overline{\Pi}_{\mathrm{K}}$. Hence by assertion (1) above, it follows that the canonical valuations v_σ and v_ρ are comparable. We notice that $v_\rho \leqslant v$, as the former valuation is a core of the latter one. We have the following case by case discussion:

- Suppose that $v_\sigma \leqslant v_\rho$: Then $v_\sigma \leqslant v_\rho \leqslant v$, hence $\mathrm{Z}_v \subseteq \mathrm{Z}_{v_\rho} \subseteq \mathrm{Z}_{v_\sigma}$, and therefore, $\mathrm{Z}^1_v \subseteq \mathrm{Z}^1_{v_\rho} \subseteq \mathrm{Z}^1_{v_\sigma}$. Since $\tau \in \mathrm{Z}_{\sigma,\tau} \subseteq \mathrm{Z}^1_v$, we finally get $\tau \in \mathrm{Z}^1_{v_\sigma}$, as claimed.

- Suppose that $v_\sigma > v_\rho$: Then in the notations from above, $\Lambda_\sigma v'_\rho \not\supseteq (\mathrm{K}v_\rho)'$, hence T_σ is mapped isomorphically into the residual Galois group $\mathrm{Gal}(\mathrm{K}'v'_\rho|\Lambda_\sigma v'_\rho)$. In particular, since $\rho \in \mathrm{T}_{v_\rho}$, it follows that $\mathrm{T}_\sigma \cap \mathrm{T}_\rho$ is trivial, hence σ, ρ generate $\mathrm{Z}_{\sigma,\tau}$. On the other hand, $v_\sigma > v_\rho$ implies $\mathrm{T}_{v_\rho} \subseteq \mathrm{T}_{v_\sigma} \subseteq \mathrm{Z}_{v_\sigma}$, hence $\mathrm{T}^1_{v_\rho} \subseteq \mathrm{T}^1_{v_\sigma} \subseteq \mathrm{Z}^1_{v_\sigma}$ too. Since $\rho \in \mathrm{T}^1_{v_\rho}$, we finally get $\rho \in \mathrm{Z}^1_{v_\sigma}$. Hence finally $\tau \in \mathrm{Z}^1_{v_\sigma}$.

Combining both cases conclude the proof of the claim.

Now recall that by (1), the valuations v_σ are comparable, and that v denotes their supremum. By general decomposition theory for valuation, since $\mathrm{Z} \subseteq \mathrm{Z}^1_{v_\sigma}$ for all σ, one has $\mathrm{Z} \subseteq \mathrm{Z}^1_v$ too. In order to show that $\Sigma_\mathrm{Z} = \mathrm{Z} \cap \mathrm{T}^1_v$, we proceed as follows. First, reasoning as above, by the Key Fact 9, we see that T^1_v, Z^1_v is c.l., thus so is the pair of groups $\mathrm{Z} \cap \mathrm{T}^1_v$, Z. Hence by the maximality of Σ_Z, it follows that

$$\mathrm{Z} \cap \mathrm{T}^1_v \subseteq \Sigma_\mathrm{Z}.$$

And reasoning as above, it follows that $\Sigma_\mathrm{Z} \subseteq \mathrm{Z} \cap \mathrm{T}^1_v$, thus finally $\Sigma_\mathrm{Z} = \mathrm{Z} \cap \mathrm{T}^1_v$, as claimed. □

10.3.2 Inertia Elements and the c.l. Property

Proposition 11 has the following consequence.

Proposition 12. *(1) Let Δ be a c.l. subgroup of $\overline{\Pi}_\mathrm{K}$. Then Δ contains a subgroup Σ consisting of inertia elements such that Δ/Σ is pro-cyclic (maybe trivial). In particular, there exists a valuation $v := v_\Sigma$ such that $\Delta \subseteq \mathrm{Z}_v$, and $\Delta \cap \mathrm{T}_v = \Sigma$, and v equals its $\mathbb{Z}/\ell$-abelian Λ-core, where Λ is the fixed field of Σ in K'.*

(2) Let $\mathrm{Z} \subseteq \overline{\Pi}_\mathrm{K}$ be a closed subgroup, and Σ_Z a maximal subgroup of Z such that the following are satisfied:

(★) Σ_Z and $(\Sigma_\mathrm{Z}, \mathrm{Z})$ are c.l.

Suppose that $1 \neq \Sigma_\mathrm{Z} \neq \mathrm{Z}$. Then Σ_Z is the unique maximal subgroup of Z satisfying (★), and it consists of all the inertia elements σ in Z such that (σ, Z) is c.l. And there exists a unique valuation v such that $\mathrm{Z} \subseteq \mathrm{Z}^1_v$,

$$\Sigma_\mathrm{Z} = \mathrm{Z} \cap \mathrm{T}^1_v,$$

and v equals its $\mathbb{Z}/\ell$-abelian Λ-core, where Λ is the fixed field of Σ_Z in K'.

Proof. (1) When all $\sigma \in \Delta$ are inertia elements, then $\Sigma = \Delta$ and the conclusion follows by applying Proposition 11 (1).

Now we assume that there is an element $\sigma_0 \in \Delta$ which is not an inertia element. Then for each $\sigma'_i \in \Delta$ such that the closed subgroup $\mathrm{Z}_{\sigma_0,\sigma'_i}$ generated by σ_0, σ'_i is not pro-cyclic, the following holds. Since (σ'_i, σ_0) is by hypothesis c.l., it follows that there exists a valuation v_i of K such that $\mathrm{Z}_{\sigma'_i,\sigma_0} \subseteq \mathrm{Z}^1_{v_i}$, and $\mathrm{T}_i := \mathrm{Z}^1_{\sigma'_i,\sigma_0} \cap \mathrm{T}^1_{v_i}$ is not trivial. Moreover, since σ_0 is by assumption not an inertia element, it follows that denoting by σ_i a generator of T_i, we have: σ_0, σ_i generate topologically $\mathrm{Z}_{\sigma'_i,\sigma_0}$, and σ_i is an inertia element in Δ. In particular, if v_{σ_i} is the canonical valuation attached to the inertia element σ_i, then $v_{\sigma_i} \leqslant v_i$. Hence we have $\sigma_0 \in \mathrm{Z}^1_{v_i} \subseteq \mathrm{Z}^1_{v_{\sigma_i}}$. We next apply Proposition 11 and get the valuation

$$v := v_\Sigma = \sup_i v_{\sigma_i}.$$

Since $\sigma_0 \in \mathrm{Z}^1_{v_{\sigma_i}}$ for all i, by general valuation theory one has $\sigma_0 \in \mathrm{Z}^1_{v_\Sigma}$.

(2) We first claim that Σ_Z consists of inertia elements. By contradiction, suppose that this is not the case, and let $\sigma_0 \in \Sigma_Z$ be a non-inertia element. Reasoning as in (1) above and in the above notations we have: For every $\sigma'_i \in Z$ such that the closed subgroup $Z_{\sigma'_i,\sigma_0}$ generated by σ_0, σ'_i is not pro-cyclic, there exists an inertia element $\sigma_i \in Z_{\sigma'_i,\sigma_0}$, such that $\sigma'_i \in Z^1_{v_{\sigma_i}}$, and the closed subgroup generated by σ_i, σ_0 equals $Z_{\sigma'_i,\sigma_0}$. Then if v_{σ_i} is the canonical valuation for σ_i, we have $Z_{\sigma'_i,\sigma_0} \subseteq Z^1_{v_{\sigma_i}}$, and $Z_{\sigma'_i,\sigma_0} \cap T^1_{v_{\sigma_i}}$ is generated by σ_i. In particular, if Λ_0 is the fixed field of σ_0 in K′, then for every σ_i one has $\Lambda_0 v'_{\sigma_i} \not\supseteq (Kv_{\sigma_i})'$, hence v_{σ_i} equals its $\mathbb{Z}/\ell$-abelian Λ_0-core. Now taking into account that for all subscripts i one has $K^{Z^1}_{v_{\sigma_i}} \subset \Lambda_0$, it follows by Proposition 7, (1), that $(v_{\sigma_i})_i$ is a family of pairwise comparable valuations. Let $v = \sup_i v_{\sigma_i}$ be the supremum of all these valuations. Since $v_{\sigma_i} \leqslant v$ for all σ_i, it follows that $T_{v_{\sigma_i}} \subseteq T_v$ and $T^1_{v_{\sigma_i}} \subseteq T^1_v$ for all σ_i and $\sigma_0 \notin T_v$. Moreover, all the σ_i together with σ_0 generate Z. Hence $T := Z \cap T_v$ together with σ_0 topologically generate Z. But then (σ_0, T) is c.l., and therefore Z is c.l., contradiction.

Hence we conclude that Σ_Z consists of inertia elements only, and by hypothesis, Σ_Z is c.l. We next let $v := v_{\Sigma_Z}$ be the valuation constructed in Proposition 11 (1), for the c.l. family consisting of all the elements from Σ_Z.

Claim. $Z \subseteq Z^1_v$.

Indeed, let $\sigma_0 \in Z \backslash \Sigma_Z$ be a fixed element $\neq 1$, and $\sigma'_i \in \Sigma_Z$. Then (σ'_i, σ_0) is c.l. by hypothesis of the Proposition. Reasoning as in (1) above, we obtain σ_i and v_{σ_i} as there. On the other hand, $\sigma'_i \in \Sigma_Z$ is itself an inertia element, as Σ_Z consists of inertia elements only by the discussion above. Since (σ'_i, Z) is c.l. by hypothesis, it follows that (σ'_i, σ_i) is c.l. Since they are inertia elements, it follows by Proposition 11 (1), that v_{σ_i} and $v_{\sigma'_i}$ are comparable. And further note that $\sigma_0 \in Z^1_{v_{\sigma_i}}$. We have the following case by case discussion.

- If $v_{\sigma_i} \geqslant v_{\sigma'_i}$ for all i, then $v_0 := \sup_i v_{\sigma_i} \geqslant \sup_i v_{\sigma'_i} = v_{\Sigma_Z} = v$. Hence $Z_{v_0} \subseteq Z_v$ and $Z^1_{v_0} \subseteq Z^1_v$. Since $\sigma_0 \in Z^1_{v_i}$ for all σ_i, it follows that $\sigma_0 \in Z^1_{v_0}$. Thus we finally get $\sigma_0 \in Z^1_v$, as claimed.

- If $v_{\sigma_i} < v_{\sigma'_i}$ for some σ'_i, then $\sigma'_i \notin T_{v_{\sigma_i}}$, hence (σ_i, σ'_i) and (σ_0, σ'_i) generate the same closed subgroup. Or equivalently, σ_0 is contained in the subgroup generated by (σ_i, σ'_i). On the other hand, $v_{\sigma_i} < v_{\sigma'_i} \leqslant v$ implies $T_{v_{\sigma_i}} \subseteq T_{v_{\sigma'_i}} \subseteq T_v$, and therefore

$$T^1_{v_{\sigma_i}} \subseteq T^1_{v_{\sigma'_i}} \subseteq T^1_v$$

too. Hence $\sigma_i, \sigma'_i \in T^1_v$. But then $\sigma_0 \in T^1_v$, and in particular, $\sigma_0 \in Z^1_v$, as claimed.

Finally to prove that $\Sigma_Z = Z \cap T^1_v$, apply Proposition 11 (2). □

10.4 Almost Quasi r-Divisorial Subgroups

In this section we will prove a more general form of the main result announced in the Introduction. We begin by quickly recalling some basic facts about (transcendence) defectless valuations, see e.g. Pop [Pop06] Appendix for more details.

10.4.1 Generalized Quasi Divisorial Valuations

Let $\mathrm{K}|k$ be a function field over the algebraically closed field k with $\mathrm{char}(k) \neq \ell$. For every valuation v on K, or on any algebraic extension of K, like for instance K' or K'', since k is algebraically closed, the group vk is a totally ordered $\mathbb{Q}$-vector space, which is trivial, if the restriction of v to k is trivial. We will denote by r_v the rational rank of the torsion free group $v\mathrm{K}/vk$, and by abuse of language call it the **rational rank of** v. Next notice that the residue field kv is algebraically closed too, and $\mathrm{K}v|kv$ is some field extension, but not necessarily a function field. We will denote $\mathrm{td}_v = \mathrm{td}(\mathrm{K}v\,|\,kv)$ and call it the **residual transcendence degree**. By general valuation theory, see e.g. [Bou64] Chap. 6 Sect. 10.3,

$$r_v + \mathrm{td}_v \leqslant \mathrm{td}(\mathrm{K}|k)$$

and we will say that v has **no (transcendence) defect**, or that v is **(transcendence) defectless**, if the above inequality is an equality.

Using Fact 5.4 from [Pop06], it follows that for a valuation v of K and $r \leqslant \mathrm{td}(\mathrm{K}|k)$ the following are equivalent:

(1) v is minimal among the valuations w satisfying $r_w = r$ and $\mathrm{td}_w = \mathrm{td}(\mathrm{K}|k) - r$.

(2) v has no relative defect, and $r_v = r$, and $r_{v'} < r$ for any proper coarsening v' of v.

Definition 13. A valuation of K with the equivalent properties (i) and (ii), above is called **almost quasi r-divisorial**, or an **almost quasi r-divisor** of $\mathrm{K}|k$, or simply a **generalized almost quasi divisor**, if the rank r is not relevant for the context.

Remark 14. In the above context, one has:

(1) The additivity of the rational rank $r_{(\cdot)}$, see [Pop06] Fact 5.4 (1), implies, if v is almost quasi r-divisorial on $\mathrm{K}|k$, and v_0 is almost quasi r_0-divisorial on $\mathrm{K}v\,|\,kv$, then the compositum $v_0 \circ v$ is an almost quasi $(r+r_0)$-divisor of K.

(2) Similar to [Pop06] Appendix, Fact 5.5 (2) (b), if v is almost quasi r-divisorial, then $\mathrm{K}v|kv$ is a function field with $\mathrm{td}(\mathrm{K}v\,|\,kv) = \mathrm{td}(\mathrm{K}|k) - r$, and $v\mathrm{K}/vk \cong \mathbb{Z}^r$.

(3) A prime divisor of $\mathrm{K}|k$ is an almost quasi 1-divisor. And conversely, an almost quasi 1-divisor v of K is a prime divisor if and only if v is trivial on k.

Proposition 15. *In the above context, suppose that* $d = \mathrm{td}(\mathrm{K}|k) > 1$. *Let* v *be an almost quasi* r*-divisor of* $\mathrm{K}|k$, *and* $\mathrm{T}_v \subset \mathrm{Z}_v$ *be the corresponding almost quasi divisorial subgroups of* $\overline{\Pi}_{\mathrm{K}}$, *and* $\mathrm{T}^1_v \subset \mathrm{Z}^1_v$ *be the minimized ones as usual. Then the following hold:*

(1) $\mathrm{T}_v^1 \cong (\mathbb{Z}/\ell)^r$ *and canonically via Kummer theory* $\mathrm{Z}_v^1/\mathrm{T}_v^1 \cong \mathrm{Hom}(\mathrm{K}v)/\ell,\mu_l)$.
(2) *The following hold:*

(a) Z_v^1 *contains c.l. subgroups* $\cong (\mathbb{Z}/\ell)^d$ *and* Z_v^1 *is maximal among the subgroups* Z^1 *of* $\overline{\Pi}_\mathrm{K}$ *which contain subgroups* $\mathrm{T}^1 \cong (\mathbb{Z}/\ell)^r$ *with* $(\mathrm{T}^1,\mathrm{Z}^1)$ *c.l.*
(b) $\mathrm{T} := \mathrm{T}_v^1$ *is the unique maximal subgroup of* $\mathrm{Z} := \mathrm{Z}_v^1$ *with* (T,Z) *c.l. Equivalently, if* $\mathrm{T}'' \subseteq \mathrm{Z}''$ *are the preimages of* $\mathrm{T} \subseteq \mathrm{Z}$ *under the canonical projection* $\overline{\Pi}^c_\mathrm{K} \to \overline{\Pi}_\mathrm{K}$, *then* T'' *is the center of* Z''.

Proof. Assertion (1) follows immediately from the behavior of the decomposition and inertia groups in towers of algebraic extensions, and Fact 2.1 (2) (3) of [Pop06].

(2a) First let us prove that Z_v contains c.l. subgroups $\Delta \cong (\mathbb{Z}/\ell)^d$. Since v is an almost quasi r-divisor, we have $\mathrm{td}(\mathrm{K}v\,|\,kv) = \mathrm{td}_v = \mathrm{td}(\mathrm{K}|k) - r$. Now if $\mathrm{td}_v = 0$, then we are done by assertion (1) above. If $\mathrm{td}_v > 0$, then we consider any almost quasi td_v-divisor v_0 on the residue field $\mathrm{K}v$. Then denoting by $w := v_0 \circ v$ the refinement of v by v_0, we have $v < w$, hence $\mathrm{T}_v \subseteq \mathrm{T}_w$ and $\mathrm{T}_v^1 \subseteq \mathrm{T}_w^1$. And

$$d = \mathrm{td}(\mathrm{K}|k) = r + \mathrm{td}_v = r_w,$$

hence w is an almost quasi d-divisor of K. An easy verification shows that $\mathrm{K}'|\mathrm{K}_w^{\mathrm{T}^1}$ has Galois group isomorphic to $(\mathbb{Z}/\ell)^d$, thus $\mathrm{T}_w^1 \cong (\mathbb{Z}/\ell)^d$. By Key Fact 9, it follows that $\mathrm{T}_w^1 \cong (\mathbb{Z}/\ell)^d$ is c.l. Second, reasoning as in the case of w, it follows that $\mathrm{T}_v^1 \cong (\mathbb{Z}/\ell)^r$, and by applying Key Fact 9, it follows that $(\mathrm{T}_v^1, \mathrm{Z}_v^1)$ are c.l.

We next show that $\mathrm{Z} = \mathrm{Z}_v^1$ is the maximal subgroup of $\overline{\Pi}_\mathrm{K}$ such that (T,Z) is c.l., where $\mathrm{T} = \mathrm{T}_v^1$. Let namely Z be such that $\mathrm{Z}_v^1 \subseteq \mathrm{Z}$ and (T,Z) is c.l. We claim that $\mathrm{Z} = \mathrm{Z}_v$. Let Σ be a maximal c.l. subgroup of Z such that $\mathrm{T} \subseteq \Sigma$, and (Σ,Z) is c.l. too. Applying Proposition 12, let $w := v_\Sigma$ be the resulting valuation from loc.cit. Hence we have $\mathrm{Z} \subseteq \mathrm{Z}_w^1$, and $\Sigma = \mathrm{Z} \cap \mathrm{T}_w$.

Claim. $w \geqslant v$.

Indeed, suppose by contradiction that $w < v$. Then by the fact that v is an almost quasi r-divisor, it follows that $r_w < r = r_v$ and w has no defect. But then

$$\dim(w\mathrm{K}/\ell) = r_w < r = \dim(v\mathrm{K}/\ell),$$

hence $\mathrm{T}_w^1 \cong (\mathbb{Z}/\ell)^{r_w}$. Since $\mathrm{T} \cong (\mathbb{Z}/\ell)^r$ and $\mathrm{T} \subseteq \Sigma \cong (\mathbb{Z}/\ell)^{r_w}$, we get a contradiction and the claim is proved. Therefore $\mathrm{Z}_w \subseteq \mathrm{Z}_v$, thus $\mathrm{Z}_w^1 \subseteq \mathrm{Z}_v^1$; and since $\mathrm{Z} \subseteq \mathrm{Z}_w^1$, we finally have $\mathrm{Z} \subseteq \mathrm{Z}_v^1$, as claimed.

(2b) By assertion (2a), it follows that $(\mathrm{T}_v^1,\mathrm{Z}_v^1)$ is c.l. We show that T_v^1 is the unique maximal (closed) subgroup of Z_v^1 with this property. Indeed, let T be a closed subgroup of Z_v^1 as in (2a). Then denoting by Σ the closed subgroup of Z_v generated by T_v^1 and T, since $(\mathrm{T}_v^1,\mathrm{Z}_v^1)$ and $(\mathrm{T},\mathrm{Z}_v^1)$ are c.l., it follows that (Σ,Z_v^1) are c.l. too. Thus w.l.o.g. we may assume that $\mathrm{T}_v^1 \subseteq \mathrm{T}$, and that T is maximal with the properties from (2b). Now if $r = d$, then $\mathrm{Z}_v^1 = \mathrm{T}_v^1$, and there is nothing to prove. Thus suppose that $r < d$. Let v_Σ be the unique valuation of K given by Proposition 12.

Claim. $v_\Sigma \leqslant v$.

Indeed, suppose by contradiction that $v_\Sigma > v$. Since v is a quasi r-divisor of K, it follows that $\mathrm{K}v|kv$ is a function field with

$$\mathrm{td}(\mathrm{K}v|kv) = \mathrm{td}_v = d - r > 0.$$

Since the valuation $v_0 := v_\Sigma / v$ on $\mathrm{K}v$ is non-trivial, it follows that $\mathrm{Z}^1_{v_0} \subseteq \mathrm{Gal}((\mathrm{K}v)'|\mathrm{K}v)$ is a proper subgroup of $\overline{\Pi}_\mathrm{K}$. Taking into account that $\mathrm{Z}^1_{v_\Sigma}$ is the preimage of

$$\mathrm{Z}^1_{v_0} \subseteq \mathrm{Gal}((\mathrm{K}v)'|\mathrm{K}v)$$

under the canonical projection $\mathrm{Z}_v \to \mathrm{Gal}((\mathrm{K}v)'|\mathrm{K}v)$, it follows that $\mathrm{Z}^1_{v_\Sigma}$ is strictly contained in Z^1_v, contradiction!

By the claim, we have $v_\Sigma \leqslant v$ and so $\mathrm{T}_{v_\Sigma} \subseteq \mathrm{T}_v$, and therefore, $\mathrm{T}^1_{v_\Sigma} \subseteq \mathrm{T}^1_v$. Hence one has $\mathrm{T} \subseteq \Sigma \subseteq \mathrm{T}^1_{v_\Sigma} \subseteq \mathrm{T}^1_v$, thus finally $\mathrm{T} \subseteq \mathrm{T}^1_v$. We conclude that $\mathrm{T} := \mathrm{T}^1_v$ is the unique maximal subgroup of $\mathrm{Z} := \mathrm{Z}^1_v$ such that (T,Z) is c.l., as claimed. □

10.4.2 Characterizing Almost Quasi r-Divisorial Subgroups

We keep the notations from the previous section.

Definition 16. We say that a closed subgroup Z of $\overline{\Pi}_\mathrm{K}$ is an **almost quasi r-divisorial subgroup**, if there exists an almost quasi r-divisor v of $\mathrm{K}|k$ such that $\mathrm{Z} = \mathrm{Z}_v$.

Below we give a characterization of the almost quasi r-divisorial subgroups of $\overline{\Pi}_\mathrm{K}$, thus of the almost quasi r-divisors of K, in terms of the group theoretical information encoded in the Galois group $\overline{\Pi}^c_\mathrm{K}$ alone, provided $r < \mathrm{td}(\mathrm{K}|k)$.

Theorem 17. *Let* $\mathrm{K}|k$ *be a function field over the algebraically closed field* k *with* $\mathrm{char}(k) \neq \ell$. *Let* $pr : \overline{\Pi}^c_\mathrm{K} \to \overline{\Pi}_\mathrm{K}$ *be the canonical projection, and for subgroups* T, Z, Δ *of* $\overline{\Pi}_\mathrm{K}$, *let* T'', Z'', Δ'' *denote their preimages in* $\overline{\Pi}^c_\mathrm{K}$. *Then the following hold:*

(1) *The transcendence degree* $d = \mathrm{td}(\mathrm{K}|k)$ *is the maximal integer* d *such that there exists a closed subgroup* $\Delta \cong (\mathbb{Z}/\ell)^d$ *of* $\overline{\Pi}_\mathrm{K}$ *with* Δ'' *abelian.*

(2) *Suppose that* $d := \mathrm{td}(\mathrm{K}|k) > r > 0$. *Let* $\mathrm{T} \subseteq \mathrm{Z}$ *be closed subgroups of* $\overline{\Pi}_\mathrm{K}$. *Then* $\mathrm{T} \subseteq \mathrm{Z}$ *is a minimized almost quasi r-divisorial subgroup of* $\overline{\Pi}_\mathrm{K}$ *if and only if* Z *is maximal in the set of closed subgroups of* $\overline{\Pi}_\mathrm{K}$ *which satisfy:*

 (i) Z *contains a closed subgroup* $\Delta \cong (\mathbb{Z}/\ell)^d$ *such that* Δ'' *is Abelian.*

 (ii) $\mathrm{T} \cong (\mathbb{Z}/\ell)^r$, *and* T'' *is the center of* Z''.

Proof. Recall from Fact/Definition 8, especially the points (3)–(5), that Δ'' being abelian is equivalent to Δ being c.l., and T'' being the center of Z'' is equivalent to T being the maximal subgroup of Z such that (T,Z) is c.l. We will use the c.l. terminology from now on.

(1) Let Δ be any c.l. closed non-procyclic subgroup of $\overline{\Pi}_{\mathrm{K}}$. Then by Proposition 12 (1), it follows that there exists a valuation v of K such that $\Delta \subseteq \mathrm{Z}_v$, and setting $\mathrm{T}_\Delta := \Delta \cap \mathrm{T}_v$, it follows that Δ/T_Δ is pro-cyclic (maybe trivial). Hence we have the following cases.

Case $\mathrm{T}_\Delta = \Delta$. Then $\mathrm{T}_\Delta \cong (\mathbb{Z}/\ell)^\delta$, and by Lemma 2 (3), it follows that $\delta_v \geqslant \delta$. Since $\mathrm{td}(\mathrm{K}|k) \geqslant r_v \geqslant \delta_v$, we finally get $\mathrm{td}(\mathrm{K}|k) \geqslant \delta$.

Case Δ/T_Δ is non-trivial. Then $\Delta/\mathrm{T}_\Delta \cong \mathbb{Z}/\ell$ and $\mathrm{T}_\Delta \cong (\mathbb{Z}/\ell)^{\delta-1}$. So the image of Δ in $\overline{\Pi}_{\mathrm{K}v}$ is non-trivial, hence $\overline{\Pi}_{\mathrm{K}}$ is non-trivial. Since kv is algebraically closed, we must have $\mathrm{K}v \neq kv$. Equivalently, $\mathrm{td}_v > 0$. Proceeding as above, we also have $r_v \geqslant (\delta - 1)$, hence finally: $\mathrm{td}(\mathrm{K}|k) \geqslant r_v + \mathrm{td}_v > (\delta - 1)$ and $\mathrm{td}(\mathrm{K}|k) \geqslant \delta$.

We now show the converse inequality. Using Pop [Pop06] Fact 5.6, one constructs valuations v of K such that $r_v = \mathrm{td}(\mathrm{K}|k) =: d$. If v is such a valuation, then $\dim(v\mathrm{K}/\ell) = d$. Hence by Lemma 2 (3), $\mathrm{T}^1_v \cong (\mathbb{Z}/\ell)^d$; and T^1_v is a c.l. closed subgroup of $\overline{\Pi}_{\mathrm{K}}$.

(2) By Proposition 15, it follows that if v is an almost quasi-prime r-divisor, then $\mathrm{T}^1_v, \mathrm{Z}^1_v$ satisfy the properties asked for T,Z in (2). For the converse assertion, let $\mathrm{T} \subseteq \mathrm{Z}$ be closed subgroups of $\overline{\Pi}_{\mathrm{K}}$ satisfying the conditions (i) and (ii). Then $\mathrm{T} \cong (\mathbb{Z}/\ell)^r$, and Z contains c.l. subgroups $\Delta \cong (\mathbb{Z}/\ell)^d$ with $d = \mathrm{tr.deg}(\mathrm{K}|k) > r$. And notice that the fact that T'' is the center of Z'' is equivalent to the fact that T is the unique maximal subgroup of Z such that T and (T,Z) are c.l.

Step 1. Consider a maximal c.l. subgroup $\Delta \cong (\mathbb{Z}/\ell)^d$ of Z –which exists by the hypothesis. Since T is by hypothesis a c.l. subgroup of Z such that (T,Z) is c.l. too, it follows that the closed subgroup T_1 of Z generated by T and Δ is a c.l. closed subgroup of Z. Hence by the maximality of Δ it follows that $\mathrm{T}_1 \subseteq \Delta$, hence $\mathrm{T} \subseteq \Delta$.

Step 2. Applying Proposition 12, let v_0 be the valuation of K deduced from the data (T,Z). Hence $\mathrm{Z} \subseteq \mathrm{Z}^1_{v_0}$, and $\mathrm{T} = \mathrm{Z} \cap \mathrm{T}_{v_0}$. Let Λ be the fixed field of T in K′. Then we obviously have $\mathrm{K}^{\mathrm{T}^1}_{v_0} \subseteq \Lambda$. Let v be the $\mathbb{Z}/\ell$-abelian Λ-core of v_0. We will eventually show that v is an almost quasi r-divisor of K, and that $\mathrm{Z} = \mathrm{Z}^1_v$, $\mathrm{T} = \mathrm{T}^1_v$, thus concluding the proof. Note that $\mathrm{T} \subseteq \mathrm{T}_{v_0}$ implies $\mathrm{K}^{\mathrm{T}^1}_{v_0} \subseteq \Lambda$, thus $\Lambda v'_0 \supseteq (\mathrm{K}v_0)'$. Hence by Proposition 6 (1), the same is true for the $\mathbb{Z}/\ell$-abelian Λ-core v of v_0. Therefore we have $\Lambda v' \supseteq (\mathrm{K}v)'$.

Step 3. By Proposition 12 applied to Δ, it follows that there exists a valuation v_Δ of K such that $\Delta \subseteq \mathrm{Z}^1_{v_\Delta}$. Moreover, by the discussion in the proof of assertion (1) above, it follows that v_Δ is defectless, and one of the following holds: Either $r_{v_\Delta} = d$, and moreover, in this case $\Delta = \mathrm{T}^1_{v_\Delta}$, thus Δ consists of inertia elements only. Or $r_{v_\Delta} = d - 1$, and in this case we have $\mathrm{T}^1_{v_\Delta} \cong (\mathbb{Z}/\ell)^{d-1}$; and since Δ is a c.l. subgroup of $\overline{\Pi}_{\mathrm{K}}$, and $\mathrm{T} \subseteq \Delta$ consists of inertia elements only, it follows that $\mathrm{T} \subseteq \mathrm{T}^1_{v_\Delta}$. Finally, Δ contains non inertia elements of $\overline{\Pi}_{\mathrm{K}}$.

Let w be the $\mathbb{Z}/\ell$-abelian Λ-core of v_Δ. Since $\mathrm{T} \subseteq \mathrm{T}^1_{v_\Delta}$, we must have $\mathrm{K}^{\mathrm{T}^1}_{v_\Delta} \subseteq \Lambda$. Hence $\Lambda v'_\Delta \subseteq \mathrm{K}'v'_\Delta$ is purely inseparable. But then by Proposition 6 (1), the same is true for w, i.e., $\Lambda w' \subseteq \mathrm{K}'w'$ is purely inseparable.

Claim. $v = w$

We argue by contradiction. Suppose that $v \neq w$, and first suppose that $v > w$. Since by the conclusion of Step 2 we have that $\Lambda v' \subseteq \mathrm{K}'v'$ is purely inseparable, and v equals its $\mathbb{Z}/\ell$-abelian Λ-core, it follows by Proposition 6 (1) that $\Lambda w' \subseteq \mathrm{K}'w'$ is not purely inseparable. But this contradicts the conclusion of Step 3.

Second, suppose that $v < w$. Then reasoning as above, we contradict the fact that $\Lambda v' \subseteq \mathrm{K}'v'$ is purely inseparable. The claim is proved.

Now since w is a coarsening of v_Δ, and the latter valuation is defectless, the same is true for w, thus for $v = w$. We next claim that v is an almost quasi r-divisor. Indeed, we have:

- First, $\mathrm{T} \subseteq \mathrm{T}^1_w = \mathrm{T}^1_v$, and since $\mathrm{T} \cong (\mathbb{Z}/\ell)^r$, it follows that $\mathrm{T}^1_v \cong (\mathbb{Z}/\ell)^\delta$ for some $r \leqslant \delta \leqslant d$.
- Second, since $v \leqslant v_0$, it follows that $\mathrm{Z}_{v_0} \subseteq \mathrm{Z}_v$ and $\mathrm{Z}^1_{v_0} \subseteq \mathrm{Z}^1_v$. Hence $\mathrm{Z} \subseteq \mathrm{Z}^1_v$, as $\mathrm{Z} \subseteq \mathrm{Z}^1_{v_0}$ by the definition of v_0.
- Moreover, $(\mathrm{T}^1_v, \mathrm{Z}^1_v)$ is c.l. In particular, $(\mathrm{T}, \mathrm{Z}^1_v)$ is a c.l. pair too.

The maximality of Z and T implies that $\mathrm{Z} = \mathrm{Z}^1_v$ and $\mathrm{T} = \mathrm{T}^1_v$, as claimed. □

10.5 A Characterization of Almost r-Divisorial Subgroups

Definition 18. We say that an almost quasi r-divisorial valuation of $\mathrm{K}|k$ is an **almost r-divisorial valuation** or an **almost prime r-divisor** of $\mathrm{K}|k$, if v is trivial on k. We will further say that a closed subgroup Z of $\overline{\Pi}_{\mathrm{K}}$ endowed with a closed subgroup $\mathrm{T} \subset \mathrm{Z}$ is an **almost divisorial r-subgroup** of $\overline{\Pi}_{\mathrm{K}}$, if there exists an almost prime r-divisor v of $\mathrm{K}|k$ such that $\mathrm{T} = \mathrm{T}^1_v$ and $\mathrm{Z} = \mathrm{Z}^1_v$.

We now show that using the information encoded in *sufficiently many* 1-dimensional projections, one can characterize the almost r-divisorial subgroups among all the almost quasi r-divisorial subgroups of $\overline{\Pi}_{\mathrm{K}}$. See Pop [Pop06], especially Fact 4.5 for more details. Let us recall that for $t \in \mathrm{K}$ a non-constant function, we denote by κ_t the relative algebraic closure of $k(t)$ in K, and that the canonical (surjective) projection $pr_t : \overline{\Pi}_{\mathrm{K}} \to \overline{\Pi}_{\kappa_t}$ is called a **one dimensional projection** of $\overline{\Pi}_{\mathrm{K}}$.

Theorem 19. *Let* $\mathrm{K}|k$ *be a function field as usual with* $\mathrm{td}(\mathrm{K}|k) > r > 0$. *Then for a given almost quasi r-divisorial subgroup* $\mathrm{Z} \subseteq \overline{\Pi}_{\mathrm{K}}$, *the following assertions are equivalent:*

(i) *Z is an almost r-divisorial subgroup of* $\overline{\Pi}_{\mathrm{K}}$.
(ii) *There is an element* $t \in \mathrm{K}\backslash k$ *such that* $p_t(\mathrm{Z}) \subseteq \overline{\Pi}_{\kappa_t}$ *is an open subgroup.*

Proof. The proof is word-by-word identical with the one of Pop [Pop06], Proposition 5.6, and therefore we will omit the proof here. □

References

[AEJ87] J. K. Arason, R. Elman, and B. Jacob. Rigid elements, valuations, and realization of Witt rings. *J. Algebra*, 110:449–467, 1987.

[Bog91] F. A. Bogomolov. On two conjectures in birational algebraic geometry. In A. Fujiki et al., editors, *Algebraic Geometry and Analytic Geometry, ICM-90 Satellite Conference Proceedings*. Springer Verlag, Tokyo, 1991.

[Bou64] N. Bourbaki. *Algèbre commutative*. Hermann Paris, 1964.

[BT02] F. A. Bogomolov and Yu. Tschinkel. Commuting elements in Galois groups of function fields. In F. A. Bogomolov and L. Katzarkov, editors, *Motives, Polylogarithms and Hodge theory*, pages 75–120. International Press, 2002.

[BT08] F. Bogomolov and Yu. Tschinkel. Reconstruction of function fields. *Geometric And Functional Analysis*, 18:400–462, 2008.

[EE77] O. Endler and A. J. Engler. Fields with Henselian valuation rings. *Math. Z.*, 152:191–193, 1977.

[Gro83] A. Grothendieck. Brief an Faltings (27/06/1983). In L. Schneps and P. Lochak, editors, *Geometric Galois Action 1*, volume 242 of *LMS Lecture Notes*, pages 49–58. Cambridge, 1997.

[Gro84] A. Grothendieck. Esquisse d'un programme. In L. Schneps and P. Lochak, editors, *Geometric Galois Action 1*, volume 242 of *LMS Lecture Notes*, pages 5–48. Cambridge, 1997.

[Koe01] J. Koenigsmann. Solvable absolute Galois groups are metabelian. *Inventiones Math.*, 144:1–22, 2001.

[MMS04] L. Mahé, J. Mináč, and T. L. Smith. Additive structure of multiplicative subgroups of fields and Galois theory. *Doc. Math.*, 9:301–355, 2004.

[Neu69] J. Neukirch. Kennzeichnung der p-adischen und endlichen algebraischen Zahlkörper. *Inventiones Math.*, 6:269–314, 1969.

[NSW08] J. Neukirch, A. Schmidt, and K. Wingberg. *Cohomology of number fields*, volume 323 of *Grundlehren der Mathematischen Wissenschaften*. Springer-Verlag, Berlin, 2nd edition, 2008.

[Pop94] F. Pop. On Grothendieck's conjecture of birational anabelian geometry. *Ann. of Math.*, 138:145–182, 1994.

[Pop98] F. Pop. Glimpses of Grothendieck's anabelian geometry. In L. Schneps and P. Lochak, editors, *Geometric Galois Action 1*, volume 242 of *LMS Lecture Notes*, pages 113–126. Cambridge, 1997.

[Pop06] F. Pop. Pro-ℓ Galois theory of Zariski prime divisors. In Débès and others, editor, *Luminy Proceedings Conference, SMF No 13*. Hérmann, Paris, 2006.

[Pop10] F. Pop. Pro-ℓ abelian-by-central Galois theory of Zariski prime divisors. *Israel J. Math.*, 180:43–68, 2010.

[Pop11] F. Pop. On the birational anabelian program initiated by Bogomolov I. *Inventiones Math.*, pages 1–23, 2011.

[Sza04] T. Szamuely. Groupes de Galois de corps de type fini (d'après Pop). *Astérisque*, 294:403–431, 2004.

[Top11] A. Topaz. $\mathbb{Z}/\ell$ commuting liftable pairs. Manuscript.

[Uch79] K. Uchida. Isomorphisms of Galois groups of solvably closed Galois extensions. *Tôhoku Math. J.*, 31:359–362, 1979.

[War81] R. Ware. Valuation rings and rigid elements in fields. *Can. J. Math.*, 33:1338–1355, 1981.

Chapter 11
On ℓ-adic Pro-algebraic and Relative Pro-ℓ Fundamental Groups

Jonathan P. Pridham*

Abstract We recall ℓ-adic relative Malcev completions and relative pro-ℓ completions of pro-finite groups and homotopy types. These arise when studying unipotent completions of fibres or of normal subgroups. Several new properties are then established, relating to ℓ-adic analytic moduli and comparisons between relative Malcev and relative pro-ℓ completions. We then summarise known properties of Galois actions on the pro-$\mathbb{Q}_\ell$-algebraic geometric fundamental group and its big Malcev completions. For smooth varieties in finite characteristics different from ℓ, these groups are determined as Galois representations by cohomology of semisimple local systems. Olsson's non-abelian étale-crystalline comparison theorem gives slightly weaker results for varieties over ℓ-adic fields, since the non-abelian Hodge filtration cannot be recovered from cohomology.

11.1 Introduction

Given a topological group Γ, an affine group scheme R over $\mathbb{Q}_\ell$, and a continuous Zariski-dense representation

$$\rho : \Gamma \to \mathrm{R}(\mathbb{Q}_\ell),$$

the relative Malcev completion $\Gamma^{\mathrm{R,Mal}}$ is the universal pro-unipotent extension $\Gamma^{\mathrm{R,Mal}} \to \mathrm{R}$ equipped with a continuous homomorphism

$$\Gamma \to \Gamma^{\mathrm{R,Mal}}(\mathbb{Q}_\ell)$$

J.P. Pridham (✉)
DPMMS, Centre for Mathematical Sciences, University of Cambridge, Wilberforce Road, Cambridge, CB3 0WB, UK
e-mail: J.P.Pridham@dpmms.cam.ac.uk

* The author was supported during this research by the Engineering and Physical Sciences Research Council, grant number EP/F043570/1.

J. Stix (ed.), *The Arithmetic of Fundamental Groups*, Contributions in Mathematical and Computational Sciences 2, DOI 10.1007/978-3-642-23905-2_11,

extending ρ. Relative Malcev completion was introduced by Hain in [Hai98] for discrete groups, and was extended to pro-finite groups in [Pri09], although the similar notion of weighted completion had already appeared in [HM03b]. In Sect. 11.2, we recall its main properties and establish several new results.

Relative Malcev completion simultaneously generalises both (unipotent) Malcev completion (take $\mathrm{R} = 1$) and the deformation theory of $\mathbb{Q}_\ell$-representations (by restricting the unipotent extensions – see 11.2.1.3). As an example of its power, consider a semi-direct product $\Gamma = \Delta \ltimes \Lambda$. Unless the action of Δ on Λ is nilpotent, the Malcev completion $\Gamma^{\mathrm{R,Mal}}$ of Γ can destroy Λ. However, Example 19 shows that for suitable R we have

$$\Gamma^{\mathrm{R,Mal}} = \Delta^{\mathrm{R,Mal}} \ltimes \Lambda^{1,\mathrm{Mal}},$$

where $\Lambda^{1,\mathrm{Mal}}$ is the unipotent Malcev completion of Λ.

Although relative Malcev completion is right-exact, it is not left-exact. However, there is a theory of relative Malcev homotopy types and higher homotopy groups, as developed in [Pri11] and summarised in 11.2.2. There is a long exact sequence of homotopy (Theorem 18), allowing us to describe the Malcev completion of the kernel of a surjection $\Gamma \twoheadrightarrow \Delta$ in terms of the relative Malcev homotopy types of Γ and Δ. In 11.2.2.3 we establish criteria for these higher homotopy groups to vanish.

In Sect. 11.3, we introduce a new notion, that of relative Malcev completion

$$\Gamma^{\rho,\mathrm{Mal}}_{\mathbb{Z}_\ell}$$

over $\mathbb{Z}_\ell$. This is a canonical $\mathbb{Z}_\ell$-form of the $\mathbb{Q}_\ell$-scheme $\Gamma^{\rho,\mathrm{Mal}}$, and is a strictly finer invariant from which we can recover analytic moduli spaces of Γ-representations over $\mathbb{Q}_\ell$ (Proposition 41), rather than just local deformations.

A similar notion to relative Malcev completion is the relative pro-ℓ completion $\Gamma^{(\ell),\rho}$ of [HM09]. For a surjective homomorphism $\bar{\rho} : \Gamma \to \bar{\mathrm{R}}$ of pro-finite groups, the pro-finite group $\Gamma^{(\ell),\bar{\rho}}$ is the universal pro-ℓ extension of $\bar{\mathrm{R}}$ equipped with a continuous homomorphism

$$\Gamma \to \Gamma^{(\ell),\rho}$$

extending $\bar{\rho}$. It turns out (Proposition 34) that $\Gamma^{(\ell),\rho}$ is in fact the relative $\mathbb{F}_\ell$-Malcev completion of Γ over $\bar{\mathrm{R}}$, where the latter is regarded as a pro-finite group scheme over $\mathbb{F}_\ell$.

However, specialisation of $\Gamma^{\rho,\mathrm{Mal}}_{\mathbb{Z}_\ell}$ to $\mathbb{F}_\ell$ does not recover $\Gamma^{(\ell),\rho}$ in general. Instead, we need to look at the specialisation of the relative Malcev homotopy type over $\mathbb{Z}_\ell$, with a universal coefficient theorem giving the required data (Proposition 45). In this sense, the homotopy type over $\mathbb{Z}_\ell$ acts as a bridge between relative Malcev and relative pro-ℓ completions.

In the final section, we summarise the main implications of [Pri09], [Ols11] and [Pri11] for relative $\mathbb{Q}_\ell$-Malcev completions

$$\pi_1^{\text{ét}}(\mathrm{X},\bar{x})^{\mathrm{R,Mal}}$$

of geometric fundamental groups. For smooth quasi-projective varieties in finite characteristic, these can be recovered as Galois representations from cohomology of semisimple local systems (Propositions 48 and 49). Over ℓ-adic local fields, there are similar results (Theorem 53) using Olsson's non-abelian étale-crystalline comparison theorem, but for a full description it is necessary to understand the non-abelian Hodge filtration as well. Over global fields K, we just have a weight filtration on $\pi_1^{\text{ét}}(X,\bar{x})^{R,\text{Mal}}$, which splits $\text{Gal}(\bar{K}_{\mathfrak{p}}/K_{\mathfrak{p}})$-equivariantly for each prime $\mathfrak{p}$ of good reduction (11.4.4).

Notation. For any affine scheme Z, we write $O(Z) := \Gamma(Z, \mathscr{O}_Z)$.

11.2 Relative Malcev Completion

Fix a topological group Γ, an affine group scheme R over $\mathbb{Q}_\ell$, and a continuous Zariski-dense representation

$$\rho : \Gamma \to R(\mathbb{Q}_\ell),$$

where $R(\mathbb{Q}_\ell)$ is given the ℓ-adic topology. Explicitly, [DMOS82, Chap. II] shows that R can be expressed as a filtered inverse limit

$$R = \varprojlim R_\alpha$$

of linear algebraic groups. Each $R_\alpha(\mathbb{Q}_\ell)$ has a canonical ℓ-adic topology induced by any embedding $R_\alpha \hookrightarrow GL_n$. We define the topology on $R(\mathbb{Q}_\ell)$ by

$$R(\mathbb{Q}_\ell) = \varprojlim R_\alpha(\mathbb{Q}_\ell).$$

The following definition appears in [Hai98] for Γ discrete, and in [Pri09] for Γ pro-finite.

Definition 1. The **Malcev completion** $\Gamma^{\rho,\text{Mal}}$ (or $\Gamma^{R,\text{Mal}}$) of Γ relative to ρ is defined to be the universal diagram

$$\Gamma \xrightarrow{\breve{\rho}} \Gamma^{\rho,\text{Mal}}(\mathbb{Q}_\ell) \xrightarrow{p} R(\mathbb{Q}_\ell),$$

with $p : \Gamma^{\rho,\text{Mal}} \to R$ a pro-unipotent extension, and with $p \circ \breve{\rho} = \rho$.

If R^{red} is the maximal pro-reductive quotient of R, then $R \to R^{\text{red}}$ is a pro-unipotent extension, so there is a morphism $\Gamma^{R^{\text{red}},\text{Mal}} \to R$. This must itself be a pro-unipotent extension, so we see that

$$\Gamma^{R^{\text{red}},\text{Mal}} = \Gamma^{R,\text{Mal}}.$$

For this reason, from now on we will (unless otherwise stated) assume that R is pro-reductive.

Example 2. If $R = 1$, then $\Gamma^{1,\mathrm{Mal}}$ is the (pro-unipotent) Malcev completion of Γ.

Example 3. Take Γ^{red} to be universal among Zariski-dense morphisms $\Gamma \to R(\mathbb{Q}_\ell)$ to pro-reductive affine group schemes, and set $R = \Gamma^{\mathrm{red}}$. Then $\Gamma^{R,\mathrm{Mal}} = \Gamma^{\mathrm{alg}}$, the pro-algebraic (or Hochschild–Mostow) completion of Γ. The morphism

$$\Gamma \to \Gamma^{\mathrm{alg}}(\mathbb{Q}_\ell)$$

is universal among continuous morphisms from Γ to affine group schemes over $\mathbb{Q}_\ell$.

In fact, we can describe $O(\Gamma^{\mathrm{red}})$ explicitly: if T is the set of isomorphism classes of irreducible representations of Γ over $\overline{\mathbb{Q}}_\ell$, then

$$O(\Gamma^{\mathrm{red}}) \otimes \overline{\mathbb{Q}}_\ell \cong \bigoplus_{V \in T} \mathrm{End}(V)$$

as a vector space. For example,

$$O(\hat{\mathbb{Z}}^{\mathrm{red}}) = \overline{\mathbb{Q}}_\ell[\overline{\mathbb{Z}}_\ell^{\,*}]^{\mathrm{Gal}(\overline{\mathbb{Q}}_\ell/\mathbb{Q}_\ell)}.$$

Note that since $\mathbb{Q}_\ell$ is of characteristic 0, there is a Levi decomposition

$$\Gamma^{R,\mathrm{Mal}} \cong R \ltimes R_u(\Gamma^{R,\mathrm{Mal}}),$$

unique up to conjugation by $R_u(\Gamma^{R,\mathrm{Mal}})$, where

$$R_u(\Gamma^{R,\mathrm{Mal}}) = \ker(\Gamma^{R,\mathrm{Mal}} \to R)$$

is the pro-unipotent radical.

Lemma 4. *The affine group scheme* $R_u(\Gamma^{R,\mathrm{Mal}})$ *is determined by its tangent space*

$$\mathfrak{r}_u(\Gamma^{R,\mathrm{Mal}})$$

at 1, *regarded as a pro-(finite dimensional nilpotent) Lie algebra.*

Proof. This is true for all pro-unipotent group schemes U. Given an inverse system $V = \{V_\alpha\}$ of vector spaces, write $V \hat{\otimes} A := \varprojlim_\alpha (V_\alpha \otimes A)$. We may thus regard the Lie algebra $\mathfrak{u}$ of U as an affine scheme, with A-valued points given by

$$\mathfrak{u}(A) := \mathfrak{u} \hat{\otimes} A,$$

for any $\mathbb{Q}_\ell$-algebra A. The Lie algebra structure of $\mathfrak{u}(A)$ over A then allows us to define a group

$$\exp(\mathfrak{u})(A) = \exp(\mathfrak{u}(A))$$

with the same elements as $\mathfrak{u}(A)$, but with multiplication given by the Baker–Campbell–Hausdorff formula. Exponentiation then gives a canonical isomorphism

$$U \cong \exp(\mathfrak{u})$$

of group schemes. For unipotent U this is standard, and the pro-unipotent case follows by taking inverse limits. □

11.2.1 Properties of Relative Malcev Completion

We now summarise various properties of relative Malcev completion.

11.2.1.1 Representations

The category of finite-dimensional $\Gamma^{R,Mal}$-representations

$$\mathrm{FDRep}(\Gamma^{R,Mal})$$

has a natural forgetful functor to the category of continuous finite-dimensional Γ-representations

$$\mathrm{FDRep}(\Gamma)\,.$$

Lemma 5. *The natural forgetul functor*

$$\mathrm{FDRep}(\Gamma^{R,Mal}) \to \mathrm{FDRep}(\Gamma)$$

is full and faithful, and a Γ-representation V lies in the essential image of this functor if and only if its semisimplification V^{ss} is an R-representation – in other words, if the morphism $\Gamma \to GL(V^{ss})$ factors through $\rho : \Gamma \to R$.

Proof. Take an algebraic morphism $\Gamma^{R,Mal} \to GL(V)$, and define a decreasing filtration S^pV on V by

$$S^pV = (R_u\Gamma^{R,Mal} - 1)^pV\,.$$

Since $R_u\Gamma^{R,Mal}$ is pro-unipotent, either $S^pV = 0$ or $\dim S^{p+1}V < \dim S^pV$. Thus the filtration is Hausdorff, and $\mathrm{gr}_S V$ is the semisimplification of V. Since $\mathrm{gr}_S V$ is an R-representation, this establishes essential surjectivity.

Full faithfulness just follows because the map $\Gamma \to \Gamma^{R,Mal}(\mathbb{Q}_\ell)$ is Zariski-dense. If it were not, then the Zariski closure of its image would have the same universal property, giving a contradiction. □

In particular, the category $\mathrm{FDRep}(\Gamma^{red})$ consists of continuous semisimple finite-dimensional Γ-representations, while $\mathrm{FDRep}(\Gamma^{alg})$ consists of all continuous finite-dimensional Γ-representations. An arbitrary $\Gamma^{R,Mal}$-representation V is by definition an $O(\Gamma^{R,Mal})$-comodule. This is the same as saying that V is a sum of finite-dimensional $\Gamma^{R,Mal}$-representations, as is true for all affine group schemes, see [DMOS82, Chap. II].

11.2.1.2 Cohomology

The following is [Pri09, Lemma 2.3].

Lemma 6. *For any finite-dimensional* R*-representation* V*, the canonical maps*

$$H^i(\Gamma^{R,Mal}, V) \to H^i(\Gamma, V),$$

are bijective for $i = 0, 1$ *and injective for* $i = 2$.

Note that the long exact sequence of cohomology then implies that the same is true for all finite-dimensional $\Gamma^{R,Mal}$-representations V.

We may regard O(R) as an R-representation via left multiplication. Applying the Hochschild-Serre spectral sequence, as in [Pri09, Lemma 2.6], to the morphism $\Gamma^{R,Mal} \to R$ then gives canonical isomorphisms

$$H^i(R_u\Gamma^{R,Mal}, \mathbb{Q}_\ell) \cong H^i(\Gamma^{R,Mal}, O(R)).$$

As observed in [Pri09, Lemma 2.7], there are canonical isomorphisms

$$H^*(R_u\Gamma^{R,Mal}, \mathbb{Q}_\ell) \cong H^*(\mathfrak{r}_u\Gamma^{R,Mal}, \mathbb{Q}_\ell).$$

These results combine to show that there is a presentation of $\mathfrak{r}_u\Gamma^{R,Mal}$ with generators dual to

$$H^1(\Gamma^{R,Mal}, O(R))$$

and relations dual to

$$H^2(\Gamma^{R,Mal}, O(R)).$$

If $H^*(\Gamma, -)$ commutes with filtered direct limits, we then have a presentation with generators $H^1(\Gamma, \rho^{-1}O(R))^\vee$ and relations $H^2(\Gamma, \rho^{-1}O(R))^\vee$, where $\rho^{-1}O(R)$ is the Γ-representation induced by the R-representation O(R) above.

11.2.1.3 Deformations

We now show how relative Malcev completions naturally encode all the information about framed deformations of a representation. We take a representation

$$\rho : \Gamma \to GL(V)$$

and consider the formal scheme F_ρ, defined for any Artinian local $\mathbb{Q}_\ell$-algebra A with residue field $\mathbb{Q}_\ell$ by

$$F_\rho(A) = \mathrm{Hom}(\Gamma, GL(V \otimes A)) \times_{\mathrm{Hom}(\Gamma, GL(V))} \{\rho\}.$$

Now, with $\mathfrak{m}(A)$ the maximal ideal of A, we have

$$GL(V \otimes A) = GL(V) \ltimes (1 + \mathrm{End}(V) \otimes \mathfrak{m}(A)).$$

If R is the Zariski closure of the image of ρ (which need not be reductive), then

$$\begin{aligned} F_\rho(A) &= \mathrm{Hom}(\Gamma, R(\mathbb{Q}_\ell) \ltimes (1+\mathrm{End}(V)\otimes \mathfrak{m}(A))) \times_{\mathrm{Hom}(\Gamma, R(\mathbb{Q}_\ell))} \rho \\ &= \mathrm{Hom}_{\mathbb{Q}_\ell}(\Gamma^{R,\mathrm{Mal}}, R \ltimes (1+\mathrm{End}(V)\otimes \mathfrak{m}(A))) \times_{\mathrm{Hom}(\Gamma^{R,\mathrm{Mal}}, R)} \rho , \end{aligned}$$

since $R \ltimes (1+\mathrm{End}(V)\otimes \mathfrak{m}(A))$ is a unipotent extension of R. Applying the logarithm

$$\log : 1+\mathrm{End}(V)\otimes \mathfrak{m}(A) \to \mathrm{End}(V)\otimes \mathfrak{m}(A),$$

we see that F_ρ is a formal subscheme contained in the germ at 0 of

$$O(\Gamma^{R,\mathrm{Mal}}) \otimes \mathrm{End}(V),$$

defined by the conditions

$$\exp(f)\cdot \rho \in F_\rho(A) \quad \Longleftrightarrow \quad f(a\cdot b) = f(a) \star (\mathrm{ad}_{\rho(a)}(f(b)))$$

for $a, b \in \Gamma^{R,\mathrm{Mal}}$, where ad denotes the adjoint action, and $\star$ is the Baker–Campbell–Hausdorff product

$$a \star b = \log(\exp(a)\cdot \exp(b)).$$

Remark 7. Note that the same formulae hold if we replace R with any larger quotient of Γ^{alg}. In particular, this means that we can recover F_ρ directly from Γ^{alg}.

Remark 8. There is also a natural conjugation action of the group

$$1+\mathrm{End}(V)\otimes \mathfrak{m}(A)$$

on $O(\Gamma^{R,\mathrm{Mal}}) \otimes \mathrm{End}(V) \otimes \mathfrak{m}(A)$, so we can even recover the formal stack

$$A \mapsto [F_\rho(A)/(1+\mathrm{End}(V)\otimes \mathfrak{m}(A))]$$

of representations modulo infinitesimal inner automorphisms.

Remark 9. If we wished to consider representations to an arbitrary linear algebraic group G, then the formulae above adapt, replacing End(V) with the Lie algebra $\mathfrak{g}$, and $1+\mathrm{End}(V)\otimes \mathfrak{m}(A)$ with

$$\ker(G(A) \to G(\mathbb{Q}_\ell)) = \exp(\mathfrak{g} \otimes \mathfrak{m}(A)).$$

11.2.2 Higher Homotopy Groups

Relative Malcev completion was developed in [Pri11] for any pointed pro-finite homotopy type (X, x). Examples of pro-finite homotopy types are the classifying space $B\Gamma$ of a pro-finite group Γ, or Artin and Mazur's pointed étale homotopy type

$(Y_{\text{ét}}, \bar{y})$ of a connected Noetherian scheme Y as in [AM69] or [Fri82]. In particular, note that:

(1) $\pi_1(B\Gamma) = \Gamma$.
(2) $\pi_n(B\Gamma) = 0$ for all $n > 1$.
(3) $\pi_1(Y_{\text{ét}}, \bar{y}) = \pi_1^{\text{ét}}(Y, \bar{y})$.
(4) $H^*(Y_{\text{ét}}, F) = H^*_{\text{ét}}(Y, F)$ for all $\pi_1^{\text{ét}}(Y, \bar{y})$-representations F in finite abelian groups.

The first stage in the construction of relative Malcev completion is to form, as in [Pri11, §1], a simplicial pro-finite group

$$\widehat{G(X,x)},$$

based on Kan's loop group construction [Kan58]. It has the following properties:

(1) $\pi_n(\widehat{G(X,x)}) = \pi_{n+1}(X,x)$ for all $n \geq 0$.
(2) $H^*(\widehat{G(X,x)}, F) = H^*(X, F)$ for all finite $\pi_1(X,x)$-modules F.
(3) For all n, the pro-finite group $\widehat{G(X,x)}_n$ is freely generated.

In particular, this means that $\widehat{G(B\Gamma)}$, for any pro-finite group Γ, is a free simplicial resolution of Γ in pro-finite groups.

Definition 10. The **relative Malcev homotopy type** $(X,x)^{R,\text{Mal}}$ (or $(X,x)^{\rho,\text{Mal}}$) of a pointed pro-finite homotopy type (X,x) relative to a pro-reductive affine group scheme R over $\mathbb{Q}_\ell$ and to a Zariski-dense map

$$\rho : \pi_1(X,x) \to R(\mathbb{Q}_\ell)$$

is defined to be the simplicial affine group scheme over $\mathbb{Q}_\ell$ given by

$$\widehat{G(X,x)}^{R,\text{Mal}}$$

as in [Pri11, Definition 3.20 and Lemma 1.17]. The **relative Malcev homotopy groups** $\varpi_n(X,x)^{R,\text{Mal}}$ are defined by

$$\varpi_n(X,x)^{R,\text{Mal}} := \pi_{n-1}\widehat{G(X,x)}^{R,\text{Mal}}.$$

Relative Malcev homotopy types have the following properties:

(1) $\varpi_1(X,x)^{R,\text{Mal}} = \pi_1(X,x)^{R,\text{Mal}}$,
(2) For $n > 1$, $\varpi_n(X,x)^{R,\text{Mal}}$ is a commutative pro-unipotent group scheme,
(3) For any finite-dimensional $\pi_1(X,x)^{R,\text{Mal}}$-representation V, the map

$$H^*(X^{R,\text{Mal}}, V) \to H^*(X, V)$$

coming from the morphism $\widehat{G(X,x)} \to \widehat{G(X,x)}^{R,\text{Mal}}$ is an isomorphism,
(4) There is a conjugation action of $\varpi_1(X,x)^{R,\text{Mal}}$ on $\varpi_n(X,x)^{R,\text{Mal}}$,
(5) For $m,n > 1$, there is a graded Lie bracket – the Whitehead bracket

$$[-,-] : \varpi_m(X,x)^{R,\text{Mal}} \times \varpi_n(X,x)^{R,\text{Mal}} \to \varpi_{m+n-1}(X,x)^{R,\text{Mal}}.$$

Remark 11. In [Pri08], a category $s\mathcal{E}(\mathrm{R})$ was introduced to model relative Malcev homotopy types over R. Its objects are simplicial diagrams $\mathrm{G} = \mathrm{G}_\bullet$ of pro-unipotent extensions $\mathrm{G}_n \to \mathrm{R}$. A morphism $f : \mathrm{G} \to \mathrm{H}$ in $s\mathcal{E}(\mathrm{R})$ is said to be a weak equivalence if it induces isomorphisms $\pi_n\mathrm{G} \to \pi_n\mathrm{H}$ on homotopy groups for all n.

The category obtained by formally inverting all weak equivalences in $s\mathcal{E}(\mathrm{R})$ forms the homotopy category

$$\mathrm{Ho}_*(s\mathcal{E}(\mathrm{R}))$$

of pointed relative Malcev homotopy types, as studied in [Pri10, Theorem 3.28].

For unpointed relative Malcev homotopy types, we define $\mathrm{Ho}(s\mathcal{E}(\mathrm{R}))$ to have the same objects as $s\mathcal{E}(\mathrm{R})$, but with

$$\mathrm{Hom}_{\mathrm{Ho}(s\mathcal{E}(\mathrm{R}))}(\mathrm{G},\mathrm{H}) := \mathrm{Hom}_{\mathrm{Ho}_*(s\mathcal{E}(\mathrm{R}))}(\mathrm{G},\mathrm{H})/(\mathrm{R_u}\mathrm{H}_0)\,,$$

where $\mathrm{R_u}\mathrm{H}_0$ acts by conjugation. As we will see in Theorem 17, the functor

$$(\mathrm{X},x) \mapsto \widehat{\mathrm{G}(\mathrm{X},x)}^{\mathrm{R,Mal}}$$

descends to a functor from unpointed pro-finite homotopy types to $\mathrm{Ho}(s\mathcal{E}(\mathrm{R}))$.

A related result is [Pri08, Corollary 3.57], which shows that $\mathrm{Ho}(s\mathcal{E}(\mathrm{R}))$ forms a full subcategory of Toën's unpointed schematic homotopy types [Toë06] over BR. The same argument shows that $\mathrm{Ho}_*(s\mathcal{E}(\mathrm{R}))$ forms a full subcategory within pointed schematic homotopy types over BR.

Definition 12. Let Lie(V) be the **free graded Lie algebra** generated by a graded vector space V. Here, the bracket follows usual graded conventions, so for a of degree i and b of degree j, we have

$$[a,b] = (-1)^{ij+1}[b,a]\,.$$

The grading on Lie(V) is given by setting any bracket of length r of homogeneous elements a_i of degrees d_i to have degree $\sum_{i=1}^r d_i$.

Let $\mathrm{Lie}_r(\mathrm{V}) \subset \mathrm{Lie}(\mathrm{V})$ be the graded subspace consisting of elements of bracket length r in V, so $\mathrm{Lie}(\mathrm{V}) = \bigoplus_{r>0} \mathrm{Lie}_r(\mathrm{V})$.

For the ease of stating the next two results, we introduce the notation that Π_1 is the Lie algebra of $\mathrm{R_u}\varpi_1(\mathrm{X},x)^{\mathrm{R,Mal}}$, while for all $n > 1$

$$\Pi_n := \varpi_n(\mathrm{X},x)^{\mathrm{R,Mal}}\,.$$

These are pro-finite-dimensional vector spaces – note that taking continuous duals $(\varprojlim_\alpha \mathrm{V}_\alpha)^\vee = \varinjlim_\alpha \mathrm{V}_\alpha^\vee$ gives a contravariant equivalence from pro-finite-dimensional vector spaces to arbitrary vector spaces.

Proposition 13. *There is a convergent Adams spectral sequence in pro-finite-dimensional vector spaces*

$$\mathrm{E}^1_{pq} = (\mathrm{Lie}_{-p}(\tilde{\mathrm{H}}^{*+1}(\mathrm{X},\rho^{-1}\mathrm{O}(\mathrm{R}))^\vee))_{p+q} \Longrightarrow \Pi_{p+q}\,,$$

where $\tilde{\mathrm{H}}$ *denotes reduced cohomology. Moreover, the differential*

$$\begin{aligned} d^1_{-1,q} : \tilde{\mathrm{H}}^q(\mathrm{X},\rho^{-1}\mathrm{O(R)})^\vee &\to (\bigwedge^2 \tilde{\mathrm{H}}^{*+1}(\mathrm{X},\rho^{-1}\mathrm{O(R)})^\vee)_{q-2} \\ &= ((\mathrm{Sym}^2\tilde{\mathrm{H}}^*(\mathrm{X},\rho^{-1}\mathrm{O(R)}))^q)^\vee \end{aligned}$$

is dual to the cup product on $\tilde{\mathrm{H}}^*(\mathrm{X},\rho^{-1}\mathrm{O(R)})$.

Proof. This is [Pri08, Proposition 4.37]. The spectral sequence is induced by studying the lower central series filtration on the pro-unipotent radical

$$\mathrm{R_u}\widehat{\mathrm{G(X},x)}^{\mathrm{R,Mal}}.$$

Beware that $\rho^{-1}\mathrm{O(R)}$ is here regarded as an ind-object of finite-dimensional local systems, with cohomology calculated accordingly. This is only an issue if $\mathrm{H}^*(\mathrm{X},-)$ does not preserve filtered colimits. □

Theorem 14. *There is a canonical convergent reverse Adams spectral sequence*

$$\mathrm{E}_1^{pq} = (\mathrm{Sym}^p(\Pi_{*-1}))^{p+q} \implies \mathrm{H}^{p+q}(\mathrm{X},\rho^{-1}\mathrm{O(R)}),$$

where Sym *is the symmetric functor on graded vector spaces.*

Proof. This is [Pri08, Theorem 1.53]. □

Finally, these combine to give a Hurewicz theorem.

Corollary 15. *Let* V *be the (ind-)local system on* X *corresponding to the* $\pi_1(\mathrm{X},x)$-*representation* $\mathrm{O}(\varpi_1(\mathrm{X},x)^{\mathrm{R,Mal}})$. *Then for* $n \geq 1$, *the following conditions are equivalent*

(a) $\varpi_i(\mathrm{X})^{\mathrm{R,Mal}} = 0$ *for all* $2 \leq i < n$,
(b) $\mathrm{H}^i(\mathrm{X},\mathrm{V}) = 0$ *for all* $2 \leq i < n$,

and if either of these conditions holds, then

$$\varpi_n(\mathrm{X})^{\mathrm{R,Mal}} \cong \mathrm{H}^n(\mathrm{X},\mathrm{V})^\vee.$$

In particular, this always holds for $n = 2$.

Proof. If $\varpi_1(\mathrm{X}^{\mathrm{R,Mal}}) = 1$, then $\mathrm{V} = \mathbb{Q}_\ell$ and these results follow by studying the Adams and reverse Adams spectral sequences. For the general result, we replace $\mathrm{X}^{\mathrm{R,Mal}}$ with its universal cover

$$\widetilde{\mathrm{X}^{\mathrm{R,Mal}}}.$$

Explicitly, $\widetilde{\mathrm{X}^{\mathrm{R,Mal}}}$ is the homotopy fibre of $\mathrm{X}^{\mathrm{R,Mal}}$ over $\varpi_1(\mathrm{X},x)^{\mathrm{R,Mal}}$, given by taking any free resolution of the kernel

$$\ker(\widehat{\mathrm{G(X},x)}^{\mathrm{R,Mal}} \to \varpi_1(\mathrm{X},x)^{\mathrm{R,Mal}}).$$

Now, $\varpi_1(\widetilde{X^{R,Mal}}) = 1$, and the Hochschild-Serre spectral sequence gives

$$H^*(\widetilde{X^{R,Mal}}, \mathbb{Q}_\ell) = H^*(X, V),$$

so the general results follow from the simply connected case. □

Beware that $\varpi_n(B\Gamma)^{R,Mal}$ can be non-zero for $n > 1$. Determining when this happens is the purpose of 11.2.2.3.

11.2.2.1 Equivariant Cochains

Given a pro-finite homotopy type (X,x) with $\pi_1(X,x) = \Gamma$, and a continuous Γ-representation Λ in finite rank $\mathbb{Z}_\ell$-modules, [Pri11, Definition 1.21] constructs a cosimplicial ℓ-adic sheaf $\mathscr{C}^\bullet(\Lambda)$. This is an acyclic resolution of Λ, so gives a cosimplicial $\mathbb{Z}_\ell$-module

$$C^\bullet(X,\Lambda) := \Gamma(X, \mathscr{C}^\bullet(\Lambda))$$

with the property that $H^*C^\bullet(X,\Lambda) = H^*(X,\Lambda)$.

In particular, if X is the étale homotopy type $Y_{\text{ét}}$ of a Noetherian scheme Y, then

$$C^\bullet(Y_{\text{ét}}, \Lambda)$$

is a model for the ℓ-adic étale Godement resolution of Y with coefficients in Λ. If $X = B\Gamma$, then $C^\bullet(B\Gamma,\Lambda)$ is just the continuous group cohomology complex

$$C^n(B\Gamma,\Lambda) = \mathrm{Hom}_{\mathrm{cts}}(\Gamma^n, \Lambda),$$

with its usual operations.

Definition 16. Take a pro-finite homotopy type X with $\pi_1(X,x) = \Gamma$, an affine group scheme R over $\mathbb{Q}_\ell$, and a representation $\rho : \Gamma \to R(\mathbb{Q}_\ell)$. Given a finite-dimensional R-representation V, choose a Γ-equivariant $\mathbb{Z}_\ell$-lattice $\Lambda \subset V$, and define the cosimplicial vector space $C^\bullet(X,\rho^{-1}V)$ by

$$C^\bullet(X,\rho^{-1}V) := C^\bullet(X,\rho^{-1}\Lambda) \otimes_{\mathbb{Z}_\ell} \mathbb{Q}_\ell .$$

If $U = \bigcup_\alpha U_\alpha$ is a nested union of finite-dimensional R-representations, define

$$C^\bullet(X,\rho^{-1}U) := \bigcup_\alpha C^\bullet(X,\rho^{-1}U_\alpha).$$

In particular this applies when $U = O(R)$, in which case Proposition 27 will provide us with a canonical choice $O(R_{\mathbb{Z}_\ell})$ of lattice.

Theorem 17. *The relative Malcev homotopy type* $(X,x)^{R,Mal}$ *is determined up to pointed homotopy (i.e., up to unique isomorphism in the category* $\mathrm{Ho}_*(s\mathcal{E}(R))$ *of Remark 11) by the quasi-isomorphism class of the augmented* R*-equivariant cosimplicial algebra*

$$\mathrm{C}^\bullet(\mathrm{X},\rho^{-1}\mathrm{O}(\mathrm{R})) \xrightarrow{x^*} \mathrm{O}(\mathrm{R}).$$

Up to unpointed homotopy, i.e., up to unique isomorphism in the category $\mathrm{Ho}(s\mathcal{E}(\mathrm{R}))$ *of Remark 11,* $(\mathrm{X},x)^{\mathrm{R,Mal}}$ *is determined by the quasi-isomorphism class of*

$$\mathrm{C}^\bullet(\mathrm{X},\rho^{-1}\mathrm{O}(\mathrm{R})).$$

In particular, the relative Malcev homotopy groups $\varpi_n(\mathrm{X},x)^{\mathrm{R,Mal}}$ *are functorially determined by the augmented cosimplicial algebra, while the unaugmented algebra determines the* $\varpi_n(\mathrm{X},x)^{\mathrm{R,Mal}}$ *up to conjugation by* $\mathrm{R}_\mathrm{u}\varpi_1(\mathrm{X},x)^{\mathrm{R,Mal}}$.

Proof. This is [Pri11, Theorem 3.30], and makes use of a bar construction from cosimplicial algebras to simplicial Lie algebras.

We now sketch a demonstration of how to recover $\varpi_1(\mathrm{X},x)^{\mathrm{R,Mal}}$ in the unpointed case. First note that for a unipotent R-equivariant group scheme U,

$$\mathrm{Hom}(\varpi_1(\mathrm{X},x)^{\mathrm{R,Mal}},\mathrm{R}\ltimes\mathrm{U})_\mathrm{R}/\mathrm{U},$$

where $\mathrm{Hom}(-,-)_\mathrm{R}$ denotes morphisms over R, is the set of isomorphism classes of $(\mathrm{R}\ltimes\mathrm{U})(\mathbb{Q}_\ell)$-torsors T on X for which

$$\mathrm{T}\times_{(\mathrm{R}\ltimes\mathrm{U})(\mathbb{Q}_\ell)}\mathrm{R}(\mathbb{Q}_\ell)$$

is the $\mathrm{R}(\mathbb{Q}_\ell)$-torsor T_0 associated to ρ.

Now, the construction of $\mathrm{C}^\bullet(\mathrm{X},\mathrm{G})$ extends to non-abelian pro-finite groups G, with group cohomology $\mathrm{H}^1(\mathrm{X},\mathrm{G})$ given by

$$\mathrm{H}^1(\mathrm{X},\mathrm{G}) = \mathrm{Z}^1(\mathrm{X},\mathrm{G})/\mathrm{C}^0(\mathrm{X},\mathrm{G}),$$

where the 1-cocycles are

$$\mathrm{Z}^1(\mathrm{X},\mathrm{G}) = \{\omega\in\mathrm{C}^1(\mathrm{X},\mathrm{G})\,;\,\partial^1\omega = (\partial^2\omega)(\partial^0\omega)\in\mathrm{C}^2(\mathrm{X},\mathrm{G})\},$$

and $\mathrm{C}^0(\mathrm{X},\mathrm{G})$ acts by setting

$$g(\omega) = (\partial^1 g)\omega(\partial^0 g)^{-1}.$$

These formulae can be extended from pro-finite groups to ℓ-adic Lie groups, and the set of torsors we want is then

$$\{\omega\in\mathrm{Z}^1(\mathrm{X},\mathrm{R}\ltimes\mathrm{U})\,;\,\omega\mapsto\mathrm{T}_0\in\mathrm{Z}^1(\mathrm{X},\mathrm{R})\}/\mathrm{C}^0(\mathrm{X},\mathrm{U}).$$

If $\mathfrak{u}$ is the Lie algebra of U, regarded as an R-representation, then it turns out that this is just

$$\{\omega\in\exp(\mathrm{C}^1(\mathrm{X},\rho^{-1}\mathfrak{u}))\,;\,\partial^1\omega = (\partial^2\omega)\star(\partial^0\omega)\}/\exp(\mathrm{C}^0(\mathrm{X},\rho^{-1}\mathfrak{u})),$$

where $\star$ is the Baker–Campbell–Hausdorff product. Since

$$C^\bullet(X,\rho^{-1}u) = C^\bullet(X,\rho^{-1}O(R))\otimes^R u\,,$$

we have recovered

$$\mathrm{Hom}(\varpi_1(X,x)^{R,\mathrm{Mal}},R\ltimes U)_R/U$$

from $C^\bullet(X,\rho^{-1}O(R))$ functorially in U, which amounts to determining

$$\varpi_1(X,x)^{R,\mathrm{Mal}}$$

up to conjugation by $R_u\varpi_1(X,x)^{R,\mathrm{Mal}}$. □

11.2.2.2 The Long Exact Sequence of Homotopy

Theorem 18. *Take a morphism $f : (X,x) \to (Y,y)$ of pro-finite homotopy types which is surjective on fundamental groups. Assume that the homotopy fibre F of f over $\{y\}$ has finite-dimensional cohomology groups $H^i(F,\mathbb{Q}_\ell)$, and let R be the reductive quotient of the Zariski closure of the homomorphism*

$$\pi_1(Y,y) \to \prod_i GL(H^i(F,\mathbb{Q}_\ell))\,.$$

Then the unipotent Malcev homotopy type $(F,x)^{1,\mathrm{Mal}}$ is the homotopy fibre of

$$(X,x)^{R,\mathrm{Mal}} \to (Y,y)^{R,\mathrm{Mal}}\,.$$

In particular, there is a long exact sequence

$$\ldots \to \varpi_n(F,x)^{1,\mathrm{Mal}} \to \varpi_n(X,x)^{R,\mathrm{Mal}} \to \varpi_n(Y,y)^{R,\mathrm{Mal}} \to \varpi_{n-1}(F,x)^{1,\mathrm{Mal}} \to$$
$$\ldots \to \varpi_1(F,x)^{1,\mathrm{Mal}} \to \varpi_1(X,x)^{R,\mathrm{Mal}} \to \varpi_1(Y,y)^{R,\mathrm{Mal}} \to 1\,.$$

Proof. This is a special case of [Pri11, Theorem 3.32]. □

If f is a fibration, then the homotopy fibre is just the fibre. One case when this happens is for nerves

$$B\Delta \to B\Gamma$$

of surjections of pro-finite groups, in which case the homotopy fibre is just

$$B\ker(\Gamma \to \Delta)\,.$$

Other cases are when f is the étale homotopy type of a geometric fibration of schemes in the sense of [Fri82, Definition 11.4]. These include smooth projective morphisms, and smooth quasi-projective morphisms where the divisor is transverse to f.

Example 19. If $\Gamma = \Delta \ltimes \Lambda$ is a semi-direct product of pro-finite groups with $H^*(\Lambda,\mathbb{Q}_\ell)$ finite-dimensional, then we may apply the theorem with $X = B\Gamma$, $Y = B\Delta$ and $F = B\Lambda$. Since $\Gamma \to \Delta$ has a section, the connecting homomorphism

$$\varpi_2(\Delta)^{\mathrm{R,Mal}} \to \Lambda^{1,\mathrm{Mal}}$$

is necessarily 0, so we get

$$\Gamma^{\mathrm{R,Mal}} \cong \Delta^{\mathrm{R,Mal}} \ltimes \Lambda^{1,\mathrm{Mal}}.$$

In fact, this even remains true if we take R to be the reductive quotient of the Zariski closure of the homomorphism $\Delta \to \mathrm{GL}(\mathrm{H}^1(\Lambda,\mathbb{Q}_\ell))$, see [Pri07, Lemma 4.6].

Example 20. For a case where higher homotopy can affect fundamental groups, consider the symplectic group $\mathrm{Sp}_g(\mathbb{Z}_\ell)$ for $g \geq 2$. This has

$$(\mathrm{Sp}_g(\mathbb{Z}_\ell))^{\mathrm{alg}} = \mathrm{Sp}_g,$$

which is reductive. Letting $\mathrm{R} = \mathrm{Sp}_g$, we get

$$\mathrm{H}^2(\mathrm{Sp}_g(\mathbb{Z}_\ell), \mathrm{O(R)}) \cong \mathbb{Q}_\ell$$

as effectively calculated in [Hai93, Hai97] and [HM09]. Thus Corollary 15 implies that

$$\varpi_2(\mathrm{BSp}_g(\mathbb{Z}_\ell))^{\mathrm{R,Mal}} = \mathbb{G}_a.$$

For any surjective map $\Gamma \twoheadrightarrow \mathrm{Sp}_g(\mathbb{Z}_\ell)$ whose kernel Λ has $\mathrm{H}^1(\Lambda,\mathbb{Q}_\ell)$ finite-dimensional, this gives us an exact sequence

$$\mathbb{G}_a \to \Lambda^{1,\mathrm{Mal}} \to \Gamma^{\mathrm{R,Mal}} \to \mathrm{Sp}_g \to 1,$$

confirming the observation in [HM09, Proposition 6.2] that $\ker(\Lambda^{1,\mathrm{Mal}} \to \Gamma^{\mathrm{R,Mal}})$ is at most 1-dimensional, and proving that it is indeed central. Examples of this form arise from taking Γ to be a group (such as a Galois group or the mapping class group) acting on cohomology of a genus g curve.

Example 21. If we set $\mathrm{Y} = \mathrm{B}\pi_1(\mathrm{X},x)$, then F will be the universal covering space of X, a simply connected space with $\pi_n(\mathrm{F},x) = \pi_n(\mathrm{X},x)$ for $n \geq 2$ (if X is an étale homotopy type, these are Artin–Mazur étale homotopy groups). When each $\mathrm{H}^i(\mathrm{F},\mathbb{Q}_\ell)$ is finite-dimensional, Theorem 18 gives a long exact sequence

$$\begin{array}{l}
\ldots \longrightarrow \pi_n(\mathrm{X},x)\otimes_{\hat{\mathbb{Z}}}\mathbb{Q}_\ell \longrightarrow \varpi_n(\mathrm{X},x)^{\mathrm{R,Mal}} \longrightarrow \varpi_n(\mathrm{B}\pi_1(\mathrm{X},x))^{\mathrm{R,Mal}} \\
\longrightarrow \pi_{n-1}(\mathrm{X},x)\otimes_{\hat{\mathbb{Z}}}\mathbb{Q}_\ell \longrightarrow \ldots \longrightarrow \varpi_3(\mathrm{B}\pi_1(\mathrm{X},x))^{\mathrm{R,Mal}} \\
\longrightarrow \pi_2(\mathrm{X},x)\otimes_{\hat{\mathbb{Z}}}\mathbb{Q}_\ell \longrightarrow \varpi_2(\mathrm{X},x)^{\mathrm{R,Mal}} \longrightarrow \varpi_2(\mathrm{B}\pi_1(\mathrm{X},x))^{\mathrm{R,Mal}} \longrightarrow 0,
\end{array}$$

see [Pri11, Theorem 3.40] for a refinement.

11.2.2.3 Relative Goodness

We now establish criteria for the higher relative Malcev homotopy groups of $B\Gamma$ to vanish.

Definition 22. Say that a pro-finite group Γ is B_n relative to a continuous Zariski-dense map $\rho : \Gamma \to R(\mathbb{Q}_\ell)$ if

$$\varpi_i(B\Gamma)^{R,\mathrm{Mal}} = 0$$

for all $1 < i \leq n$. We say that Γ is **good relative to** ρ if it is B_n for all n.

By [Pri11, Examples 3.38], the following are good relative to all representations: free pro-finite groups, finitely generated nilpotent pro-finite groups, and étale fundamental groups of smooth projective curves over algebraically closed fields.

Proposition 23. *For $\rho : \Gamma \to R(\mathbb{Q}_\ell)$ as above, the following are equivalent:*

(a) Γ *is* B_n *relative to* ρ.
(b) $H^i(\Gamma, \rho^{-1}O(\Gamma^{R,\mathrm{Mal}})) = 0$ *for all* $0 < i \leq n$*, where* $\rho^{-1}O(\Gamma^{R,\mathrm{Mal}})$ *is interpreted as an ind-finite-dimensional representation.*
(c) For all finite-dimensional $\Gamma^{R,\mathrm{Mal}}$*-representations,* $H^i(\Gamma^{R,\mathrm{Mal}}, V) \to H^i(\Gamma, V)$ *is an isomorphism for all* $i \leq n$*, and injective for* $i = n+1$.
(d) For all finite-dimensional $\Gamma^{R,\mathrm{Mal}}$*-representations,* $H^i(\Gamma^{R,\mathrm{Mal}}, V) \to H^i(\Gamma, V)$ *is surjective for all* $i \leq n$.
(e) For all finite-dimensional $\Gamma^{R,\mathrm{Mal}}$*-representations* V*, all* $1 < i \leq n$*, and any element* $x \in H^i(\Gamma, V)$*, there exists an embedding* $V \hookrightarrow W_x$ *of finite-dimensional* $\Gamma^{R,\mathrm{Mal}}$*-representations, with* x *lying in the kernel of* $H^i(\Gamma, V) \to H^i(\Gamma, W_x)$.

Proof. This is based on [KPT09, Lemma 4.15]. The Hurewicz theorem, Corollary 15, implies that (a) and (b) are equivalent. So let us assume (b). As a $\Gamma^{R,\mathrm{Mal}}$-representation,

$$V \otimes O(\Gamma^{R,\mathrm{Mal}})$$

is injective, so there is a cosimplicial injective resolution $V \otimes O(W\Gamma^{R,\mathrm{Mal}})$, as in [Pri08, Example 1.45], given by $O(W\Gamma^{R,\mathrm{Mal}}) = O(\Gamma^{R,\mathrm{Mal}})^{\otimes n+1}$ in level n. This gives us spectral sequences

$$E_1^{ij} = H^i(\Gamma, V \otimes O(W\Gamma^{R,\mathrm{Mal}})^j) \implies H^{i+j}(\Gamma, V).$$

By hypothesis (b), E_1^{ij} for $0 < i \leq n$. Since $V \otimes O(W\Gamma^{R,\mathrm{Mal}})$ is an injective resolution, and

$$H^0(\Gamma, -) = H^0(\Gamma^{R,\mathrm{Mal}}, -),$$

the complex $E_1^{0\bullet}$ computes $H^*(\Gamma^{R,\mathrm{Mal}}, V)$. Thus

$$E_2^{0j} = H^j(\Gamma^{R,\mathrm{Mal}}, V),$$

and $E_2^{ij} = 0$ for $0 < i \leq n$, implying (c).

That (c) implies (d) is immediate, and, since $O(\Gamma^{R,Mal})$ is injective as a $\Gamma^{R,Mal}$-representation, assumption (d) implies that $H^i(\Gamma, O(\Gamma^{R,Mal})) = 0$ for $0 < i \leq n$ and thus (b).

It remains to show the equivalence of (d) and (e). Let us assume condition (d). Then $x \in H^i(\Gamma, V)$ lifts to $\tilde{x} \in H^i(\Gamma^{R,Mal}, V)$. Write

$$V \otimes O(\Gamma^{R,Mal}) = \varinjlim_{\alpha} W_\alpha$$

as a union of finite-dimensional subrepresentations. Thus the image of $\tilde{x}$ in

$$\varinjlim_{\alpha} H^i(\Gamma^{R,Mal}, W_\alpha)$$

is 0, so for some W_α, the image of x in $H^i(\Gamma, W_\alpha)$ is 0, and (e) holds.

Conversely, let us assume that (e) holds. We prove (d) by induction on i, the case $i = 0$ being trivial. Choosing some $x \in H^i(\Gamma, V)$, it follows from the long exact sequence of cohomology that x lies in the image of the connecting homomorphism

$$H^{i-1}(\Gamma, W_x/V) \to H^i(\Gamma, V).$$

Let y lie in the pre-image of x. By induction, there exists $\tilde{y} \in H^{i-1}(\Gamma^{R,Mal}, W_x/V)$ lying over y. Thus the image of $\tilde{y}$ in $H^i(\Gamma^{R,Mal}, V)$ lies over x, giving the required surjectivity. □

Thus super rigid groups Γ cannot be good for any representation. This is because we necessarily have $\Gamma^{R,Mal} = R$ for any R as above, so

$$H^*(\Gamma^{R,Mal}, \mathbb{Q}_\ell) = \mathbb{Q}_\ell,$$

whereas $H^*(\Gamma, \mathbb{Q}_\ell)$ has non-trivial higher cohomology. Examples of super rigid groups are $Sp_g(\mathbb{Z}_\ell)$ for $g \geq 2$, and $SL_n(\mathbb{Z}_\ell)$ for $n \geq 3$. For these examples, the respective pro-algebraic completions are Sp_g and SL_n, since every $Sp_g(\mathbb{Z}_\ell)$-representation is an algebraic Sp_g-representation, and likewise for SL_n.

11.2.3 Weighted Completion

A closely related notion to relative Malcev completion is that of the weighted completion of $\rho : \Gamma \to R(\mathbb{Q}_\ell)$, developed in [HM03b]. This assumes the extra data of a cocharacter

$$\mathbb{G}_m \to R,$$

and agrees with the relative completion if $R_u(\Gamma^{R,Mal})^{ab}$ is of strictly negative weights for the $\mathbb{G}_m$-action. If not, the weighted completion is the largest quotient G of $\Gamma^{R,Mal}$ on which $R_u(G)$ is of strictly negative weights.

As shown in [HM03a, §7], representations of the weighted completion correspond to Γ-representations equipped with a well-behaved weight filtration.

Example 24. Let Γ be the Galois group of the maximal algebraic extension of $\mathbb{Q}$ unramified outside ℓ. Then for the cyclotomic character

$$\xi : \Gamma \to \mathbb{G}_m(\mathbb{Q}_\ell),$$

Hain and Matsumoto proved [HM03b, Theorem 7.3] that the pro-unipotent radical $\mathrm{R_u(G)}$ of weighted completion G of Γ is freely generated by Soulé elements

$$s_1, s_3, s_5, \ldots .$$

In Sect. 11.4, we will be establishing weight filtrations on relative completions

$$\varpi_1(\mathrm{X}_{\text{ét}}, \bar{x})^{\mathrm{R,Mal}}$$

of geometric fundamental groups, with the pro-unipotent radical being of strictly negative weights. These will therefore correspond to weighted completions whenever the $\mathbb{G}_m$-action is an inner action.

In general, relative completion and weighted completion tend to be applied to different types of groups. As we saw in Theorem 18, relative completions of geometric fundamental groups arise when studying fibrations. Galois actions respect the weight filtrations of 11.4, and often the action on the graded group

$$\mathrm{gr_W}(\varpi_1(\mathrm{X}_{\text{ét}}, \bar{x})^{\mathrm{R,Mal}})$$

is pro-reductive and algebraic, so we can set S to be the Zariski closure of

$$\mathrm{Gal} \to \mathrm{Aut}(\mathrm{gr_W}(\varpi_1(\mathrm{X}_{\text{ét}}, \bar{x})^{\mathrm{R,Mal}})).$$

Assuming that the canonical weight map

$$\mathbb{G}_m \to \mathrm{Aut}(\mathrm{gr_W}(\varpi_1(\mathrm{X}_{\text{ét}}, \bar{x})^{\mathrm{R,Mal}}))$$

factors through S, it then follows that the weighted completion of Gal over S acts on $\varpi_1(\mathrm{X}_{\text{ét}}, \bar{x})^{\mathrm{R,Mal}}$.

Remark 25. An alternative way to look at weighted completion is to use affine monoid schemes rather than group schemes. Tannakian theory shows that for any exact tensor category C (not necessarily containing duals) fibred over $\mathbb{Q}_\ell$ vector spaces, there is an affine monoid scheme M such that FDRep(M)is equivalent to C. For instance, for the subcategory of FDRep($\mathbb{G}_m$) generated by $\{\mathbb{Q}(n)\}_{n\leq 0}$, M is just the multiplicative monoid $\mathbb{A}^1$.

Since we just want to work with the category generated by $\{\mathbb{Q}(n)\}_{n<0}$, we can go further, and require that our monoid M contains an element 0, with the property that $0 \cdot g = g \cdot 0 = 0$ for all $g \in \mathrm{M}$. Then we define M-representations to be multiplicative morphisms $\mathrm{M} \to \mathrm{End(V)}$ preserving 0 and 1, so $\mathbb{A}^1$-representations are strictly negatively weighted vector spaces. For affine group schemes G, the corresponding monoid is then just $G \sqcup \{0\}$.

In the scenario of Theorem 18, we would then replace R with the Zariski closure R′ of $\mathbb{A}^1 \cup R$ in $\mathrm{End}(H^{>0}(F, \mathbb{Q}_\ell))$ (for $\mathbb{A}^1$ acting according to the weights on cohomology). Weighted completion of $\pi_1(Y, y)$ can then be interpreted as a kind of (monoidal) relative completion over R′.

An even more efficient choice would be to set R′ as the Zariski closure of the monoid $\{0\} \cup \pi_1(Y, y)$, so FDRep(R′) would be the exact tensor subcategory of $\mathrm{FDRep}(\pi_1(Y, y))$ generated by $H^{>0}(F, \mathbb{Q}_\ell)$, meaning that we only consider local systems of geometric origin (and not their duals).

11.3 Relative Malcev Completion over $\mathbb{Z}_\ell$ and $\mathbb{F}_\ell$

In this section, we introduce canonical $\mathbb{Z}_\ell$-forms for relative Malcev completion, and show how this recovers finer invariants of the fundamental group. Beware that unlike most models over $\mathbb{Z}_\ell$, our affine group scheme $G_{\mathbb{Z}_\ell}$ will seldom be of finite type, even when G is so, see Example 30.

11.3.1 Forms Defined over $\mathbb{Z}_\ell$

Now assume that our topological group Γ is compact, e.g. pro-finite.

Definition 26. Given a set S, a scheme X over a ring A, and a map $f : S \to X(A)$, say that f is **Z-dense** if there is no closed subscheme $Y \subsetneq X$ with $f(S) \subset Y(A)$. If A is a field, note that this is equivalent to saying that X is reduced and f is Zariski-dense.

Proposition 27. *Given a continuous Zariski-dense group homomorphism*

$$\phi : \Gamma \to G(\mathbb{Q}_\ell)$$

to an affine group scheme G *over* $\mathbb{Q}_\ell$*, there is a model* $G_{\mathbb{Z}_\ell}$ *for* G *over* $\mathbb{Z}_\ell$*, unique subject to the conditions*

(i) $\phi : \Gamma \to G_{\mathbb{Z}_\ell}(\mathbb{Q}_\ell)$ *factors through* $G_{\mathbb{Z}_\ell}(\mathbb{Z}_\ell)$.

(ii) If we set $\bar{G} := G_{\mathbb{Z}_\ell} \otimes_{\mathbb{Z}_\ell} \mathbb{F}_\ell$*, then the morphism* $\bar{\phi} : \Gamma \to \bar{G}(\mathbb{F}_\ell)$*, given by reduction modulo* ℓ*, is Z-dense.*

Proof. Define a valuation on O(G) by setting

$$\|f\| = \max_{\gamma \in \Gamma} |f(\phi\gamma)|,$$

noting that this is well-defined because Γ is compact. The first condition above says that for all $f \in O(G_{\mathbb{Z}_\ell})$, we find $\|f\| \leq 1$.

The second condition says that the morphism $\psi : O(G_{\mathbb{Z}_\ell}) \to \mathrm{Hom}_{\mathrm{cts}}(\Gamma, \mathbb{F}_\ell)$ to the set of continuous maps $\Gamma \to \mathbb{F}_\ell$ is injective. Considering $\ker\psi$, this is equivalent to saying that

$$\ker\psi = \{f \in O(G_{\mathbb{Z}_\ell}) \, ; \, \|f\| < 1\} \subset \ell O(G_{\mathbb{Z}_\ell}) .$$

The conditions thus force us to set

$$O(G_{\mathbb{Z}_\ell}) = \{f \in O(G) \, ; \, \|f\| \leq 1\} ,$$

and it is straightforward to check that this is indeed a Hopf algebra over $\mathbb{Z}_\ell$. □

Note that we may apply this construction to the representation $\rho : \Gamma \to R(\mathbb{Q}_\ell)$ considered earlier, and even to the universal representation $\breve{\rho} : \Gamma \to \Gamma^{R,\mathrm{Mal}}(\mathbb{Q}_\ell)$. As the topology on $G(\mathbb{Q}_\ell)$ is totally disconnected, the image of ϕ is pro-finite, so these maps all factor through the pro-finite completion $\hat{\Gamma}$ of the topological group Γ.

Lemma 28. *Given $\phi : \Gamma \to G(\mathbb{Q}_\ell)$ as above and an affine group scheme* H *over $\mathbb{Z}_\ell$, morphisms $G_{\mathbb{Z}_\ell} \to H$ correspond to morphisms $\psi : G \to H \otimes_{\mathbb{Z}_\ell} \mathbb{Q}_\ell$ for which $\psi\phi(\Gamma) \subset H(\mathbb{Z}_\ell)$.*

Proof. A Hopf algebra map $\psi^\sharp : O(H) \to O(G_{\mathbb{Z}_\ell})$ is determined by the corresponding map $\psi^\sharp : O(H) \otimes_{\mathbb{Z}_\ell} \mathbb{Q}_\ell \to O(G)$. Now, $\psi^\sharp$ preserves the $\mathbb{Z}_\ell$-models if and only if $\|\psi^\sharp(f)\| \leq 1$ for all $f \in O(H)$. This is equivalent to saying that $f(\psi\phi\gamma) \in \mathbb{Z}_\ell$ for all $\gamma \in \Gamma$, or equivalently that $\psi\phi(\Gamma) \subset H(\mathbb{Z}_\ell)$. □

In particular, if we take $H = GL_n$, this describes $G_{\mathbb{Z}_\ell}$-representations in finite free $\mathbb{Z}_\ell$-modules. It also implies that $\Gamma^{R,\mathrm{Mal}}_{\mathbb{Z}_\ell} \to R_{\mathbb{Z}_\ell}$ is the universal pro-unipotent extension under Γ, i.e., the relative Malcev $\mathbb{Z}_\ell$-completion.

Proposition 29. *Assume we have a $\mathbb{Z}_\ell$-model* H *for an affine group scheme* G *over $\mathbb{Q}_\ell$, and a* surjective *continuous group homomorphism $\phi : \Gamma \to H(\mathbb{Z}_\ell)$ for which the induced map $\phi : \Gamma \to H(\mathbb{Q}_\ell) = G(\mathbb{Q}_\ell)$ is Zariski-dense.*

Then $G_{\mathbb{Z}_\ell}$ is the affine scheme over $\mathbb{Z}_\ell$ given on $\mathbb{Z}_\ell$-algebras A *by*

$$G_{\mathbb{Z}_\ell}(A) = H(W(A)) \times_{H(A)^{\mathbb{N}_0}} H(A) ,$$

where $W = W_{\ell^\infty}$ is the Witt vector functor, $w : W(A) \to A^{\mathbb{N}_0}$ is the ghost component morphism, and $H(A) \to H(A)^{\mathbb{N}_0}$ is the diagonal map.

Explicitly, $O(G_{\mathbb{Z}_\ell})$ is the smallest $\mathbb{Z}_\ell$-subalgebra of O(G) *containing* O(H) *and closed under the operations*

$$f \mapsto w^{-1}(f, f, \ldots)_n$$

for all $n \geq 0$, where $w^{-1} : O(G)^{\mathbb{N}_0} \to W(O(G))$ is inverse to w.

Proof. First observe that the functor $G_{\mathbb{Z}_\ell}$ above preserves arbitrary limits. Write $H = \varprojlim H_\alpha$ as a filtered limit of finitely generated affine group schemes, and set

$$G_{\alpha,n,\mathbb{Z}_\ell}(A) := H_\alpha(W_n(A)) \times_{H_\alpha(A)^{[0,n]}} H_\alpha(A)\,.$$

Thus the functor $G_{\alpha,n,\mathbb{Z}_\ell}$ commutes with filtered colimits and arbitrary limits, so is represented by a finitely generated affine group scheme. Since

$$G_{\mathbb{Z}_\ell} = \varprojlim_{\alpha,n} G_{\alpha,n,\mathbb{Z}_\ell}\,,$$

it is also an affine group scheme over $\mathbb{Z}_\ell$.

We need to show that $G_{\mathbb{Z}_\ell}$ satisfies the conditions of Proposition 27. The first observation to make is that for $\mathbb{Q}_\ell$-algebras A, the map $w : W(A) \to A^{\mathbb{N}_0}$ is an isomorphism, so $G_{\mathbb{Z}_\ell}(A) = H(A) = G(A)$ and

$$G_{\mathbb{Z}_\ell} \otimes_{\mathbb{Z}_\ell} \mathbb{Q}_\ell = G\,.$$

We now have to check that $\phi(\Gamma) \subset G_{\mathbb{Z}_\ell}(\mathbb{Z}_\ell)$ in $G(\mathbb{Q}_\ell)$. We know that

$$W(\mathbb{Z}_\ell) \times_{\mathbb{Z}_\ell^{\mathbb{N}_0}} \mathbb{Z}_\ell = \mathbb{Z}_\ell\,,$$

since on the one hand ℓ is not a zero divisor in $\mathbb{Z}_\ell$ and the map $w : W(\mathbb{Z}_\ell) \to \mathbb{Z}_\ell^{\mathbb{N}_0}$ is injective, and on the other hand $(b,b,b,\ldots)$ lies in the image of $w : W(\mathbb{Z}_\ell) \to \mathbb{Z}_\ell^{\mathbb{N}_0}$ by the ghost component integrality lemma, e.g. [Haz78, Lemma 17.6.1]. We conclude that

$$G_{\mathbb{Z}_\ell}(\mathbb{Z}_\ell) = H(\mathbb{Z}_\ell)\,.$$

The last check is that $\phi(\Gamma) \to \bar{G}(\mathbb{F}_\ell)$ is Z-dense. For this, we first make the definition

$$\bar{G}_{\alpha,n} := G_{\alpha,n,\mathbb{Z}_\ell} \otimes_{\mathbb{Z}_\ell} \mathbb{F}_\ell\,,$$

for $G_{\alpha,n,\mathbb{Z}_\ell}$ as above, and note that for $n \geq 1$ and A an $\mathbb{F}_\ell$-algebra,

$$W_n(A) \times_{(A)^{[0,n]}} A = W_n(\{a \in A : a^\ell = a\})\,.$$

If Spec A is connected, this is just $W_n(\mathbb{F}_\ell) = \mathbb{Z}/\ell^n$, so $\bar{G}_{\alpha,n}$ is the finite group scheme $H_\alpha(\mathbb{Z}/\ell^n)$. This means that $\bar{G}(\mathbb{F}_\ell)$ is pro-finite, so $\phi(\Gamma) \to \bar{G}(\mathbb{F}_\ell)$ is Z-dense if and only if it is surjective. But $\bar{G}(\mathbb{F}_\ell) = H(W(\mathbb{F}_\ell)) = H(\mathbb{Z}_\ell)$, and we have surjectivity by hypothesis.

Finally, for the description of $O(G_{\mathbb{Z}_\ell})$, let $D_n(f) := w^{-1}(f,f,\ldots)_n$, and let B be the smallest $\mathbb{Z}_\ell$-subalgebra of O(G) containing O(H) and closed under the operations D_n. Then B is a Hopf algebra over $\mathbb{Z}_\ell$, and for any ring homomorphism $f : O(H) \to \mathbb{Z}_\ell$, there is a unique compatible homomorphism $\tilde{f} : B \to \mathbb{Z}_\ell$, determined by the conditions that $\tilde{f}(D_n a) = D_n f(a)$. Thus $H' := \operatorname{Spec} B$ satisfies the same conditions as H. Now, there is a canonical element

$$\omega \in H'(W(B)) \times_{H'(B)^{\mathbb{N}_0}} H'(B) = G_{\mathbb{Z}_\ell}(B)\,,$$

given by $a \mapsto (\underline{D}a, a)$ for $a \in \mathrm{B}$. This amounts to giving a section of $\mathrm{G}_{\mathbb{Z}_\ell} \to \mathrm{H}'$, so we must have $\mathrm{H}' = \mathrm{G}_{\mathbb{Z}_\ell}$. □

Example 30. For $\Gamma = \mathbb{Z}_\ell$ and $\mathrm{G} = \mathbb{G}_a$, with $\phi : \mathbb{Z}_\ell \to \mathbb{G}_a(\mathbb{Q}_\ell)$ the standard inclusion $\mathbb{Z}_\ell \hookrightarrow \mathbb{Q}_\ell$, the $\mathbb{Z}_\ell$ form is given by

$$\mathrm{G}_{\mathbb{Z}_\ell} = \mathbb{W} \times_{\mathbb{G}_a^{\mathbb{N}_0}} \mathbb{G}_a \,,$$

where $\mathbb{W}$ is the Witt vector group scheme $\mathbb{W}(\mathrm{A}) = \mathrm{W}(\mathrm{A})$.

11.3.2 Relative Malcev Completion over $\mathbb{F}_\ell$

In fact, relative Malcev completion can be defined over any field, and we now replace $\mathbb{Q}_\ell$ with $\mathbb{F}_\ell$. Assume that we have an affine group scheme $\bar{\mathrm{R}}$ over $\mathbb{F}_\ell$ and a continuous Z-dense representation $\bar{\rho} : \Gamma \to \bar{\mathrm{R}}(\mathbb{F}_\ell)$, where $\bar{\mathrm{R}}(\mathbb{F}_\ell)$ is given the pro-discrete topology. Explicitly, [DMOS82, Chap. II] shows that $\bar{\mathrm{R}}$ can be expressed as a filtered inverse limit $\bar{\mathrm{R}} = \varprojlim \bar{\mathrm{R}}_\alpha$ of linear algebraic groups. Each $\bar{\mathrm{R}}_\alpha(\mathbb{F}_\ell)$ is given the discrete topology, and we define $\bar{\mathrm{R}}(\mathbb{F}_\ell)$ to be the topological space $\varprojlim \mathrm{R}_\alpha(\mathbb{F}_\ell)$. In particular, this implies that $\bar{\mathrm{R}}(\mathbb{F}_\ell)$ is a pro-finite topological group.

Definition 31. Define the **Malcev completion** $\Gamma^{\bar{\rho},\mathrm{Mal}}$ (or $\Gamma^{\bar{\mathrm{R}},\mathrm{Mal}}$) of Γ relative to $\bar{\rho}$ to be the universal diagram

$$\Gamma \to \Gamma^{\bar{\rho},\mathrm{Mal}}(\mathbb{F}_\ell) \xrightarrow{p} \bar{\mathrm{R}}(\mathbb{F}_\ell)\,,$$

with $p : \Gamma^{\bar{\rho},\mathrm{Mal}} \to \bar{\mathrm{R}}$ a pro-unipotent extension of affine group schemes over $\mathbb{F}_\ell$, and the composition equal to ρ.

There are various ways to prove that this universal object exists. Since the analogue of Lemma 5 must also hold over $\mathbb{F}_\ell$, we can characterise the category of $\Gamma^{\bar{\rho},\mathrm{Mal}}$-representations in terms of Γ and $\bar{\rho}$. The Tannakian formalism of [DMOS82, Chap. II] then uniquely determines $\Gamma^{\bar{\rho},\mathrm{Mal}}$.

Definition 32. Given a pro-finite group $\Gamma = \varprojlim \Gamma_\alpha$, define the **associated affine group scheme** $\Gamma_{\mathbb{F}_\ell}$ over $\mathbb{F}_\ell$ by

$$\Gamma_{\mathbb{F}_\ell} := \varprojlim (\Gamma_\alpha \times \operatorname{Spec} \mathbb{F}_\ell)\,.$$

Explicitly, $\mathrm{O}(\Gamma_{\mathbb{F}_\ell})$ consists of continuous functions from Γ to $\mathbb{F}_\ell$, and $\Gamma_{\mathbb{F}_\ell}(\mathrm{U}) = \Gamma$ for any connected affine scheme U.

Definition 33. We say that an affine group scheme G over $\mathbb{F}_\ell$ is **pro-finite** if the map $\mathrm{G}(\mathbb{F}_\ell)_{\mathbb{F}_\ell} \to \mathrm{G}$ is an isomorphism.

This is equivalent to saying that G is a filtered inverse limit of group schemes of the form $\mathrm{F} \times \operatorname{Spec} \mathbb{F}_\ell$, for F finite.

From now on, assume that Γ is compact.

Proposition 34. *The group schemes* $\bar{\mathrm{R}}$ *and* $\Gamma^{\bar{\rho},\mathrm{Mal}}$ *are pro-finite, and*

$$\Gamma^{\bar{\rho},\mathrm{Mal}}(\mathbb{F}_\ell) \to \bar{\mathrm{R}}(\mathbb{F}_\ell)$$

is the relative pro-ℓ *completion* $\Gamma^{(\ell),\rho}$ *of* Γ *over* $\bar{\mathrm{R}}(\mathbb{F}_\ell)$*, in the sense of [HM09].*

Proof. The image of $\bar{\rho} : \Gamma \to \bar{\mathrm{R}}(\mathbb{F}_\ell)$ is compact and totally disconnected, hence pro-finite. Thus $\bar{\rho}(\Gamma)_{\mathbb{F}_\ell}$ is a closed subscheme of $\bar{\mathrm{R}}$ containing the image of Γ, so must equal $\bar{\mathrm{R}}$, since $\bar{\rho}$ is Z-dense. The same holds for $\Gamma^{\bar{\rho},\mathrm{Mal}}$ (where the corresponding map is Z-dense by universality).

Now, observe that a pro-finite group scheme over $\mathbb{F}_\ell$ is pro-unipotent if and only if it is a pro-ℓ group. Thus

$$\Gamma \to \Gamma^{\bar{\rho},\mathrm{Mal}}(\mathbb{F}_\ell) \xrightarrow{p} \bar{\mathrm{R}}(\mathbb{F}_\ell)$$

is the universal such diagram with p a pro-ℓ extension; in other words, this says that

$$\Gamma^{\bar{\rho},\mathrm{Mal}}(\mathbb{F}_\ell) = \Gamma^{(\ell),\rho}\,.$$

□

Proposition 35. *Given an affine group scheme* $\mathrm{G}_{\mathbb{Z}_\ell}$ *over* $\mathbb{Z}_\ell$*, arising from a continuous Zariski-dense group homomorphism* $\phi : \Gamma \to \mathrm{G}(\mathbb{Q}_\ell)$ *as in Proposition 27, set* $\bar{\mathrm{G}} := \mathrm{G}_{\mathbb{Z}_\ell} \otimes_{\mathbb{Z}_\ell} \mathbb{F}_\ell$*. Then* $\phi : \Gamma \to \mathrm{G}_{\mathbb{Z}_\ell}(\mathbb{Z}_\ell)$ *is surjective, and* $\bar{\mathrm{G}} = \phi(\Gamma)_{\mathbb{F}_\ell}$*.*

Proof. By Proposition 27, $\bar{\phi} : \Gamma \to \bar{\mathrm{G}}(\mathbb{F}_\ell)$ is Z-dense, so Proposition 34 implies that $\bar{\mathrm{G}} = \mathrm{G}_{\mathbb{Z}_\ell}(\mathbb{F}_\ell)_{\mathbb{F}_\ell}$. We therefore need to show that the maps

$$\phi(\Gamma) \to \mathrm{G}_{\mathbb{Z}_\ell}(\mathbb{Z}_\ell) \to \mathrm{G}_{\mathbb{Z}_\ell}(\mathbb{F}_\ell)$$

are isomorphisms. Since the first map is injective and the composition surjective, it suffices to show that $\mathrm{G}_{\mathbb{Z}_\ell}(\mathbb{Z}_\ell) \to \mathrm{G}_{\mathbb{Z}_\ell}(\mathbb{F}_\ell)$ is injective.

If $\epsilon^2 = 0$, there is a ring isomorphism

$$\begin{aligned}(\mathbb{Z}/\ell^{n+1}) \times_{\mathbb{F}_\ell} \mathbb{F}_\ell[\epsilon] \;&\cong\; (\mathbb{Z}/\ell^{n+1}) \times_{(\mathbb{Z}/\ell^n)} (\mathbb{Z}/\ell^{n+1})\\ a + b\epsilon &\mapsto (a, a + \ell^n b)\,.\end{aligned}$$

Thus

$$\mathrm{G}_{\mathbb{Z}_\ell}(\mathbb{Z}/\ell^{n+1}) \times_{\mathrm{G}_{\mathbb{Z}_\ell}(\mathbb{Z}/\ell^n)} \mathrm{G}_{\mathbb{Z}_\ell}(\mathbb{Z}/\ell^{n+1}) \cong \mathrm{G}_{\mathbb{Z}_\ell}(\mathbb{Z}/\ell^{n+1}) \times_{\bar{\mathrm{G}}(\mathbb{F}_\ell)} \bar{\mathrm{G}}(\mathbb{F}_\ell[\epsilon])\,.$$

Now, since $\bar{\mathrm{G}}$ is pro-finite, $\bar{\mathrm{G}}(\mathbb{F}_\ell[\epsilon]) \cong \bar{\mathrm{G}}(\mathbb{F}_\ell)$, so we have shown that

$$\mathrm{G}_{\mathbb{Z}_\ell}(\mathbb{Z}/\ell^{n+1}) \to \mathrm{G}_{\mathbb{Z}_\ell}(\mathbb{Z}/\ell^n)$$

is injective. Since $\mathrm{G}_{\mathbb{Z}_\ell}(\mathbb{Z}_\ell) = \varprojlim_n \mathrm{G}_{\mathbb{Z}_\ell}(\mathbb{Z}/\ell^n)$, this completes the proof. □

Proposition 36. *If* Γ *is a freely generated pro-finite group, then the natural morphism*

$$\Gamma^{(\ell),\rho} \to \Gamma^{\rho,\mathrm{Mal}}(\mathbb{Q}_\ell)$$

is injective.

Proof. Let K be the kernel of $\Gamma^{(\ell),\rho} \to \rho(\Gamma)$, and form the lower central series $\mathrm{L}^n\mathrm{K}$ by setting $\mathrm{L}^1\mathrm{K} = \mathrm{K}$, and

$$\mathrm{L}^{n+1}\mathrm{K} = [\mathrm{K}, \mathrm{L}^n\mathrm{K}]\,.$$

Then $\Gamma^{(\ell),\rho} = \varprojlim_n \Gamma/\mathrm{L}^n\mathrm{K}$, and this maps to

$$(\Gamma^{\rho,\mathrm{Mal}}/\mathrm{L}^n\mathrm{R}_\mathrm{u}(\Gamma^{\rho,\mathrm{Mal}}))(\mathbb{Q}_\ell)\,.$$

It therefore suffices to show that the associated graded map

$$\prod_{n\geq 1} \mathrm{gr}^n_\mathrm{L}\mathrm{K} \to \prod_{n\geq 1} \mathrm{gr}^n_\mathrm{L}\mathrm{R}_\mathrm{u}(\Gamma^{\rho,\mathrm{Mal}})$$

is injective. Now,

$$\mathrm{gr}_\mathrm{L}\mathrm{R}_\mathrm{u}(\Gamma^{\rho,\mathrm{Mal}}) \cong \mathrm{gr}_\mathrm{L}\mathfrak{r}_u(\Gamma^{\rho,\mathrm{Mal}})\,,$$

which is the free pro-(nilpotent finite-dimensional) Lie algebra generated by

$$\mathrm{H}^1(\Gamma, \rho^{-1}\mathrm{O}(\mathrm{R}))^\vee\,.$$

Meanwhile, $\mathrm{gr}_\mathrm{L}\mathrm{K}$ is a Lie ring (topologically) generated by $\mathrm{K}/[\mathrm{K},\mathrm{K}]$. Thus it suffices to show that the map

$$\mathrm{K}/[\mathrm{K},\mathrm{K}] \to \mathrm{H}^1(\Gamma^{\rho,\mathrm{Mal}}, \rho^{-1}\mathrm{O}(\mathrm{R}))^\vee$$

is injective. It follows from the Hochschild–Serre spectral sequence for $\Gamma \to \rho(\Gamma)$ that

$$\mathrm{K}/[\mathrm{K},\mathrm{K}] = \mathrm{H}_1(\Gamma^{(\ell),\rho}, \mathbb{Z}_\ell[\rho(\Gamma)])\,,$$

which is just $\mathrm{H}_1(\Gamma, \mathbb{Z}_\ell[\rho(\Gamma)])$. Meanwhile, if we write $\mathrm{O}(\mathrm{R}) = \varinjlim \mathrm{V}_\alpha$ for V_α finite-dimensional, then

$$\mathrm{H}^1(\Gamma^{\rho,\mathrm{Mal}}, \rho^{-1}\mathrm{O}(\mathrm{R})) = \varinjlim_\alpha \mathrm{H}^1(\Gamma, \rho^{-1}\mathrm{V}_\alpha)\,,$$

so

$$\mathrm{H}^1(\Gamma^{\rho,\mathrm{Mal}}, \rho^{-1}\mathrm{O}(\mathrm{R}))^\vee = \varprojlim_\alpha \mathrm{H}_1(\Gamma, \rho^{-1}\mathrm{V}^\vee_\alpha) = \mathrm{H}_1(\Gamma, \rho^{-1}\mathrm{O}(\mathrm{R})^\vee)\,,$$

where we regard $\rho^{-1}\mathrm{O}(\mathrm{R})^\vee$ as a pro-finite-dimensional Γ-representation.

Since $\mathbb{Z}_\ell[\rho(\Gamma)]$ embeds into $\mathrm{O}(\mathrm{R})^\vee$, we now apply the long exact sequence of homology with coefficients in pro-abelian groups. As Γ is free, H_2 is identically 0, so

$$\mathrm{H}_1(\Gamma^{(\ell),\rho}, \mathbb{Z}_\ell[\rho(\Gamma)]) \hookrightarrow \mathrm{H}_1(\Gamma, \mathrm{O}(\mathrm{R})^\vee)$$

is injective, as required. □

Corollary 37. *If Γ is a freely generated pro-finite group, then*

$$\Gamma^{\rho,\mathrm{Mal}}_{\mathbb{Z}_\ell} \otimes_{\mathbb{Z}_\ell} \mathbb{F}_\ell = \Gamma^{\bar{\rho},\mathrm{Mal}} .$$

Proof. By Proposition 35, $\Gamma^{\rho,\mathrm{Mal}}_{\mathbb{Z}_\ell} \otimes_{\mathbb{Z}_\ell} \mathbb{F}_\ell = \rho(\Gamma)_{\mathbb{F}_\ell}$. By Proposition 36, this is $\Gamma^{(\ell),\rho}_{\mathbb{F}_\ell}$, which is $\Gamma^{\bar{\rho},\mathrm{Mal}}$ by Proposition 34. □

11.3.3 The ℓ-adic Analytic Moduli Space of Representations

The space $\mathrm{Hom}(\Gamma, \mathrm{GL(V)})$ has the structure of an ℓ-adic analytic space. As a set, it is just $\mathrm{Hom}(\Gamma^{\mathrm{alg}}, \mathrm{GL(V)})$, and 11.2.1.3 shows how infinitesimal neighbourhoods in this space can be recovered from Γ^{alg}. The purpose of this section is to show how the full analytic structure can be recovered from the $\mathbb{Z}_\ell$-form of Γ^{alg}.

Definition 38. Given a Zariski-dense morphism $\phi : \Gamma \to \mathrm{G}(\mathbb{Q}_\ell)$, define $\widehat{\mathrm{O(G)}}$ to be the completion of $\mathrm{O(G)}$ with respect to the valuation $\|.\|$ from the proof of Proposition 27. Explicitly,

$$\widehat{\mathrm{O(G)}} = \varprojlim_n \mathrm{O(G)}/\ell^n \mathrm{O}(\mathrm{G}_{\mathbb{Z}_\ell}) .$$

Lemma 39. *The canonical morphism $\widehat{\mathrm{O(G)}} \to \mathrm{Hom}_{\mathrm{cts}}(\phi(\Gamma), \mathbb{Q}_\ell)$ is an isomorphism.*

Proof. We can rewrite this ring homomorphism as

$$(\varprojlim_n \mathrm{O}(\mathrm{G}_{\mathbb{Z}_\ell})/\ell^n \mathrm{O}(\mathrm{G}_{\mathbb{Z}_\ell})) \otimes_{\mathbb{Z}_\ell} \mathbb{Q}_\ell \to (\varprojlim_n \mathrm{Hom}_{\mathrm{cts}}(\phi(\Gamma), \mathbb{Z}/\ell^n)) \otimes_{\mathbb{Z}_\ell} \mathbb{Q}_\ell ,$$

since Γ is compact. It therefore suffices to show that the maps

$$\ell^n \mathrm{O}(\mathrm{G}_{\mathbb{Z}_\ell})/\ell^{n+1} \mathrm{O}(\mathrm{G}_{\mathbb{Z}_\ell}) \to \mathrm{Hom}_{\mathrm{cts}}(\phi(\Gamma), \ell^n \mathbb{Z}/\ell^{n+1}\mathbb{Z})$$

are isomorphisms. But Proposition 35 gives

$$\mathrm{O}(\mathrm{G}_{\mathbb{Z}_\ell}) \otimes_{\mathbb{Z}_\ell} \mathbb{F}_\ell = \mathrm{Hom}_{\mathrm{cts}}(\phi(\Gamma), \mathbb{F}_\ell) ,$$

yielding the required isomorphisms. □

Definition 40. Define $\mathcal{R}(\mathrm{G}_{\mathbb{Z}_\ell}, \mathrm{GL(V)})$ to be the subset of $\widehat{\mathrm{O(G)}} \otimes \mathrm{End(V)}$ consisting of f such that

$$\begin{aligned} \mu(f) &= m(f \otimes f) \in \widehat{\mathrm{O(G \times G)}} \otimes \mathrm{End(V)} \\ \varepsilon(f) &= 1 \in \mathrm{End(V)} , \end{aligned}$$

where $\mu : \mathrm{O(G)} \to \mathrm{O(G)} \otimes \mathrm{O(G)}$ is the comultiplication, $\varepsilon : \mathrm{O(G)} \to \mathbb{Q}_\ell$ the co-unit, and $m : \mathrm{End(V)} \otimes \mathrm{End(V)} \to \mathrm{End(V)}$ multiplication.

Proposition 41. *$\mathcal{R}(G_{\mathbb{Z}_\ell}, GL(V))$ has the natural structure of an ℓ-adic analytic space, isomorphic to* $\mathrm{Hom}(\phi(\Gamma), GL(V))$.

Proof. This follows immediately from Lemma 39. □

Remark 42. Note that $\mathrm{Hom}(G, GL(V))$ is just

$$\mathcal{R}(G_{\mathbb{Z}_\ell}, GL(V)) \cap (O(G) \otimes \mathrm{End}(V)) .$$

If $G = \Gamma^{\mathrm{alg}}$ (or any affine group scheme, such as $\hat{\Gamma}$, for which the map $\hat{\Gamma} \to G(\mathbb{Q}_\ell)$ is injective), then this shows that the analytic spaces

$$\mathrm{Hom}(\Gamma, GL(V))$$

can be recovered directly from $G_{\mathbb{Z}_\ell}$. If $G = \Gamma^{R,\mathrm{Mal}}$, then for any

$$\psi \in \mathrm{Hom}(\Gamma^{R,\mathrm{Mal}}, GL(V)) ,$$

the results from 11.2.1.3 show that the space $\mathcal{R}(G_{\mathbb{Z}_\ell}, GL(V))$ contains an open neighbourhood of ψ in $\mathrm{Hom}(\Gamma, GL(V))$.

11.3.4 Homotopy Types over $\mathbb{F}_\ell$

For any field k, [KPT09] develops a theory of schematic homotopy types over k, using simplicial affine group schemes over k. In many respects, these behave like schematic homotopy types over fields of characteristic 0, except that we no longer have Levi decompositions or the correspondence between unipotent group schemes and nilpotent Lie algebras. This means that although there is not an explicit analogue of Theorem 17, equivariant cochains still determine the homotopy type [KPT09, Proposition 3.26].

Definition 43. Take a pro-finite homotopy type (X, x), an affine group scheme $\bar{R}$ over $\mathbb{F}_\ell$, and a continuous Z-dense representation

$$\bar{\rho} : \pi_1(X, x) \to \bar{R}(\mathbb{F}_\ell) .$$

Define the **relative Malcev homotopy type** $(X, x)^{\bar{R},\mathrm{Mal}}$ of (X, x) over $\bar{R}$ to be the simplicial affine group scheme

$$\widehat{G(X, x)}^{\bar{R},\mathrm{Mal}} = \widehat{G(X, x)}^{(\ell),\bar{\rho}} ,$$

the identification following from Proposition 34.

Definition 44. Define **relative Malcev homotopy groups** by

$$\varpi_n(X, x)^{\bar{R},\mathrm{Mal}} := \pi_{n-1}\widehat{G(X, x)}^{\bar{R},\mathrm{Mal}} .$$

Observe that, since relative Malcev completion is right exact,

$$\varpi_1(X,x)^{\bar{R},\mathrm{Mal}} = \pi_1(X,x)^{\bar{R},\mathrm{Mal}} .$$

Theorem 18 is also true for relative Malcev homotopy types over $\mathbb{F}_\ell$, since the proof only involves the Hochschild-Serre spectral sequence. Thus the long exact sequence of homotopy allows us to interpret the failure of relative completion to be left exact, as observed in [HM09], in terms of the non-vanishing of $\varpi_2(B\Gamma)^{\bar{R},\mathrm{Mal}}$.

Now take a Zariski-dense continuous homomorphism $\rho : \pi_1(X,x) \to R(\mathbb{Q}_\ell)$, form $R_{\mathbb{Z}_\ell}$ as in Proposition 27, and set $\bar{R} := R_{\mathbb{Z}_\ell} \otimes_{\mathbb{Z}_\ell} \mathbb{F}_\ell$. We cannot recover $\pi_1(X,x)^{(\ell),\bar{\rho}}$ from the relative Malcev completion over $\mathbb{Z}_\ell$, since the latter annihilates elements of $\ker\rho$ which are not infinitely ℓ-divisible. However, the following proposition implies that we can recover $\pi_1(X,x)^{(\ell),\bar{\rho}}$ from the $\mathbb{Z}_\ell$ form

$$\widehat{G(X,x)}^{R,\mathrm{Mal}}_{\mathbb{Z}_\ell}$$

of the homotopy type, given by applying Proposition 27 levelwise. This can be interpreted as saying that information about non-divisible elements of $\ker\rho$ is encoded by higher homotopy over $\mathbb{Z}_\ell$.

Proposition 45. *For $\rho : \pi_1(X,x) \to R(\mathbb{Q}_\ell)$ as above,*

$$\widehat{G(X,x)}^{\bar{R},\mathrm{Mal}} = \widehat{G(X,x)}^{R,\mathrm{Mal}}_{\mathbb{Z}_\ell} \otimes_{\mathbb{Z}_\ell} \mathbb{F}_\ell .$$

This gives an exact sequence

$$\begin{aligned} 0 \to O(\varpi_1(X,x)^{R,\mathrm{Mal}}_{\mathbb{Z}_\ell}) \otimes_{\mathbb{Z}_\ell} \mathbb{F}_\ell \to O(\pi_1(X,x)^{\bar{R},\mathrm{Mal}}) \\ \to H^1(O(\widehat{G(X,x)}^{R,\mathrm{Mal}}_{\mathbb{Z}_\ell})) \xrightarrow{\ell} H^1(O(\widehat{G(X,x)}^{R,\mathrm{Mal}}_{\mathbb{Z}_\ell})). \end{aligned}$$

Proof. Since $\widehat{G_n(X,x)}$ is freely generated as a pro-finite group, it satisfies the hypotheses of Corollary 37, giving $\widehat{G_n(X,x)}^{\bar{R},\mathrm{Mal}} = \widehat{G_n(X,x)}^{R,\mathrm{Mal}}_{\mathbb{Z}_\ell} \otimes_{\mathbb{Z}_\ell} \mathbb{F}_\ell$ for all n.

This gives an exact sequence

$$0 \to O(\widehat{G(X,x)}^{R,\mathrm{Mal}}_{\mathbb{Z}_\ell}) \xrightarrow{\ell} O(\widehat{G(X,x)}^{R,\mathrm{Mal}}_{\mathbb{Z}_\ell}) \to O(\widehat{G(X,x)}^{\bar{R},\mathrm{Mal}}) \to 0 .$$

Applying the long exact sequence of cohomology gives the required result, since

$$H^0(O(\widehat{G(X,x)}^{\bar{R},\mathrm{Mal}}) = O(\pi_1(X,x)^{\bar{R},\mathrm{Mal}}) .$$

□

Note that Proposition 14 relates $H^*(O(\widehat{G(X,x)}^{R,\mathrm{Mal}}_{\mathbb{Z}_\ell})) \otimes_{\mathbb{Z}_\ell} \mathbb{Q}_\ell$ to the homotopy groups $\varpi_*(X,x)^{R,\mathrm{Mal}}$.

11.4 Geometric Fundamental Groups

In this section, we will describe geometric fundamental groups as Galois representations. All relative Malcev completions will be over $\mathbb{Q}_\ell$.

X_0 will be a connected variety over a field k, with $\bar{k}$ an algebraic closure of k, and we write $X = X_0 \otimes_k \bar{k}$. Assume that we have a point $x \in X_0(k)$, with associated geometric point $\bar{x} \in X(\bar{k})$.

11.4.1 Weight Filtrations

If X_0 is smooth and quasi-projective, with smooth compactification $j : X_0 \to \bar{X}_0$, then [Pri11, Definition 4.37] gives an associated Leray filtration W (there denoted by J) on $\varpi_1^{\text{ét}}(X,\bar{x})^{R,\text{Mal}}$, and indeed on $\varpi_n^{\text{ét}}(X,\bar{x})^{R,\text{Mal}}$ for all n. Explicitly (as in [Pri11, Corollary 6.15]), we have a sequence

$$\ldots \le W_{-r}\varpi_*^{\text{ét}}(X,\bar{x})^{R,\text{Mal}} \le \ldots \le W_0\varpi_*^{\text{ét}}(X,\bar{x})^{R,\text{Mal}} = \varpi_*^{\text{ét}}(X,\bar{x})^{R,\text{Mal}}$$

of closed subgroup schemes, with

$$[W_{-r}\varpi_m^{\text{ét}}(X,\bar{x})^{R,\text{Mal}}, W_{-s}\varpi_n^{\text{ét}}(X,\bar{x})^{R,\text{Mal}}] \le W_{-r-s}\varpi_{m+n-1}^{\text{ét}}(X,\bar{x})^{R,\text{Mal}} .$$

If R is a quotient of $\varpi_1^{\text{ét}}(X,\bar{x})$, then the filtration W has the additional property that

$$W_{-1}\varpi_n^{\text{ét}}(X,\bar{x})^{R,\text{Mal}} = \ker(\varpi_n^{\text{ét}}(X,\bar{x})^{R,\text{Mal}} \to \varpi_n^{\text{ét}}(\bar{X},\bar{x})^{R,\text{Mal}}) .$$

The construction of W is based on the idea that the R-equivariant cosimplicial algebra $C^\bullet(X,\rho^{-1}O(R))$ of 11.2.2.1 is quasi-isomorphic to the diagonal of the bicosimplicial algebra

$$C^\bullet(\bar{X}, j_*\mathscr{C}^\bullet(\rho^{-1}O(R))) ,$$

and that good truncations of $j_*\mathscr{C}^\bullet(\rho^{-1}O(R))$ give an increasing Leray filtration

$$W_0 = C^\bullet(\bar{X}, j_*\rho^{-1}O(R)) \subset \ldots \subset W_\infty \simeq C^\bullet(X,\rho^{-1}O(R))$$

by R-equivariant cosimplicial complexes, with $W_i \cdot W_j \subset W_{i+j}$. This filtration is essentially the same as the weight filtration of [Del71, Proposition 3.1.8].

In [Pri11, Theorem 4.22], the bar construction is used to transfer this filtration to a filtration $W_0 \ge W_{-1} \ge \ldots$ by (simplicial) subgroup schemes on the homotopy type $(X_{\text{ét}},\bar{x})^{R,\text{Mal}}$ and homotopy groups $\varpi_n^{\text{ét}}(X,\bar{x})^{R,\text{Mal}}$, satisfying the conditions above. The rough idea is to adapt Theorem 17 to give a functor on negatively filtered Lie algebras, replacing the cosimplicial Lie algebra $C^\bullet(X,\rho^{-1}\mathfrak{u})$ with

$$W_0C^\bullet(X, j_*\mathscr{C}^\bullet(\rho^{-1}\mathfrak{u})) := \sum_{i\ge 0} W_iC^\bullet(X, j_*\mathscr{C}^\bullet(\rho^{-1}O(R))) \otimes^R W_{-i}\mathfrak{u} .$$

Studying the spectral sequence of Proposition 13 shows that when R is a quotient of $\varpi_1^{\text{ét}}(X,\bar{x})$, the Leray filtration on $\varpi_1^{\text{ét}}(X,\bar{x})^{R,\text{Mal}}$ is given by

$$\begin{aligned} W_{-1}\varpi_1^{\text{ét}}(X,\bar{x})^{R,\text{Mal}} &= \ker(\varpi_1^{\text{ét}}(X,\bar{x})^{R,\text{Mal}} \to \varpi_1^{\text{ét}}(\bar{X},\bar{x})^{R,\text{Mal}}) \\ W_{-n}\varpi_1^{\text{ét}}(X,\bar{x})^{R,\text{Mal}} &= [W_{-1}\varpi_1^{\text{ét}}(X,\bar{x})^{R,\text{Mal}}, W_{1-n}\varpi_1^{\text{ét}}(X,\bar{x})^{R,\text{Mal}}] \end{aligned}$$

for $n \geq 2$.

In fact, décalage gives another filtration, Dec W, on $C^\bullet(\bar{X}, j_*\mathscr{C}^\bullet(\rho^{-1}O(R)))$, and this is the true weight filtration, cf. [Mor78] or [Del75], in the sense that

$$H^a(\bar{X}, \mathbf{R}^b j_*\rho^{-1}O(R))$$

has weight $a+2b$ with respect to Dec W, but only weight b with respect to W.

Via the bar construction, this also induces a filtration

$$(\text{Dec}\, W)_0 \geq (\text{Dec}\, W)_{-1} \geq \dots$$

on the homotopy type and homotopy groups. Beware, however, that décalage does not commute with the bar construction. Studying the spectral sequence of Proposition 13 then gives that when R is a quotient of $\varpi_1^{\text{ét}}(X,\bar{x})$,

$$\begin{aligned} (\text{Dec}\, W)_{-1}\varpi_1^{\text{ét}}(X,\bar{x})^{R,\text{Mal}} &= R_u\varpi_1^{\text{ét}}(X,\bar{x})^{R,\text{Mal}} \\ (\text{Dec}\, W)_{-2}\varpi_1^{\text{ét}}(X,\bar{x})^{R,\text{Mal}} &= \ker(R_u\varpi_1^{\text{ét}}(X,\bar{x})^{R,\text{Mal}} \to (R_u\varpi_1^{\text{ét}}(\bar{X},\bar{x})^{R,\text{Mal}})^{\text{ab}}), \end{aligned}$$

with the lower terms determined inductively by the condition that for $n \geq 3$, the subgroup $(\text{Dec}\, W)_{-n}\varpi_1^{\text{ét}}(X,\bar{x})^{R,\text{Mal}}$ is the smallest closed normal subgroup containing $[(\text{Dec}\, W)_{1-n}, (\text{Dec}\, W)_{-1}]$ and $[(\text{Dec}\, W)_{2-n}, (\text{Dec}\, W)_{-2}]$. This filtration is analogous to the weight filtration of [Pri10, Theorem 5.14], which however is only defined for smooth proper complex varieties.

The following sections give circumstances in which the filtration Dec W splits canonically.

11.4.2 Finite Characteristic, $\ell \neq p$

In this section, we assume that k is a finite field of characteristic $p \neq \ell$. Fix a Galois-equivariant Zariski-dense representation $\rho : \pi_1^{\text{ét}}(X,\bar{x}) \to R(\mathbb{Q}_\ell)$, where R is a pro-reductive affine group scheme equipped with an algebraic $\text{Gal}(\bar{k}/k)$-action. In other words, $\text{Gal}(\bar{k}/k)^{\text{alg}} \ltimes R$ is a quotient of $\varpi_1^{\text{ét}}(X_0,\bar{x})$.

Example 46. To see how such groups R arise naturally, assume that $f_0 : Y_0 \to X_0$ is a smooth proper morphism with connected fibres. Let R be the Zariski closure of the map

$$\pi_1^{\text{ét}}(X,\bar{x}) \to \prod_n \text{Aut}((\mathbf{R}^n f_{\text{ét},*}\mathbb{Q}_\ell)_{\bar{x}}),$$

then R is a pro-reductive affine group scheme satisfying the hypotheses.

Example 47. The universal case is given by letting G be the image of the homomorphism $\pi_1^{\text{ét}}(X,\bar{x})^{\text{alg}} \to \pi_1^{\text{ét}}(X_0,\bar{x})^{\text{alg}}$, then setting $R := G^{\text{red}}$, the pro-reductive quotient. In that case, [Pri09, Lemma 1.3] implies that $G = \pi_1^{\text{ét}}(X,\bar{x})^{R,\text{Mal}}$, and that

$$\pi_1^{\text{ét}}(X_0,\bar{x})^{\text{alg}} = G \rtimes \text{Gal}(\bar{k}/k)^{\text{alg}} .$$

The following is [Pri09, Theorem 2.10]; see [Pri11, Theorem 6.10] for a generalisation to higher homotopy groups.

Proposition 48. *If* X *is smooth and proper over* $\bar{k}$, *then there is a unique Galois-equivariant isomorphism*

$$\varpi_1^{\text{ét}}(X,\bar{x})^{R,\text{Mal}} \cong R \ltimes \exp(\text{Fr}(H^1(X,\rho^{-1}O(R))^\vee)/\sim),$$

where Fr *is the free pro-(finite-dimensional nilpotent) Lie algebra functor, and* $\sim$ *is generated by*

$$H^2(X,\rho^{-1}O(R))^\vee \xrightarrow{\cup^\vee} \bigwedge^2 H^1(X,\rho^{-1}O(R))^\vee ,$$

the map dual to the cup product.

Proof (sketch). The Galois action on R gives the sheaf $\rho^{-1}O(R)$ the natural structure of a sheaf on X_0. Lafforgue's Theorem [Laf02, Thm. VII.6 and Cor. VII.8], combined with the description of Example 3, shows that the sheaf $\rho^{-1}O(R)$ is pure of weight 0.

By [Del80, Corollaries 3.3.4–3.3.6], the group $H^n(X,\rho^{-1}O(R))$ is thus pure of weight n, so the spectral sequence of Proposition 13 thus degenerates at E_2. This gives a description of all homotopy groups in terms of $H^*(X,\rho^{-1}O(R))$.

Explicitly, write $\mathfrak{u}$ for the Lie algebra of $R_u\varpi_1^{\text{ét}}(X,\bar{x})^{R,\text{Mal}}$, and note that

$$\mathfrak{u}^{ab} \cong H^1(X,\rho^{-1}O(R))^\vee ,$$

which is pure of weight 1. Now,

$$H^2(\mathfrak{u},\mathbb{Q}_\ell) \cong H^2(X,\rho^{-1}O(R)) ,$$

which is pure of weight 2, so the only possible relation defining $\mathfrak{u}$ is

$$\cup^\vee : H^2(X,\rho^{-1}O(R))^\vee \to \bigwedge^2 H^1(X,\rho^{-1}O(R))^\vee \subset \text{Fr}(H^1(X,\rho^{-1}O(R))^\vee) . \qquad \square$$

This can be used to construct examples of groups which cannot be fundamental groups of any smooth proper variety in finite characteristic, e.g. [Pri09, Ex 2.30].

The following specialises [Pri11, Theorem 6.15] and Corollary 6.16 to the case of fundamental groups.

Proposition 49. *Assume that* $X = \bar{X} - D$ *for* $\bar{X}$ *smooth and proper over* $\bar{k}$, *with* $D \subset \bar{X}$ *a divisor locally of normal crossings. If* ρ *has tame monodromy around the components of* D, *then there is a Galois-equivariant isomorphism*

$$\varpi_1^{\text{ét}}(X,\bar{x})^{R,\mathrm{Mal}} \cong R \ltimes \exp(\mathrm{Fr}(H^1(\bar{X}, j_*\rho^{-1}O(R))^\vee \oplus H^0(\bar{X}, \mathbf{R}^1 j_*\rho^{-1}O(R))^\vee)/\sim),$$

where $\sim$ *is generated by the images of the maps*

$$H^2(\bar{X}, j_*\rho^{-1}O(R))^\vee \xrightarrow{(d_2^\vee,\cup^\vee)} H^0(\bar{X}, \mathbf{R}^1 j_*\rho^{-1}O(R))^\vee \oplus \bigwedge^2 H^1(X, j_*\rho^{-1}O(R))^\vee,$$

$$H^1(\bar{X}, \mathbf{R}^1 j_*\rho^{-1}O(R))^\vee \xrightarrow{(\cup^\vee)} H^0(\bar{X}, \mathbf{R}^1 j_*\rho^{-1}O(R))^\vee \otimes H^1(X, j_*\rho^{-1}O(R))^\vee,$$

$$H^0(\bar{X}, \mathbf{R}^2 j_*\rho^{-1}O(R))^\vee \xrightarrow{(\cup^\vee)} \bigwedge^2 H^0(\bar{X}, \mathbf{R}^1 j_*\rho^{-1}O(R))^\vee.$$

Here $\cup^\vee$ *is the map dual to the cup product, and* $d_2^\vee$ *is dual to the differential* d_2 *on the* E_2 *sheet of the Leray spectral sequence.*

Proof (sketch). Again, the Frobenius action on $\rho^{-1}O(R)$ is pure of weight 0, so [Del80, Corollaries 3.3.4–3.3.6] imply that $H^a(\bar{X}, \mathbf{R}^b j_*\rho^{-1}O(R))$ is pure of weight $a+2b$. This means that the Leray spectral sequence

$$E_2^{ab} = H^a(\bar{X}, \mathbf{R}^b j_*\rho^{-1}O(R)) \implies H^{a+b}(X, \rho^{-1}O(R))$$

degenerates at E_3.

Substituting the terms $H^a(\bar{X}, \mathbf{R}^b j_*\rho^{-1}O(R))^\vee$ into the Adams spectral sequence of Proposition 13, the relations above turn out to be the only maps compatible with both the Frobenius weights and the Leray filtration W on $\varpi_1^{\text{ét}}(X,\bar{x})^{R,\mathrm{Mal}}$. □

Note that the filtration W (resp. Dec W) on $\varpi_1^{\text{ét}}(X,\bar{x})^{R,\mathrm{Mal}}$ is then determined by the conditions that $W_{-1}R = (\mathrm{Dec}\,W)_{-1}R = 1$, and that $H^a(\bar{X}, \mathbf{R}^b j_*\rho^{-1}O(R))^\vee$ is contained in W_{-b} (resp. $(\mathrm{Dec}\,W)_{-a-2b}$), but not in W_{-b-1} (resp. $(\mathrm{Dec}\,W)_{-a-2b-1}$). Thus Dec W is precisely the filtration by weights of Frobenius on the Lie algebra of $\varpi_1^{\text{ét}}(X,\bar{x})^{R,\mathrm{Mal}}$.

Proposition 49 can be used to construct examples of groups which cannot be fundamental groups of any smooth quasi-projective variety in finite characteristic, e.g. [Pri09, Example 2.31].

11.4.3 Mixed Characteristic, $\ell = p$

In this section, we will assume that X_0 is a connected variety of good reduction over a local field K, with residue field k.

Explicitly, let V be a complete discrete valuation ring, with residue field k (finite, of characteristic p), and fraction field K (of characteristic 0). Let $\bar{k}, \bar{K}$ be the algebraic closures of k, K respectively, and $\bar{V}$ the algebraic closure of V in $\bar{K}$. Write K_0 for the fraction field of $W(k)$.

Assume that we have a scheme $X_V = \bar{X}_V - D_V$ over V, with $\bar{X}_V$ smooth and proper, D_V a normal crossings divisor, and $X_0 = X_V \otimes_V K$. Also fix a basepoint $x_V \in X_V(V)$, giving $x \in X_V(K) = X_0(K)$ and $\bar{x} \in X_0(\bar{K})$. Write $X := X_0 \otimes_K \bar{K}$.

11.4.3.1 Crystalline Étale Sheaves

We now introduce crystalline étale sheaves, as in [Fal89] V(f), [Ols11, §6.13], or [AI09].

Definition 50. Say that a smooth $\mathbb{Q}_p$-sheaf $\mathbb{V}$ on X_K is **crystalline** if it is associated to a filtered convergent F-isocrystal on $(\bar{X}_V, D_V)$.

This means that there exists a filtered convergent F-isocrystal E, and a collection of isomorphisms

$$\iota_U : \mathbb{V} \otimes_{\mathbb{Q}_p} B_{cris}(\hat{U}) \to E(B_{cris}(\hat{U}))$$

for $U \to X_V$ étale, compatible with the filtrations and semi-linear Frobenius automorphisms, and with morphisms over X, so that ι becomes an isomorphism of étale presheaves. Here, $B_{cris}(\hat{U})$ is formed by applying Fontaine's construction to the p-adic completion $\hat{U}$ of U.

By [Pri11, Proposition 7.8], the category of crystalline $\mathbb{Q}_p$-sheaves is closed under extensions and subquotients, and the isocrystal associated to a crystalline $\mathbb{Q}_p$-sheaf is essentially unique. More precisely, association gives a fully faithful functor D^X_{cris} from crystalline $\mathbb{Q}_p$-sheaves to filtered convergent F-isocrystals.

11.4.3.2 Structure of Fundamental Groups

Now fix a Galois-equivariant Zariski-dense representation $\rho : \pi_1^{ét}(X, \bar{x}) \to R(\mathbb{Q}_\ell)$, where R is a pro-reductive affine group scheme equipped with an algebraic $\mathrm{Gal}(\bar{k}/k)$-action.

Assume that $D^X_{cris}\rho^{-1}O(R)$ is an ind-object in the category of ι-pure overconvergent F-isocrystals. This is equivalent to saying that for every R-representation V, the corresponding sheaf $\mathbb{V}$ on $X_{\bar{K}}$ can be embedded in the pullback of a crystalline étale sheaf $\mathbb{U}$ on X_K, associated to an ι-pure overconvergent F-isocrystal on $(\bar{X}_k, D_k)/K$. Also note that this implies that O(R) is a crystalline Galois representation for which the Frobenius action on $D_{cris}O(R)$ is ι-pure.

Example 51. To see how these hypotheses arise naturally, assume that $f_0 : Y_0 \to X_0$ is a smooth proper morphism with connected components, for Y of good reduction. Let R be the Zariski closure of the map

$$\pi_1^{ét}(X, \bar{x}) \to \prod_n \mathrm{Aut}((\mathbf{R}^n f_{ét,*}\mathbb{Q}_p)_{\bar{x}}),$$

so R is a pro-reductive affine group scheme. By [Fal89], $\mathbf{R}^n f_{ét,*}\mathbb{Q}_p$ is associated to $\mathbf{R}^n f^{cris}_{\bar{k},*}\mathscr{O}_{Y_{\bar{k}},cris}$, which by [Ked06, Theorem 6.6.2] is ι-pure. Thus the semisimplifications of the R-representations $(\mathbf{R}^n f_{ét,*}\mathbb{Q}_p)_{\bar{x}}$ are direct sums of ι-pure representations. Since these generate the Tannakian category of R-representations, the hypotheses are satisfied.

We may write $F := Y \times_{f,X,\bar{x}} \mathrm{Spec}\,\bar{K}$, and Theorem 18 then shows that the homotopy fibre of

$$Y^{R,\mathrm{Mal}}_{\acute{e}t} \to X^{R,\mathrm{Mal}}_{\acute{e}t}$$

over $\bar{x}$ is the unipotent Malcev homotopy type $F^{1,\mathrm{Mal}}_{\acute{e}t}$.

Definition 52. From now on, write $B := B_{\mathrm{cris}}(V)$, with $B^{\sigma} \subset B$ the invariants under Frobenius.

Theorem 53. *For* R *as above, there is a Galois-equivariant isomorphism*

$$\varpi_1^{\acute{e}t}(X,\bar{x})^{R,\mathrm{Mal}} \otimes_{\mathbb{Q}_p} B^{\sigma}$$
$$\cong (R \ltimes \exp(\mathrm{Fr}(H^1(\bar{X}, j_*\rho^{-1}O(R))^{\vee} \oplus H^0(\bar{X}, \mathbf{R}^1 j_*\rho^{-1}O(R))^{\vee})/\sim)) \otimes_{\mathbb{Q}_p} B^{\sigma},$$

of affine group schemes over B^{σ}, *where the relations* $\sim$ *are defined as in Proposition 49.*

Proof. This is [Pri11, Theorem 7.35], or alternatively [Ols11, Theorem 7.22] when X is projective, which also has corresponding results for higher homotopy groups and indeed the whole homotopy type. [Ols11, 6.8] introduces a ring $\tilde{B} \supset B_{\mathrm{cris}}(V)$ equipped with a Hodge filtration and Galois action. The proof then proceeds by using the results of [Ols11], which give a weak equivalence

$$X^{R,\mathrm{Mal}}_{\bar{K},\acute{e}t} \otimes_{\mathbb{Q}_p} \tilde{B} \sim X^{D_{\mathrm{cris}}R,\mathrm{Mal}}_{\bar{k},\mathrm{cris}} \otimes_{K_0} \tilde{B},$$

preserving Hodge filtrations and Galois actions. Here, $X^{D_{\mathrm{cris}}R,\mathrm{Mal}}_{\bar{k},\mathrm{cris}}$ is a relative Malcev crystalline homotopy type over K_0; representations of its fundamental group are isocrystals, and its cohomology is crystalline cohomology.

This implies that $O(\varpi_1^{\acute{e}t}(X,\bar{x})^{R,\mathrm{Mal}})$ is crystalline as a Galois representation, and that

$$D_{\mathrm{cris}} O(\varpi_1^{\acute{e}t}(X,\bar{x})^{R,\mathrm{Mal}}) = O(\varpi_1^{\mathrm{cris}}(X_{\bar{k}},\bar{x})^{D_{\mathrm{cris}}R,\mathrm{Mal}}).$$

Now, if we write $\mathscr{E}(R) := D^{X}_{\mathrm{cris}}\rho^{-1}O(R)$, then replacing [Del80] with [Ked06], Proposition 49 adapts to show that

$$\varpi_1^{\mathrm{cris}}(X_{\bar{k}},\bar{x})^{D_{\mathrm{cris}}R,\mathrm{Mal}}$$
$$\cong D_{\mathrm{cris}}R \ltimes \exp(\mathrm{Fr}(H^1_{\mathrm{cris}}(\bar{X}, j_*\mathscr{E}(R))^{\vee} \oplus H^0(\bar{X}, \mathbf{R}^1 j_*\mathscr{E}(R))^{\vee})/\sim),$$

for $\sim$ defined as in Proposition 49. This isomorphism is Frobenius-equivariant, but need not respect the Hodge filtration.

The final step is to tensor this isomorphism with B_{cris} and to take Frobenius-invariants, using the comparison above to replace crystalline fundamental groups and cohomology with étale fundamental groups and cohomology. □

In fact, [Pri11, Theorem 7.35] also shows that the isomorphism of Theorem 53 also holds without having to tensor with B^{σ}, but at the expense of Galois-equivariance.

Remark 54. Although Theorem 53 is weaker than Proposition 49, it is more satisfactory in one important respect. Proposition 49 effectively shows that relative Malcev fundamental groups over $\mathbb{Q}_\ell$ carry no more information than cohomology, whereas to recover relative Malcev fundamental groups over $\mathbb{Q}_p$, we still need to identify

$$\varpi_1^{\text{ét}}(X,\bar{x})^{R,\text{Mal}} \subset \varpi_1^{\text{ét}}(X,\bar{x})^{R,\text{Mal}} \otimes_{\mathbb{Q}_p} B^\sigma .$$

This must be done by describing the Hodge filtration on $\varpi_1^{\text{cris}}(X_{\bar{k}},\bar{x})^{D_{\text{cris}}R,\text{Mal}}$, which is not determined by cohomology (since it is not Frobenius-equivariant). Thus the Hodge filtration is the only really new structure on the relative Malcev fundamental group.

Remark 55. There is a similar Archimedean phenomenon established in [Pri10, §2]. If X is a smooth proper variety over $\mathbb{C}$, with R a real affine group scheme and

$$\rho : \pi_1(X(\mathbb{C}),x) \to R(\mathbb{R})$$

Zariski-dense, then we can study the relative Malcev completion $\varpi_1(X,x)^{R,\text{Mal}}$ of the topological fundamental group $\pi_1(X(\mathbb{C}),x)$. If all R-representations underlie variations of Hodge structure, then [Pri10, Theorems 5.14 and 4.20] show that the Hopf algebra $O(\varpi_1(X,x)^{R,\text{Mal}})$ is a sum of real mixed Hodge structures.

If we define $B(\mathbb{R}) := \mathbb{C}[t]$ to be of weight 0, with Hodge filtration given by

$$\text{Fil}^n B(\mathbb{R}) = (t-i)^n B(\mathbb{R}) ,$$

and with σ denoting complex conjugation, then by [Pri10, Theorem 4.21], there is an equivariant isomorphism

$$\varpi_1(X(\mathbb{C}),x)^{R,\text{Mal}} \otimes_{\mathbb{R}} B(\mathbb{R})^\sigma \cong (R \ltimes \exp(\text{Fr}(H^1(X(\mathbb{C}),\rho^{-1}O(R))^\vee)/\sim)) \otimes_{\mathbb{R}} B(\mathbb{R})^\sigma$$

preserving Hodge and weight filtrations, for $\sim$ as in Theorem 48.

If X is the complex form of a real variety X_0, then (by [Pri10, Remark 2.15]) this isomorphism is moreover $\text{Gal}(\mathbb{C}/\mathbb{R})$-equivariant, where the non-trivial element of $\text{Gal}(\mathbb{C}/\mathbb{R})$ acts on $B(\mathbb{R})$ as the $\mathbb{C}$-algebra homomorphism determined by $t \mapsto -t$.

11.4.4 Global Fields

We now summarise how the previous sections provide information over global fields. Given a smooth quasi-projective variety $X_0 = \bar{X}_0 - D_0$ over a number field K, Sect. 11.4.1 gives a filtration $\text{Dec}\,W$ on $\varpi_1^{\text{ét}}(X,\bar{x})^{R,\text{Mal}}$. Assume that $\text{Gal}(\bar{K}/K)$ acts algebraically on R, and that the Zariski-dense representation

$$\rho : \pi_1^{\text{ét}}(X,\bar{x}) \to R(\mathbb{Q}_\ell)$$

is Galois-equivariant.

Theorem 56. *For each prime $\mathfrak{p} \nmid \ell$ of $\bar{K}$ at which $(\bar{X},D)$ has potentially good reduction and tame monodromy round the divisor, there is a weight decomposition*

$$\varpi_1^{\text{ét}}(X,\bar{x})^{R,\text{Mal}} = \prod_{n\le 0} {}_{\mathfrak{p}}\mathcal{W}_n \varpi_1^{\text{ét}}(X,\bar{x})^{R,\text{Mal}},$$

splitting the true weight filtration Dec W. *These decompositions are conjugate under the action of* $\text{Gal}(\bar{K}/K)$*, in the sense that*

$$g({}_{\mathfrak{p}}\mathcal{W}_*) = {}_{g\mathfrak{p}}\mathcal{W}_*.$$

If $\mathfrak{p} \mid \ell$ *is a prime at which* $(\bar{X}, D)$ *has potentially good reduction, and* $\rho^{-1}O(R)$ *is a potentially crystalline* $\mathbb{Q}_\ell$*-sheaf associated to a sum of* ι*-pure overconvergent* F*-isocrystals, then there is a weight decomposition*

$$\varpi_1^{\text{ét}}(X,\bar{x})^{R,\text{Mal}} \otimes_{\mathbb{Q}_p} B_{\text{cris}}^{\sigma} = \prod_{n\le 0} {}_{\mathfrak{p}}\mathcal{W}_n \varpi_1^{\text{ét}}(X,\bar{x})^{R,\text{Mal}} \otimes_{\mathbb{Q}_p} B_{\text{cris}}^{\sigma},$$

of affine schemes over B_{cris}^{σ}*, splitting the true weight filtration* Dec W. *These decompositions are conjugate under the action of* $\text{Gal}(\bar{K}/K)$.

Proof. This combines Proposition 49, using smooth specialisation to compare special and generic fibres, and Theorem 53, assigning R the weight 0, then

$$H^1(\bar{X}, j_*\rho^{-1}O(R))^\vee$$

the weight -1 and $H^0(\bar{X}, \mathbf{R}^1 j_*\rho^{-1}O(R))^\vee$ the weight -2. □

References

[AI09] F. Andreatta and A. Iovita. Comparison isomorphisms for smooth formal schemes. 2009, www.mathstat.concordia.ca/faculty/iovita/research.html.

[AM69] M. Artin and B. Mazur. *Etale homotopy*. Number 100 in Lecture Notes in Mathematics. Springer, Berlin, 1969.

[Del71] P. Deligne. Théorie de Hodge. II. *Inst. Hautes Études Sci. Publ. Math.*, 40:5–57, 1971.

[Del75] P. Deligne. Poids dans la cohomologie des variétés algébriques. In *Proceedings of the International Congress of Mathematicians (Vancouver, B. C., 1974), Vol. 1*, pages 79–85. Canad. Math. Congress., Montreal, Que., 1975.

[Del80] P. Deligne. La conjecture de Weil. II. *Inst. Hautes Études Sci. Publ. Math.*, 52:137–252, 1980.

[DMOS82] P. Deligne, J. S. Milne, A. Ogus, and K. Y. Shih. *Hodge cycles, motives, and Shimura varieties*, volume 900 of *Lecture Notes in Mathematics*. Springer, Berlin, 1982.

[Fal89] G. Faltings. *Crystalline cohomology and p-adic Galois-representations*, pages 25–80. Johns Hopkins Univ. Press, Baltimore, MD, 1989.

[Fri82] E. M. Friedlander. *Étale homotopy of simplicial schemes*, volume 104 of *Annals of Mathematics Studies*. 1982.

[Hai93] R. M. Hain. Completions of mapping class groups and the cycle $C - C^-$. In *Mapping class groups and moduli spaces of Riemann surfaces (Göttingen, 1991/Seattle, WA, 1991)*, volume 150 of *Contemp. Math.*, pages 75–105. Amer. Math. Soc., Providence, RI, 1993.

[Hai97] R. Hain. Infinitesimal presentations of the Torelli groups. *J. Amer. Math. Soc.*, 10(3):597–651, 1997.

[Hai98] R. M. Hain. The Hodge de Rham theory of relative Malcev completion. *Ann. Sci. École Norm. Sup. (4)*, 31(1):47–92, 1998.

[Haz78] M. Hazewinkel. *Formal groups and applications*, volume 78 of *Pure and Applied Mathematics*. Academic Press Inc. [Harcourt Brace Jovanovich Publishers], New York, 1978.

[HM03a] R. Hain and M. Matsumoto. Tannakian fundamental groups associated to Galois groups. In *Galois groups and fundamental groups*, volume 41 of *Math. Sci. Res. Inst. Publ.*, pages 183–216. Cambridge Univ. Press, Cambridge, 2003.

[HM03b] R. Hain and M. Matsumoto. Weighted completion of Galois groups and Galois actions on the fundamental group of $\mathbb{P}^1 - \{0,1,\infty\}$. *Compositio Math.*, 139(2):119–167, 2003.

[HM09] R. Hain and M. Matsumoto. Relative pro-ℓ completions of mapping class groups. *J. Algebra*, 321(11):3335–3374, 2009.

[Kan58] Daniel M. Kan. On homotopy theory and c.s.s. groups. *Ann. of Math. (2)*, 68:38–53, 1958.

[Ked06] K. S. Kedlaya. Fourier transforms and p-adic 'Weil II'. *Compos. Math.*, 142(6):1426–1450, 2006.

[KPT09] L. Katzarkov, T. Pantev, and B. Toën. Algebraic and topological aspects of the schematization functor. *Compos. Math.*, 145(3):633–686, 2009.

[Laf02] L. Lafforgue. Chtoucas de Drinfeld et correspondance de Langlands. *Invent. Math.*, 147(1):1–241, 2002.

[Mor78] J. W. Morgan. The algebraic topology of smooth algebraic varieties. *Inst. Hautes Études Sci. Publ. Math.*, 48:137–204, 1978.

[Ols11] M. C. Olsson. Towards non-abelian p-adic Hodge theory in the good reduction case. *Mem. Amer. Math. Soc.*, 220(990), 2011.

[Pri07] J. P. Pridham. The pro-unipotent radical of the pro-algebraic fundamental group of a compact Kähler manifold. *Ann. Fac. Sci. Toulouse Math. (6)*, 16(1):147–178, 2007.

[Pri08] J. P. Pridham. Pro-algebraic homotopy types. *Proc. London Math. Soc.*, 97(2):273–338, 2008.

[Pri09] J. P. Pridham. Weight decompositions on étale fundamental groups. *Amer. J. Math.*, 131(3):869–891, 2009.

[Pri10] J. P. Pridham. Formality and splitting of real non-abelian mixed Hodge structures. 2010, arXiv:math.AG/0902.0770v2. submitted.

[Pri11] J. P. Pridham. Galois actions on homotopy groups. *Geom. Topol.*, 15:1:501-607, 2011.

[Toë06] B. Toën. Champs affines. *Selecta Math. (N.S.)*, 12(1):39–135, 2006.

Chapter 12
On 3-Nilpotent Obstructions to π_1 Sections for $\mathbb{P}^1_{\mathbb{Q}} - \{0, 1, \infty\}$

Kirsten Wickelgren*

Abstract We study which rational points of the Jacobian of $\mathbb{P}^1_k - \{0, 1, \infty\}$ can be lifted to sections of geometrically 3-nilpotent quotients of étale π_1 over the absolute Galois group. This is equivalent to evaluating certain triple Massey products of elements of $k^* \subseteq \mathrm{H}^1(\mathrm{G}_k, \hat{\mathbb{Z}}(1))$ or $\mathrm{H}^1(\mathrm{G}_k, \mathbb{Z}/2\mathbb{Z})$. For $k = \mathbb{Q}_p$ or $\mathbb{R}$, we give a complete mod 2 calculation. This permits some mod 2 calculations for $k = \mathbb{Q}$. These are computations of obstructions of Jordan Ellenberg.

12.1 Introduction

The generalized Jacobian of a pointed smooth curve can be viewed as its abelian approximation. It is natural to consider non-abelian nilpotent approximations. Grothendieck's anabelian conjectures predict that smooth hyperbolic curves over certain fields are controlled by their étale fundamental groups. In particular, approximating π_1 should be similar to approximating the curve. We study the effect of 2 and 3-nilpotent quotients of the étale fundamental group of $\mathbb{P}^1 - \{0, 1, \infty\}$ on its rational points, using obstructions of Jordan Ellenberg.

More specifically, a pointed smooth curve X embeds into its generalized Jacobian via the Abel-Jacobi map. Applying π_1 to the Abel-Jacobi map gives the abelianization of the étale fundamental group of X. Quotients by subgroups in the lower central series lie between $\pi_1(\mathrm{X})$ and its abelianization, giving rise to obstructions to a rational point of the Jacobian lying in the image of the Abel-Jacobi map. These obstructions were defined by Ellenberg in [Ell00].

K. Wickelgren (✉)
Harvard University, Cambridge MA, USA
e-mail: `wickelgren@post.harvard.edu`

* Supported by an NSF Graduate Research Fellowship, a Stanford Graduate Fellowship, and an American Institute of Math Five Year Fellowship

J. Stix (ed.), *The Arithmetic of Fundamental Groups*, Contributions in Mathematical and Computational Sciences 2, DOI 10.1007/978-3-642-23905-2_12,

For simplicity, first assume that X is a proper, smooth, geometrically connected curve over a field k. The absolute Galois group of k will be denoted by

$$G_k = \mathrm{Gal}(\bar{k}/k)$$

where $\bar{k}$ denotes an algebraic closure of k. Assume that X is equipped with a k point, denoted b and used as a base point. A k-variety will be said to be **pointed** if it is equipped with a k-point. The point b gives rise to an Abel-Jacobi map

$$\alpha : X \to \mathrm{Jac}\, X$$

from X to its Jacobian, sending b to the identity. Applying π_1 to $\alpha \otimes \bar{k}$ produces the abelianization of $\pi_1(X_{\bar{k}})$. For any pointed variety Z over k, there is a natural map

$$\kappa : Z(k) \to H^1(G_k, \pi_1(Z_{\bar{k}}))$$

where $Z(k)$ denotes the k points of Z. In particular, we have the commutative diagram

$$\begin{array}{ccc} \mathrm{Jac}(X)(k) & \longrightarrow & H^1(G_k, \pi_1(X_{\bar{k}})^{ab}) \\ \uparrow & & \uparrow \\ X(k) & \longrightarrow & H^1(G_k, \pi_1(X_{\bar{k}})) \end{array} \tag{12.1}$$

Any k point of Jac(X) which is in the image of the Abel-Jacobi map satisfies the condition that its associated element of $H^1(G_k, \pi_1(X_{\bar{k}})^{ab})$ lifts through the map

$$H^1(G_k, \pi_1(X_{\bar{k}})) \to H^1(G_k, \pi_1(X_{\bar{k}})^{ab}). \tag{12.2}$$

Therefore, showing that the associated element of $H^1(G_k, \pi_1(X_{\bar{k}})^{ab})$ does not admit such a lift obstructs this point of the Jacobian from lying on the curve. Ellenberg's obstructions are obstructions to lifting through the map (12.2). Since they obstruct a conjugacy class of sections of π_1 of $\mathrm{Jac}\, X \to \mathrm{Spec}\, k$ from being the image of a conjugacy class of sections of π_1 of $X \to \mathrm{Spec}\, k$, they are being called **obstructions to π_1 sections** in the title. They arise from the lower central series and are defined in Sect. 12.2.

More specifically, Ellenberg's obstruction δ_n is the $H^1 \to H^2$ boundary map in G_k cohomology for the extension

$$1 \to [\pi]_n/[\pi]_{n+1} \to \pi/[\pi]_{n+1} \to \pi/[\pi]_n \to 1 \tag{12.3}$$

where π is the étale fundamental group of $X_{\bar{k}}$, and

$$\pi = [\pi]_1 \supset [\pi]_2 \supset [\pi]_3 \supset \ldots$$

denotes the lower central series of π. The obstruction δ_n is regarded as a multi-valued function on $\mathrm{H}^1(\mathrm{G}_k,\pi^{\mathrm{ab}})$ via $\mathrm{H}^1(\mathrm{G}_k,\pi/[\pi]_n) \to \mathrm{H}^1(\mathrm{G}_k,\pi^{\mathrm{ab}})$ and also on $\mathrm{Jac}\,\mathrm{X}(k)$ via κ.

Now assume that $\mathrm{X} = \mathbb{P}^1_k - \{0,1,\infty\}$ and that k is a subfield of $\mathbb{C}$ or a completion of a number field. By replacing the Jacobian by the generalized Jacobian and enlarging $\mathrm{X}(k)$ to include k rational tangential base points, we obtain a commutative diagram generalizing (12.1). The same obstructions to lifting through (12.2) define obstructions δ_n for X. As there is an isomorphism

$$\pi \cong \langle x,y\rangle^{\wedge}$$

between π and the profinite completion of the topological fundamental group of $\mathbb{P}^1_{\mathbb{C}} - \{0,1,\infty\}$, bases of

$$[\pi]_n/[\pi]_{n+1} \cong \hat{\mathbb{Z}}(n)^{\mathrm{N}(n)}$$

can be specified by order n commutators of x and y, decomposing the obstructions δ_n into multi-valued, partially defined maps

$$\mathrm{H}^1(\mathrm{G}_k,\hat{\mathbb{Z}}(1)\oplus\hat{\mathbb{Z}}(1)) \dashrightarrow \mathrm{H}^2(\mathrm{G}_k,\hat{\mathbb{Z}}(n))\,.$$

Section 12.3 expresses δ_2 and δ_3 in terms of cup products and Massey products. For a in k^*, let $\{a\}$ denote the image of a in $\mathrm{H}^1(\mathrm{G}_k,\hat{\mathbb{Z}}(1))$ under the Kummer map. For (b,a) in $\mathrm{Jac}\,\mathrm{X}(k) \cong (\mathbb{G}_m\times\mathbb{G}_m)(k)$, the obstruction δ_2 is given by $\delta_2(b,a) = \{b\}\cup\{a\}$, see [Ell00]. It is a charming observation of Jordan Ellenberg that this computation shows that the cup product factors through $\mathrm{K}_2(k)$, see Remark 28. The obstruction δ_3 is computed by Theorem 19 as

$$\delta_{3,[[x,y],x]}(b,a) = \langle\{-b\},\{b\},\{a\}\rangle$$

$$\delta_{3,[[x,y],y]}(b,a) = -\langle\{-a\},\{a\},\{b\}\rangle - f\cup\{a\}\,,$$

where $f \in \mathrm{H}^1(\mathrm{G}_k,\hat{\mathbb{Z}}(2))$ is associated to the monodromy between 0 and 1. The indeterminacy of the Massey product and the conditions required for its definition coincide with the multiple values assumed by δ_3 and the condition for its definition.

Section 12.4 contains computations of δ_2, and its mod 2 reduction. In particular, Sect. 12.4.4 provides points on which to evaluate δ_3, which can be phrased as the failure of a 2-nilpotent section conjecture for $\mathbb{P}^1_k - \{0,1,\infty\}$. Tate's computation of $\mathrm{K}_2(\mathbb{Q})$ gives a finite algorithm for determining whether or not $\delta_2(b,a) = 0$ for $k = \mathbb{Q}$ described in Sect. 12.4.5.

The main results of this paper are in Sect. 12.5. The mod 2 reduction of δ_3 for a finite extension k_v of $\mathbb{Q}_p$ with p odd is computed as follows.

Theorem 36. *Suppose that* $\delta_2^{\mathrm{mod}\,2}(b,a) = 0$. *Then* $\delta_3^{\mathrm{mod}\,2}(b,a) \neq 0$ *if and only if one of the following holds:*

- $\{-b\} = 0$ *and* $\{2\sqrt{-b}\}\cup\{a\} \neq 0$.
- $\{-a\} = 0$ *and* $\{2\sqrt{-a}\}\cup\{b\} + \{2\}\cup\{a\} \neq 0$.
- $\{b\} = \{a\}$ *and* $\{2\sqrt{b}\sqrt{a}\}\cup\{a\} \neq 0$.

Here, equalities such as $\{-b\} = 0$ take place in $H^1(G_{k_v}, \mathbb{Z}/2\mathbb{Z})$ and non-equalities such as $\{2\sqrt{-b}\} \cup a \neq 0$ take place in $H^2(G_{k_v}, \mathbb{Z}/2\mathbb{Z})$. The cocycle

$$f : G_k \to [\pi]_2/([\pi]_3([\pi]_2)^2) \cong \mathbb{Z}/2\mathbb{Z}$$

described above is known due to contributions of Anderson, Coleman, Deligne, Ihara, Kaneko, and Yukinari, and its computation is required for Theorem 36. For points

$$(b,a) \in (\mathbb{Z} - \{0\}) \times (\mathbb{Z} - \{0\}) \subset \operatorname{Jac} X(\mathbb{Q}_p)$$

such that p divides ab exactly once, the vanishing of $\delta_3^{\mathrm{mod}\,2}$ for $\mathbb{Q}_p$ can be expressed in terms of the congruence conditions of Corollary 38:

- $\delta_2^{\mathrm{mod}\,2}(b,a) = 0 \iff a+b$ is a square mod p.
- When $\delta_2^{\mathrm{mod}\,2}(b,a) = 0$, then $\delta_3^{\mathrm{mod}\,2}(b,a) = 0 \iff a+b$ is a fourth power mod p.

As the image of X in its Jacobian consists of (b,a) such that $b+a=1$, and the image of the tangential points of X are (b,a) such that $b+a=0$, or $b=1$, or $a=1$, we see in Corollary 38 that $\delta_3^{\mathrm{mod}\,2}$ vanishes on the points and tangential points of X. Of course, $\delta_3^{\mathrm{mod}\,2}$ is constructed to satisfy this property, but here it is visible that $\delta_2^{\mathrm{mod}\,2}$ and $\delta_3^{\mathrm{mod}\,2}$ are increasingly accurate approximations to X inside its Jacobian.

The obstruction $\delta_3^{\mathrm{mod}\,2}$ for $k = \mathbb{R}$ is computed in Sect. 12.5.3. Consider $k = \mathbb{Q}$. Although an element of $H^2(G_{\mathbb{Q}}, \mathbb{Z}/2\mathbb{Z})$ is 0 if and only if its restriction to all places, or all but one place, vanishes, the previous local calculations can only be combined to produce a global calculation when the Massey products are evaluated locally using compatible defining systems. This involves the local-global comparison map on Galois cohomology with coefficients in a 2-nilpotent group, see Remark 46. In Sect. 12.5.5, such lifts are arranged and the local calculations are used to show that

$$\delta_3^{\mathrm{mod}\,2}(-p^3, p) = 0$$

for $k = \mathbb{Q}$. Proposition 48 computes $\delta_3^{\mathrm{mod}\,2}$ on a specific lift of $(-p^3, p)$ for $k = \mathbb{Q}$, which is equivalent to the calculation of the $G_{\mathbb{Q}}$ Massey products with $\mathbb{Z}/2\mathbb{Z}$ coefficients $\langle\{p^3\},\{-p^3\},\{p\}\rangle$ and $\langle\{-p\},\{p\},\{-p^3\}\rangle$ for any specified defining system.

Acknowledgements. I wish to thank Gunnar Carlsson, Jordan Ellenberg, and Mike Hopkins for many useful discussions. I thank the referee for correcting sign errors in Proposition 17 and Sect. 12.5.2. I also thank Jakob Stix for clarifying Sect. 12.4.4, shortening the proofs of Lemma 34, Propositions 32 and 45, and for extensive and thoughtful editing.

12.2 Ellenberg's Obstructions to π_1 Sections for $\mathbb{P}^1_k - \{0,1,\infty\}$

We work with fields k which are subfields of $\mathbb{C}$ or completions of a number field. In the latter case, fix an embedding of the number field into $\mathbb{C}$, as well as an algebraic closure $\overline{k}$ of k, and an embedding $\overline{\mathbb{Q}} \subset \overline{k}$, where $\overline{\mathbb{Q}}$ denotes the algebraic closure of $\mathbb{Q}$ in $\mathbb{C}$. These specifications serve to choose maps between topological and étale fundamental groups, as in (12.11).

This section defines Ellenberg's obstructions. In Sect. 12.2.1, we recall Deligne's notion of a tangential point [Del89, §15] [Nak99], and define in (12.5) the map κ from k points and tangential points to $\mathrm{H}^1(\mathrm{G}_k, \pi_1(\mathrm{X}_{\overline{k}}))$. We then specialize to $\mathrm{X} = \mathbb{P}^1_k - \{0,1,\infty\}$, give the computation of κ composed with

$$\mathrm{H}^1(\mathrm{G}_k, \pi_1(\mathrm{X}_{\overline{k}})) \to \mathrm{H}^1(\mathrm{G}_k, \pi_1(\mathrm{X}_{\overline{k}})^{\mathrm{ab}})$$

in (12.16) and Lemma 4, and define Ellenberg's obstructions in Sect. 12.2.3.

12.2.1 Tangential Base Points, Path Torsors, and the Galois Action

Let X be a smooth, geometrically connected curve over k with smooth compactification $\mathrm{X} \subseteq \overline{\mathrm{X}}$ and $x \in \overline{\mathrm{X}}(k)$, so in particular, x could be in $(\overline{\mathrm{X}} - \mathrm{X})(k)$. A local parameter z at x gives rise to an isomorphism

$$k[[z]] \xrightarrow{\cong} \widehat{\mathcal{O}}_{\overline{\mathrm{X}},x},$$

where $\widehat{\mathcal{O}}_{\overline{\mathrm{X}},x}$ denotes the completion of the local ring of x. Let $\overline{k}$ be a fixed algebraic closure of k. Since we assume that k has characteristic 0, the field of Puiseux series

$$\overline{k}((z^{\mathbb{Q}})) := \cup_{n \in \mathbb{Z}_{>0}} \overline{k}((z^{1/n}))$$

is algebraically closed. The composition

$$\mathrm{Spec}\,\overline{k}((z^{\mathbb{Q}})) \to \mathrm{Spec}\,k[[z]] \cong \mathrm{Spec}\,\widehat{\mathcal{O}}_{\overline{\mathrm{X}},x} \to \overline{\mathrm{X}}$$

factors through the generic point of $\overline{\mathrm{X}}$ and thus defines a geometric point of X

$$b_z : \mathrm{Spec}\,\overline{k}((z^{\mathbb{Q}})) \to \mathrm{X}$$

that will be called the **tangential base point** of X at x in the direction of z. The tangential base point b_z determines an embedding

$$k(\mathrm{X}) \subset k((z)) \subset \overline{k}((z^{\mathbb{Q}})).$$

The coefficientwise action of G_k on $\bar{k}((z^{\mathbb{Q}}))$ gives a splitting of $G_{k((z))} \to G_k$. Combined with the embedding $k(X) \subset k((z))$, this splitting gives a splitting of $G_{k(X)} \to G_k$ and therefore a splitting of

$$\pi_1^{et}(X, b_z) \to G_k\,,$$

see [SGA1, V Prop 8.2], and a G_k action on $\pi_1^{et}(X_{\bar{k}}, b_z)$.

A **geometric point associated to a k point or tangential point** will mean

$$x : \operatorname{Spec}\Omega_x \to X$$

where Ω_x is an algebraically closed extension of $\bar{k}$, such that either x arises as a tangential base point as described above or x has a k point as its image. Such a geometric point determines a canonical geometric point of $X_{\bar{k}}$, and the associated fiber functor has a canonical G_k action. A path between two such geometric points b and x is a natural transformation of the associated fiber functors, and the set of paths

$$\pi_1(X_{\bar{k}}; b, x)$$

from b to x form a trivial $\pi_1^{et}(X_{\bar{k}}, b)$ torsor whose G_k action determines an element

$$[\pi_1(X_{\bar{k}}; b, x)] \in H^1(G_k, \pi_1^{et}(X_{\bar{k}}, b))$$

represented by the cocycle

$$g \mapsto \gamma^{-1} \circ g(\gamma) \tag{12.4}$$

where γ is any path from b to x. Composition of paths is written right to left so that $\gamma^{-1} \circ g(\gamma)$ is the path formed by first traversing $g(\gamma)$ and then γ^{-1}.

A local parameter z at a point x of $\overline{X}$ determines a tangent vector

$$\operatorname{Spec} k[[z]]/\langle z^2 \rangle \to \overline{X}\,.$$

For b or x a tangential base point, the associated element of $H^1(G_k, \pi_1^{et}(X_{\bar{k}}, b))$ only depends on the choice of local parameter up to the associated tangent vector. Furthermore, if x is a k tangential point which comes from a tangent vector at a point p of X, then

$$[\pi_1(X_{\bar{k}}; b, x)] = [\pi_1(X_{\bar{k}}; b, p)]\,.$$

This describes a map κ, which is often called the **non-abelian Kummer map**,

$$\kappa = \kappa_{(X,b)} : X(k) \cup \bigcup_{x \in \overline{X} - X} (T_x\overline{X}(k) - \{0\}) \longrightarrow H^1(G_k, \pi_1(X_{\bar{k}}, b))\,. \tag{12.5}$$

Example 1. The boundary map for the Kummer sequence

$$1 \to \mathbb{Z}/n\mathbb{Z}(1) \to \mathbb{G}_m \xrightarrow{n\cdot} \mathbb{G}_m \to 1 \tag{12.6}$$

yields in the limit over all n the Kummer map

$$k^* \to \mathrm{H}^1(\mathrm{G}_k, \hat{\mathbb{Z}}(1)) \tag{12.7}$$

which is represented on the level of cocycles by

$$\sigma \mapsto \{z\}(\sigma) = \left(\sigma(\sqrt[n]{z})/\sqrt[n]{z}\right)_n \tag{12.8}$$

for any compatible choice of nth roots of $z \in \overline{k}^*$. This cocycle, or by abuse of notation also the class it represents, will also be denoted by z, or denoted by $\{z\}$ if there is possible confusion.

When $n = 2$, both choices of square root of z produce the same cocycle. Furthermore, canonically $\mu_2 = \mathbb{Z}/2\mathbb{Z}$ and thus we have a well-defined homomorphism

$$k^* \to \mathrm{C}^1(\mathrm{G}_k, \mathbb{Z}/2\mathbb{Z}),$$

where $\mathrm{C}^1(\mathrm{G}_k, \mathbb{Z}/2\mathbb{Z})$ denotes the group of continuous 1-cocycles of G_k with values in $\mathbb{Z}/2\mathbb{Z}$.

For $(\mathrm{X}, b) = (\mathbb{G}_m, 1)$, the map κ is the Kummer map: for

$$x \in \mathbb{G}_{\mathrm{m},k}(k) = k^*,$$

choose compatible nth roots $\sqrt[n]{x}$ of x, and choose 1 as the nth root of unity for each $n \in \mathbb{Z}_{>0}$. These choices determine a path γ from 1 to x as follows. On the degree n cover

$$p_n : \mathbb{G}_{\mathrm{m},\bar{k}} \to \mathbb{G}_{\mathrm{m},\bar{k}}$$

given by $t \mapsto t^n$, the path γ maps

$$\gamma : p_n^{-1}(1) \to p_n^{-1}(x)$$

by multiplication by $\sqrt[n]{x}$. For $g \in \mathrm{G}_k$, the path $g\gamma$ is the path sending $g1$ to $g(\sqrt[n]{x})$, and thus multiplies by $g(\sqrt[n]{x})$. We conclude

$$\kappa(x) = \gamma^{-1} \circ g(\gamma) = g(\sqrt[n]{x})/\sqrt[n]{x} = \{x\}(g).$$

Identifying the choice of path from 1 to x with the choice of compatible n^{th} roots of x, there is an equality of cocycles $\kappa(x) = \{x\}$.

Similarly, for $w \in \mathrm{T}_0\mathbb{P}^1_k(k) - \{0\} = k^*$, a compatible choice of nth roots $\sqrt[n]{w}$ of w determines a path γ from 1 to $\overrightarrow{0w}$, where $\overrightarrow{0w}$ is the k tangential point

$$\operatorname{Spec} \bar{k}((z^{\mathbb{Q}})) \to \operatorname{Spec} k[t, t^{-1}] \tag{12.9}$$

given by $t \mapsto wz$, by defining γ to map $\zeta \in \mu_n(\bar{k}) = p_n^{-1}(1)$ to the point of $p_n^{-1}(\overrightarrow{0w})$ given by (12.9) and $t \mapsto \sqrt[n]{w}\zeta z^{1/n}$. For any $g \in \mathrm{G}_k$, we have $g(\gamma)(\zeta) = g(\gamma(g^{-1}\zeta))$ is the path given by (12.9) and $t \mapsto (g\sqrt[n]{w})\zeta z^{1/n}$, whence $\gamma((g\sqrt[n]{w})\zeta/\sqrt[n]{w}) = g(\gamma)(\zeta)$. We conclude

$$\kappa(\overrightarrow{0w}) = \gamma^{-1} \circ g(\gamma) = g(\sqrt[n]{w})/\sqrt[n]{w} = \{w\}(g) = \kappa(w).$$

Example 2. The map $\kappa_{(X,b)}$ depends on the choice of base point, even when $\pi_1(X_{\bar{k}}, b)$ is abelian and is therefore independent of b. In this case, if b_1 and b_2 are two geometric points associated to a k point or tangential point of X, a straightforward cocycle manipulation shows that

$$\kappa_{(X,b_2)}(x) = \kappa_{(X,b_1)}(x) - \kappa_{(X,b_1)}(b_2)$$

for any k point or tangential point x. In particular, Example 1 implies that

$$\kappa_{(\mathbb{G}_m,\overrightarrow{01})} = \kappa_{(\mathbb{G}_m,1)}\ .$$

For higher dimensional geometrically connected varieties Z over k, we will not need the notion of a tangential base point, but we will use the map

$$\kappa = \kappa_{(Z,b)}\ :\ Z(k) \to H^1(G_k, \pi_1(Z_{\bar{k}}, b)) \tag{12.10}$$

defined precisely as in the case of curves above.

Let $X = \mathbb{P}^1_k - \{0,1,\infty\} = \operatorname{Spec} k[t, \frac{1}{t}, \frac{1}{1-t}]$, and base X at $\overrightarrow{01}$ as in Example 1 (12.9). We fix an isomorphism

$$\pi = \pi_1^{et}(\mathbb{P}^1_{\bar{k}} - \{0,1,\infty\}, \overrightarrow{01}) \cong \langle x, y\rangle^\wedge \tag{12.11}$$

between π and the profinite completion of the free group on two generators as follows: recall that we assume that k is a subfield of $\mathbb{C}$ or the completion of a number field at a place, and we have fixed $\mathbb{C} \supset \overline{\mathbb{Q}} \subseteq \bar{k}$, see 12.2. The morphisms $\mathbb{C} \supset \overline{\mathbb{Q}} \subseteq \bar{k}$ and the Riemann existence theorem give an isomorphism $\pi \cong \pi_1^{top}(\mathbb{P}^1_{\mathbb{C}} - \{0,1,\infty\}, \overrightarrow{01})^\wedge$, where the base point for the topological fundamental group, also denoted $\overrightarrow{01}$, is the tangent vector at 0 pointing towards 1. Let x be a small counterclockwise loop around 0 based at $\overrightarrow{01}$. Let y' be the pushforward of x by the automorphism of $\mathbb{P}^1_{\mathbb{C}} - \{0,1,\infty\}$ given by $1 \mapsto 1-t$, so in particular, y' is a small loop around 1 based at $\overrightarrow{10}$, where $\overrightarrow{10}$ is the tangent vector at 1 pointing towards 0. Conjugating y' by the direct path along the real axis between $\overrightarrow{10}$ and $\overrightarrow{01}$ produces a loop y, and an isomorphism $\pi_1^{top}(\mathbb{P}^1_{\mathbb{C}} - \{0,1,\infty\}, \overrightarrow{01}) = \langle x, y\rangle$, giving (12.11).

An element $\sigma \in G_k$ acts on π by

$$\begin{aligned} \sigma(x) &= x^{\chi(\sigma)} \\ \sigma(y) &= \mathfrak{f}(\sigma)^{-1} y^{\chi(\sigma)} \mathfrak{f}(\sigma) = [\mathfrak{f}(\sigma)^{-1}, y^{\chi(\sigma)}] y^{\chi(\sigma)}\ , \end{aligned} \tag{12.12}$$

where $\mathfrak{f}\ :\ G_k \to [\pi]_2$ is a cocycle with values in the commutator subgroup $[\pi]_2$ of π coming from the monodromy of the above path from $\overrightarrow{01}$ to $\overrightarrow{10}$, and $\chi\ :\ G_k \to \hat{\mathbb{Z}}^*$ denotes the cyclotomic character, see [Iha94].

12.2.2 The Abel-Jacobi map for $\mathbb{P}^1_k - \{0,1,\infty\}$

Let $\mathrm{X} \subseteq \overline{\mathrm{X}}$ denote a smooth curve over k inside its smooth compactification. The generalized Jacobian $\mathrm{Jac}\,\mathrm{X}$ of X is the algebraic group of equivalence classes of degree 0 divisors of X where two divisors are considered equivalent if they differ by $\mathrm{Div}(\phi)$ for a rational function ϕ such that $\phi(p) = 1$ for all p in $\overline{\mathrm{X}} - \mathrm{X}$. It follows that $\mathrm{Jac}(\mathrm{X})$ is an extension of $\mathrm{Jac}(\overline{\mathrm{X}})$ by the torus

$$\mathbb{T} = \Big(\prod_{p \in \overline{\mathrm{X}} - \mathrm{X}} \mathrm{Res}_{k(p)/k}\, \mathbb{G}_{\mathrm{m},k(p)} \Big) / \mathbb{G}_{\mathrm{m},k}$$

where p ranges over the closed points of $\overline{\mathrm{X}} - \mathrm{X}$ with residue field $k(p)$, the torus $\mathrm{Res}_{k(p)/k}\, G_{m,k(p)}$ denotes the restriction of scalars of $\mathbb{G}_{\mathrm{m},k(p)}$ to k, and where $\mathbb{G}_{\mathrm{m},k}$ acts diagonally. For more information on generalized Jacobians see [Ser88].

For $\mathrm{X} = \mathbb{P}^1_k - \{0,1,\infty\}$, the Jacobian of $\overline{\mathrm{X}} = \mathbb{P}^1_k$ is trivial. The complement of X in $\overline{\mathrm{X}}$ consists of three rational points and

$$\mathrm{Jac}(\mathbb{P}^1_k - \{0,1,\infty\}) \cong \mathbb{G}_{\mathrm{m},k} \times \mathbb{G}_{\mathrm{m},k} \,. \tag{12.13}$$

Since the fundamental group of a connected group is abelian, the fundamental group does not depend on base points. We find

$$\pi_1(\mathrm{Jac}(\mathbb{P}^1_{\bar{k}} - \{0,1,\infty\})) = \pi_1(\mathbb{G}_{\mathrm{m},\bar{k}}, 1) \times \pi_1(\mathbb{G}_{\mathrm{m},\bar{k}}, 1) = \hat{\mathbb{Z}}(1) \oplus \hat{\mathbb{Z}}(1)\,. \tag{12.14}$$

We choose an isomorphism (12.13) by sending $\mathrm{Div}(f)$ for a rational function f on $\mathbb{P}^1$ to $f(0)/f(\infty) \times f(1)/f(\infty)$ in $\mathbb{G}_{\mathrm{m}} \times \mathbb{G}_{\mathrm{m}}$. The **Abel-Jacobi map** based at $\overrightarrow{01}$

$$\alpha : \mathbb{P}^1_k - \{0,1,\infty\} \to \mathrm{Jac}(\mathbb{P}^1_k - \{0,1,\infty\}) = \mathbb{G}_{\mathrm{m},k} \times \mathbb{G}_{\mathrm{m},k}$$

$$t \mapsto (t, 1-t)$$

induces the abelianization

$$\pi = \pi_1(\mathbb{P}^1_{\bar{k}} - \{0,1,\infty\}, \overrightarrow{01}) \twoheadrightarrow \pi^{\mathrm{ab}} = \pi_1(\mathrm{Jac}(\mathbb{P}^1_{\bar{k}} - \{0,1,\infty\})). \tag{12.15}$$

Remark 3. From (12.11) and (12.12), we have fixed an isomorphism

$$\pi^{ab} \cong \hat{\mathbb{Z}}(\chi)x \oplus \hat{\mathbb{Z}}(\chi)y\,.$$

The isomorphism $\pi^{ab} \cong \hat{\mathbb{Z}}(1) \oplus \hat{\mathbb{Z}}(1)$ of (12.14) and (12.15) above is the composition of the former with the isomorphism

$$\hat{\mathbb{Z}}(\chi) \cong \hat{\mathbb{Z}}(1) := \varprojlim_n \mu_{n,\bar{k}}$$

corresponding to the compatible choice of roots of unity given by the action of the loop x on the fiber over $\overrightarrow{01}$ on the finite étale covers of $\mathbb{G}_{m,\bar{k}}$, i.e., to the choice $(\zeta_n)_{n\in\mathbb{Z}_{>0}}$, with $\zeta_n = e^{2\pi i/n}$, of compatible primitive nth roots of unity. We will hereafter identify

$$\hat{\mathbb{Z}}(1) = \hat{\mathbb{Z}}(\chi)$$

by this isomorphism, and for typographical reasons, we will use the notation $\hat{\mathbb{Z}}(n)$ for $\hat{\mathbb{Z}}(\chi^n)$, although the group law will be written additively.

By Example 1, the map κ for the k scheme $\mathbb{G}_{\mathrm{m}} \times \mathbb{G}_{\mathrm{m}}$ pointed by $(1,1)$ is two copies of the Kummer map:

$$k^* \times k^* \to \mathrm{H}^1(\mathrm{G}_k, \hat{\mathbb{Z}}(1)) \times \mathrm{H}^1(\mathrm{G}_k, \hat{\mathbb{Z}}(1))$$

$$b \times a \mapsto \{b\} \times \{a\} .$$

By functoriality of κ and Example 2, the following diagram is commutative:

$$\begin{array}{ccc} \mathrm{H}^1(\mathrm{G}_k, \pi) & \longrightarrow & \mathrm{H}^1(\mathrm{G}_k, \pi^{ab}) \\ \uparrow \kappa_{\overrightarrow{01}} & & \uparrow \kappa_{(1,1)} \\ (\mathbb{P}^1_k - \{0,1,\infty\})(k) & \xrightarrow{t \mapsto (t,1-t)} & (\mathbb{G}_m \times \mathbb{G}_m)(k) \end{array} \tag{12.16}$$

We can similarly compute the image of the k tangential points of $\mathbb{P}^1_k - \{0,1,\infty\}$ in $\mathrm{H}^1(\mathrm{G}_k, \pi^{ab})$, which is what we now do, see also [Ell00]. We define a map

$$\alpha : \cup_{t=0,1,\infty} (\mathrm{T}_t\mathbb{P}^1 - \{0\})(k) \to (\mathbb{G}_m \times \mathbb{G}_m)(k)$$

$$\alpha(\overrightarrow{0w}) = (w,1), \quad \alpha(\overrightarrow{1w}) = (1,-w), \quad \alpha(\overrightarrow{\infty w}) = (w^{-1}, -w^{-1})$$

where $w \in k^*$, the tangent vector $\overrightarrow{1w}$ is the pushforward of $\overrightarrow{0w}$ under $t \mapsto t+1$, and $\overrightarrow{\infty w}$ is the pushforward of $\overrightarrow{0w}$ under $t \mapsto 1/t$.

Lemma 4. *The following diagram commutes:*

$$\begin{array}{ccc} \mathrm{H}^1(\mathrm{G}_k, \pi) & \longrightarrow & \mathrm{H}^1(\mathrm{G}_k, \pi^{\mathrm{ab}}) \\ \uparrow & & \uparrow \\ \bigcup_{t=0,1,\infty} (\mathrm{T}_t\mathbb{P}^1 - \{0\})(k) & \xrightarrow{\alpha} & (\mathbb{G}_m \times \mathbb{G}_m)(k) . \end{array}$$

Proof. The commutativity of the two diagrams

$$\begin{array}{ccc} X & \xrightarrow{\alpha} & \mathrm{Jac}\,X \\ \big\downarrow{\scriptstyle t\mapsto 1-t} & & \big\downarrow{\scriptstyle (b,a)\mapsto(a,b)} \\ X & \xrightarrow{\alpha} & \mathrm{Jac}\,X \end{array} \qquad \begin{array}{ccc} X & \xrightarrow{\alpha} & \mathrm{Jac}\,X \\ \big\downarrow{\scriptstyle t\mapsto 1/t} & & \big\downarrow{\scriptstyle (b,a)\mapsto(\frac{1}{b},\frac{-a}{b})} \\ X & \xrightarrow{\alpha} & \mathrm{Jac}\,X \end{array}$$

reduces the lemma for $t = 1$ or ∞ (respectively) to the case $t = 0$.

By functoriality of κ applied to the Abel-Jacobi map $t \mapsto (t, 1-t)$, we have that the image of $\kappa_{(\mathbb{P}^1_k - \{0,1,\infty\}, \overrightarrow{01})}(\overrightarrow{0w})$ in $\mathrm{H}^1(\mathrm{G}_k, \pi^{\mathrm{ab}})$ is

$$\kappa_{(\mathbb{G}_m \times \mathbb{G}_m, \overrightarrow{01} \times 1)}(\overrightarrow{0w}, \overrightarrow{1(-w)}) = \kappa_{(\mathbb{G}_m, \overrightarrow{01})}(\overrightarrow{0w}) \times \kappa_{(\mathbb{G}_m,1)}(\overrightarrow{1(-w)}) .$$

The geometric point $\overrightarrow{1(-w)}$ factors through $1 - w\frac{\partial}{\partial t}$: $\mathrm{Spec}\,\overline{k}[[z]] \to \mathbb{G}_{m,\overline{k}}$

$$t \mapsto 1 - wz .$$

The quotient map $\mathrm{Spec}\,\overline{k} \to \mathrm{Spec}\,\overline{k}[[z]]$ given by $z \mapsto 0$ gives a bijection between the fiber over $1 - w\frac{\partial}{\partial t}$ and the fiber over 1 of the multiplication by n cover

$$p_n : \mathbb{G}_{m,\overline{k}} \to \mathbb{G}_{m,\overline{k}} .$$

These bijections determine a Galois equivariant path between 1 and $\overrightarrow{1(-w)}$ showing that

$$\kappa_{(\mathbb{G}_m,1)}(\overrightarrow{1(-w)}) = \kappa_{(\mathbb{G}_m,1)}(1) .$$

The lemma now follows from Examples 1 and 2. □

12.2.3 Ellenberg's Obstructions

Let π be $\pi_1(\mathbb{P}^1_{\overline{k}} - \{0,1,\infty\}, \overrightarrow{01})$ or more generally π can be any profinite group with a continuous G_k action, e.g., the étale fundamental group of a k-variety after base change to $\overline{k}$. The lower central series of π is the filtration of closed characteristic subgroups

$$\pi = [\pi]_1 \supset [\pi]_2 \supset \ldots \supset [\pi]_n \supset \ldots$$

where the commutator is defined $[x,y] = xyx^{-1}y^{-1}$, and $[\pi]_{n+1} = \overline{[\pi, [\pi]_n]}$ is the closure of the subgroup generated by commutators of elements of $[\pi]_n$ with elements of π. The central extension

$$1 \to [\pi]_n/[\pi]_{n+1} \to \pi/[\pi]_{n+1} \to \pi/[\pi]_n \to 1$$

gives rise to a boundary map in continuous group cohomology

$$\delta_n : \mathrm{H}^1(\mathrm{G}_k, \pi/[\pi]_n) \to \mathrm{H}^2(\mathrm{G}_k, [\pi]_n/[\pi]_{n+1})$$

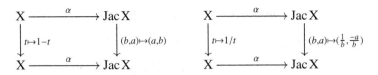

that is part of an exact sequence of pointed sets (see for instance [Ser02, I 5.7]),

$$\begin{aligned}1 \to &([\pi]_n/[\pi]_{n+1})^{G_k} \to (\pi/[\pi]_{n+1})^{G_k} \to (\pi/[\pi]_n)^{G_k} \\ \to &H^1(G_k,[\pi]_n/[\pi]_{n+1}) \to H^1(G_k,\pi/[\pi]_{n+1}) \to H^1(G_k,\pi/[\pi]_n) \\ \to &H^2(G_k,[\pi]_n/[\pi]_{n+1}).\end{aligned}$$

The δ_n give a series of obstructions to an element of $H^1(G_k,\pi/[\pi]_2)$ being the image of an element of $H^1(G_k,\pi)$, thereby also providing a series of obstructions to a rational point of the Jacobian coming from a rational point of the curve: to a given element x of $H^1(G_k,\pi/[\pi]_2)$, if $\delta_2(x) \neq 0$, then x is not the image of an element of $H^1(G_k,\pi)$. Otherwise, x lifts to $H^1(G_K,\pi/[\pi]_3)$. Apply δ_3 to all the lifts of x. If δ_3 is never 0, then x is not the image of an element of $H^1(G_k,\pi)$. Otherwise, x lifts to $H^1(G_k,\pi/[\pi]_4)$, and so on.

Definition 5. For x in $H^1(G_k,\pi/[\pi]_2)$, say that $\delta_n x = 0$ if x is in the image of

$$H^1(G_k,\pi/[\pi]_{n+1}) \to H^1(G_k,\pi/[\pi]_2)\,. \tag{12.17}$$

Otherwise, say $\delta_n x \neq 0$.

Let $X = \mathbb{P}^1_k - \{0,1,\infty\}$, or more generally X could be a smooth, geometrically connected, pointed curve over k, with an Abel-Jacobi map $X \to \operatorname{Jac} X$. As we are interested in obstructing points of the Jacobian from lying on X, it is convenient to identify a rational point of Jac X with its image under κ cf. (12.5).

Definition 6. For a k-point x of Jac X, say that $\delta_n x = 0$ if $\kappa_{(\operatorname{Jac} X,0)} x$ is in the image of (12.17), where 0 denotes the identity of Jac X. Otherwise, say $\delta_n x \neq 0$.

For k a number field, and K the completion of k at a place v, the obstruction δ_n for K will sometimes be denoted δ_n^v, and applied to elements of $H^1(G_k,\pi/[\pi]_n)$; it is to be understood that one first restricts to $H^1(G_K,\pi/[\pi]_n)$. In other words, given x in $H^1(G_k,\pi/[\pi]_2)$, the meaning of $\delta_n^v x = 0$ is that there exists x_{n+1} in $H^1(G_K,\pi/[\pi]_{n+1})$ lifting the restriction of x to $H^1(G_K,\pi/[\pi]_2)$. For x a point of Jac X(k), the meaning of $\delta_n^v x = 0$ is that $\delta_n^v \kappa_{(\operatorname{Jac} X,0)} x = 0$. This is equivalent to taking the image of x under $\operatorname{Jac} X(k) \to \operatorname{Jac} X_K(K)$ and applying Definition 6 with K as the base field.

Any filtration of π by characteristic subgroups such that successive quotients give rise to central extensions produces an analogous sequence of obstructions. For instance, consider the lower exponent 2 central series

$$\pi = [\pi]_1^2 \supset [\pi]_2^2 \supset \ldots \supset [\pi]_n^2 \supset \ldots ,$$

defined inductively by

$$[\pi]_{n+1}^2 = \overline{[\pi,[\pi]_n^2] \cdot ([\pi]_n^2)^2}$$

where $[\pi,[\pi]_n^2] \cdot ([\pi]_n^2)^2$ denotes the subgroup generated by the indicated commutators and the squares of elements of $[\pi]_n^2$. The resulting obstructions are denoted δ_n^2, and will also be evaluated on Jac X(k) in the following manner: π^{ab} maps to $(\pi/[\pi]_{n+1}^2)^{ab}$, giving a map

$$\mathrm{H}^1(\mathrm{G_K},\pi^{\mathrm{ab}}) \to \mathrm{H}^1(\mathrm{G_K},(\pi/[\pi]^2_{n+1})^{\mathrm{ab}}),$$

where either $\mathrm{K} = k$ or K is the completion of a number field $k \subset \mathbb{C}$ at a place v as above. Precomposing with κ for the Jacobian (12.5) gives a map

$$\mathrm{Jac(X)}(k) \to \mathrm{H}^1(\mathrm{G_K},(\pi/[\pi]^2_{n+1})^{\mathrm{ab}}).$$

For x in Jac(X)(k), say $\delta_n^2 x = 0$ (respectively $\delta_n^{(2,v)} x = 0$) if there exists x_{n+1} in $\mathrm{H}^1(\mathrm{G_K},\pi/[\pi]^2_{n+1})$ such that x and x_{n+1} have equal image in $\mathrm{H}^1(\mathrm{G_K},(\pi/[\pi]^2_{n+1})^{ab})$. Otherwise, say $\delta_n^2 x \neq 0$ (respectively $\delta_n^{(2,v)} x \neq 0$). The obstruction δ_n^v for v the place 2 will not be considered, so the notation δ_n^2 will not be ambiguous. Obstructions δ_n^m corresponding to the lower exponent m central series,

$$[\pi]^m_{n+1} = \overline{[\pi,[\pi]^m_n]\cdot([\pi]^m_n)^m}$$

are defined similarly.

As one final note of caution, $\mathrm{H}^1(\mathrm{G_K},\pi/[\pi]_n)$ is in general only a pointed set. Furthermore, even for $n = 2$, the map δ_2 is not a homomorphism, see Proposition 7.

12.3 The Obstructions δ_2 and δ_3 as Cohomology Operations

We express δ_2 and δ_3 for $\mathbb{P}^1_k - \{0,1,\infty\}$ in terms of cohomology operations, where k is a subfield of $\mathbb{C}$ or the completion of a number field at a place; in the latter case, fix an embedding of the number field into $\mathbb{C}$ and an embedding $\overline{\mathbb{Q}} \subset \overline{k}$, giving the isomorphism

$$\pi = \pi_1^{et}(\mathbb{P}^1_{\overline{k}} - \{0,1,\infty\},\overrightarrow{01}) \cong \langle x,y\rangle^\wedge$$

of (12.11). We will use the following notation.

For elements x and y of a group, let $[x,y] = xyx^{-1}y^{-1}$ denote their commutator. For a profinite group G and a profinite abelian group A with a continuous action of G, let $(\mathrm{C}^*(\mathrm{G,A}),\mathrm{D})$ be the complex of inhomogeneous cochains of G with coefficients in A as in [NSW08, I.2 p. 14]. For $c \in \mathrm{C}^p(\mathrm{G,A})$ and $d \in \mathrm{C}^q(\mathrm{G,A}')$, where A′ is a profinite abelian group with a continuous action of G, let $c \cup d$ denote the cup product $c \cup d \in \mathrm{C}^{p+q}(\mathrm{G,A}\otimes\mathrm{A}')$

$$(c\cup d)(g_1,\ldots,g_{p+q}) = c(g_1,\ldots,g_p)\otimes g_1\cdots g_p d(g_{p+1},\ldots,g_{p+q}),$$

which induces a well defined map on cohomology, and gives $\mathrm{C}^*(\mathrm{G,A})$ the structure of a differential graded algebra, for A a commutative ring, via $\mathrm{A}\otimes\mathrm{A} \to \mathrm{A}$. For a profinite group Q, no longer assumed to be abelian, the continuous 1-cocycles

$$\mathrm{Z}^1(\mathrm{G}_k,\mathrm{Q}) = \{s : \mathrm{G}_k \to \mathrm{Q}\ ;\ s \text{ is continuous, } s(gh) = s(g)gs(h)\}$$

of G_k with values in Q form a subset of the set of continuous inhomogeneous cochains

$$C^1(G_k, Q) = \{s : G_k \to Q \,;\, s \text{ is continuous}\} .$$

See [Ser02, I §5] for instance. For $s \in C^1(G_k, Q)$, let

$$Ds : G_k \times G_k \to Q$$

denote the function $Ds(g,h) = s(g)gs(h)s(gh)^{-1}$.

12.3.1 The Obstruction δ_2 as a Cup Product

For any based curve X over k with fundamental group $\pi = \pi_1(X_{\bar{k}})$, [Zar74, Thm p. 242] or [Ell00, Prop. 1] show that

$$\delta_2(p+q) - \delta_2(p) - \delta_2(q) = [-,-]_* p \cup q ,$$

where $[-,-]_*$ is the map on H^2 induced by the commutator

$$[-,-] : \pi^{ab} \otimes \pi^{ab} \to [\pi]_2/[\pi]_3$$

defined by

$$[\overline{\gamma}, \overline{\ell}] \mapsto \gamma \ell \gamma^{-1} \ell^{-1} ,$$

where $\overline{\gamma} \in \pi^{ab}$ is the image of $\gamma \in \pi/[\pi]_3$ and similarly for ℓ. It follows that δ_2 is the sum of a cup product term and a linear term, after inverting 2.

For $X = \mathbb{P}^1_k - \{0, 1, \infty\}$ based at $\overrightarrow{01}$, the linear term vanishes and we can avoid inverting 2 by slightly changing what is meant by the cup-product term. This was shown by Ellenberg, who gave a complete calculation of δ_2 in this case [Ell00, p. 11]. Here is an alternative calculation of this δ_2, showing the same result. Using the notation of (12.11) we identify $\pi/[\pi]_2$ with $\hat{\mathbb{Z}}(1) \oplus \hat{\mathbb{Z}}(1)$ using the basis x, y, and identify $[\pi]_2/[\pi]_3$ with $\hat{\mathbb{Z}}(2)$ using the basis $[x, y]$. So δ_2 is identified with a map

$$H^1(G_k, \hat{\mathbb{Z}}(1) \oplus \hat{\mathbb{Z}}(1)) \to H^2(G_k, \hat{\mathbb{Z}}(2)) .$$

Proposition 7. *Let $p(g) = y^{a(g)} x^{b(g)}$ be a 1-cocycle of G_k with values in $\pi/[\pi]_2$, so*

$$b, a : G_k \to \hat{\mathbb{Z}}(1)$$

are the cocycles produced by the isomorphism $\pi/[\pi]_2 \cong \hat{\mathbb{Z}}(1)x \oplus \hat{\mathbb{Z}}(1)y$. Then

$$\delta_2 p = b \cup a .$$

Proof. Sending $y^a x^b \in \pi/[\pi]_2$ to $y^a x^b \in \pi/[\pi]_3$ determines a set-theoretic section s of the quotient map $\pi/[\pi]_3 \to \pi/[\pi]_2$. Then $\delta_2(p)$ is represented by the cocycle

$$(g,h) \mapsto \delta_2(p)(g,h) = s(p(g))gs(p(h))s(p(gh))^{-1}$$

Using (12.12) and since $\mathfrak{f}(g) \in [\pi]_2$ is mapped to a central element in $\pi/[\pi]_3$ we find

$$\begin{aligned}
\delta_2(p)(g,h) &= (y^{a(g)}x^{b(g)})\big((\mathfrak{f}(g)^{-1}y^{\chi(g)}\mathfrak{f}(g))^{a(h)}x^{\chi(g)b(h)}\big)(y^{a(gh)}x^{b(gh)})^{-1} \\
&= (y^{a(g)}x^{b(g)})(y^{\chi(g)a(h)}x^{\chi(g)b(h)})(x^{-b(g)-\chi(g)b(h)}y^{-a(g)-\chi(g)a(h)}) \\
&= y^{a(g)}[x^{b(g)},y^{\chi(g)a(h)}]y^{-a(g)} = [x^{b(g)},y^{\chi(g)a(h)}] = [x,y]^{b(g)\cdot\chi(g)a(h)} \\
&= [x,y]^{(b\cup a)(g,h)}
\end{aligned}$$

giving the desired result. □

Proposition 7 characterizes the lifts to $\mathrm{H}^1(\mathrm{G}_k,\pi/[\pi]_3)$ of an element of

$$\mathrm{H}^1(\mathrm{G}_k,\pi^{ab}) \cong \mathrm{H}^1(\mathrm{G}_k,\hat{\mathbb{Z}}(1)) \oplus \mathrm{H}^1(\mathrm{G}_k,\hat{\mathbb{Z}}(1)) .$$

Let $b,a : \mathrm{G}_k \to \hat{\mathbb{Z}}(1)$ be cochains. For any $c \in \mathrm{C}^1(\mathrm{G}_k,\hat{\mathbb{Z}}(2))$, define

$$(b,a)_c : \mathrm{G}_k \to \pi/[\pi]_3 \quad \text{by} \quad (b,a)_c(g) = y^{a(g)}x^{b(g)}[x,y]^{c(g)} . \tag{12.18}$$

Corollary 8. *Let $p(g) = y^{a(g)}x^{b(g)}$ be a 1-cocycle of G_k with values in $\pi/[\pi]_2$. The lifts of p to a cocycle in $\mathrm{C}^1(\mathrm{G}_k,\pi/[\pi]_3)$ are in bijection with the set of cochains $c \in \mathrm{C}^1(\mathrm{G}_k,\hat{\mathbb{Z}}(2))$ such that*

$$\mathrm{D}c = -b \cup a$$

by

$$c \leftrightarrow (b,a)_c .$$

Proof. $\mathrm{D}((b,a)_c) = \delta_2(p) + \mathrm{D}c$, where $\mathrm{D}((b,a)_c)$ is as above (cf. 12.3), and $\delta_2(p)$ denotes its cocycle representative given in the proof of Proposition 7. □

12.3.2 The Obstruction δ_3 as a Massey Product

Let χ be the cyclotomic character. Note that $\chi(g) - 1$ is divisible by 2 in $\hat{\mathbb{Z}}$ for any $g \in \mathrm{G}_k$. This allows us to define

$$\frac{\chi-1}{2} : \mathrm{G}_k \to \hat{\mathbb{Z}}(\chi) \quad \text{by} \quad g \mapsto \frac{\chi(g)-1}{2} \in \hat{\mathbb{Z}}(\chi) .$$

For any compatible system of primitive nth roots of unity in $\hat{\mathbb{Z}}(1)$ giving an identification $\hat{\mathbb{Z}}(1) = \hat{\mathbb{Z}}(\chi)$, and in particular for (ζ_n) determined by Remark 3, we have

$$\{-1\} = \frac{\chi-1}{2} \tag{12.19}$$

in $\mathrm{H}^1(\mathrm{G}_k,\hat{\mathbb{Z}}(1))$, where $\{-1\}$ denotes the image of -1 under the Kummer map. The equality (12.19) holds as an equality of cocycles in $\mathrm{C}^1(\mathrm{G}_k,\hat{\mathbb{Z}}(1))$ when $\{-1\}$

is considered as the cocycle $G_k \to \hat{\mathbb{Z}}(1)$ given by choosing as the nth root of -1, the chosen primitive $(2n)$th root of unity. This is shown by the following calculation.

$$\{-1\}(g) = (g(\zeta_{2n})/\zeta_{2n})_n = (\zeta_{2n}^{\chi(g)-1})_n = (\zeta_n^{\frac{\chi(g)-1}{2}})_n\,,$$

where for an element $a \in \hat{\mathbb{Z}}(1)$, the reduction of a in $\mathbb{Z}/n(1)$ is denoted $(a)_n$.

Definition 9. Profinite binomial coefficients are the maps $\binom{}{m} : \hat{\mathbb{Z}} \to \hat{\mathbb{Z}}$ for $m \geqslant 0$ defined for $a \in \hat{\mathbb{Z}}$ with $a \equiv a_n \mod n$ by

$$\binom{a}{m} \equiv a_{m!n}(a_{m!n} - 1)(a_{m!n} - 2)\ldots(a_{m!n} - m + 1)/m! \mod n$$

for every $n \in \mathbb{N}$.

Example 10. For a cocycle $b \in C^1(G_k, \hat{\mathbb{Z}}(\chi))$, let $\binom{b}{2}$ in $C^1(G_k, \hat{\mathbb{Z}}(\chi^2))$ denote the cochain given by

$$g \mapsto \binom{b(g)}{2}.$$

We have

$$D\binom{b}{2} = -(b + \frac{\chi - 1}{2}) \cup b\,,$$

as shown by the computation:

$$D\binom{b}{2}(g,h) = \binom{b(g)}{2} + \chi(g)^2\binom{b(h)}{2} - \binom{b(gh)}{2} = \frac{b(g)(b(g)-1)}{2}$$
$$+ \frac{\chi(g)^2 b(h)(b(h)-1)}{2} - \frac{(b(g)+\chi(g)b(h))(b(g)+\chi(g)b(h)-1)}{2}$$
$$= -b(g)\chi(g)b(h) - \frac{\chi(g)^2 b(h) - \chi(g)b(h)}{2}$$
$$= -(b \cup b)(g,h) - (\frac{\chi - 1}{2} \cup b)(g,h)\,.$$

Example 11. Let b be an element of k^* with compatibly chosen nth roots $\sqrt[n]{b}$ in $\bar{k}$, giving a cocycle $b : G_k \to \hat{\mathbb{Z}}(1)$ via the Kummer map. Identify $\hat{\mathbb{Z}}(1) = \hat{\mathbb{Z}}(\chi)$ with $(\zeta_n)_{n \in \mathbb{Z}_{>0}}$ from Remark 3. When restricted to an element of $C^1(G_{k(\sqrt{b})}, \mathbb{Z}/2\mathbb{Z})$,

$$\binom{b}{2} = \{\sqrt{b}\} \quad \in C^1(G_{k(\sqrt{b})}, \mathbb{Z}/2\mathbb{Z})\,,$$

where $\{\sqrt{b}\}$ denotes the image of $\sqrt{b}$ under the Kummer map, which is independent of the choice of $\sqrt{\sqrt{b}}$. To see this, note that $(\{b\}(g))_4 = 2(\{\sqrt{b}\}(g))_4$ is even, whence $(\{b\}(g)-1)_2 = 1$, and the value of $\frac{1}{2}((\{b\}(g))_4)$ in $\mathbb{Z}/2\mathbb{Z}$ is $(\{\sqrt{b}\}(g))_2$. Here, as above, the reduction mod n of an element $a \in \hat{\mathbb{Z}}$ is denoted $(a)_n$.

Remark 12. Note that $\binom{b}{2}$ is a cochain taking values $\hat{\mathbb{Z}}(\chi^2)$, but Example 11 identifies its image in $\mathrm{C}^1(\mathrm{G}_{k(\sqrt{b})}, \mathbb{Z}/2\mathbb{Z})$ with a cocycle taking values in $\mathbb{Z}/2\mathbb{Z}(1)$. Furthermore, the cohomology class of the image of $\binom{b}{2}$ in $\mathrm{H}^1(\mathrm{G}_{k(\sqrt{b})}, \mathbb{Z}/2\mathbb{Z})$ depends on the choice of $\sqrt{b}$ used to define $b : \mathrm{G}_k \to \hat{\mathbb{Z}}(1)$. Nevertheless, $\binom{b}{2}$ appears in the expressions for δ_3 which will be given in Proposition 17; it is involved in expressions which make the choice of $\sqrt[n]{b}$ irrelevant, cf. Remark 37. In writing down elements of π in terms of x and y, we have identified $\hat{\mathbb{Z}}(1)$ and $\hat{\mathbb{Z}}(\chi)$ because monodromy around x distinguishes a compatible system of roots of unity.

Identify $\hat{\mathbb{Z}}(1)$ and $\hat{\mathbb{Z}}(\chi)$ using (ζ_n) as in Remark 3. In particular, we can apply profinite binomial coefficients to elements of $\hat{\mathbb{Z}}(1)$ or any $\hat{\mathbb{Z}}(n)$.

We define a 1-cocycle $f(\sigma) \in \mathrm{C}^1(\mathrm{G}_k, \hat{\mathbb{Z}}(2))$ by

$$\mathfrak{f}(\sigma) = [x,y]^{f(\sigma)} \mod [\pi]_3 \tag{12.20}$$

where $\mathfrak{f}(\sigma)$ is the 1-cocycle from (12.12) that describes the Galois action on π.

The basis $[[x,y],x], [[x,y],y]$ for $[\pi]_3/[\pi]_4$ as a $\hat{\mathbb{Z}}$ module decomposes δ_3 into two obstructions

$$\delta_{3,[[x,y],x]}, \delta_{3,[[x,y],y]} : \mathrm{H}^1(\mathrm{G}_k, \pi/[\pi]_3) \to \mathrm{H}^2(\mathrm{G}_k, \hat{\mathbb{Z}}(3)) .$$

Since an arbitrary element of $\pi/[\pi]_3$ can be written uniquely in the form $y^a x^b [x,y]^c$ for $a,b,c \in \hat{\mathbb{Z}}$, an arbitrary element of $\mathrm{C}^1(\mathrm{G}_k, \pi/[\pi]_3)$ is of the form $(b,a)_c$, as in (12.18). The obstruction δ_3 is therefore computed by the following.

Proposition 13. *Let $(b,a)_c$ be a 1-cocycle for G_k with values in $\pi/[\pi]_3$. Then:*

(1) $\delta_{3,[[x,y],x]}(b,a)_c$ is represented by the cocycle that maps (g,h) to

$$c(g)\chi(g)b(h) + \binom{b(g)+1}{2}\chi(g)a(h) + b(g)\chi(g)^2 a(h)b(h) - \frac{\chi(g)-1}{2}\chi(g)^2 c(h) ,$$

(2) $\delta_{3,[[x,y],y]}(b,a)_c$ is represented by the cocycle that maps (g,h) to

$$c(g)\chi(g)a(h) + b(g)\binom{\chi(g)a(h)+1}{2} - \frac{\chi(g)-1}{2}\chi(g)^2 c(h) - f(g)\chi(g)a(h) .$$

Proof. We have the following equality in $\pi/[\pi]_4$:

$$x^b y^a = y^a x^b [x,y]^{ab} [[x,y],y]^{b\binom{a+1}{2}} [[x,y],x]^{a\binom{b+1}{2}} . \tag{12.21}$$

Replacing b and a by $-a$ in (12.21), yields:

$$[x^a, y^a] = [x,y]^{a^2} [[x,y],x]^{-a\binom{a}{2}} [[x,y],y]^{-a\binom{a}{2}} \tag{12.22}$$

For any $g \in \mathrm{G}_k$, a straightforward computation using (12.12), (12.22) and (12.20) shows that:

$$g(y^a x^b [x,y]^c) = y^{\chi(g)a} x^{\chi(g)b} [x,y]^{\chi(g)^2 c} [[x,y],x]^{-\frac{\chi(g)-1}{2}\chi(g)^2 c} \times [[x,y],y]^{-\frac{\chi(g)-1}{2}\chi(g)^2 c - f(g)\chi(g)a} \qquad (12.23)$$

An arbitrary element of $\pi/[\pi]_3$ can be written uniquely in the form $y^a x^b [x,y]^c$ for $a,b,c \in \hat{\mathbb{Z}}$. Sending $y^a x^b [x,y]^c \in \pi/[\pi]_3$ to $y^a x^b [x,y]^c \in \pi/[\pi]_4$ determines a section s of the quotient map $\pi/[\pi]_4 \to \pi/[\pi]_3$. Let $p = (b,a)_c$. $\delta_3 p$ is represented by the cocycle

$$(g,h) \mapsto s(p(g)) g s(p(h)) s(p(gh))^{-1}$$

which gives cocycles representing $\delta_{3,[[x,y],x]} p$ and $\delta_{3,[[x,y],y]} p$. Combining (12.21) and (12.23) gives the desired result. □

We give a formula for δ_3 in terms of triple Massey products.

Definition 14. The **triple Massey product** $\langle \alpha,\beta,\gamma \rangle$ for 1-cocycles α,β,γ such that $\alpha \cup \beta = 0$ and $\beta \cup \gamma = 0$ is described by choosing cochains A, B such that $\mathrm{DA} = \alpha \cup \beta$ and $\mathrm{DB} = \beta \cup \gamma$, and setting

$$\langle \alpha,\beta,\gamma \rangle = \mathrm{A} \cup \gamma + \alpha \cup \mathrm{B}\,.$$

The choice $\{\mathrm{A},\mathrm{B}\}$ is the called the **defining system**. The triple Massey product determines a partially defined multivalued product on H^1.

Remark 15. Results of Dwyer and Stallings [Dwy75] relate the element of

$$\mathrm{H}^2(\pi/[\pi]_n, [\pi]_n/[\pi]_{n+1})$$

classifying the central extension

$$1 \to [\pi]_n/[\pi]_{n+1} \to \pi/[\pi]_{n+1} \to \pi/[\pi]_n \to 1$$

to nth order Massey products. The computation of δ_3 is equivalent to computing the element of $\mathrm{H}^2(\pi/[\pi]_3 \rtimes \mathrm{G}_k, [\pi]_3/[\pi]_4)$ classifying

$$1 \to [\pi]_3/[\pi]_4 \to \pi/[\pi]_4 \rtimes \mathrm{G}_k \to \pi/[\pi]_3 \rtimes \mathrm{G}_k \to 1\,.$$

Because $\mathrm{D}c = -b \cup a$ and $\mathrm{D}\binom{b+1}{2} = -(b - \frac{\chi-1}{2}) \cup b$, where the latter equality follows by the argument of Example 10, the expression for $\delta_{3,[[x,y],x]}(b,a)_c$ given in Proposition 13 looks similar to a triple Massey product $\pm\langle\{\pm b\},\{\pm b\},a\rangle$, except the $c \cup b$ term should be $b \cup c$. The cup product on cohomology is graded commutative, and the analogue on the level of cochains, given below in Lemma 16, allows us to change the order of c and b, which will express $\delta_{3,[[x,y],x]}(b,a)_c$ as a Massey product.

For cochains $c \in \mathrm{C}^1(\mathrm{G}_k, \hat{\mathbb{Z}}(n))$ and $b \in \mathrm{C}^1(\mathrm{G}_k, \hat{\mathbb{Z}}(m))$, define

$$cb : \mathrm{G}_k \to \hat{\mathbb{Z}}(n+m) \quad \text{by} \quad (cb)(g) = c(g)b(g)\,.$$

Lemma 16. *Let $c \in \mathrm{C}^1(\mathrm{G}_k, \hat{\mathbb{Z}}(n))$ be an arbitrary cochain, and b in $\mathrm{C}^1(\mathrm{G}_k, \hat{\mathbb{Z}}(m))$ be a cocycle. Then*

$$(\mathrm{D}(cb) + b \cup c + c \cup b)(g,h) = \mathrm{D}c(g,h)b(g) + \mathrm{D}c(g,h)\chi(g)^m b(h)\,.$$

Proof. By definition of D, for any $g,h \in \mathrm{G}_k$,

$$\mathrm{D}(cb)(g,h) = (cb)(g) + \chi(g)^{n+m}(cb)(h) - (cb)(gh)\,,$$
$$c(gh) = c(g) + \chi(g)^n c(h) - \mathrm{D}c(g,h)\,.$$

Since b is a cocycle, $b(gh) = b(g) + \chi(g)^m b(h)$. Combining equations, we have

$$\begin{aligned}\mathrm{D}(cb)(g,h) =& c(g)b(g) + \chi(g)^{n+m}c(h)b(h)\\ &- (c(g) + \chi(g)^n c(h) - \mathrm{D}c(g,h))(b(g) + \chi(g)^m b(h))\\ =& \mathrm{D}c(g,h)b(g) + \mathrm{D}c(g,h)\chi(g)^m b(h) - (c \cup b(g,h) + b \cup c(g,h))\end{aligned}$$

which proves the lemma. □

Proposition 17. *Let* $p = (b,a)_c \in \mathrm{C}^1(\mathrm{G}_k, \pi/[\pi]_3)$ *be a* 1*-cocycle, where* $(b,a)_c$ *is as in the notation of* (12.18)*. Then the following holds:*

(1) $\delta_{3,[[x,y],x]}(p) = -(b + \frac{\chi-1}{2}) \cup c - \binom{b}{2} \cup a\,,$

(2) $\delta_{3,[[x,y],y]}(p) = (a + \frac{\chi-1}{2}) \cup (ab - c) + \binom{a}{2} \cup b - f \cup a\,.$

Proof. By Corollary 8 and Lemma 16, we have that $-\mathrm{D}(cb)(g,h) =$

$$c(g)\chi(g)b(h) + b(g)\chi(g)^2 c(h) + b(g)^2\chi(g)a(h) + b(g)\chi(g)^2 a(h)b(h)\,.$$

Subtracting this expression for $-\mathrm{D}(cb)$ from the cocycle representing $\delta_{3,[[x,y],x]}p$ given in Proposition 13 shows that $\delta_{3,[[x,y],x]}p$ is represented by the cocycle sending (g,h) to

$$-b(g)\chi(g)^2 c(h) - b(g)^2\chi(g)a(h) + \binom{b(g)+1}{2}\chi(g)a(h) - \frac{\chi(g)-1}{2}\chi(g)^2 c(h)\,.$$

Note that

$$\begin{aligned}-b(g)^2\chi(g)a(h) + \binom{b(g)+1}{2}\chi(g)a(h) &= \frac{b(g)(-b(g)+1)}{2}\chi(g)a(h)\\ &= -(\binom{b}{2} \cup a)(g,h)\,.\end{aligned}$$

Therefore

$$\delta_{3,[[x,y],x]}p = -b \cup c - \binom{b}{2} \cup a - \frac{\chi-1}{2} \cup c\,,$$

giving the claimed expression for $\delta_{3,[[x,y],x]}p$.

The claimed expression for $\delta_{3,[[x,y],y]}p$ follows from this formula for $\delta_{3,[[x,y],x]}p$ and a symmetry argument. Consider the action of G_k on the profinite completion of the free group on two generators $\mathrm{F}_2^\wedge = \langle x,y \rangle^\wedge$ given by

$$g(x) = x^{\chi(g)} \quad \text{and} \quad g(y) = y^{\chi(g)} .$$

The G_k action on π described by (12.12) would reduce to this action on $F_2^\wedge$ if $\mathfrak{f}$ were in the center of π. In particular, sending x to x and y to y induces isomorphisms of profinite groups with G_k actions

$$\pi/[\pi]_3 \cong F_2^\wedge/[F_2^\wedge]_3$$

$$[\pi]_3/[\pi]_4 \cong [F_2^\wedge]_3/[F_2^\wedge]_4$$

Furthermore, viewing f as a formal variable in the proof of Proposition 13, we see that Proposition 13 implies that these isomorphisms fit into the commutative diagram

$$\begin{array}{ccc} H^1(G_k, \pi/[\pi]_3) & \xrightarrow{\delta_3 + \mathfrak{f} \cup a} & H^2(G_k, [\pi]_3/[\pi]_4) \\ \downarrow \cong & & \uparrow \cong \\ H^1(G_k, F_2^\wedge/[F_2^\wedge]_3) & \xrightarrow{\delta_3} & H^2(G_k, [F_2^\wedge]_3/[F_2^\wedge]_4) \end{array} \tag{12.24}$$

Let $i : F_2^\wedge \to F_2^\wedge$ be the G_k equivariant involution defined by

$$i(x) = y$$

$$i(y) = x .$$

Note that i induces an endomorphism of the short exact sequence of G_k modules

$$1 \to [F_2^\wedge]_3/[F_2^\wedge]_4 \to F_2^\wedge/[F_2^\wedge]_4 \to F_2^\wedge/[F_2^\wedge]_3 \to 1 .$$

Thus we have a commutative diagram

$$\begin{array}{ccc} H^1(G_k, F_2^\wedge/[F_2^\wedge]_3) & \xrightarrow{\delta_3} & H^2(G_k, [F_2^\wedge]_3/[F_2^\wedge]_4) \\ i_* \uparrow & & \uparrow i_* \\ H^1(G_k, F_2^\wedge/[F_2^\wedge]_3) & \xrightarrow{\delta_3} & H^2(G_k, [F_2^\wedge]_3/[F_2^\wedge]_4) \end{array}$$

Since i is an involution, so is i_*, whence

$$\delta_3 = i_* \delta_3 i_* .$$

With respect to the decomposition

$$H^2(G_k, [F_2^\wedge]_3/[F_2^\wedge]_4) = H^2(G_k, \hat{\mathbb{Z}}(3)) \cdot [[x,y],x] \oplus H^2(G_k, \hat{\mathbb{Z}}(3)) \cdot [[x,y],y] . \tag{12.25}$$

i_* acts by the matrix

$$\begin{pmatrix} 0 & -1 \\ -1 & 0 \end{pmatrix},$$

or in other terms,

$$\delta_{3,[[x,y],y]} = -\delta_{3,[[x,y],x]} i_* \,. \tag{12.26}$$

The map i_* on $\mathrm{H}^1(\mathrm{G}_k, \mathrm{F}_2^\wedge/[\mathrm{F}_2^\wedge]_3)$ we compute as

$$\begin{aligned} i_*((b,a)_c)(g) &= x^{a(g)} y^{b(g)} [y,x]^{c(g)} = [x^{a(g)}, y^{b(g)}] y^{b(g)} x^{a(g)} [x,y]^{-c(g)} \\ &= y^{b(g)} x^{a(g)} [x,y]^{a(g)b(g)-c(g)} = (a,b)_{ab-c}(g) \,. \end{aligned}$$

Combining (12.24) with (12.26) we get

$$\begin{aligned} \delta_{3,[[x,y],y]}((b,a)_c) &= \delta^{\mathrm{F}_2^\wedge}_{3,[[x,y],y]}((b,a)_c) - f \cup a = -\delta^{\mathrm{F}_2^\wedge}_{3,[[x,y],x]}(i_*(b,a)_c) - f \cup a \\ &= -\delta^{\mathrm{F}_2^\wedge}_{3,[[x,y],x]}((a,b)_{ab-c}) - f \cup a = -\delta_{3,[[x,y],x]}((a,b)_{ab-c}) - f \cup a \\ &= (a + \frac{\chi - 1}{2}) \cup (ab - c) + \binom{a}{2} \cup b - f \cup a \end{aligned}$$

as claimed by the proposition. In the above manipulation, we have marked obstructions corresponding to $\mathrm{F}_2^\wedge$ with a superscript to avoid confusion. □

Remark 18. The above symmetry argument combined with the explicit cocycle for $\delta_{3,[[x,y],y]}$ given in Proposition 13 gives unexploited computational information.

Theorem 19. *Let p be an element of* $\mathrm{H}^1(\mathrm{G}_k, \pi/[\pi]_3)$*, so p is represented by a cocycle* $(b,a)_c \in \mathrm{C}^1(\mathrm{G}_k, \pi/[\pi]_3)$ *in the notation of* (12.18)*. Then*

$$\delta_{3,[[x,y],x]}(p) = \langle (b + \frac{\chi-1}{2}), b, a \rangle \qquad \textit{defining system:} \{-\binom{b}{2}, -c\}$$

$$\delta_{3,[[x,y],y]}(p) = -\langle (a + \frac{\chi-1}{2}), a, b \rangle - f \cup a \qquad \textit{defining system:} \{-\binom{a}{2}, c - ab\} \,.$$

In particular, for $(b,a) \in \mathrm{Jac}(\mathbb{P}^1_k - \{0,1,\infty\})(k) = k^* \times k^*$*, we have*

$$\begin{aligned} \delta_{3,[[x,y],x]}(b,a) &= \langle \{-b\}, b, a \rangle \\ \delta_{3,[[x,y],y]}(b,a) &= -\langle \{-a\}, a, b \rangle - f \cup a \,. \end{aligned}$$

Remark 20. (i) As above, an element of k^* also denotes its image in $\mathrm{H}^1(\mathrm{G}_k, \hat{\mathbb{Z}}(1))$ under the Kummer map in the last two equations. The brackets in the notation $\{-b\}, \{-a\}$ serve to distinguish between the additive inverse of b in $\mathrm{H}^1(\mathrm{G}_k, \hat{\mathbb{Z}}(1))$ and the image of $-b$ under the Kummer map. We note the obvious remark that the Kummer map is a homomorphism, because this will appear in (ii) of this remark on the level of cocycles; namely given $a, b \in k^*$ with compatibly chosen n^{th} roots $\sqrt[n]{a}$ and $\sqrt[n]{b}$, then $\{a\} + \{b\} : \mathrm{G}_k \to \hat{\mathbb{Z}}(1)$ is the image under the Kummer map of ab with $\sqrt[n]{a}\sqrt[n]{b}$ chosen as the nth root of ab. This means that if one has chosen compatible primitive roots of -1, as is the case by (12.19) and Remark 3, then the cocycle $\{-1\} + \{a\}$ is different from $-\{-1\} + \{a\}$ although both give the same class in cohomology, namely the class $\{-a\}$.

(ii) Expressing δ_3 in terms of Massey products reduces the dependency on c to the choices of the defining systems. In fact, after restricting to defining systems of the appropriate form, the choice of these defining systems and the choice of lift are equivalent. More explicitly, we will say that a choice for the defining systems $\{A,B\}$ of $\langle\{-x\},x,y\rangle$ and $\{C,D\}$ of $\langle\{-y\},y,x\rangle$ is **compatible** if $B+D=-xy$, $A=-\binom{x}{2}$, and $C=-\binom{y}{2}$. The choice of defining system also encompasses choosing cocycle representatives for the cohomology classes involved; the cocycle representative for $\{-x\}$ is the representative for x plus $\frac{\chi-1}{2}$, and similarly for $\{-y\}$, as in (i). Then choosing a lift of (b,a) in $H^1(G_k,\pi/[\pi]_2)$ to $(b,a)_c$ in $H^1(G_k,\pi/[\pi]_3)$ is equivalent to choosing compatible defining systems for $\langle\{-b\},b,a\rangle$ and $\langle\{-a\},a,b\rangle$, by Corollary 8 and Theorem 19. As we are ultimately interested in obstructing points of the Jacobian from lying on the curve, it is natural to suppress both the defining system in the Massey product and the choice of lift, and view δ_3 and triple Massey products as multivalued functions on $H^1(G_k,\pi/[\pi]_2)$.

Proof. Comparing the expression for $\delta_{3,[[x,y],x]}(b,a)_c$ of Proposition 17, and the equations $D(-\binom{b}{2})=(b+\frac{\chi-1}{2})\cup b$ of Example 10, and $D(-c)=b\cup a$ of Corollary 8 with the definition of the triple Massey product shows

$$\delta_{3,[[x,y],x]}(b,a)_c=\langle(b+\frac{\chi-1}{2}),b,a\rangle$$

with the defining system $\{-\binom{b}{2},-c\}$. By Lemma 16,

$$-D(ab)=a\cup b+b\cup a,$$

whence $D(c-ab)=a\cup b$ by Corollary 8. Comparing $D(-\binom{a}{2})=(a+\frac{\chi-1}{2})\cup a$, $D(c-ab)=a\cup b$, and the expression for $\delta_{3,[[x,y],y]}(b,a)_c$ of Proposition 17 with the definition of the triple Massey product shows

$$\delta_{3,[[x,y],y]}(b,a)_c=-\langle(a+\frac{\chi-1}{2}),a,b\rangle-f\cup a$$

with defining system $\{-\binom{a}{2},c-ab\}$. By (12.19), we have $a+\frac{\chi-1}{2}=\{-a\}$ and $b+\frac{\chi-1}{2}=\{-b\}$, showing the theorem. □

12.4 Evaluating δ_2 on Jac(k)

Let k be a subfield of $\mathbb{C}$ or a completion of a number field equipped with $\mathbb{C}\supset\overline{\mathbb{Q}}\subseteq\overline{k}$ as in 12.2. Let $X=\mathbb{P}^1_k-\{0,1,\infty\}$ and recall that in 12.2.2 we fixed an isomorphism $\mathrm{Jac}(X)=\mathbb{G}_{m,k}\times\mathbb{G}_{m,k}$. Let $\pi=\pi_1(X_{\overline{k}},\overrightarrow{01})$. In (12.11), we specified an isomorphism $\pi=\langle x,y\rangle^{\wedge}$.

The obstruction δ_2 is given on (b,a) in Jac(X)(k) by $\delta_2(b,a)=b\cup a$, so evaluating δ_2 is equivalent to evaluating the cup product

$$\mathrm{H}^1(\mathrm{G}_k,\hat{\mathbb{Z}}(1))\otimes\mathrm{H}^1(\mathrm{G}_k,\hat{\mathbb{Z}}(1))\to\mathrm{H}^2(\mathrm{G}_k,\hat{\mathbb{Z}}(2))\,. \tag{12.27}$$

Evaluating the obstruction δ_2^2 coming from the lower exponent 2 central series, cf. Sect. 12.2.3, is equivalent to evaluating the mod 2 cup product

$$\mathrm{H}^1(\mathrm{G}_k,\mathbb{Z}/2\mathbb{Z})\otimes\mathrm{H}^1(\mathrm{G}_k,\mathbb{Z}/2\mathbb{Z})\to\mathrm{H}^2(\mathrm{G}_k,\mathbb{Z}/2\mathbb{Z})\,, \tag{12.28}$$

and this evaluation is recalled for $k=\mathbb{Q}_p$, $\mathbb{R}$, and $\mathbb{Q}$ in Sects. 12.4.1–12.4.3. The remainder of Sect. 12.4 gives evaluation results for the obstruction δ_2 itself. From the bilinearity of δ_2, Proposition 24 finds infinite families of points of Jac(k) which are unobstructed by δ_2, but which are not the image of a rational point or tangential point under the Abel-Jacobi map. This is rephrased as "the 2-nilpotent section conjecture is false" in Sect. 12.4.4 and Proposition 25. These families provide a certain supply of points on which to evaluate δ_3. Section 12.4.5 contains a finite algorithm for determining if δ_2 is zero or not for $k=\mathbb{Q}$, using Tate's calculation of $\mathrm{K}_2(\mathbb{Q})$, and gives Jordan Ellenberg's geometric proof that the cup product factors through K_2.

12.4.1 The mod 2 *cup Product for* k_v

Let p be an odd prime. Let k_v be a finite extension of $\mathbb{Q}_p$ with valuation $v: k_v^*\twoheadrightarrow\mathbb{Z}$, ring of integers O_v and residue field $\mathbb{F}_v$. Let $\mathfrak{p}$ be a uniformizer of O_v and $u\in O_v^*$ be a unit and not a square, so $\mathfrak{p},u$ is a basis for the $\mathbb{F}_2$ vector space

$$\mathrm{H}^1(\mathrm{G}_{k_v},\mathbb{Z}/2\mathbb{Z})=k_v^*/(k_v^*)^2\,.$$

By Hilbert 90 and the local invariant map, see e.g. [CF67, VI], we have

$$\mathrm{H}^2(\mathrm{G}_{k_v},\mathbb{Z}/2\mathbb{Z})=\mathrm{Br}(k_v)[2]=\frac{1}{2}\mathbb{Z}/\mathbb{Z}\,.$$

The mod 2 cup product (12.28) is given by the table

$\cup$	u	$\mathfrak{p}$
u	0	$1/2$
$\mathfrak{p}$	$1/2$	$\{-1\}\cup\mathfrak{p}$

where $\{-1\}\cup\mathfrak{p}=0$ if -1 is a square in $\mathbb{F}_v$ and $\{-1\}\cup\mathfrak{p}=1/2$ otherwise.

We include a derivation of this well-known calculation: as $\mathrm{H}^2(\mathrm{G}_{k_v},\mathbb{Z}/2\mathbb{Z})$ injects into $\mathrm{H}^2(\mathrm{G}_{k_v},\overline{k_v}^*)$, we may calculate in the Brauer group. For $(a,b)\in k_v^*\oplus k_v^*$, define $\mathrm{E}_{(\sqrt{a},b)}\in\mathrm{C}^1(\mathrm{G}_{k_v},\overline{k_v}^*)$ by

$$\mathrm{E}_{(\sqrt{a},b)}(\sigma)=(\sqrt{a})^{b(\sigma)}\,,$$

where $b:\mathrm{G}_{k_v}\to\{0,1\}$ is defined by $(-1)^{a(\sigma)}=(\sigma\sqrt{a})/\sqrt{a}$. A short calculation shows that

$$\mathrm{DE}_{(\sqrt{a},b)}(\sigma,\tau)=(-1)^{a(\sigma)b(\tau)}a^{b(\sigma)b(\tau)}\,,$$

whence $a \cup b$ in $\mathrm{H}^2(\mathrm{G}_{k_v}, \overline{k_v}^*)$ is represented by

$$(\sigma, \tau) \mapsto a^{b(\sigma)b(\tau)}. \tag{12.29}$$

Let k_v^{nr} denote the maximal unramified extension of k_v, and $v : (k_v^{nr})^* \to \mathbb{Z}$ the extension of the valuation. For $b = u$, the cocycle (12.29) factors through $\mathrm{Gal}(k_v^{nr}/k_v)^2$, and by [CF67, Chap VI 1.1 Thm 2 pg 130],

$$v_* : \mathrm{H}^2(\mathrm{Gal}(k_v^{nr}/k_v), k_v^{nr,*}) \to \mathrm{H}^2(\mathrm{Gal}(k_v^{nr}/k_v), \mathbb{Z})$$

is an isomorphism, showing that $\{u\} \cup \{u\} = 0$, and the invariant of $\{\mathfrak{p}\} \cup \{u\}$ is $1/2$ as claimed, see [CF67, pg 130].

To compute $\mathfrak{p} \cup \mathfrak{p}$, note that for $a = -1$, the cochain $(\sigma, \tau) \mapsto a^{b(\sigma)b(\tau)}$ equals $b \cup b$. By the above, $a \cup b$ is also represented by $(\sigma, \tau) \mapsto a^{b(\sigma)b(\tau)}$, so it follows that

$$\{-1\} \cup \mathfrak{p} = \mathfrak{p} \cup \mathfrak{p}\,.$$

It follows that δ_2 is non-trivial: let $\mathrm{X} = \mathbb{P}^1_{\mathbb{Q}} - \{0, 1, \infty\}$. Let $\delta_2^{(2,p)}$ denote δ_2^2 for $k = \mathbb{Q}_p$ and $\mathrm{X}_{\mathbb{Q}_p}$. Consider $\delta_2^{(2,p)}$ as a function on $\mathrm{Jac(X)}(\mathbb{Q})$ by evaluating $\delta_2^{(2,p)}$ on the corresponding $\mathbb{Q}_p$ point of $\mathrm{Jac}(\mathrm{X}_{\mathbb{Q}_p})$.

Corollary 21. *Choose x, y in $\mathbb{Q}^*$. Let p be an odd prime and u be an integer which is not a quadratic residue mod p. Then $\delta_2^{(2,p)}(uy^2, px^2) \neq 0$, and therefore $\delta_2^2(uy^2, px^2) \neq 0$ and $\delta_2(uy^2, px^2) \neq 0$.*

12.4.2 The mod 2 *Cohomology of* $\mathrm{G}_{\mathbb{R}}$

By [Hat02, III Example 3.40 p. 250] or [Bro94, III.1 Ex 2 p.58 p.108], there is a ring isomorphism

$$\mathrm{H}^*(\mathrm{G}_{\mathbb{R}}, \mathbb{Z}/2\mathbb{Z}) \cong \mathrm{H}^*(\mathbb{Z}/2\mathbb{Z}, \mathbb{Z}/2\mathbb{Z}) \cong \mathbb{Z}/2\mathbb{Z}[\alpha]$$

where α is the nontrivial class in degree 1, namely

$$\alpha = \{-1\} \in \mathbb{R}^*/(\mathbb{R}^*)^2 = \mathrm{H}^1(\mathrm{G}_{\mathbb{R}}, \mu_2) = \mathrm{H}^1(\mathrm{G}_{\mathbb{R}}, \mathbb{Z}/2\mathbb{Z})\,.$$

In particular, the cup product is an isomorphism

$$\mathrm{H}^1(\mathrm{G}_{\mathbb{R}}, \mathbb{Z}/2\mathbb{Z}(1)) \otimes \mathrm{H}^1(\mathrm{G}_{\mathbb{R}}, \mathbb{Z}/2\mathbb{Z}(1)) \to \mathrm{H}^2(\mathrm{G}_{\mathbb{R}}, \mathbb{Z}/2\mathbb{Z}(2))\,.$$

The map $\mathrm{C}^2(\mathrm{G}_{\mathbb{R}}, \mathbb{Z}/2\mathbb{Z}) \to \mathbb{Z}/2\mathbb{Z}$ given by evaluating a cochain at (τ, τ) for τ the non-trivial element of $\mathrm{G}_{\mathbb{R}}$ determines an isomorphism $\mathrm{H}^2(\mathrm{G}_{\mathbb{R}}, \mathbb{Z}/2\mathbb{Z}) \to \mathbb{Z}/2\mathbb{Z}$.

Let $\mathrm{X} = \mathbb{P}^1_{\mathbb{Q}} - \{0, 1, \infty\}$, and let $\delta_2^{(2,\mathbb{R})}$ denote δ_2^2 for $k = \mathbb{R}$ and $\mathrm{X}_{\mathbb{R}}$. Consider $\delta_2^{(2,\mathbb{R})}$ as a function on $\mathrm{Jac(X)}(\mathbb{Q})$ by evaluating $\delta_2^{(2,\mathbb{R})}$ on the corresponding $\mathbb{R}$ point of $\mathrm{Jac}(\mathrm{X}_{\mathbb{R}})$.

Corollary 22. *Let b,a be elements of $\mathbb{Q}^*$. Then $\delta_2^{(2,\mathbb{R})}(b,a) \neq 0$ if and only if $a,b<0$. If $a,b<0$, then $\delta_2^2(b,a) \neq 0$ and $\delta_2(b,a) \neq 0$.*

12.4.3 The mod 2 Cup Product for $\mathrm{G}_\mathbb{Q}$

There is a finite algorithm for computing the cup product of two Kummer classes

$$\mathbb{Q}^* \otimes \mathbb{Q}^* \to \mathrm{H}^1(\mathrm{G}_\mathbb{Q}, \mathbb{Z}/2\mathbb{Z}(1)) \otimes \mathrm{H}^1(\mathrm{G}_\mathbb{Q}, \mathbb{Z}/2\mathbb{Z}(1)) \to \mathrm{H}^2(\mathrm{G}_\mathbb{Q}, \mathbb{Z}/2\mathbb{Z}(2)) .$$

The phrase "computing an element of $\mathrm{H}^2(\mathrm{G}_\mathbb{Q}, \mathbb{Z}/2\mathbb{Z}(2))$" means the following. By the local-global principle for the Brauer group, the theorem of Hasse, Brauer, and Noether, see [NSW08, 8.1.17 Thm p. 436], we have an isomorphism

$$\mathrm{H}^2(\mathrm{G}_\mathbb{Q}, \mathbb{Z}/2\mathbb{Z}(2)) = \mathrm{Br}(\mathbb{Q})_2 \otimes \mu_2 = \ker(\bigoplus_v \mathrm{Br}(\mathbb{Q}_v)_2 \xrightarrow{\sum_v \mathrm{inv}_v} \mathbb{Q}/\mathbb{Z}) \otimes \mu_2$$

where A_2 is the 2-torsion of an abelian group A and $\mathrm{inv}_v : \mathrm{Br}(k_v) \hookrightarrow \mathbb{Q}/\mathbb{Z}$ is the local invariant map for the Brauer group of the local field k_v, see [CF67, VI], and v ranges over all places of $\mathbb{Q}$. The class in $\mathrm{H}^2(\mathrm{G}_\mathbb{Q}, \mathbb{Z}/2\mathbb{Z}(2))$ is therefore completely determined by its restrictions to local cohomology groups with the freedom to ignore one place by reciprocity. We will benefit from this freedom by ignoring the prime 2 for which we did not describe the mod 2 cup product computation above.

Given b and a in $\mathbb{Q}^*$, we give a finite algorithm for computing $\mathrm{inv}_v(\{b\} \cup \{a\})$ for every $v \neq 2$. By Sect. 12.4.1, for an odd prime p with $p \nmid ab$ we have $\{b\} \cup \{a\} = 0$. It therefore remains to evaluate

$$\mathrm{inv}_v(\{b\} \cup \{a\})$$

for $v = \mathbb{R}$ and the finitely many odd primes $v = p$ with $p \mid ab$. This is accomplished in finitely many steps by Sects. 12.4.1 and 12.4.2.

In fact, given any field extension $\mathbb{Q} \subset \mathrm{E}$ and any cocycle in $\mathrm{C}^2(\mathrm{Gal}(\mathrm{E}/\mathbb{Q}), \mathbb{Z}/2\mathbb{Z})$, for instance the ones given in Proposition 17, there is a finite algorithm for computing the associated element of $\mathrm{H}^2(\mathrm{G}_\mathbb{Q}, \mathbb{Z}/2\mathbb{Z})$.

12.4.4 The 2*-Nilpotent Section Conjecture for Number Fields Is False*

We describe families of points of $\mathrm{Jac}(\mathbb{P}^1_k - \{0,1,\infty\})(k)$ such that δ_2 vanishes.

Example 23. (1) The map δ_2 vanishes on the k points and tangential points of the curve $\mathbb{P}^1_k - \{0,1,\infty\}$ by design. Therefore, the k points of

$$\mathrm{Jac}(\mathbb{P}^1_k - \{0,1,\infty\}) = \mathbb{G}_{m,k} \times \mathbb{G}_{m,k}$$

of the form $(x, 1-x)$ or $(-x, x)$ satisfy $\delta_2 = 0$ by (12.16) and Lemma 4.

From a more computational point of view, the vanishing of δ_2 on $(-x, x)$ follows from the the calculation in Lemma 10 identifying the cochain whose boundary is $\{-x\} \cup \{x\}$. I do not presently see a specific cochain in $C^1(G_k, \hat{\mathbb{Z}}(2))$ whose boundary is $\{x\} \cup \{1-x\}$, although I would not be surprised if such a cochain could be written down explicitly.

(2) Since $\delta_2(b,a) = \{b\} \cup \{a\}$ by Proposition 7, the map δ_2 is bilinear in both coordinates of

$$k^* \oplus k^* = (\mathbb{G}_m \times \mathbb{G}_m)(k) = \mathrm{Jac}(\mathbb{P}^1_k - \{0,1,\infty\})(k)$$

Therefore, for any (b,a) in $k^* \oplus k^*$ such that $\delta_2(b,a) = 0$, the points of the form (b^m, a^n), for integers m and n, satisfy $\delta_2(b^m, a^n) = 0$ as well. Likewise for any (b,a), (b,c) such that $\delta_2(b,a) = \delta_2(b,c) = 0$, the point of the Jacobian (b, ac) also satisfies $\delta_2(b, ac) = 0$. As was pointed out by Jordan Ellenberg, this produces many families of k points of $\mathrm{Jac}(\mathbb{P}^1_k - \{0,1,\infty\})$ which are unobstructed by δ_2. A special case of this is mentioned in Proposition 24 below.

Proposition 24. *Let* $X = \mathbb{P}^1_k - \{0,1,\infty\}$. *For any x in k^*, the map δ_2 vanishes on*

$$(x, (1-x)^m), \quad ((-x)^m, x), \quad ((1-x)(-x), x),$$

$$(x^{n_1}(1-((1-x)^{n_2}(-x)^{n_3})), (1-x)^{n_2}(-x)^{n_3}).$$

Proof. This follows immediately from the discussion in Example 23 above. □

Say the "n-nilpotent section conjecture" for a smooth curve X over a field k holds if the natural map from k points and tangential points to conjugacy classes of sections of

$$1 \to \pi_1(X_{\bar{k}})^{ab} \to \pi_1(X)/[\pi_1(X_{\bar{k}})]_2 \to G_k \to 1 \tag{12.30}$$

which arise from k points of Jac X, and lift to sections of

$$1 \to \pi_1(X_{\bar{k}})/[\pi_1(X_{\bar{k}})]_{n+1} \to \pi_1(X)/[\pi(X_{\bar{k}})]_{n+1} \to G_k \to 1 \tag{12.31}$$

is a bijection. This is similar to the notion of "minimalistic" birational section conjectures introduced by Florian Pop [Pop10]. The notion given here has the disadvantage that it mentions the points of the Jacobian, and is therefore not entirely group theoretic. The reason for this is that finite nilpotent groups decompose as a product of p-groups, so the sections of (12.30) and (12.31) decompose similarly, allowing for sections which at different primes arise from different rational points of X. Restricting to a single prime will not give a "minimalistic" section conjecture by results of Hoshi [Hos10].

Because the conjugacy classes of sections of a split short exact sequence of profinite groups

$$1 \to Q \to Q \rtimes G \to G \to 1$$

are in natural bijective correspondence with the elements of $H^1(G, Q)$, the n-nilpotent section conjecture for a curve with a rational point is equivalent to: *the* *natural map from k points and tangential points to the kernel of δ_n is a bijection,* where the kernel of δ_n is considered as a subset of Jac X. More precisely, a smooth, pointed curve X gives rise to a commutative diagram

$$\begin{array}{ccc}
\mathrm{H}^1(\mathrm{G}_k, \pi/[\pi]_n) & \xrightarrow{\delta_n} & \mathrm{H}^2(\mathrm{G}_k, [\pi]_n/[\pi]_{n+1}) \\
\downarrow{\scriptstyle pr} & & \\
\mathrm{H}^1(\mathrm{G}_k, \pi/[\pi]_2) & \xleftarrow[\alpha]{} & \mathrm{X}(k) \cup \bigcup_{x\in\overline{\mathrm{X}}-\mathrm{X}} (\mathrm{T}_x\overline{\mathrm{X}}(k) - \{0\}) \\
\uparrow{\scriptstyle \kappa} & & \\
\mathrm{Jac}(\mathrm{X})(k) & &
\end{array}$$

where $\overline{\mathrm{X}}$ denotes the smooth compactification of X, and π denotes the fundamental group of $\mathrm{X}_{\overline{k}}$. The n-nilpotent section conjecture is the claim α induces a bijection

$$\mathrm{X}(k) \cup \bigcup_{x\in\overline{\mathrm{X}}-\mathrm{X}} (\mathrm{T}_x\overline{\mathrm{X}}(k) - \{0\}) \to \kappa(\mathrm{Jac}(\mathrm{X})(k)) \cap pr(\ker(\delta_n)) .$$

Say the "n-nilpotent section conjecture" holds for a field k, if for all smooth, hyperbolic curves over k, the n-nilpotent section conjecture holds.

Proposition 25. *The 2-nilpotent section conjecture fails for any subfield k of $\mathbb{C}$ or k the completion of a number field. In fact, the 2-nilpotent section conjecture does not hold for $\mathbb{P}^1_k - \{0, 1, \infty\}$.*

Proof. Choose $x \neq 1$ in k^* such that $(1-x)^2$ does not equal $-x$. Then the k point of $\mathrm{Jac}(\mathbb{P}^1_k - \{0, 1, \infty\})$ given by $(x, (1-x)^2)$ is not the image of a k point or tangential point of $\mathbb{P}^1_k - \{0, 1, \infty\}$ by (12.16) and Lemma 4. By Proposition 24, the section of (12.30) determined by $(x, (1-x)^2)$ lifts to a section of (12.31) for $n = 2$. □

While the failure of a "minimalistic section conjecture" is not surprising at all, some do hold. Moreover, the "minimalistic section conjectures" in [Pop10] [Wic10] do not mention the points of the Jacobian and so control the rational points of X group theoretically.

12.4.5 The Obstruction δ_2 over $\mathbb{Q}$

By [Tat76, Theorem 3.1], the cup product (12.27) composed with the Kummer map (12.7)

$$k^* \otimes k^* \to \mathrm{H}^2(\mathrm{G}_k, \hat{\mathbb{Z}}(2)) \tag{12.32}$$

factors through the Milnor K_2-group of k

$$\mathrm{K}_2(k) = k^* \otimes_{\mathbb{Z}} k^* / \langle x \otimes (1-x) : x \in k^* \rangle$$

mapping to $H^2(G_k,\hat{\mathbb{Z}}(2))$ by the Galois symbol $h_k : K_2(k) \to H^2(G_k,\hat{\mathbb{Z}}(2))$.

Proposition 26. *Let k be a finite extension of $\mathbb{Q}$, and $X = \mathbb{P}^1_k - \{0,1,\infty\}$. For any $(b,a) \in \mathrm{Jac}(X)(k) = k^* \times k^*$ we have $\delta_2(b,a) = 0$ if and only if $b \otimes a = 0$ in $K_2(k)$.*

Proof. As $\delta_2(b,a) = b \cup a$ by Proposition 7, the map δ_2 factors through h_k, which is an isomorphism onto the torsion subgroup of $H^2(G_k,\hat{\mathbb{Z}}(2))$ by [Tat76, Thm. 5.4]. □

Tate's computation of $K_2(\mathbb{Q})$ thus gives an algorithm for computing δ_2 for the curve $\mathbb{P}^1_{\mathbb{Q}} - \{0,1,\infty\}$ on any rational point (b,a) of $\mathrm{Jac}(\mathbb{P}^1_{\mathbb{Q}} - \{0,1,\infty\})(\mathbb{Q}) = \mathbb{Q}^* \times \mathbb{Q}^*$. By [Mil71, Thm. 11.6], there is an isomorphism

$$K_2(\mathbb{Q}) \cong \mu_2 \oplus_{p \text{ odd prime}} \mathbb{F}_p^*$$

given by the tame symbols: for odd p (with p-adic valuation v_p)

$$K_2(\mathbb{Q}) \to \mathbb{F}_p^*$$
$$x \otimes y \mapsto (x,y)_p = (-1)^{v_p(x)v_p(y)} x^{v_p(y)} y^{-v_p(x)} \in \mathbb{F}_p^*,$$

and a map at the prime 2

$$K_2(\mathbb{Q}) \to \mu_2$$
$$x \otimes y \mapsto (x,y)_2 = (-1)^{i\mathrm{I}+j\mathrm{K}+k\mathrm{J}}$$

where $x = (-1)^i 2^j 5^k u$ and $y = (-1)^{\mathrm{I}} 2^{\mathrm{J}} 5^{\mathrm{K}} u'$ with $k, \mathrm{K} = 0$ or 1, and u, u' quotients of integers congruent to 1 mod 8.

By Proposition 26, we have $\delta_2(b,a) = 0$ if and only if $(b,a)_p = 0$ for $p = 2$ and for p equal to all odd primes dividing a or b.

Example 27. As an example of this algorithm, consider $\delta_2(p,-p)$ for p an odd prime: for q different from p and 2, we have $(p,-p)_q = 1$ because $v_q(p)$ and $v_q(-p)$ vanish. At the prime p, we have $(p,-p)_p = (-1)^1 p/(-p) = 1$. For the prime 2, express p in the form $p = (-1)^i 2^j 5^k u$ with $k = 0,1$ and u a quotient of integers congruent to 1 mod 8. Then $-p = (-1)^{i+1} 2^j 5^k u$, and

$$(p,-p)_2 = (-1)^{i(i+1)+2jk} = 1 .$$

Thus $\delta_2(p,-p) = 0$, as also followed from Lemma 4, as $(p,-p)$ is the image of a tangential point of $\mathbb{P}^1_{\mathbb{Q}} - \{0,1,\infty\}$, or can be deduced from the Steinberg relation.

Remark 28. The computation of δ_2 for $\mathbb{P}^1_k - \{0,1,\infty\}$ gives a geometric proof that (12.32) factors through $K_2(k)$, as observed by Jordan Ellenberg. Namely, by construction δ_2 vanishes on the image of

$$\mathbb{P}^1_k - \{0,1,\infty\}(k) \to \mathrm{Jac}(\mathbb{P}^1_k - \{0,1,\infty\})(k) = k^* \times k^* .$$

By (12.16), this image is $(x, 1-x)$, and by Proposition 7, the obstruction δ_2 is given by $\delta_2(b,a) = b \cup a$. Thus, the map (12.32) vanishes on $x \otimes (1-x)$, and therefore factors through $K_2(k)$.

12.5 Evaluating Quotients of δ_3

We keep the notation k, $X = \mathbb{P}^1_k - \{0,1,\infty\}$, and $\pi = \pi_1(\mathbb{P}^1_{\bar{k}} - \{0,1,\infty\}, \overrightarrow{01})$ from Sect. 12.4. In particular, we have a chosen isomorphism $\pi = \langle x,y\rangle^\wedge$ as above. To evaluate δ_3 on points in $\mathrm{Ker}\,\delta_2$ requires Galois cohomology computations with coefficients in $[\pi]_3/[\pi]_4 \cong \hat{\mathbb{Z}}(3)\oplus\hat{\mathbb{Z}}(3)$. As computing in $\mathrm{H}^2(\mathrm{G}_k,\hat{\mathbb{Z}}(3))$ seems difficult, we will evaluate a quotient of δ_3 which can be computed in $\mathrm{H}^2(\mathrm{G}_k,\mathbb{Z}/2\mathbb{Z}(3)) = \mathrm{H}^2(\mathrm{G}_k,\mathbb{Z}/2\mathbb{Z})$. This quotient is denoted $\delta_3^{\mathrm{mod}\,2}$ and can be described as "the reduction of δ_3 mod 2" as well as "the 3-nilpotent piece of δ_3^2," where δ_3^2 is the obstruction coming from the lower exponent 2 central series as in Sect. 12.2.3.

12.5.1 Definition of $\delta_3^{\mathrm{mod}\,2}$

Composing the obstruction

$$\delta_3 : \mathrm{H}^1(\mathrm{G}_k,\pi/[\pi]_3) \to \mathrm{H}^2(\mathrm{G}_k,[\pi]_3/[\pi]_4)$$

with the map on H^2 induced from the quotient

$$\begin{array}{c}[\pi]_3/[\pi]_4 \cong \hat{\mathbb{Z}}(3)[[x,y],x]\oplus\hat{\mathbb{Z}}(3)[[x,y],y] \\ \big\downarrow \\ [\pi]_3/[\pi]_4([\pi]_3)^2 \cong \mathbb{Z}/2\mathbb{Z}[[x,y],x]\oplus\mathbb{Z}/2\mathbb{Z}[[x,y],y]\end{array}$$

gives a map $\mathrm{H}^1(\mathrm{G}_k,\pi/[\pi]_3) \to \mathrm{H}^2(\mathrm{G}_k,[\pi]_3/[\pi]_4([\pi]_3)^2)$ which factors through

$$\mathrm{H}^1(\mathrm{G}_k,\pi/[\pi]_3) \to \mathrm{H}^1(\mathrm{G}_k,\pi/[\pi]_3^2)$$

as follows from Proposition 17 where $[\pi]_n^2$ denotes the nth subgroup of the lower exponent 2 central series, cf. Sect. 12.2.3. The resulting map

$$\delta_3^{\mathrm{mod}\,2} : \mathrm{H}^1(\mathrm{G}_k,\pi/[\pi]_3^2) \to \mathrm{H}^1(\mathrm{G}_k,[\pi]_3/[\pi]_4([\pi]_3)^2)$$

is defined as $\delta_3^{\mathrm{mod}\,2}$. The basis $[[x,y],x],[[x,y],y]$ decomposes $\delta_3^{\mathrm{mod}\,2}$ into two obstructions

$$\delta^2_{3,[[x,y],x]},\delta^2_{3,[[x,y],y]} : \mathrm{H}^1(\mathrm{G}_k,\pi/[\pi]_3^2) \to \mathrm{H}^2(\mathrm{G}_k,\mathbb{Z}/2\mathbb{Z})$$

which are compatible with the previously defined $\delta_{3,[[x,y],x]}$, $\delta_{3,[[x,y],y]}$ in the obvious manner. In words, $\delta_3^{\mathrm{mod}\,2}$ is δ_3 reduced mod 2.

The obstruction $\delta_3^{\mathrm{mod}\,2}$ can also be constructed from a central extension extension of groups with an action of G_k. To see this, we recall certain well-known results on the lower exponent p central series of free groups: for a free group F, the successive quotient $[\mathrm{F}]_n/[\mathrm{F}]_{n+1}$ of the lower central series is isomorphic to the homogeneous degree n component of the free Lie algebra on the same generators. The Lie basis theorem gives bases for $[\mathrm{F}]_n/[\mathrm{F}]_{n+1}$ explicitly via bases for the free Lie algebra [MKS04, Thm 5.8]. For the free group on 2 generators x and y, the following are bases in the respective degree n:

$$\begin{array}{ll} n=1 & x,y \\ n=2 & [x,y] \\ n=3 & [[x,y],x],[[x,y],y] \\ n=4 & [[[x,y],x],x],[[[x,y],y],y],[[[x,y],y],x] \end{array}$$

Results of [MKS04] and [Laz54] can be used to show that if β_i is a basis for $[\mathrm{F}]_i/[\mathrm{F}]_{i+1}$ for $i=1,\ldots,n$ and $\beta_i^{p^{n-i}}$ denotes the set whose elements are the elements of β_i raised to the p^{n-i}, then

$$\beta_1^{p^{n-1}} \cup \beta_2^{p^{n-2}} \cup \beta_3^{p^{n-3}} \cup \ldots \cup \beta_n$$

is a basis for $[\mathrm{F}]_n^p/[\mathrm{F}]_{n+1}^p$, where $[\mathrm{F}]_n^p$ denotes the nth subgroup of the lower exponent p central series. Thus, as a G_k-module we have

$$[\pi]_3^p/[\pi]_4^p = \mathbb{Z}/p\mathbb{Z}(3)\cdot[[x,y],x] \oplus \mathbb{Z}/p\mathbb{Z}(3)\cdot[[x,y],y]$$

$$\oplus \mathbb{Z}/p\mathbb{Z}(2)\cdot[x,y]^2 \oplus \mathbb{Z}/p\mathbb{Z}(1)\cdot x^4 \oplus \mathbb{Z}/p\mathbb{Z}(1)\cdot y^4\,.$$

It follows that

$$1 \to [\pi]_3/[\pi]_4([\pi]_3)^p \to \pi/[\pi]_4^p([\pi]_2^p)^p \to \pi/[\pi]_3^p \to 1 \tag{12.33}$$

is an exact sequence, and $\delta_3^{\mathrm{mod}\,2}$ is also the boundary map in G_k cohomology of (12.33) for $p=2$.

Since we view Ellenberg's obstructions as constraints for points of the Jacobian, we will define a *lift* of a point of Jac(X)(k) to $\mathrm{H}^1(\mathrm{G}_k,\pi/[\pi]_3^2)$ and then define

$$\delta_3^{\mathrm{mod}\,2} : \mathrm{Ker}(\delta_2^2 : \mathrm{Jac}(\mathbb{P}_k^1 - \{0,1,\infty\})(k) \to \mathrm{H}^2(\mathrm{G}_k,[\pi]_2^2/[\pi]_3^2)) \to \{0,1\}$$

which assigns to (b,a) in $k^* \times k^*$ the element 0 if there exists a lift of (b,a) such that $\delta_3^{\mathrm{mod}\,2}=0$, and assigns to (b,a) the element 1 if there does not exist such a lift. If $\delta_3^{\mathrm{mod}\,2}(b,a)\neq 0$, it follows that (b,a) is not the image of a k point or tangential point under the Abel-Jacobi map.

For (b,a) in $k^*\times k^* = \mathrm{Jac}(\mathbb{P}_k^1 - \{0,1,\infty\})(k)$, let (b,a) also denote the associated element of $\mathrm{H}^1(\mathrm{G}_k,\pi^{ab})$. For a characteristic closed subgroup $\mathrm{N}<\pi$, a **lift** of (b,a) to $\mathrm{H}^1(\mathrm{G}_k,\pi/\mathrm{N})$ is an element whose image under

$$\mathrm{H}^1(\mathrm{G}_k,\pi/\mathrm{N}) \to \mathrm{H}^1(\mathrm{G}_k,(\pi/\mathrm{N})^{\mathrm{ab}})$$

equals the image of (b,a) under

$$\mathrm{Jac(X)}(k) \to \mathrm{H}^1(\mathrm{G}_k, \pi^{\mathrm{ab}}) \to \mathrm{H}^1(\mathrm{G}_k, (\pi/\mathrm{N})^{\mathrm{ab}}).$$

For example, the lifts of (b,a) to $\mathrm{H}^1(\mathrm{G}_k, \pi/[\pi]_3^2)$ can be described as follows. The fixed embeddings $\mathbb{C} \supset \overline{\mathbb{Q}} \subseteq \overline{k}$ and resulting identification $\pi = \langle x,y\rangle^\wedge$ give canonical identifications

$$(\pi/[\pi]_3^2)^{\mathrm{ab}} = \mathbb{Z}/4\mathbb{Z}(1)\cdot x \oplus \mathbb{Z}/4\mathbb{Z}(1)\cdot y,$$

$$\mathrm{Ker}(\pi/[\pi]_3^2 \to (\pi/[\pi]_3^2)^{\mathrm{ab}}) = \mathbb{Z}/2\mathbb{Z}(2)\cdot [x,y].$$

We choose fourth roots of b and a, that give cocycles via the Kummer map

$$b, a : \mathrm{G}_k \to \mathbb{Z}/4\mathbb{Z}(1),$$

and $g \mapsto y^{a(g)}x^{b(g)}$ represents (b,a) in $\mathrm{H}^1(\mathrm{G}_k, (\pi/[\pi]_3^2)^{\mathrm{ab}})$. The obstruction to lifting (b,a) to $\mathrm{H}^1(\mathrm{G}_k, \pi/[\pi]_3^2)$ is

$$\delta_2^2(b,a) = b \cup a \in \mathrm{H}^1(\mathrm{G}_k, \mathbb{Z}/2\mathbb{Z}(2)).$$

For (b,a) such that $\delta_2^2(b,a) = 0$, the lifts of (b,a) are in bijection with the set of cochains $c \in \mathrm{C}^1(\mathrm{G}_k, \mathbb{Z}/2\mathbb{Z})$ such that $\mathrm{D}c = -b \cup a$ (note the minus sign) up to coboundary by

$$c \leftrightarrow (b,a)_c,$$

where

$$(b,a)_c(g) = y^{a(g)}x^{b(g)}[x,y]^{c(g)}.$$

The set of these lifts is a $\mathrm{H}^1(\mathrm{G}_k, \mathbb{Z}/2\mathbb{Z}(2))$ torsor. We could have equivalently considered the set of cochains c such that $\mathrm{D}c = -b \cup a$, as the G_k action on $\mathbb{Z}/2\mathbb{Z}(2)$ is trivial, cf. Corollary 8.

Example 29. (1) Let $(b,a) \in \mathrm{Jac(X)}(k)$ be the image of a rational point or a rational tangential point of $\mathrm{X} = \mathbb{P}_k^1 - \{0,1,\infty\}$. For $x,y \in k^*$ then (bx^4, ay^4) is unobstructed by $\delta_2^{\mathrm{mod}\,2}$ and $\delta_3^{\mathrm{mod}\,2}$ since (b,a) is unobstructed and defines the same class in $\mathrm{H}^1(\mathrm{G}_k, \pi^{\mathrm{ab}} \otimes \mathbb{Z}/4\mathbb{Z})$.

(2) Similarly, the mod m obstructions $\delta_n^{\mathrm{mod}\,m}$ coming from the lower m-central series of π are m-adically continuous in the sense that the obstruction map $\delta_{n+1}^{\mathrm{mod}\,m}$ is constant on cosets by $(k^*)^{m^n}$.

(3) $(b,(1-b)^4)$ determines the same element of $\mathrm{H}^1(\mathrm{G}_{\mathrm{K}}, \pi/([\pi]_2\pi^4))$ as $(b,1)$ which is the image of a k tangential point at 0 (Lemma 4). Therefore

$$\delta_3^{\mathrm{mod}\,2}(b,(1-b)^4) = 0.$$

This is an example of a point of the Jacobian unobstructed both by δ_2, see Proposition 24, and by $\delta_3^{\mathrm{mod}\,2}$.

We can also compute $\delta_3^{\mathrm{mod}\,2}(b,(1-b)^4)$ directly using Proposition 13. We include the calculation. Let $[\pi]_4^m$ denote the subgroup of the lower exponent m central series, cf. Sect. 12.2.3.

Proposition 30. *The conjugacy class of the section of*

$$1 \to \pi^{ab}/(\pi)^{m^3} \to \pi_1(\mathbb{P}^1_k - \{0,1,\infty\}, \overrightarrow{01})/([\pi]_2[\pi]_4^m) \to G_k \to 1 \tag{12.34}$$

determined by $(b,(1-b)^{m^2})$ *lifts to a section of*

$$1 \to \pi/[\pi]_4^m \to \pi_1(\mathbb{P}^1_k - \{0,1,\infty\}, \overrightarrow{01})/[\pi]_4^m \to G_k \to 1\,.$$

Proof. Let $(b,a) = (b,(1-b)^{m^2})$. Choose compatible systems of nth roots of b and $1-b$, giving cocycles $b, 1-b : G_k \to \hat{\mathbb{Z}}(1)$, and let $a : G_k \to \hat{\mathbb{Z}}(1)$ be $m^2(1-b)$. Conjugacy classes of sections of (12.34) are in bijection with $H^1(G_k, \pi^{ab}/(\pi)^{m^3})$ using $\overrightarrow{01}$ as the marked splitting. We claim that

$$g \mapsto y^{a(g)} x^{b(g)} \tag{12.35}$$

defines a cocycle $G_k \to \pi/[\pi]_4^m$, showing the proposition. Note that the reduction mod m^2 of a is 0, showing that $b \cup a = 0 \bmod m^2$, whence (12.35) is a cocycle with values in $\pi/[\pi]_3^m$. By Proposition 13, with $a = 0 \bmod m$ and $c = 0$, we have that $\delta^m_{3,[[x,y],x]} = 0$. An easy algebraic manipulation shows that $\binom{\chi(g)a(h)+1}{2}$ is 0 mod m^2 for m odd, and 0 mod m for any m, so the expression for $\delta_{3,[[x,y],y]}$ of Proposition 13 with $a = c = 0 \bmod m$ shows that $\delta^m_{3,[[x,y],y]} = 0$. □

12.5.2 Evaluating the 2-Nilpotent Quotient of $\mathfrak{f}$

The cocycle $\mathfrak{f} : G_k \to [\pi]_2$ in the description given in (12.12) of the Galois action on $\pi_1(\mathbb{P}^1_{\bar{k}} - \{0,1,\infty\}, \overrightarrow{01})$ records the monodromy of the standard path from $\overrightarrow{01}$ to $\overrightarrow{10}$. To evaluate $\delta^2_{[[x,y],y]}$, we will need to evaluate the 2-nilpotent quotient

$$f : G_k \to \hat{\mathbb{Z}}(2) \cdot [x,y]$$

of $\mathfrak{f}$ that was introduced in (12.20), or more precisely its mod 2 reduction. Work of Anderson [And89], Coleman [Col89], Deligne, Ihara [Iha91, 6.3 Thm p.115], Kaneko, and Yukinari [IKY87] gives the formula

$$f(\sigma) = \frac{1}{24}(\chi(\sigma)^2 - 1) \tag{12.36}$$

where we recall that $\chi : G_k \to \hat{\mathbb{Z}}^*$ denotes the cyclotomic character. For the convenience of the reader, we introduce enough of the notation of [Iha91] to check (12.36) from the statement given in loc. cit. 6.3 Thm p.115.

Let $\hat{\mathbb{Z}}\langle\langle \xi, \eta \rangle\rangle$ denote the non-commutative power series algebra in two variables over $\hat{\mathbb{Z}}$. The Magnus embedding of the free group on two generators in $\hat{\mathbb{Z}}\langle\langle \xi, \eta \rangle\rangle$ gives rise to an injective map

$$\mathcal{M} : \pi \cong \langle x,y\rangle^{\wedge} \hookrightarrow \hat{\mathbb{Z}}\langle\langle \xi,\eta\rangle\rangle$$

defined by

$$\mathcal{M}(x) = 1+\xi\,,$$

$$\mathcal{M}(y) = 1+\eta\,.$$

By [MKS04, Cor 5.7], $\mathcal{M}$ takes $[\pi]_n$ to elements of the form $1+\sum_{m\geqslant n} u_m$, where u_m is homogeneous of degree m. By [MKS04, §5.5] Lemma 5.4, for any $j \in \hat{\mathbb{Z}}$,

$$\mathcal{M}([x,y]^j) = 1 + j(\xi\eta - \eta\xi) + \sum_{m>2} u_m$$

where u_m is some homogeneous element of degree m depending on j. More generally, to lowest order in $\hat{\mathbb{Z}}\langle\langle \xi,\eta\rangle\rangle$, $\mathcal{M}$ takes the commutator of the group π to the Lie bracket of the associative algebra $\hat{\mathbb{Z}}\langle\langle \xi,\eta\rangle\rangle$, in the manner made precise by loc. cit. §5.5 Lemma 5.4 (7). Thus

$$\mathfrak{f}(\sigma) = 1 + f(\sigma)(\xi\eta - \eta\xi) + O(3) \tag{12.37}$$

where $O(3)$ is a sum of monomials of degree $\geqslant 3$.

As a $\hat{\mathbb{Z}}$ module, $\hat{\mathbb{Z}}\langle\langle \xi,\eta\rangle\rangle$ is the direct sum

$$\hat{\mathbb{Z}}\langle\langle \xi,\eta\rangle\rangle \cong \hat{\mathbb{Z}} \oplus \hat{\mathbb{Z}}\langle\langle \xi,\eta\rangle\rangle\xi \oplus \hat{\mathbb{Z}}\langle\langle \xi,\eta\rangle\rangle\eta\,.$$

Define $\psi : G_k \to \hat{\mathbb{Z}}\langle\langle \xi,\eta\rangle\rangle$ to be the projection of $(\mathfrak{f})^{-1}$ onto the direct summand $\hat{\mathbb{Z}} \oplus \hat{\mathbb{Z}}\langle\langle \xi,\eta\rangle\rangle\xi$ i.e.,

$$(\mathfrak{f}(\sigma))^{-1} = 1 + a_1\xi + a_2\eta$$

$$\psi(\sigma) = 1 + a_1\xi\,.$$

By (12.37), we have

$$(\mathfrak{f}(\sigma))^{-1} = 1 - f(\sigma)(\xi\eta - \eta\xi) + O(3)\,,$$

so $f(\sigma)$ is the coefficient of $\eta\xi$ in $\psi(\sigma)$. As the only degree 2 terms that $\psi(\sigma)$ can contain are $\hat{\mathbb{Z}}$ linear combinations of $\eta\xi$ and ξ^2, the cocycle f is determined by the degree 2 terms of the projection of $\psi(\sigma)$ to the commutative power series ring. More explicitly, let $\psi^{ab} : G_k \to \hat{\mathbb{Z}}[[\xi,\eta]]$ denote the composition of ψ with the quotient

$$\hat{\mathbb{Z}}\langle\langle \xi,\eta\rangle\rangle \to \hat{\mathbb{Z}}[[\xi,\eta]]$$

where $\hat{\mathbb{Z}}[[\xi,\eta]]$ denotes the commutative power series ring. Then

$$\psi^{ab}(\sigma) = 1 + f(\sigma)\eta\xi + r$$

where r is a sum a monomial of the form $b\xi^2$ with $b \in \hat{\mathbb{Z}}$ and monomials of degree greater than one.

The formula in [Iha91, 6.3 Thm p.115] expresses $\psi^{ab}(\sigma)$ in terms of the Bernoulli numbers and the variables $X = \log(1+\xi)$, and $Y = \log(1+\eta)$. This formula gives

$$\psi^{ab}(\sigma) = 1 - \frac{1}{2}b_2(1-\chi(\sigma)^2)\eta\xi + O(3)$$

where $O(3)$ is a sum of monomial terms in the variables η,ξ of degree $\geqslant 3$, and $b_2 = \frac{1}{12}$. This implies (12.36).

We denote the mod 2 reduction of f by

$$\overline{f} : G_k \to \hat{\mathbb{Z}}(2) \to \mathbb{Z}/2\mathbb{Z}(2) = \mathbb{Z}/2\mathbb{Z}\,.$$

Lemma 31. *The class represented by $\overline{f}$ in $H^1(G_k, \mathbb{Z}/2\mathbb{Z})$ is the class of 2 under the Kummer map.*

Proof. We may assume that $k = \mathbb{Q}$ by functoriality. The value of $\frac{1}{24}(1-\chi(\sigma)^2)$ mod 2 is determined by $1-\chi(\sigma)^2$ mod 48, which is determined by $\chi(\sigma)$ mod 8. A direct check shows that $\overline{f}(\sigma) = 1$ when $\chi(\sigma)$ is ± 3 mod 8, and $\overline{f}(\sigma) = 0$ otherwise. Thus $\overline{f}$ corresponds to the quadratic extension $k \subset k(\zeta_8 + \zeta_8^{-1}) = k(\sqrt{2})$ inside $k \subset k(\zeta_8)$. □

12.5.3 Local 3-Nilpotent Obstructions mod 2 at $\mathbb{R}$

We compute $\delta_3^{\mathrm{mod}\,2}$ for $k = \mathbb{R}$. For a point

$$(b,a) \in \mathbb{R}^* \times \mathbb{R}^* \cong \mathrm{Jac}(\mathbb{P}^1_{\mathbb{R}} - \{0,1,\infty\})(\mathbb{R})\,,$$

we have the associated element of $H^1(G_{\mathbb{R}}, \pi^{ab})$. Recall the notation $(b,a)_c$ and characterization of the lifts of this element to $H^1(G_{\mathbb{R}}, \pi/[\pi]_3^2)$ of Sect. 12.5.1. Recall as well that $\{-1\}$ in $C^1(G_k, \mathbb{Z}/2\mathbb{Z})$ denotes the image of -1 under the Kummer map, and note that given c in $C^1(G_{\mathbb{R}}, \mathbb{Z}/2\mathbb{Z}(2))$ such that $Dc = -b \cup a$, we have that $c + \{-1\}$ is another cochain such that $D(c + \{-1\}) = -b \cup a$. Thus $(b,a)_c$ and $(b,a)_{c+\{-1\}}$ are two lifts of (b,a) to $H^1(G_{\mathbb{R}}, \pi/[\pi]_3^2)$, and they are the only two as $H^1(G_{\mathbb{R}}, \mathbb{Z}/2\mathbb{Z}(2)) = \mathbb{Z}/2\mathbb{Z}$.

Proposition 32. *For any $(b,a)_c$ in $H^1(G_{\mathbb{R}}, \pi/[\pi]_3^2)$, either*

$$\delta_3^{\mathrm{mod}\,2}(b,a)_c = 0 \quad or \quad \delta_3^{\mathrm{mod}\,2}(b,a)_{c+\{-1\}} = 0\,.$$

Proof. Since $\overline{f} = 2 \in H^1(G_{\mathbb{R}}, \mathbb{Z}/2\mathbb{Z})$ vanishes by Lemma 31, it suffices by Theorem 19 to show that the triple Massey products

$$\delta^{\mathrm{mod}\,2}_{3,[[x,y],x]}(b,a) = \langle \{-b\}, b, a \rangle \in H^2(G_{\mathbb{R}}, \mathbb{Z}/2\mathbb{Z})$$
$$\delta^{\mathrm{mod}\,2}_{3,[[x,y],y]}(b,a) = -\langle \{-a\}, a, b \rangle - \overline{f} \cup b = \langle \{-a\}, a, b \rangle \in H^2(G_{\mathbb{R}}, \mathbb{Z}/2\mathbb{Z})$$

admit the value 0 for a compatible choice of the defining systems as in Remark 20 (ii). Changing c by a 1-cocycle $\epsilon \in C^1(G_{\mathbb{R}}, \mathbb{Z}/2\mathbb{Z})$ has the effect by Proposition 17 that

$$\delta_3^{\mathrm{mod}\,2}(b,a)_c - \delta_3^{\mathrm{mod}\,2}(b,a)_{c+\epsilon} = \{-b\} \cup \epsilon \cdot [[x,y],x] + \{-a\} \cup \epsilon \cdot [[x,y],y] \,.$$

Since $a \cup b = 0$ we conclude that at most one of a, b is negative.

Case $a, b > 0$: we may choose trivial defining systems, i.e., the defining system for $\langle \{-b\}, b, a \rangle$ is $\{\binom{b}{2}, 0\} = \{0,0\}$ and the defining system for $\langle \{-a\}, a, b \rangle$ is the pair $\{\binom{a}{2}, ab + 0\} = \{0,0\}$, so the obstruction vanishes.

Case $a > 0$ *and* $b < 0$: regardless of the defining system, the Massey product $\langle \{-b\}, b, a \rangle$ always vanishes. The other Massey product can be adjusted if necessary by $\epsilon = -1$ since $\{-1\} \cup \{-1\}$ generates $\mathrm{H}^2(\mathrm{G}_{\mathbb{R}}, \mathbb{Z}/2\mathbb{Z})$.

Case $a < 0$ *and* $b > 0$: regardless of the defining system, the Massey product $\langle \{-a\}, a, b \rangle$ always vanishes. The other Massey product can be adjusted if necessary by $\epsilon = -1$ since $\{-1\} \cup \{-1\}$ generates $\mathrm{H}^2(\mathrm{G}_{\mathbb{R}}, \mathbb{Z}/2\mathbb{Z})$. □

12.5.4 *Local* 3*-Nilpotent Obstructions mod* 2 *Above Odd Primes*

Let k be a number field embedded into $\mathbb{C}$. Let $p \in \mathbb{Z}$ be an odd prime, k_v the completion of k at a prime v above p, and choose an embedding $\overline{\mathbb{Q}} \subset \overline{k_v}$, where $\overline{\mathbb{Q}}$ is the algebraic closure of $\mathbb{Q}$ in $\mathbb{C}$.

We compute $\delta_3^{(\mathrm{mod}\,2,v)}$ for all k_v-points of the Jacobian of $\mathbb{P}^1_{k_v} - \{0,1,\infty\}$ in the kernel of $\delta_2^{(\mathrm{mod}\,2,v)}$. Note that Sect. 12.4.1 and

$$\delta_2^{\mathrm{mod}\,2,v}(b,a) = a \cup b \in \mathrm{H}^2(\mathrm{G}_{k_v}, \mathbb{Z}/2\mathbb{Z}(2))$$

characterize this kernel, and recall the notation $(b,a)_c$ for lifts of $(b,a) \in \mathrm{Ker}\,\delta_2^{\mathrm{mod}\,2,v}$ to $\mathrm{H}^1(\mathrm{G}_{k_v}, \pi/[\pi]_3^2)$ given in Sect. 12.5.1.

Lemma 33. *Let* (b,a) *in* $\mathrm{Jac}(\mathbb{P}^1_{k_v} - \{0,1,\infty\})(k_v)$ *be in the kernel of* $\delta_2^{\mathrm{mod}\,2,v}$ *and let* $(b,a)_c$ *be a lift of* (b,a) *to* $\mathrm{H}^1(\mathrm{G}_{k_v}, \pi/[\pi]_3^2)$.

(1) *If* $\{-b\} = 0$ *in* $\mathrm{H}^1(\mathrm{G}_{k_v}, \mathbb{Z}/2\mathbb{Z})$, *then* $\delta^{(2,v)}_{3,[[x,y],x]}(b,a)_c = \binom{b}{2} \cup a$ *is independent of* c.

(2) *If* $\{-a\} = 0$ *in* $\mathrm{H}^1(\mathrm{G}_{k_v}, \mathbb{Z}/2\mathbb{Z})$, *then* $\delta^{(2,v)}_{3,[[x,y],y]}(b,a)_c = \binom{a}{2} \cup b + \overline{f} \cup a$ *is independent of* c.

(3) *If* $\{b\} = \{a\}$ *in* $\mathrm{H}^1(\mathrm{G}_{k_v}, \mathbb{Z}/2\mathbb{Z})$, *then*

$$\delta^{(2,v)}_{3,[[x,y],x]}(b,a)_c + \delta^{(2,v)}_{3,[[x,y],y]}(b,a)_c = (\binom{a}{2} + \binom{b}{2} + \overline{f}) \cup a$$

is independent of c.

Otherwise

$$\delta_3^{(\mathrm{mod}\,2,v)}(b,a) = 0 \,.$$

Proof. Changing the lift is equivalent to adding a 1-cocycle $\epsilon \in \mathrm{C}^1(\mathrm{G}_{k_v}, \mathbb{Z}/2\mathbb{Z})$ to c with the effect by Proposition 17 that

$$\delta_3^{\mathrm{mod}\,2}(a,b)_c - \delta_3^{\mathrm{mod}\,2}(a,b)_{c+\epsilon} = \{-b\} \cup \epsilon \cdot [[x,y],x] + \{-a\} \cup \epsilon \cdot [[x,y],y]. \quad (12.38)$$

We abbreviate the differences componentwise by introducing the notation

$$\Delta^2_{3,[[x,y],x]}(b,a,z) = \{-b\} \cup z\,,$$

$$\Delta^2_{3,[[x,y],y]}(b,a,z) = \{-a\} \cup z\,.$$

Using Proposition 17, case (1) and (2) are now immediate. In case (3), note that Proposition 17 implies that

$$\delta^{(2,v)}_{3,[[x,y],x]}(b,a)_c + \delta^{(2,v)}_{3,[[x,y],y]}(b,a)_c = (a + \frac{\chi - 1}{2}) \cup (ab) + \binom{b}{2} \cup a + \binom{a}{2} \cup b + \overline{f} \cup a$$

Since $a = b$, we have $(ab) = a^2 = a$, so

$$(a + \frac{\chi - 1}{2}) \cup (ab) = (a + \frac{\chi - 1}{2}) \cup a = 0\,.$$

This equation and $a = b$ show case (3).

If the map

$$\Delta : \mathrm{H}^1(\mathrm{G}_{k_v}, \mathbb{Z}/2\mathbb{Z}) \to \mathrm{H}^2(\mathrm{G}_{k_v}, \mathbb{Z}/2\mathbb{Z}) \oplus \mathrm{H}^2(\mathrm{G}_{k_v}, \mathbb{Z}/2\mathbb{Z}) \quad (12.39)$$

$$z \mapsto (\Delta^2_{3,[[x,y],x]}(b,a,z), \Delta^2_{3,[[x,y],y]}(b,a,z)) = (\{-b\} \cup z, \{-a\} \cup z)$$

is surjective, namely by the nondegeneracy of the cup product pairing, if $\{-a\}, \{-b\}$ forms a basis of $\mathrm{H}^1(\mathrm{G}_{k_v}, \mathbb{Z}/2\mathbb{Z})$, then for a suitable choice of correction term z, the corresponding c will lead to a vanishing

$$\delta_3^{\mathrm{mod}\,2}((b,a)_c) = 0\,.$$

Since $\mathrm{H}^1(\mathrm{G}_{k_v}, \mathbb{Z}/2\mathbb{Z}) \cong \mathbb{Z}/2\mathbb{Z} \oplus \mathbb{Z}/2\mathbb{Z}$, the elements $\{-a\}, \{-b\}$ form a basis unless at least one of cases (1), (2), or (3) holds. □

To compute when there is a local obstruction as in case (1) or (2), we need to understand the cochain

$$\binom{a}{2} : \mathrm{G}_{k_v} \to \mathbb{Z}/2\mathbb{Z}$$

when $\{-a\}$ is 0 in $\mathrm{C}^1(\mathrm{G}_{k_v}, \mathbb{Z}/2\mathbb{Z})$, cf. Examples 10 and 11.

Lemma 34. *Let* K *be a field of characteristic* $\neq 2$, *and let* ζ_4 *be a primitive fourth root of unity in a fixed algebraic closure of* K. *Let* $-x \in (\mathrm{K}^*)^2$ *be a square, and choose a fourth root* $\sqrt[4]{x}$ *of* x *giving a Kummer cocycle* $x : \mathrm{G_K} \to \mathbb{Z}/4\mathbb{Z}(1)$ *and the cochain* $\binom{x}{2} : \mathrm{G_K} \to \mathbb{Z}/2\mathbb{Z}$ *obtained by the identification* $\mathbb{Z}/4\mathbb{Z}(1) = \mathbb{Z}/4\mathbb{Z}(\chi)$ *using* ζ_4. *Then there is an equality of cocycles*

$$\binom{x}{2} = \{2\sqrt{-x}\} \; : \; G_K \to \mathbb{Z}/2\mathbb{Z}$$

where $\sqrt{-x} = \zeta_4(\sqrt[4]{x})^2$.

Proof. Note that squaring $(1+\zeta_4)\sqrt[4]{x}$ gives $2\sqrt{-x}$, so letting $\eta = 2\sqrt{-x} \in K$, we see that $K(\sqrt[4]{x}, \zeta_4) = K(\sqrt{\eta}, \zeta_4)$. Both cochains $\binom{x}{2}, \{2\sqrt{-x}\} : G_K \to \mathbb{Z}/2\mathbb{Z}$ factor through $\mathrm{Gal}(K(\sqrt[4]{x}, \zeta_4)/K)$, and there are four possibilities for the action of $g \in G_K$ on $\sqrt{\eta}$ and ζ_4. It is enough to check that $\binom{x}{2}(g)$ and $\{\eta\}(g)$ agree in each case:

$g(\sqrt{\eta})$	$g(\zeta_4)$	$x(g)$	$\binom{x(g)}{2}$
$\sqrt{\eta}$	ζ_4	$(\zeta_4)^0$	0
$\sqrt{\eta}$	$(\zeta_4)^3$	$(\zeta_4)^1$	0
$-\sqrt{\eta}$	ζ_4	$(\zeta_4)^2$	1
$-\sqrt{\eta}$	$(\zeta_4)^3$	$(\zeta_4)^3$	1

This shows the claim $\binom{x}{2} = \{\eta\} = \{2\sqrt{-x}\}$. □

To compute when there is a local obstruction as in case (3), we need to understand the cochain

$$\binom{a}{2} + \binom{b}{2} \; : \; G_{k_v} \to \mathbb{Z}/2\mathbb{Z}$$

when $a = b$ in $C^1(G_{k_v}, \mathbb{Z}/2\mathbb{Z})$.

Lemma 35. *Let* K *be a field of characteristic* $\neq 2$, *and let* ζ_4 *be a primitive fourth root of unity in a fixed algebraic closure of* K. *Let* a, b *be non-zero elements of* K *such that* a/b *is a square* $a/b \in (K^*)^2$. *We choose fourth roots of both, and let* $a, b : G_K \to \mathbb{Z}/4\mathbb{Z}(1)$ *denote the corresponding cocyles. Then*

$$\binom{a}{2} + \binom{b}{2} \; : \; G_K \to \mathbb{Z}/2\mathbb{Z}$$

equals the cocycle

$$\{\sqrt{b}/\sqrt{a}\} \; : \; G_K \to \mathbb{Z}/2\mathbb{Z},$$

where $\sqrt{b}$, $\sqrt{a}$ *are defined as the squares of the chosen fourth roots of* b, a *respectively.*

Proof. For any two elements d_1, d_2 in $\mathbb{Z}/4\mathbb{Z}$, direct calculation shows that in $\mathbb{Z}/2\mathbb{Z}$

$$\binom{d_1+d_2}{2} - \binom{d_1}{2} - \binom{d_2}{2} = d_1 d_2 .$$

Therefore for all g in G_K,

$$\binom{a(g)}{2} + \binom{b(g)}{2} = \binom{a(g)+b(g)}{2} - a(g)b(g)$$

Since mod 2, $a(g) = b(g)$, we have that $a(g)b(g) = a(g)$ mod 2. Therefore

$$\binom{a(g)}{2} + \binom{b(g)}{2} = \binom{a(g)+b(g)}{2} - a(g)$$

Since b/a is a square in K, ab is also a square in K. Let $\sqrt{a}, \sqrt{b}$ denote the squares of our chosen fourth roots of a and b respectively. By Lemma 11, and because ab is a square in K,

$$g \mapsto \binom{a(g)+b(g)}{2}$$

equals $\{\sqrt{a}\sqrt{b}\}$. Therefore

$$\binom{a}{2} + \binom{b}{2} = \{\sqrt{a}\sqrt{b}\} - a\,.$$

Since $\{\sqrt{a}\sqrt{b}\} - a = \{\sqrt{b}/\sqrt{a}\}$, the lemma is shown. □

The previous results combine to give necessary and sufficient conditions for $\delta_3^{(\mathrm{mod}\,2,v)}$ to obstruct a k_v-point of the Jacobian in the kernel of $\delta_2^{(\mathrm{mod}\,2,v)}$ from lying on the curve, i.e., from being the image of a k_v-point or tangential point of X. For any given point (b,a) of Jac X, these conditions are easy to verify using Sect. 12.4.1.

Theorem 36. *Let k_v be the completion of a number field k at a place above an odd prime. For $(b,a) \in \mathrm{Jac}(\mathbb{P}^1_{k_v} - \{0,1,\infty\})(k_v)$ such that $\delta_2^{(\mathrm{mod}\,2,v)}(b,a) = 0$, we have $\delta_3^{\mathrm{mod}\,2}(b,a) \neq 0$ if and only if one of the following holds:*

(i) $-b \in (k_v^*)^2$ *and* $\{2\sqrt{-b}\} \cup a \neq 0.$
(ii) $-a \in (k_v^*)^2$ *and* $\{2\sqrt{-a}\} \cup b + \{2\} \cup a \neq 0$
(iii) $ab \in (k_v^*)^2$ *and* $\{2\sqrt{b}\sqrt{a}\} \cup a \neq 0$

Remark 37. In case (i), the notation $\sqrt{-b}$ denotes either square root of $-b$, both of which are in k_v. The expression $\{2\sqrt{-b}\} \cup a$ denotes the corresponding element of $\mathrm{H}^2(\mathrm{G}_{k_v}, \mathbb{Z}/2\mathbb{Z})$, which is independent of the choice of square root because

$$\{-1\} \cup a = \{-b\} \cup a = -\delta_2^{(\mathrm{mod}\,2,v)}(b,a) = 0\,.$$

Similar remarks hold in cases (ii) and (iii).

Note that obstructions such as $\{2\sqrt{-b}\} \cup a$ look as though they are naturally elements of $\mathrm{H}^2(\mathrm{G}_{\mathbb{Q}_p}, \mathbb{Z}/2\mathbb{Z}(2))$, but $\delta^{(2,p)}_{3,[[x,y],x]}(b,a)_c$ is in $\mathrm{H}^2(\mathrm{G}_{\mathbb{Q}_p}, \mathbb{Z}/2\mathbb{Z}(3))$. The shift in weight happened in Lemmas 11, 34, and 35.

Proof. Fix $\mathbb{C} \supset \overline{k} \subset \overline{k_v}$. It is sufficient to show that in Lemma 33 case (i) (ii) (iii) respectively, we have

$$\binom{b}{2} \cup a = \{2\sqrt{-b}\} \cup a\,,$$

$$\binom{a}{2} \cup b + \overline{f} \cup a = \{2\sqrt{-a}\} \cup b + \{2\} \cup a\,,$$

$$(\binom{a}{2} + \binom{b}{2} + \overline{f}) \cup a = \{2\sqrt{b}\sqrt{a}\} \cup a\,.$$

For cases (i) and (ii), this follows immediately from Lemmas 34 and 31. In case (iii), Propositions 35 and 31 show that

$$(\binom{a}{2} + \binom{b}{2} + \overline{f}) \cup a = \{2\sqrt{b}/\sqrt{a}\} \cup a\,.$$

Since $\sqrt{b}/\sqrt{a}$ is in k_v, we have that $a\sqrt{b}/\sqrt{a} = \sqrt{b}\sqrt{a}$ is in k_v. As $a \cup b = 0$ and $a = b$, it follows that $\{2\sqrt{b}/\sqrt{a}\} \cup a$ is also equal to $\{2\sqrt{b}/\sqrt{a}\} \cup a + a \cup a$, which in turn equals $\{2\sqrt{b}\sqrt{a}\} \cup a$. □

Corollary 38. *Let (b,a) be a rational point of* $\mathrm{Jac}(\mathbb{P}^1_{\mathbb{Q}} - \{0,1,\infty\})$ *such that (b,a) is in $(\mathbb{Z} - \{0\}) \times (\mathbb{Z} - \{0\})$ and p divides ab exactly once. Then:*

(1) $\delta_2^{(\mathrm{mod}\,2,p)}(b,a) = 0 \iff$ *$a+b$ is a square mod p*
(2) When (1) holds, $\delta_3^{(\mathrm{mod}\,2,p)}(b,a) = 0 \iff$ *$a+b$ is a fourth power mod p*

Remark 39. (1) Note that under the hypotheses of Corollary 38, the condition that $a+b$ is congruent to a (2^n)th power mod p is equivalent to the condition that whichever of a or b not divisible by p is a (2^n)th power mod p.

(2) For the points of the Jacobian satisfying the conditions described in its statement, Corollary 38 computes $\delta_3^{(\mathrm{mod}\,2,p)}$ and $\delta_2^{(\mathrm{mod}\,2,p)}$ in terms of congruence conditions mod p. These congruence conditions allow us to see that $\delta_3^{(\mathrm{mod}\,2,p)}$ and $\delta_2^{(\mathrm{mod}\,2,p)}$ vanish on the points and tangential points of the curve which satisfy the hypotheses of the Corollary. Namely, by (12.16) and Lemma 4, the image of $\mathbb{P}^1 - \{0,1,\infty\}$ and its tangential points is the set of (b,a) such that $a+b=0$, $a+b=1$, $a=1$, or $b=1$. As 0 and 1 are fourth powers mod every prime, we see the vanishing of Ellenberg's obstructions on the points of the curve.

(3) It is tempting to hope that under certain hypotheses

- $\delta_n^{(\mathrm{mod}\,2,p)}(b,a) = 0 \iff a+b$ is a 2^{n-1} power mod p.

Since 0 and 1 are the only integers which are 2^{n-1} powers for every n, mod every prime, such a result could show that the nilpotent completion of π determines the points and tangential points of $\mathbb{P}^1 - \{0,1,\infty\}$ from those of the Jacobian. This is a mod 2, pro-nilpotent section conjecture for $\mathbb{P}^1 - \{0,1,\infty\}$, cf. Sect. 12.4.4.

Proof. (1) Note that by hypothesis, exactly one of b and a equals $\mathfrak{p}$ or $u+\mathfrak{p}$ in $\mathrm{H}^1(\mathrm{G}_{\mathbb{Q}_p}, \mathbb{Z}/2\mathbb{Z}(1))$, where the notation $\mathfrak{p}$,u is as defined in Sect. 12.4.1. By Sect. 12.4.1, it follows that the other is 0 in $\mathrm{H}^1(\mathrm{G}_{\mathbb{Q}_p}, \mathbb{Z}/2\mathbb{Z}(1))$ if and only if

$b \cup a$ vanishes. By Hensel's Lemma, this is equivalent to the other being a square mod p, which in turn is equivalent to the condition that $a+b$ is a square mod p.

(2) Exactly one of b and a is not divisible by p. Call this element r. The only case listed in Theorem 36 that can hold is $\{-r\} = 0$ in $\mathrm{H}^1(\mathrm{G}_{\mathbb{Q}_p}, \mathbb{Z}/2\mathbb{Z}(1))$. By (1), we have that r is a square mod p, whence $\{-r\} = \{-1\}$ in $\mathrm{H}^1(\mathrm{G}_{\mathbb{Q}_p}, \mathbb{Z}/2\mathbb{Z}(1))$. Note also that if $r = a$, then $\overline{f} \cup a = \{2\} \cup a = 0$, as neither 2 nor a is divisible by p. Therefore, $\delta_3^{(\mathrm{mod}\,2,p)}(b,a) \neq 0$ if and only if $p = 1$ mod 4, and $\{2\sqrt{-r}\} \cup p \neq 0$.

For $p = 1$ mod 4, and r and $-r$ squares mod p,

$$\{2\sqrt{-r}\} \cup p = \{\sqrt{r}\} \cup p$$

in $\mathrm{H}^1(\mathrm{G}_{\mathbb{Q}_p}, \mathbb{Z}/2\mathbb{Z}(1))$ where either square root of r or $-r$ in $\mathbb{Q}_p$ can be chosen. To see this: note that since -1 is a square, it is clear that changing the square root has no effect. Note that $(1+\xi_4)^2 = 2\xi_4$, for ξ_4 a primitive fourth root of unity in $\mathbb{Q}_p$, from which it follows that

$$\{\sqrt{r}\} = \{\sqrt{r}(1+\xi_4)^2\} = \{2\sqrt{-r}\}\,.$$

In the last equality $\sqrt{-r}$ is $\xi_4\sqrt{r}$, but we are free to choose either square root to see the claimed equality.

Thus, $\delta_3^{(\mathrm{mod}\,2,p)}(b,a) \neq 0$ if and only if $p = 1$ mod 4, and $\{\sqrt{r}\} \cup p \neq 0$ in $\mathrm{H}^1(\mathrm{G}_{\mathbb{Q}_p}, \mathbb{Z}/2\mathbb{Z}(1))$. Note that $\{\sqrt{r}\} \cup p \neq 0$ if and only if $\{\sqrt{r}\} \neq 0$, since r is not divisible by p. The condition $p = 1$ mod 4 and $\{\sqrt{r}\} \neq 0$ is equivalent to the condition $p = 1$ mod 4 and r is not a fourth power mod p. Since r is a square mod p, the condition that r is not a fourth power implies that $p = 1$ mod 4. Thus, $\delta_3^{(\mathrm{mod}\,2,p)}(b,a) \neq 0$ if and only if r is not a fourth power mod p. This last condition is equivalent to $a+b$ is not a fourth power mod p. □

Remark 40. The proof of Corollary 38 only uses the computation of $\overline{f}$ given in Lemma 31 to ensure that $f \in \{0, \mathrm{u}\} \subset \mathrm{H}^1(\mathrm{G}_{\mathbb{Q}_p}, \mathbb{Z}/2\mathbb{Z})$.

Definition 41. For an obstruction δ' which is defined on the vanishing locus of an obstruction δ, we say that δ' **is not redundant with** δ if δ' does not vanish identically.

Example 42. We compare the obstruction $\delta_3^{\mathrm{mod}\,2}$ with δ_2 for $k = \mathbb{Q}$.

(1) As $4 = (-1)+5$ is a square but not a fourth power mod 5, Corollary 38 implies that $\delta_2^{(\mathrm{mod}\,2,5)}(-1,5) = 0$, and

$$\delta_3^{(\mathrm{mod}\,2,5)}(-1,5) \neq 0\,.$$

In other words, the 3-nilpotent obstruction $\delta_3^{(\mathrm{mod}\,2,p)}$ is not redundant with the 2-nilpotent obstruction $\delta_2^{(\mathrm{mod}\,2,p)}$.

(2) In fact, it is easy to check that $\{-1\} \cup \{5\} = 0$ in $\mathrm{H}^2(\mathrm{G}_{\mathbb{Q}}, \mathbb{Z}/2\mathbb{Z})$ since

$$\{-1\} \cup \{5\} = \{-1\} \cup \{5\} + \{1-5\} \cup \{5\} = \{4\} \cup \{5\} = 2 \cdot (\{2\} \cup \{5\}) = 0\,.$$

Alternatively, $\{-1\}\cup\{5\}=0$ in $\mathrm{H}^2(\mathrm{G}_{\mathbb{Q}},\mathbb{Z}/2\mathbb{Z})$ because the Brauer-Severi variety

$$-u^2+5v^2=w^2$$

has the rational point $[u,v,w]=[1,1,2]$. Thus, $\delta_3^{(\mathrm{mod}\,2,p)}$ is not redundant with the global obstruction $\delta_2^{\mathrm{mod}\,2}$. It also follows that the global 3-nilpotent obstruction $\delta_3^{\mathrm{mod}\,2}$ is not redundant with the global 2-nilpotent obstruction $\delta_2^{\mathrm{mod}\,2}$.

(3) One can ask whether $\delta_3^{(\mathrm{mod}\,2,p)}$ is redundant with the global obstruction δ_2 for $k=\mathbb{Q}$. The tame symbol at p

$$b\otimes a\mapsto (b,a)_p=(-1)^{v_p(b)v_p(a)}\frac{b^{v_p(a)}}{a^{v_p(b)}}\in\mathbb{F}_p^*$$

vanishes on any (b,a) such that $\delta_2(b,a)=0$. In particular, given $b,a\in\mathbb{Z}$ such that p divides $ab\neq 0$ exactly once, we will have that

$$\frac{b^{v_p(a)}}{a^{v_p(b)}}=1 \mod p$$

and that $\frac{b^{v_p(a)}}{a^{v_p(b)}}$ equals either b or $1/a$ depending on which of b or a is divisible by p. In particular, $a+b=1 \mod p$, and thus, Corollary 38 does not show that δ_3 is not redundant with δ_2 for $k=\mathbb{Q}$.

The points (b,a) of $\mathrm{Jac}(\mathbb{P}^1_{\mathbb{Q}}-\{0,1,\infty\})(\mathbb{Q})=\mathbb{Q}^*\times\mathbb{Q}^*$ considered in Corollary 38 and satisfying $\delta_2^{(\mathrm{mod}\,2,p)}(b,a)=0$ have the property that at any finite prime p, either b or a determines the 0 element of

$$\mathrm{C}^1(\mathrm{G}_{\mathbb{Q}_p},\mathbb{Z}/2\mathbb{Z})\cong\mathrm{H}^1(\mathrm{G}_{\mathbb{Q}_p},\mathbb{Z}/2\mathbb{Z})\,.$$

Thus a lift $(b,a)_c$ of (b,a) to a class of $\mathrm{H}^1(\mathrm{G}_{\mathbb{Q}_p},\pi/[\pi]_3^2)$ is such that c is a cocycle, as opposed to a cochain. By Theorem 19, Corollary 38 consists of evaluations of Massey products where certain cup products of cochains are not only coboundaries, but 0 as cochains. Indeed, a direct proof of Corollary 38 can be given along these lines, although the methods involved are not sufficiently different from those of Theorem 36 to merit inclusion.

Let p vary through the odd primes, and let m vary through the positive integers. The points $((-p)^{2m+1},p)$ satisfy $\delta_2=0$ by Proposition 24, but both $(-p)^{2m+1}$ and p determine non-zero elements of $\mathrm{C}^1(\mathrm{G}_{\mathbb{Q}_p},\mathbb{Z}/2\mathbb{Z})$ via the Kummer map, unlike the examples computed via Corollary 38. Theorem 36 allows us to evaluate $\delta_3^{(\mathrm{mod}\,2,p)}$ on these points.

Example 43. Let p be an odd prime and m a positive integer. Then $\delta_3^{(\mathrm{mod}\,2,p)}$ vanishes on $(-p^{2m+1},p)$ and (p^{2m},p) in $\mathrm{Jac}(\mathbb{P}^1_{\mathbb{Q}}-\{0,1,\infty\})(\mathbb{Q})$. For $(-p^{2m+1},p)$ we show that neither case (i–iii) of Theorem 36 holds. Note that $\{p^{2m+1}\}$ (resp. $\{-p\}$) is nontrivial in $\mathrm{H}^1(\mathrm{G}_{\mathbb{Q}_p},\mathbb{Z}/2\mathbb{Z})$, so case (i) (resp. case (ii)) does not hold. When $p \equiv 3 \bmod 4$, the class $\{-p^{2m+1}\cdot p\} = \{-1\}$ is nontrivial in $\mathrm{H}^1(\mathrm{G}_{\mathbb{Q}_p},\mathbb{Z}/2\mathbb{Z})$ and case (iii) does not hold.

For $p \equiv 1 \bmod 4$, we have a fourth root of unity $\zeta_4 \in \mathbb{Q}_p$ and thus the product $-p^{2m+1}\cdot p$ is a square in $\mathbb{Q}_p$. In this case the first equation of (iii) is satisfied, but the second is not:

$$\{2\sqrt{-p^{2m+2}}\}\cup p = \{2p^{m+1}\zeta_4\}\cup p = \{2\zeta_4\}\cup p = \{(1+\zeta_4)^2\}\cup p = 0$$

because $p \cup p = 0$ by Sect. 12.4.1.

For (p^{2m}, p) one shows similarly that $\delta_3^{(\mathrm{mod}\,2,p)}(p^{2m}, p) = 0$. Now cases (ii) and (iii) do not apply for obvious reasons and (i) can at most apply for $p \equiv 1 \bmod 4$. But then

$$\{2\sqrt{-p^{2m}}\}\cup\{p\} = (\{2\zeta_4\} + m\{p\})\cup\{p\} = \{(1+\zeta_4)^2\}\cup p = 0$$

vanishes as well.

Example 44. Let p be a prime congruent to 3 mod 4. Let $x \in \mathbb{Z} - \{0,1\}$ be divisible by p. Then

$$\delta_3^{(\mathrm{mod}\,2,p)}((1-x)(-x), x) = 0\,.$$

Note that by Proposition 24, the point $((1-x)(-x), x)$ is in the kernel of $\delta_2^{(\mathrm{mod}\,2,p)}$, and even in the kernel of δ_2.

We again show that neither case (i)-(iii) of Theorem 36 holds. The element $1-x$ is a square in $\mathbb{Q}_p$, as $1-x \equiv 1 \bmod p$, whence

$$\{(1-x)(-x)\} = \{-x\} \in \mathrm{H}^1(\mathrm{G}_{\mathbb{Q}_p},\mathbb{Z}/2\mathbb{Z})\,.$$

Since $p \equiv 3 \bmod 4$, the class $\{-1\}$ is nonzero in $\mathrm{H}^1(\mathrm{G}_{\mathbb{Q}_p},\mathbb{Z}/2\mathbb{Z})$. Therefore case (iii) does not hold. Case (ii) holds if and only if $\{-x\} = 0$ and

$$\{2\sqrt{-x}\}\cup\{(1-x)(-x)\} + \{2\}\cup\{x\} \neq 0\,.$$

Since $(-x, x)$ is the image of a rational tangential base point by Lemma 4, the obstruction $\delta_3^{(\mathrm{mod}\,2,p)}(-x, x) = 0$ vanishes. By Theorem 36 case (ii), this implies that

$$\{2\sqrt{-x}\}\cup\{(-x)\} + \{2\}\cup\{x\} = 0$$

when $\{-x\} = 0$, so (ii) does not hold for $((1-x)(-x), x)$, because

$$\{(1-x)(-x)\} = \{(-x)\}\,.$$

Case (i) holds if and only if $\{(1-x)(x)\} = \{x\} = 0$ and

$$\{2\sqrt{(1-x)x}\}\cup\{x\} \neq 0$$

which is impossible.

12.5.5 A Global mod 2 Calculation

The local calculations of Sect. 12.5.4 and Sect. 12.5.3 allow us to evaluate the global obstruction $\delta_3^{\mathrm{mod}\,2}$ on $(-p^3, p)$ in $\mathrm{Jac}(\mathbb{P}^1_{\mathbb{Q}} - \{0,1,\infty\})(\mathbb{Q})$. This evaluation relies on the Hasse–Brauer–Noether Theorem for the Brauer group, see [NSW08, Thm. 8.1.17],

$$0 \to \mathrm{H}^2(\mathrm{G}_{\mathbb{Q}}, \mathbb{Z}/2\mathbb{Z}) \to \bigoplus_v \mathrm{H}^2(\mathrm{G}_{\mathbb{Q}_p}, \mathbb{Z}/2\mathbb{Z}) \xrightarrow{\Sigma_v \mathrm{inv}_v} \frac{1}{2}\mathbb{Z}/\mathbb{Z} \to 0 \tag{12.40}$$

but is more subtle, as the evaluation of $\delta_3^{\mathrm{mod}\,2}$ over $\mathbb{Q}$ depends on the lifts of a point of the Jacobian to $\mathrm{H}^1(\mathrm{G}_{\mathbb{Q}}, \pi/[\pi]_3^2)$, whereas each evaluation of $\delta_3^{(\mathrm{mod}\,2,p)}$ depends on the lifts to $\mathrm{H}^1(\mathrm{G}_{\mathbb{Q}_p}, \pi/[\pi]_3^2)$. One may not be able to find a global lift to restricting to some given set of local lifts.

In Proposition 30, it was shown that

$$\delta_3^{\mathrm{mod}\,2}(b, (1-b)^4) = 0$$

for k a number field and b in $k - \{0,1\} = \mathbb{P}^1_k - \{0,1,\infty\}(k)$, giving a calculation of the global obstruction $\delta_3^{\mathrm{mod}\,2}$ on a point of the Jacobian not lying on the curve or coming from a tangential base point. However, the point $(b, (1-b)^4)$ determines the same element of $\mathrm{H}^1(\mathrm{G}_{\mathbb{Q}}, \pi/([\pi]_2\pi^4))$ as the point $(b, 1)$ which is the image of a rational tangential point by Lemma 4, so this vanishing is trivial.

We now let p vary through the primes congruent to 1 mod 4, and evaluate $\delta_3^{\mathrm{mod}\,2}$ on the family of points $(-p^3, p)$. Note that $(-p^3, p)$ does not determine the same element of $\mathrm{H}^1(\mathrm{G}_{\mathbb{Q}}, \pi/([\pi]_2\pi^4))$ as a rational point or tangential point of $\mathbb{P}^1_{\mathbb{Q}} - \{0,1,\infty\}$ by (12.16) and Lemma 4, so this gives a nontrivial calculation of $\delta_3^{\mathrm{mod}\,2}$ over $\mathbb{Q}$.

Proposition 45. *Let p be a prime congruent to 5 mod 8. Consider $(-p^3, p)$ in $\mathrm{Jac}(\mathbb{P}^1_{\mathbb{Q}} - \{0,1,\infty\})(\mathbb{Q})$. Then $\delta_3^{\mathrm{mod}\,2}(-p^3, p) = 0$.*

Proof. We evaluate the obstruction as triple Massey products with compatible defining systems by Theorem 19

$$\begin{aligned} \delta^2_{3,[[x,y],x]}(-p^3, p) &= \langle p^3, \{-p^3\}, p\rangle \\ \delta^2_{3,[[x,y],y]}(-p^3, p) &= -\langle \{-p\}, p, \{-p^3\}\rangle - \{2\} \cup p \end{aligned} \tag{12.41}$$

valued in $\mathrm{H}^2(\mathrm{G}_{\mathbb{Q}}, \mathbb{Z}/2\mathbb{Z})$, using Lemma 31 to evaluate $\overline{f}$.

Let $\mathrm{S} = \{2, p, \infty\}$ and let $\mathbb{Q}_{\mathrm{S}}$ denote the maximal extension of $\mathbb{Q}$ unramified outside S. Then all classes $\{-1\}$, $\{\pm p\}$, $\{2\}$, etc. involved in (12.41) are unramified outside S, i.e., they already lie in $\mathrm{H}^1(\mathrm{Gal}(\mathbb{Q}_{\mathrm{S}}/\mathbb{Q}), \mathbb{Z}/2\mathbb{Z})$. The map

$$\mathrm{H}^2(\mathrm{Gal}(\mathbb{Q}_{\mathrm{S}}/\mathbb{Q}), \mathbb{Z}/2\mathbb{Z}) \hookrightarrow \mathrm{H}^2(\mathrm{G}_{\mathbb{Q}}, \mathbb{Z}/2\mathbb{Z}) \tag{12.42}$$

is injective. One way to see this injectivity is: let $\mathcal{O}_{\mathbb{Q},\mathrm{S}}$ denote the S integers of $\mathbb{Q}$ and let $\mathrm{U} = \mathrm{Spec}\,\mathcal{O}_{\mathbb{Q},\mathrm{S}}$. The étale cohomology groups $\mathrm{H}^*(\mathrm{U}, \mathbb{Z}/2\mathbb{Z})$ are isomorphic to the Galois cohomology groups $H^*(\mathrm{Gal}(\mathbb{Q}_S/\mathbb{Q}),\mathbb{Z}/2\mathbb{Z})$ by [Hab78, Appendix 2 Prop 3.3.1], and by the Kummer exact sequence, the sequence

$$H^1(U,\mathbb{G}_m)\to H^2(U,\mathbb{Z}/2\mathbb{Z})\to H^2(U,\mathbb{G}_m)$$

is exact in the middle. Since $O_{\mathbb{Q},S}$ is a principal ideal domain, $H^1(U,\mathbb{G}_m)=0$, and by [Mil80, III 2.22], the natural map $H^2(U,\mathbb{G}_m)\to H^2(G_{\mathbb{Q}},\mathbb{G}_m)$ is an injection. Thus (12.42) is injective. Its image consists of the classes whose image under (12.40) have vanishing local component except possibly at 2,p and ∞. It follows we can restrict to defining systems of cochains for $\mathrm{Gal}(\mathbb{Q}_S/\mathbb{Q})$. Thus the Massey product takes values in $H^2(\mathrm{Gal}(\mathbb{Q}_S/\mathbb{Q}),\mathbb{Z}/2\mathbb{Z})$ and the local components for primes not in S vanish a priori.

We will show the vanishing of a global lift at p and ∞, and deduce the vanishing at 2 from reciprocity (12.40).

Since $p\equiv 5 \bmod 8$, we have that 2 is not a quadratic residue mod p, so $\{2\}$ and $\{p\}$ span $H^1(G_{\mathbb{Q}_p},\mathbb{Z}/2\mathbb{Z})$, as in Sect. 12.4.1. Thus the set of lifts $(-p^3,p)_c$ where c varies among the cochains factoring through $\mathrm{Gal}(\mathbb{Q}_S/\mathbb{Q})$ surjects onto the set of all lifts of $(-p^3,p)$ to $H^1(G_{\mathbb{Q}_p},\pi/[\pi]_3^2)$. By Example 43, we can therefore choose such a lift such that $\delta_3^{(2,p)}(-p^3,p)_c=0$. Since p is congruent to 1 mod 4, the restriction of $\{-1\}$ to $C^1(G_{\mathbb{Q}_p},\mathbb{Z}/2\mathbb{Z})$ is 0. By Proposition 32, either $\delta_3^{\mathrm{mod}\,2}(-p^3,p)_c=0$ or $\delta_3^{\mathrm{mod}\,2}(-p^3,p)_{c+\{-1\}}=0$, so we can choose a lift factoring through $\mathrm{Gal}(\mathbb{Q}_S/\mathbb{Q})$ such that $\delta_3^{\mathrm{mod}\,2}$ vanishes at both p and ∞. □

Remark 46. (1) In the proof of Proposition 45, the vanishing of $\delta_3^{\mathrm{mod}\,2}(-p^3,p)$ was shown for $p\equiv 5 \bmod 8$ by using the local vanishing of $\delta_3^{(\mathrm{mod}\,2,v)}(-p^3,p)$ and showing that the global lifts of $(-p^3,p)$ to $H^1(G_{\mathbb{Q}},\pi/[\pi]_3^2)$ with ramification constrained to lie above $S=\{2,p,\infty\}$ surjected onto the product over the local lifts of $(-p^3,p)$ to $H^1(G_{\mathbb{Q}_v},\pi/[\pi]_3^2)$ for the places p and ∞.

(2) It would be desirable to relate $\delta_3^{\mathrm{mod}\,2}=0$ to the simultaneous vanishing of all (or all but one) $\delta_3^{(2,v)}$, where v varies over the places of a given number field k. For this, we would need to compare the set of restrictions to the k_v of lifts of (b,a) to $H^1(G_k,\pi/[\pi]_3^2)$ with the set of independently chosen lifts of (b,a) to $H^1(G_{k_v},\pi/[\pi]_3^2)$ for all v. In other words, we are interested in the map

$$H^1_{(b,a)}(G_k,\pi/[\pi]_3^2)\to\prod_v H^1_{(b,a)}(G_{k_v},\pi/[\pi]_3^2) \qquad (12.43)$$

where $H^1_{(b,a)}(G_k,\pi/[\pi]_3^2)$ denotes the subset of $H^1(G_k,\pi/[\pi]_3^2)$ of lifts of (b,a) and similarly for each $H^1_{(b,a)}(G_{k_v},\pi/[\pi]_3^2)$. A nonabelian version of Poitou-Tate duality would give information about (12.43).

We can also evaluate $\delta^2_{3,[[x,y],x]}$ and $\delta^2_{3,[[x,y],y]}$ on a specific lift of $(-p^3,p)$, which is equivalent to the calculation of the Massey products $\langle p^3,-p^3,p\rangle$ and $\langle -p,p,-p^3\rangle$ with the defining systems specified in Remark 20 (ii), and the mod 2 cup product

$\{2\} \cup \{p\}$. The cup product $\{2\} \cup \{p\}$ can be calculated with Sect. 12.4.3; it vanishes except at 2 and p, and at p, $\{2\} \cup \{p\}$ vanishes if and only if $p \equiv \pm 1$ mod 8. An arbitrary defining system for $\langle p^3, -p^3, p\rangle$ or $\langle -p, p, -p^3\rangle$ produces Massey products differing from the originals by cup products, which can also be evaluated with Sect. 12.4.3. So evaluating $\delta^2_{3,[[x,y],x]}$ and $\delta^2_{3,[[x,y],y]}$ on a specific lift allows for the computation of $\langle p^3, -p^3, p\rangle$ and $\langle -p, p, -p^3\rangle$ in $\mathrm{H}^2(\mathrm{G}_{\mathbb{Q}}, \mathbb{Z}/2\mathbb{Z})$ with any defining system. We remark that a complete computation of the triple Massey product on $\mathrm{H}^1(\mathrm{Gal}(k_S(2)/k), \mathbb{Z}/2\mathbb{Z})$ for certain maximal 2-extensions with restricted ramification $k_S(2)$ of a number field k is given in [Vog04, II §1].

Remark 47. Note that $\{-p^3\} = \{-p\}$ and $\{p^3\} = \{p\}$ in $\mathrm{H}^1(\mathrm{G}_{\mathbb{Q}}, \mathbb{Z}/2\mathbb{Z})$. The reason for distinguishing between, say, $-p^3$ and $-p$ in $\langle p^3, -p^3, p\rangle$ is that the defining systems to evaluate $\delta^2_{3,[[x,y],x]}$ and $\delta^2_{3,[[x,y],y]}$ are different for $-p^3$ and $-p$, as they depend on the image of $-p^3$ in $\mathrm{H}^1(\mathrm{G}_{\mathbb{Q}}, \mathbb{Z}/4\mathbb{Z}(1))$. However, for the discussion of evaluating triple Massey products of elements of $\mathrm{H}^1(\mathrm{G}_{\mathbb{Q}}, \mathbb{Z}/2\mathbb{Z})$ for any defining system, the distinction is of course irrelevant.

Consider the following lift of $(-p^3, p)$: choose compatible nth roots of p, and let $\{p\}$ denote the corresponding element of $\mathrm{C}^1(\mathrm{G}_{\mathbb{Q}}, \hat{\mathbb{Z}}(1))$ via the Kummer map. As above, $\{p\}$ will sometimes be abbreviated by p. Note that the chosen nth roots of p give rise to a choice of compatible nth roots of $-p^3$ such that the corresponding element of $\mathrm{C}^1(\mathrm{G}_{\mathbb{Q}}, \hat{\mathbb{Z}}(1))$ is $3(p + \frac{\chi - 1}{2})$. It is therefore consistent to let $\{-p^3\}$ and $-p^3$ denote $3(p + \frac{\chi-1}{2})$. Let $c_0 = 3\binom{p}{2}$ in $\mathrm{C}^1(\mathrm{G}_{\mathbb{Q}}, \hat{\mathbb{Z}}(2))$. Let $(-p^3, p)_{c_0}$ in $\mathrm{C}^1(\mathrm{G}_{\mathbb{Q}}, \pi/[\pi]_3)$ be as in Corollary 8, i.e., for all g in $\mathrm{G}_{\mathbb{Q}}$

$$(-p^3, p)_{c_0}(g) = y^{\{p\}(g)} x^{\{-p^3\}(g)} [x,y]^{c_0(g)} ,$$

so $(-p^3, p)_{c_0}$ is a cocycle lifting $(-p^3, p)$. The image of $(-p^3, p)_{c_0}$ under the map $\mathrm{C}^1(\mathrm{G}_{\mathbb{Q}}, \pi/[\pi]_3) \to \mathrm{C}^1(\mathrm{G}_{\mathbb{Q}}, \pi/[\pi]_3^2)$ will also be denoted $(-p^3, p)_{c_0}$.

Proposition 48. *Let p be a prime congruent to* 1 *mod* 4. *Let* $(-p^3, p)_{c_0}$ *be as above.* $\delta_3^{\mathrm{mod}\,2}(-p^3, p)_{c_0}$ *is the element of* $\mathrm{H}^2(\mathrm{G}_{\mathbb{Q}}, [\pi]_3/[\pi]_4([\pi]_3)^2)$ *determined by*

$$\delta^{(2,p)}_{3,[[x,y],x]}(-p^3, p)_{c_0} = \delta^{(2,p)}_{3,[[x,y],y]}(-p^3, p)_{c_0} = 2 \cup p = \begin{cases} \frac{1}{2} & \text{if } p \equiv 5 \mod 8, \\ 0 & \text{if } p \equiv 1 \mod 8. \end{cases}$$

$$\delta^{(2,v)}_{3,[[x,y],x]}(-p^3, p)_{c_0} = \delta^{(2,v)}_{3,[[x,y],y]}(-p^3, p)_{c_0} = 0$$

for v equal to $\mathbb{R}$ *or a finite odd prime not equal to p.*

Here, $\mathrm{H}^2(\mathrm{G}_{\mathbb{Q}_p}, \mathbb{Z}/2\mathbb{Z})$ is identified with the two torsion of $\mathbb{Q}/\mathbb{Z}$ for all finite primes p via the invariant map, and elements of $\mathrm{H}^2(\mathrm{G}_{\mathbb{Q}}, \mathbb{Z}/2\mathbb{Z})$ are identified with their images under (12.40).

Proof. Let $\mathrm{S} = \{2, p, \infty\}$ and let $\mathbb{Q}_{\mathrm{S}}$ denote the maximal extension of $\mathbb{Q}$ unramified outside S. The cocycle $(-p^3, p)_{c_0}$ factors through $\mathrm{Gal}(\mathbb{Q}_{\mathrm{S}}/\mathbb{Q})$, and it follows that

$$\delta^{(2,v)}_{3,[[x,y],x]}(-p^3,p)_{c_0} = \delta^{(2,v)}_{3,[[x,y],y]}(-p^3,p)_{c_0} = 0$$

for v equal to any prime not in S.

The obstruction $\delta_3^{(2,v)}(-p^3,p)_{c_0}$ for $v=\mathbb{R}$ decomposes into two elements

$$\delta^{(2,\mathbb{R})}_{3,[[x,y],x]}(-p^3,p)_{c_0} \quad \text{and} \quad \delta^{(2,\mathbb{R})}_{3,[[x,y],y]}(-p^3,p)_{c_0}$$

of $\mathbb{Z}/2\mathbb{Z} \cong \mathrm{H}^2(\mathrm{G}_{\mathbb{R}},\mathbb{Z}/2\mathbb{Z})$, obtained by evaluating each of the cocycles given in Proposition 13 at $(g_1,g_2)=(\tau,\tau)$, where τ denote complex conjugation in $\mathrm{G}_{\mathbb{Q}}$, c.f. Sect. 12.4.2. Note that since p is positive, the equalities $\{-p^3\}(\tau)=1$, $\{p\}(\tau)=0$, $\binom{-\{p\}(\tau)+1}{2}=0$, and $c_0(\tau)=0$ hold in $\mathbb{Z}/2\mathbb{Z}$. Substituting these equations into the cocycles in Proposition 13 shows that

$$\delta^{(2,\mathbb{R})}_{3,[[x,y],x]}(-p^3,p)_{c_0} = \delta^{(2,\mathbb{R})}_{3,[[x,y],y]}(-p^3,p)_{c_0} = 0\,.$$

By Lemma 33 case (3), we have that

$$\delta^{(2,p)}_{3,[[x,y],x]}(-p^3,p)_c + \delta^{(2,p)}_{3,[[x,y],y]}(-p^3,p)_c \tag{12.44}$$

does not depend on the choice of lift. By Example 43, there is a lift $(-p^3,p)_c$ such that

$$\delta^{(2,p)}_{3,[[x,y],x]}(-p^3,p)_c = \delta^{(2,p)}_{3,[[x,y],y]}(-p^3,p)_c = 0$$

and it follows that (12.44) vanishes. It follows from Proposition 13 that

$$\delta^{(2,p)}_{3,[[x,y],x]}(-p^3,p)_{c_0} = (\binom{p}{2} + \binom{-(p+\frac{\chi-1}{2})}{2}) \cup p. \tag{12.45}$$

To see this, note that $\frac{\chi-1}{2}=0$ and $-p^3=p$ in $\mathrm{C}^1(\mathrm{G}_{\mathbb{Q}_p},\mathbb{Z}/2\mathbb{Z})$. Thus the cocycle

$$g \mapsto p(g)3(p+\frac{\chi-1}{2})(g)$$

equals the cocycle $g \mapsto p(g)$ in $\mathrm{C}^1(\mathrm{G}_{\mathbb{Q}_p},\mathbb{Z}/2\mathbb{Z})$. Note that equating these two cocycles requires identifying $\mathbb{Z}/2\mathbb{Z}(1)$ with $\mathbb{Z}/2\mathbb{Z}(2)$, so the weight is not being respected. Substituting these equalities into Proposition 13 implies

$$\delta^{(2,p)}_{3,[[x,y],x]}(-p^3,p)_{c_0} = \binom{p}{2} \cup p + \binom{3(p+\frac{\chi-1}{2})+1}{2} \cup p + p \cup p\,.$$

Then note that in $\mathrm{C}^1(\mathrm{G}_{\mathbb{Q}_p},\mathbb{Z}/2\mathbb{Z})$,

$$\binom{3(p+\frac{\chi-1}{2})+1}{2} = \binom{3(p+\frac{\chi-1}{2})}{2} + 3(p+\frac{\chi-1}{2}) = \binom{-(p+\frac{\chi-1}{2})}{2} + p\,,$$

showing (12.45). Here it is important to distinguish between $3(p+\frac{\chi-1}{2})$ and p in the binomial coefficient as these two cocycles are not equal in $C^1(G_{\mathbb{Q}_p}, \mathbb{Z}/4\mathbb{Z}(1))$.

For any two elements d_1, d_2 in $\mathbb{Z}/4\mathbb{Z}$, direct calculation shows

$$\binom{d_1+d_2}{2} - \binom{d_1}{2} - \binom{d_2}{2} = d_1 d_2$$

in $\mathbb{Z}/2\mathbb{Z}$. Thus

$$\binom{p}{2} + \binom{-(p+\frac{\chi-1}{2})}{2} = \binom{-\frac{\chi-1}{2}}{2} + p(p+\frac{\chi-1}{2}) = \binom{-\frac{\chi-1}{2}}{2} + p\,.$$

Combining with the above, we see

$$\delta^2_{3,[[x,y],x]}(-p^3, p)_{c_0} = \binom{-\frac{\chi-1}{2}}{2} \cup p\,.$$

Since p is congruent to 1 mod 4, $\mathbb{Q}_p$ contains a primitive fourth root of unity and $\chi(g) \equiv 1 \mod 4$ for every g in $G_{\mathbb{Q}_p}$. Therefore, $\frac{\chi(g)-1}{2}$ is 0 or 2 mod 4, whence $\binom{-\frac{\chi-1}{2}}{2}$ is 0 for g fixing the eight roots of unity and 1 otherwise. It follows that

$$\delta^2_{3,[[x,y],x]}(-p^3, p)_{c_0} = \begin{cases} \frac{1}{2} & \text{if } p \equiv 5 \mod 8\,, \\ 0 & \text{if } p \equiv 1 \mod 8\,. \end{cases}$$

□

References

[And89] G. W. Anderson. The hyperadelic gamma function. *Invent. Math.*, 95(1):63–131, 1989.

[Bro94] K. S. Brown. *Cohomology of groups*, volume 87 of *Graduate Texts in Mathematics*. Springer, New York, 1994. Corrected reprint of the 1982 original.

[CF67] *Algebraic number theory*. Proceedings of an instructional conference organized by the London Mathematical Society (a NATO Advanced Study Institute) with the support of the International Mathematical Union. Edited by J. W. S. Cassels and A. Fröhlich. Academic Press, London, 1967.

[Col89] R. F. Coleman. Anderson-Ihara theory: Gauss sums and circular units. In *Algebraic number theory*, volume 17 of *Adv. Stud. Pure Math.*, pages 55–72. Academic Press, Boston, MA, 1989.

[Del89] P. Deligne. Le groupe fondamental de la droite projective moins trois points. In *Galois groups over* $\mathbb{Q}$ *(Berkeley, CA, 1987)*, volume 16 of *Math. Sci. Res. Inst. Publ.*, pages 79–297. Springer, New York, 1989.

[Dwy75] W. G. Dwyer. Homology, Massey products and maps between groups. *J. Pure Appl. Algebra*, 6(2):177–190, 1975.

[Ell00] J. Ellenberg. 2-nilpotent quotients of fundamental groups of curves. Preprint, 2000.

[Hab78] K. Haberland. *Galois cohomology of algebraic number fields*. VEB Deutscher Verlag der Wissenschaften, Berlin, 1978. With two appendices by Helmut Koch and Thomas Zink, 145.

[Hat02] A. Hatcher. *Algebraic topology*. Cambridge University Press, Cambridge, 2002.

[Hos10] Y. Hoshi. Existence of nongeometric pro-p Galois sections of hyperbolic curves. *Publ. RIMS Kyoto Univ.* 46:1–19, 2010.

[Iha91] Y. Ihara. Braids, Galois groups, and some arithmetic functions. In *Proceedings of the International Congress of Mathematicians, Vol. I, II (Kyoto, 1990)*, pages 99–120, Tokyo, 1991. Math. Soc. Japan.

[Iha94] Y. Ihara. On the embedding of Gal($\overline{\mathbf{Q}}/\mathbf{Q}$) into $\widehat{GT}$. In *The Grothendieck theory of dessins d'enfants (Luminy, 1993)*, volume 200 of *London Math. Soc. Lecture Note Ser.*, pages 289–321. Cambridge Univ. Press, Cambridge, 1994. With an appendix: the action of the absolute Galois group on the moduli space of spheres with four marked points by Michel Emsalem and Pierre Lochak.

[IKY87] Y. Ihara, M. Kaneko, and A. Yukinari. On some properties of the universal power series for Jacobi sums. In *Galois representations and arithmetic algebraic geometry (Kyoto, 1985/Tokyo, 1986)*, volume 12 of *Adv. Stud. Pure Math.*, pages 65–86. North-Holland, Amsterdam, 1987.

[Laz54] M. Lazard. Sur les groupes nilpotents et les anneaux de Lie. *Ann. Sci. Ecole Norm. Sup. (3)*, 71:101–190, 1954.

[Mil71] J. Milnor. *Introduction to algebraic* K*-theory*. Princeton University Press, Princeton, N.J., 1971. Annals of Mathematics Studies, No. 72.

[Mil80] J. S. Milne. *Étale cohomology*, volume 33 of *Princeton Mathematical Series*. Princeton University Press, Princeton, N.J., 1980.

[MKS04] W. Magnus, A. Karrass, and D. Solitar. *Combinatorial group theory*. Dover Publications Inc., Mineola, NY, second edition, 2004. Presentations of groups in terms of generators and relations.

[Nak99] H. Nakamura. Tangential base points and Eisenstein power series. In *Aspects of Galois theory (Gainesville, FL, 1996)*, volume 256 of *London Math. Soc. Lecture Note Ser.*, pages 202–217. Cambridge Univ. Press, Cambridge, 1999.

[NSW08] J. Neukirch, A. Schmidt, and K. Wingberg. *Cohomology of number fields*, volume 323 of *Grundlehren der Mathematischen Wissenschaften [Fundamental Principles of Mathematical Sciences]*. Springer-Verlag, Berlin, second edition, 2008.

[Pop10] F. Pop. On the birational p-adic section conjecture. *Compos. Math.*, 146(3):621–637, 2010.

[Ser88] J.-P. Serre. *Algebraic groups and class fields*, volume 117 of *Graduate Texts in Mathematics*. Springer, New York, 1988. Translated from the French, x+207.

[Ser02] J.-P. Serre. *Galois cohomology*. Springer Monographs in Mathematics. Springer, Berlin, english edition, 2002. Translated from the French by Patrick Ion and revised by the author.

[SGA1] A. Grothendieck. *Revêtements étale et groupe fondamental (SGA 1)*. Séminaire de géométrie algébrique du Bois Marie 1960-61, directed by A. Grothendieck, augmented by two papers by Mme M. Raynaud, *Lecture Notes in Math.* 224, Springer-Verlag, Berlin-New York, 1971. Updated and annotated new edition: *Documents Mathématiques* 3, Société Mathématique de France, Paris, 2003.

[Tat76] J. Tate. Relations between K_2 and Galois cohomology. *Invent. Math.*, 36:257–274, 1976.

[Vog04] D. Vogel. Massey products in the Galois cohomology of number fields. Thesis: Universität Heidelberg, 2004.

[Wic10] K. Wickelgren. 2-nilpotent real section conjecture. Preprint arXiv:1006.0265, 2010.

[Zar74] Yu. G. Zarkhin. Noncommutative cohomology and Mumford groups. *Mat. Zametki*, 15:415–419, 1974.

Chapitre 13
Une remarque sur les courbes de Reichardt–Lind et de Schinzel

Olivier Wittenberg

Résumé. We prove that the arithmetic fundamental group of X admits no section over the absolute Galois group of $\mathbb{Q}$ when X is the Schinzel curve, thereby confirming in this example the prediction given by Grothendieck's section conjecture.

13.1 Introduction

Les courbes de Reichardt–Lind [Rei42, Lin40] et de Schinzel [Sch84] sont deux exemples célèbres de courbes projectives et lisses X sur $\mathbb{Q}$ possédant des points réels et des points p-adiques pour tout premier p, mais pas de point rationnel ni même de diviseur de degré 1. La première est la courbe de genre 1 intersection dans $\mathbb{P}^3$ des quadriques d'équations $w^2 = xz$ et $2y^2 = x^2 - 17z^2$. La seconde est de genre 3 : il s'agit de la courbe quartique plane d'équation $x^4 - 17z^4 = 2(y^2 + 4z^2)^2$. Dans cette note, à l'aide des résultats de [EW09] concernant l'algébricité des classes de cycles de sections du groupe fondamental arithmétique sur les corps locaux, nous établissons le théorème suivant :

Théorème 1. *Soit* X *la courbe de Reichardt–Lind ou la courbe de Schinzel. La suite exacte fondamentale*

$$1 \longrightarrow \pi_1(\mathrm{X} \otimes_{\mathbb{Q}} \overline{\mathbb{Q}}) \longrightarrow \pi_1(\mathrm{X}) \longrightarrow \mathrm{Gal}(\overline{\mathbb{Q}}/\mathbb{Q}) \longrightarrow 1 \qquad (*)$$

n'est pas scindée.

O. Wittenberg (✉)
Département de mathématiques et applications, École normale supérieure,
45 rue d'Ulm, 75230 Paris Cedex 05, France
e-mail: wittenberg@dma.ens.fr

J. Stix (ed.), *The Arithmetic of Fundamental Groups*, Contributions in Mathematical and Computational Sciences 2, DOI 10.1007/978-3-642-23905-2_13,

Dans le cas de la courbe de Schinzel, le Théorème 1 confirme la prédiction donnée par la conjecture des sections de Grothendieck [Gro83]. Dans le cas de la courbe de Reichardt–Lind, il répond à une question posée par Stix [Sti11].

La courbe quartique de Schinzel est ainsi le second exemple connu d'une courbe projective et lisse X sur $\mathbb{Q}$, de genre $\geqslant 2$, telle que la suite $(*)$ ne soit pas scindée bien que $X(\mathbb{A}_{\mathbb{Q}}) \neq \emptyset$. Un premier exemple avait en effet été construit par Harari, Szamuely et Flynn [HS09]. Les arguments de [HS09] requièrent la connaissance d'informations fines sur l'arithmétique de la jacobienne de X (entre autres : finitude du groupe de Tate–Shafarevich, rang). Il n'en est pas ainsi de la preuve du Théorème 1, dont le seul ingrédient arithmétique global est la loi de réciprocité pour le groupe de Brauer de $\mathbb{Q}$.

Au paragraphe 13.2 de cette note, nous donnons un critère général pour que le groupe fondamental d'une courbe projective et lisse sur un corps de nombres n'admette pas de section, sous l'hypothèse qu'une obstruction de Brauer–Manin s'oppose à l'existence d'un diviseur de degré 1 sur cette courbe. Nous démontrons ensuite le Théorème 1, à l'aide de ce critère, au paragraphe 13.3.

Remerciements. Le contenu de cette note fut en partie exposé à Heidelberg lors de la conférence PIA 2010 organisée par Jakob Stix, que je remercie pour son hospitalité. Je remercie également Tamás Szamuely pour ses commentaires utiles sur une première version du texte.

Notations. Si M est un groupe abélien, on note M_{tors} le sous-groupe de torsion de M. Tous les groupes de cohomologie apparaissant ci-dessous sont des groupes de cohomologie étale. Si X est une courbe irréductible, projective et lisse sur un corps k, on note $\mathrm{Pic}^1(X)$ le sous-ensemble de $\mathrm{Pic}(X)$ constitué des classes de degré 1 et $\mathrm{Br}(X) = H^2(X, \mathbb{G}_m)$ le groupe de Brauer de X. Lorsque k est un corps de nombres, on désigne par Ω l'ensemble des places de k et, pour $v \in \Omega$, par k_v le complété de k en v. Dans cette situation, Manin a défini un accouplement

$$\left(\prod_{v\in\Omega} \mathrm{Pic}(X \otimes_k k_v)\right) \times \mathrm{Br}(X) \to \mathbb{Q}/\mathbb{Z}, \tag{13.1}$$

somme d'accouplements locaux

$$\mathrm{Pic}(X \otimes_k k_v) \times \mathrm{Br}(X \otimes_k k_v) \to \mathbb{Q}/\mathbb{Z}; \tag{13.2}$$

le noyau à gauche de (13.1) contient l'image diagonale de Pic(X), cf. [Sai89, (8-2)]. On dit qu'un sous-groupe $B \subset \mathrm{Br}(X)$ (resp. une classe $A \in \mathrm{Br}(X)$) est **responsable d'une obstruction de Brauer–Manin à l'existence d'un diviseur de degré** 1 **sur** X si aucun élément de $\prod_{v\in\Omega} \mathrm{Pic}^1(X \otimes_k k_v)$ n'est orthogonal à B (resp. à A) pour (13.1).

13.2 Un critère pour l'absence de sections sur un corps de nombres

Dans ce paragraphe, nous supposons donnés un corps de nombres k, une courbe X projective, lisse et géométriquement irréductible sur k, un entier $\mathrm{N} \geqslant 1$ et un sous-groupe $\Gamma \subset \mathrm{H}^2(\mathrm{X}, \boldsymbol{\mu}_{\mathrm{N}})$. Considérons les deux hypothèses suivantes :

(BM) L'ensemble $\mathrm{Pic}^1(\mathrm{X} \otimes_k k_v)$ est non vide pour tout $v \in \Omega$ et l'image de Γ par la flèche naturelle $p : \mathrm{H}^2(\mathrm{X}, \boldsymbol{\mu}_{\mathrm{N}}) \to \mathrm{Br}(\mathrm{X})$ est responsable d'une obstruction de Brauer–Manin à l'existence d'un diviseur de degré 1 sur X.

(T) Pour toute place v de k divisant N et tout $\gamma \in \Gamma$, il existe $\gamma_{v,0} \in \mathrm{H}^2(k_v, \boldsymbol{\mu}_{\mathrm{N}})$ et $\gamma_{v,1} \in \mathrm{Pic}(\mathrm{X} \otimes_k k_v)_{\mathrm{tors}}$ tels que l'image γ_v de γ dans $\mathrm{H}^2(\mathrm{X} \otimes_k k_v, \boldsymbol{\mu}_{\mathrm{N}})$ s'écrive

$$\gamma_v = \gamma_{v,0} + c(\gamma_{v,1}),$$

où c désigne l'application classe de cycle $c : \mathrm{Pic}(\mathrm{X} \otimes_k k_v) \to \mathrm{H}^2(\mathrm{X} \otimes_k k_v, \boldsymbol{\mu}_{\mathrm{N}})$.

L'hypothèse (BM) entraîne que $\mathrm{Pic}^1(\mathrm{X}) = \emptyset$ et, par conséquent, que $\mathrm{X}(k) = \emptyset$. D'après Grothendieck [Gro83], il devrait s'ensuivre, si le genre de X est $\geqslant 2$, que la suite exacte fondamentale

$$1 \longrightarrow \pi_1(\mathrm{X} \otimes_k \bar{k}) \longrightarrow \pi_1(\mathrm{X}) \longrightarrow \mathrm{Gal}(\bar{k}/k) \longrightarrow 1, \tag{13.3}$$

cf. [SGA1, Exp. IX, 6.1], n'est pas scindée. Nous montrons dans le Théorème 2 que tel est bien le cas si en outre l'hypothèse (T) est satisfaite (et ce, même si X est de genre 1).

Théorème 2. *Soit* X *une courbe projective, lisse et géométriquement irréductible, sur un corps de nombres k. Soient* $\mathrm{N} \geqslant 1$ *un entier et* $\Gamma \subset \mathrm{H}^2(\mathrm{X}, \boldsymbol{\mu}_{\mathrm{N}})$ *un sous-groupe. Si (BM) et (T) sont vérifiées, la suite (13.3) n'est pas scindée.*

Ainsi, étant données une courbe X et une classe $\mathrm{A} \in \mathrm{Br}(\mathrm{X})$ responsable d'une obstruction de Brauer–Manin à l'existence d'un diviseur de degré 1 sur X, pour que la suite (13.3) ne soit pas scindée, il suffit qu'il existe un entier $\mathrm{N} \geqslant 1$ et un relèvement $\gamma \in \mathrm{H}^2(\mathrm{X}, \boldsymbol{\mu}_{\mathrm{N}})$ de A tels que pour toute place v de k divisant N, l'image de γ dans $\mathrm{H}^2(\mathrm{X} \otimes_k k_v, \boldsymbol{\mu}_{\mathrm{N}})$ appartienne au sous-groupe engendré par $\mathrm{H}^2(k_v, \boldsymbol{\mu}_{\mathrm{N}})$ et par $c(\mathrm{Pic}(\mathrm{X} \otimes_k k_v)_{\mathrm{tors}})$.

Preuve du Théorème 2. Par l'absurde, supposons la suite (13.3) scindée et fixons-en une section $s : \mathrm{Gal}(\bar{k}/k) \to \pi_1(\mathrm{X})$. L'hypothèse (BM) entraîne que la courbe X est de genre $\geqslant 1$; c'est donc un $\mathrm{K}(\pi, 1)$ et l'on peut associer à s une classe de cohomologie étale $\alpha \in \mathrm{H}^2(\mathrm{X}, \boldsymbol{\mu}_{\mathrm{N}})$, cf. [EW09, 2.6]. Rappelons que α, dite **classe de cycle de** s, est caractérisée par la propriété suivante : il existe un revêtement étale $f : \mathrm{Y} \to \mathrm{X}$ tel que s se factorise par $\pi_1(\mathrm{Y})$ et que, notant $\mathrm{pr}_1 : \mathrm{X} \times_k \mathrm{Y} \to \mathrm{X}$ la première projection, la classe dans $\mathrm{H}^2(\mathrm{X} \times_k \mathrm{Y}, \boldsymbol{\mu}_{\mathrm{N}})$ du graphe de f soit égale à $\mathrm{pr}_1^\star \alpha$.

Pour $v \in \Omega$, notons $c : \mathrm{Pic}(\mathrm{X} \otimes_k k_v) \to \mathrm{H}^2(\mathrm{X} \otimes_k k_v, \boldsymbol{\mu}_{\mathrm{N}})$ l'application classe de cycle et α_v l'image de α dans $\mathrm{H}^2(\mathrm{X} \otimes_k k_v, \boldsymbol{\mu}_{\mathrm{N}})$.

Proposition 3. *Pour tout* $v \in \Omega$ *ne divisant pas* N, *il existe* $D_v \in \mathrm{Pic}^1(X \otimes_k k_v)$ *tel que* $\alpha_v = c(D_v)$.

Preuve. Soit $v \in \Omega$ ne divisant pas N. Notons $\bar{k}_v$ une clôture algébrique de k_v contenant $\bar{k}$. La flèche verticale de gauche du diagramme commutatif

$$\begin{array}{ccccccccc} 1 & \longrightarrow & \pi_1(X \otimes_k \bar{k}_v) & \longrightarrow & \pi_1(X \otimes_k k_v) & \longrightarrow & \mathrm{Gal}(\bar{k}_v/k_v) & \longrightarrow & 1 \\ & & \downarrow & & \downarrow & & \downarrow & & \\ 1 & \longrightarrow & \pi_1(X \otimes_k \bar{k}) & \longrightarrow & \pi_1(X) & \longrightarrow & \mathrm{Gal}(\bar{k}/k) & \longrightarrow & 1 \end{array}$$

étant un isomorphisme, cf. [SGA1, Exp. X, 1.8], la section s induit une section $s_v : \mathrm{Gal}(\bar{k}_v/k_v) \to \pi_1(X \otimes_k k_v)$ de la première ligne de ce diagramme. À l'aide de la caractérisation de la classe de cycle d'une section rappelée ci-dessus, on voit que la classe de cycle de s_v est égale à α_v. Comme v ne divise pas N, il résulte maintenant de [EW09, Cor. 3.4, Rem. 3.7 (ii)], si v est finie, ou de [EW09, Rem. 3.7 (iv)], si v est réelle, qu'il existe $D_{v,0} \in \mathrm{Pic}(X \otimes_k k_v)$ tel que $\alpha_v = c(D_{v,0})$. L'image de α_v dans $H^2(X \otimes_k \bar{k}_v, \boldsymbol{\mu}_N) = \mathbb{Z}/N\mathbb{Z}$ est égale à 1, cf. [EW09, 2.6]. Par conséquent

$$\deg(D_{v,0}) = 1 + Nm$$

pour un $m \in \mathbb{Z}$. Soit $D_{v,1}$ un élément de l'ensemble $\mathrm{Pic}^1(X \otimes_k k_v)$, non vide par hypothèse. Posons $D_v = D_{v,0} - NmD_{v,1}$. On a alors bien $\deg(D_v) = 1$ et $c(D_v) = \alpha_v$. □

Fixons, pour chaque $v \in \Omega$ divisant N, un élément arbitraire $D_v \in \mathrm{Pic}^1(X \otimes_k k_v)$, et pour chaque $v \in \Omega$ ne divisant pas N, un élément $D_v \in \mathrm{Pic}^1(X \otimes_k k_v)$ tel que $\alpha_v = c(D_v)$. Nous allons maintenant démontrer que la famille $(D_v)_{v \in \Omega}$ est orthogonale, pour l'accouplement (13.1), à l'image de Γ par $p : H^2(X, \boldsymbol{\mu}_N) \to \mathrm{Br}(X)$.

Pour toute extension K/k, notons

$$\langle -,- \rangle : H^2(X \otimes_k K, \boldsymbol{\mu}_N) \times H^2(X \otimes_k K, \boldsymbol{\mu}_N) \to \mathrm{Br}(K)$$

la composée du cup-produit

$$H^2(X \otimes_k K, \boldsymbol{\mu}_N) \times H^2(X \otimes_k K, \boldsymbol{\mu}_N) \to H^4(X \otimes_k K, \boldsymbol{\mu}_N^{\otimes 2})$$

et de la flèche

$$\delta : H^4(X \otimes_k K, \boldsymbol{\mu}_N^{\otimes 2}) \to H^2(K, H^2(X \otimes_k \overline{K}, \boldsymbol{\mu}_N^{\otimes 2})) = H^2(K, \boldsymbol{\mu}_N) \subset \mathrm{Br}(K)$$

issue de la suite spectrale de Hochschild–Serre. D'autre part, pour $v \in \Omega$, notons $\mathrm{inv}_v : \mathrm{Br}(k_v) \hookrightarrow \mathbb{Q}/\mathbb{Z}$ l'invariant de la théorie du corps de classes local et notons encore $p : H^2(X \otimes_k k_v, \boldsymbol{\mu}_N) \to \mathrm{Br}(X \otimes_k k_v)$ la flèche naturelle.

Lemme 4. *Soit* $v \in \Omega$. *Les accouplements* $\langle -,- \rangle$ *et (13.2) s'inscrivent dans un diagramme commutatif*

$$\begin{array}{ccc} \mathrm{H}^2(\mathrm{X}\otimes_k k_v,\boldsymbol{\mu}_\mathrm{N})\times \mathrm{H}^2(\mathrm{X}\otimes_k k_v,\boldsymbol{\mu}_\mathrm{N}) & \longrightarrow & \mathrm{Br}(k_v) \\ c\uparrow \qquad\qquad \downarrow p & & \downarrow \mathrm{inv}_v \\ \mathrm{Pic}(\mathrm{X}\otimes_k k_v)\times \mathrm{Br}(\mathrm{X}\otimes_k k_v) & \longrightarrow & \mathbb{Q}/\mathbb{Z}. \end{array}$$

Preuve. Soient $x \in \mathrm{X}\otimes_k k_v$ un point fermé et $i : \mathrm{Spec}(k_v(x)) \hookrightarrow \mathrm{X}\otimes_k k_v$ l'injection canonique. Pour tout $y \in \mathrm{H}^2(\mathrm{X}\otimes_k k_v,\boldsymbol{\mu}_\mathrm{N})$, on a

$$\langle c(x),y\rangle = \delta(c(x)\smile y) = \delta(i_\star i^\star y)\,,$$

où la seconde égalité résulte de la formule de projection. Compte tenu de la définition de (13.2), cf. [Sai89, p. 399], il suffit donc, pour conclure, de vérifier que l'application $\delta\circ i_\star : \mathrm{H}^2(k_v(x),\boldsymbol{\mu}_\mathrm{N}) \to \mathrm{H}^2(k_v,\boldsymbol{\mu}_\mathrm{N})$ s'identifie au morphisme de corestriction de $k_v(x)$ à k_v.

Notons $\rho : \mathrm{X}\otimes_k k_v \to \mathrm{Spec}(k_v)$ le morphisme structural et $a : i_\star\boldsymbol{\mu}_\mathrm{N} \to \boldsymbol{\mu}_\mathrm{N}^{\otimes 2}[2]$ la flèche, dans la catégorie dérivée des faisceaux étales en groupes abéliens sur X, donnant naissance au morphisme de Gysin

$$i_\star : \mathrm{H}^2(k_v(x),\boldsymbol{\mu}_\mathrm{N}) \to \mathrm{H}^4(\mathrm{X}\otimes_k k_v,\boldsymbol{\mu}_\mathrm{N}^{\otimes 2})\,.$$

La composée de $\mathrm{R}\rho_\star a$ et de la troncation

$$\mathrm{R}\rho_\star\boldsymbol{\mu}_\mathrm{N}^{\otimes 2}[2] \to \left(\tau_{\geqslant 2}\mathrm{R}\rho_\star\boldsymbol{\mu}_\mathrm{N}^{\otimes 2}\right)[2] = \boldsymbol{\mu}_\mathrm{N}$$

est une flèche entre complexes concentrés en degré 0. Elle provient donc d'un unique morphisme de modules galoisiens $b : (\rho\circ i)_\star\boldsymbol{\mu}_\mathrm{N} \to \boldsymbol{\mu}_\mathrm{N}$. La flèche obtenue en appliquant à b le foncteur $\mathrm{H}^0(\bar{k}_v,-)$ est la composée

$$\boldsymbol{\mu}_\mathrm{N}(x\otimes_{k_v}\bar{k}_v) \to \mathrm{H}^2(\mathrm{X}\otimes_k \bar{k}_v,\boldsymbol{\mu}_\mathrm{N}^{\otimes 2}) \to \boldsymbol{\mu}_\mathrm{N}$$

des applications classe de cycle et degré tordues par $\boldsymbol{\mu}_\mathrm{N}$. Par conséquent b est l'application norme. D'autre part, en appliquant à b le foncteur $\mathrm{H}^2(k_v,-)$, on retrouve le morphisme $\delta\circ i_\star$; ainsi le lemme est prouvé. □

Soit $\gamma \in \Gamma$. Pour $v \in \Omega$, notons γ_v l'image de γ dans $\mathrm{H}^2(\mathrm{X}\otimes_k k_v,\boldsymbol{\mu}_\mathrm{N})$.

Proposition 5. *Pour tout $v \in \Omega$, on a $\langle \alpha_v,\gamma_v\rangle = \langle c(\mathrm{D}_v),\gamma_v\rangle$.*

Preuve. Soit $v \in \Omega$. Si v ne divise pas N, on a même $\alpha_v = c(\mathrm{D}_v)$. Supposons donc que v divise N et reprenons les notations $\gamma_{v,0}$, $\gamma_{v,1}$ apparaissant dans l'hypothèse (T), de sorte que $\gamma_v = \gamma_{v,0} + c(\gamma_{v,1})$.

Lemme 6. *Soient $x \in \mathrm{H}^2(\mathrm{X}\otimes_k k_v,\boldsymbol{\mu}_\mathrm{N})$ et $y \in \mathrm{H}^2(k_v,\boldsymbol{\mu}_\mathrm{N})$. Si x s'annule dans le groupe $\mathrm{H}^2(\mathrm{X}\otimes_k \bar{k}_v,\boldsymbol{\mu}_\mathrm{N})$, alors $\langle x,y\rangle = 0$.*

Preuve. Notons $(\mathrm{F}^i\mathrm{H}^{2n})_{i\in\mathbb{Z}}$ la filtration décroissante de $\mathrm{H}^{2n}(\mathrm{X}\otimes_k k_v,\boldsymbol{\mu}_\mathrm{N}^{\otimes n})$ induite par la suite spectrale de Hochschild–Serre. Les hypothèses du lemme signifient que $x \in \mathrm{F}^1\mathrm{H}^2$ et $y \in \mathrm{F}^2\mathrm{H}^2$. Or le cup-produit respecte cette filtration, cf. [Jan88, Prop. 6.2], d'où $x\smile y \in \mathrm{F}^3\mathrm{H}^4$, ce qui se traduit par l'égalité $\langle x,y\rangle = 0$. □

La construction de la classe de cycle associée à une section du groupe fondamental étant fonctorielle par rapport aux coefficients, la classe α_v se relève dans $\varprojlim_{m\geqslant 1} \mathrm{H}^2(\mathrm{X}\otimes_k k_v, \boldsymbol{\mu}_m)$, cf. [EW09, 2.6]. Par conséquent $p(\alpha_v) \in \mathrm{Br}(\mathrm{X}\otimes_k k_v)$ est infiniment divisible. En particulier $p(\alpha_v)$ est orthogonal à $\mathrm{Pic}(\mathrm{X}\otimes_k k_v)_{\mathrm{tors}}$ pour l'accouplement (13.2). Or $\gamma_{v,1} \in \mathrm{Pic}(\mathrm{X}\otimes_k k_v)_{\mathrm{tors}}$ d' après l'hypothèse (T), le Lemme 4 implique donc que $\langle \alpha_v, c(\gamma_{v,1})\rangle = 0$.

D'autre part, il résulte du Lemme 4 que $\langle c(\mathrm{D}_v), c(\gamma_{v,1})\rangle = 0$, puisque $p\circ c = 0$. Enfin, le Lemme 6 entraîne que $\langle \alpha_v - c(\mathrm{D}_v), \gamma_{v,0}\rangle = 0$. Vu la décomposition

$$\langle \alpha_v - c(\mathrm{D}_v), \gamma_{v,0} + c(\gamma_{v,1})\rangle = \langle \alpha_v - c(\mathrm{D}_v), \gamma_{v,0}\rangle + \langle \alpha_v, c(\gamma_{v,1})\rangle - \langle c(\mathrm{D}_v), c(\gamma_{v,1})\rangle,$$

la Proposition 5 est maintenant établie. □

Nous sommes en position d'achever la preuve du Théorème 2. D'après la loi de réciprocité globale, la somme des invariants de $\langle \alpha, \gamma\rangle \in \mathrm{Br}(k)$ est nulle. Par conséquent $\sum_{v\in\Omega} \mathrm{inv}_v \langle \alpha_v, \gamma_v\rangle = 0$. Il s'ensuit, grâce à la Proposition 5, que

$$\sum_{v\in\Omega} \mathrm{inv}_v \langle c(\mathrm{D}_v), \gamma_v\rangle = 0\,,$$

ce qui signifie, compte tenu du Lemme 4, que la famille $(\mathrm{D}_v)_{v\in\Omega}$ est orthogonale à $p(\gamma)$ pour l'accouplement (13.1). L'élément $\gamma \in \Gamma$ étant quelconque, l'hypothèse (BM) est ainsi contredite. □

13.3 Les courbes de Reichardt–Lind et de Schinzel

Il existe un morphisme évident de la courbe de Schinzel vers la courbe de Reichardt–Lind, à savoir $[x:y:z] \mapsto [xz : x^2 : y^2+4z^2 : z^2]$. Or la suite exacte $(*)$ dépend fonctoriellement de X. Pour établir le Théorème 1, il suffit donc de traiter le cas de la courbe de Reichardt–Lind.

Soit X la courbe de Reichardt–Lind, définie comme la compactification lisse de la courbe affine d'équation $2y^2 = w^4 - 17$ sur $\mathbb{Q}$. Pour montrer que la suite $(*)$ n'est pas scindée, nous allons appliquer le Théorème 2 avec $\mathrm{N} = 2$ et $\Gamma = \{0, \gamma\}$ où $\gamma \in \mathrm{H}^2(\mathrm{X}, \mathbb{Z}/2\mathbb{Z})$ est une classe à préciser. Le reste du présent paragraphe est consacré à la construction de γ et à la vérification des hypothèses du Théorème 2.

Le diviseur de la fonction rationnelle y sur X s'écrit $\mathrm{P} - 2\mathrm{Q}$, où $\mathrm{P}, \mathrm{Q} \in \mathrm{X}$ sont des points fermés de degrés respectifs 4 et 2 sur $\mathbb{Q}$. Posons $\mathrm{V} = \mathrm{X} \setminus \{\mathrm{P}\}$. Le diviseur de y sur V étant un double, le revêtement de V obtenu en extrayant une racine carrée de y est étale. Notons $[y]$ sa classe dans $\mathrm{H}^1(\mathrm{V}, \mathbb{Z}/2\mathbb{Z})$. Notons d'autre part $[17]$ l'image de 17 par l'application naturelle

$$\mathbb{Q}^\star/\mathbb{Q}^{\star 2} = \mathrm{H}^1(\mathbb{Q}, \mathbb{Z}/2\mathbb{Z}) \to \mathrm{H}^1(\mathrm{V}, \mathbb{Z}/2\mathbb{Z})\,.$$

Lemme 7. *La flèche de restriction* $\mathrm{H}^2(\mathrm{X}, \mathbb{Z}/2\mathbb{Z}) \to \mathrm{H}^2(\mathrm{V}, \mathbb{Z}/2\mathbb{Z})$ *est injective.*

Preuve. En effet, dans la suite exacte de localisation

$$\mathrm{H}^1(\mathrm{V},\mathbb{Z}/2\mathbb{Z}) \longrightarrow \mathrm{H}^2_{\mathrm{P}}(\mathrm{X},\mathbb{Z}/2\mathbb{Z}) \longrightarrow \mathrm{H}^2(\mathrm{X},\mathbb{Z}/2\mathbb{Z}) \longrightarrow \mathrm{H}^2(\mathrm{V},\mathbb{Z}/2\mathbb{Z})\,,$$

la première flèche est surjective puisque $\mathrm{H}^2_{\mathrm{P}}(\mathrm{X},\mathbb{Z}/2\mathbb{Z}) = \mathrm{H}^0(\mathrm{P},\mathbb{Z}/2\mathbb{Z}) = \mathbb{Z}/2\mathbb{Z}$ et que $[y] \in \mathrm{H}^1(\mathrm{V},\mathbb{Z}/2\mathbb{Z})$ s'envoie sur $1 \in \mathbb{Z}/2\mathbb{Z}$. □

Le résidu de $[y] \smile [17] \in \mathrm{H}^2(\mathrm{V},\mathbb{Z}/2\mathbb{Z})$ en P est nul puisque 17 est un carré dans $\mathbb{Q}(\mathrm{P})$. Par conséquent $[y] \smile [17]$ est la restriction d'un élément de $\mathrm{H}^2(\mathrm{X},\mathbb{Z}/2\mathbb{Z})$. D'après le Lemme 7, celui-ci est uniquement déterminé. Nous le noterons γ.

L'image de γ dans Br(X) est la classe de l'algèbre de quaternions $(y,17)$. Il est bien connu que cette classe est responsable d'une obstruction de Brauer–Manin à l'existence d'un diviseur de degré 1 sur X et que $\mathrm{X}(\mathbb{Q}_v) \neq \emptyset$ pour toute place v de $\mathbb{Q}$. Voir [Sti11, §5], où l'obstruction de Brauer–Manin à l'existence d'un point rationnel est discutée ; noter que $\mathrm{Pic}^1(\mathrm{X} \otimes_{\mathbb{Q}} \mathrm{K}) = \mathrm{X}(\mathrm{K})$ pour toute extension $\mathrm{K}/\mathbb{Q}$ puisque X est une courbe de genre 1 ; ainsi l'obstruction de Brauer–Manin à l'existence d'un diviseur de degré 1 sur X est équivalente à l'obstruction de Brauer–Manin à l'existence d'un point rationnel.

L'hypothèse (BM) du paragraphe 13.2 est donc satisfaite. Il reste à vérifier l'hypothèse (T).

Afin de simplifier les notations, posons $\overline{\mathrm{X}} = \mathrm{X} \otimes_{\mathbb{Q}} \overline{\mathbb{Q}}$, $\overline{\mathrm{V}} = \mathrm{V} \otimes_{\mathbb{Q}} \overline{\mathbb{Q}}$ et enfin

$$\mathrm{H}^2(\mathrm{X} \otimes_{\mathbb{Q}} \mathrm{K},\mathbb{Z}/2\mathbb{Z})^0 = \mathrm{Ker}\Big(\mathrm{H}^2(\mathrm{X} \otimes_{\mathbb{Q}} \mathrm{K},\mathbb{Z}/2\mathbb{Z}) \to \mathrm{H}^2(\mathrm{X} \otimes_{\mathbb{Q}} \overline{\mathrm{K}},\mathbb{Z}/2\mathbb{Z})\Big)$$

pour toute extension $\mathrm{K}/\mathbb{Q}$, où $\overline{\mathrm{K}}$ désigne une clôture algébrique de K.

Proposition 8. *L'image de* γ *dans* $\mathrm{H}^2(\overline{\mathrm{X}},\mathbb{Z}/2\mathbb{Z})$ *est nulle.*

Preuve. Le diviseur de la fonction rationnelle $w^2 - \sqrt{17}$ sur $\overline{\mathrm{X}}$ est un double ; celle-ci définit donc une classe $[w^2 - \sqrt{17}] \in \mathrm{H}^1(\overline{\mathrm{X}},\mathbb{Z}/2\mathbb{Z})$. Cette classe est invariante sous l'action de $\mathrm{Gal}(\overline{\mathbb{Q}}/\mathbb{Q})$ puisque $(w^2 - \sqrt{17})(w^2 + \sqrt{17}) = w^4 - 17 = 2y^2$ est un carré dans $\overline{\mathbb{Q}}(\mathrm{X})$.

Lemme 9. *Les images de* $[w^2 - \sqrt{17}] \times [-2]$ *et de* $[y] \times [17]$ *par le cup-produit*

$$\mathrm{H}^0(\mathbb{Q},\mathrm{H}^1(\overline{\mathrm{V}},\mathbb{Z}/2\mathbb{Z})) \times \mathrm{H}^1(\mathbb{Q},\mathbb{Z}/2\mathbb{Z}) \to \mathrm{H}^1(\mathbb{Q},\mathrm{H}^1(\overline{\mathrm{V}},\mathbb{Z}/2\mathbb{Z})) \tag{13.4}$$

coïncident.

Preuve. Il suffit de montrer que le cocycle $a\,:\,\mathrm{Gal}(\overline{\mathbb{Q}}/\mathbb{Q}) \to \mathrm{H}^1(\overline{\mathrm{V}},\mathbb{Z}/2\mathbb{Z})$ défini par

$$a(\sigma) = \chi_{-2}(\sigma)[w^2 - \sqrt{17}] - \chi_{17}(\sigma)[y]$$

est un cobord, où $\chi_q\,:\,\mathrm{Gal}(\overline{\mathbb{Q}}/\mathbb{Q}) \to \mathbb{Z}/2\mathbb{Z}$ désigne le caractère quadratique associé à $q \in \mathbb{Q}^\star$. Soit $f = w^2 - \sqrt{17} - \sqrt{-2}y$. Le diviseur de f sur $\overline{\mathrm{V}}$ est un double. D'où une classe $[f] \in \mathrm{H}^1(\overline{\mathrm{V}},\mathbb{Z}/2\mathbb{Z})$. On vérifie aisément que $a(\sigma) = \sigma[f] - [f]$ pour tout σ tel que l'un au moins de $\chi_{-2}(\sigma)$ et de $\chi_{17}(\sigma)$ soit nul. Or a est un cocycle ; par conséquent $a(\sigma) = \sigma[f] - [f]$ pour tout $\sigma \in \mathrm{Gal}(\overline{\mathbb{Q}}/\mathbb{Q})$ et a est donc un cobord. □

Comme $H^2(\overline{V}, \mathbb{Z}/2\mathbb{Z}) = 0$, la suite spectrale de Hochschild–Serre et les flèches de restriction fournissent un diagramme commutatif

$$\begin{array}{ccccc} H^2(X,\mathbb{Z}/2\mathbb{Z})^0 & \xrightarrow{\delta_X} & H^1(\mathbb{Q},H^1(\overline{X},\mathbb{Z}/2\mathbb{Z})) & \longrightarrow & H^3(\mathbb{Q},H^0(\overline{X},\mathbb{Z}/2\mathbb{Z})) \\ \downarrow & & \downarrow & & \downarrow \wr \\ H^2(V,\mathbb{Z}/2\mathbb{Z}) & \xrightarrow{\delta_V} & H^1(\mathbb{Q},H^1(\overline{V},\mathbb{Z}/2\mathbb{Z})) & \xrightarrow{d} & H^3(\mathbb{Q},H^0(\overline{V},\mathbb{Z}/2\mathbb{Z})) \end{array}$$

dont les lignes sont exactes et dont la flèche verticale de droite est un isomorphisme. Cette suite spectrale étant compatible au cup-produit, cf. [Jan88, Prop. 6.2], l'image de $[y] \times [17]$ par (13.4) est égale, à un signe près, à $\delta_V([y] \smile [17])$; en particulier appartient-elle à $\mathrm{Ker}(d)$. Il s'ensuit, grâce au Lemme 9 et à une chasse au diagramme, que l'image de $[w^2 - \sqrt{17}] \times [-2]$ par le cup-produit

$$H^0(\mathbb{Q},H^1(\overline{X},\mathbb{Z}/2\mathbb{Z})) \times H^1(\mathbb{Q},\mathbb{Z}/2\mathbb{Z}) \to H^1(\mathbb{Q},H^1(\overline{X},\mathbb{Z}/2\mathbb{Z}))$$

s'écrit $\delta_X(\gamma')$ pour un $\gamma' \in H^2(X,\mathbb{Z}/2\mathbb{Z})^0$.

La flèche naturelle $\mathrm{Ker}(\delta_X) \to \mathrm{Ker}(\delta_V)$ est surjective puisque tout élément de $\mathrm{Ker}(\delta_V)$ provient de $H^2(\mathbb{Q},\mathbb{Z}/2\mathbb{Z})$. Quitte à modifier γ', on peut donc supposer que la restriction de γ' à V coïncide avec $[y] \smile [17] \in H^2(V,\mathbb{Z}/2\mathbb{Z})$. Le Lemme 7 entraîne maintenant que $\gamma = \gamma'$. D'où finalement $\gamma \in H^2(X,\mathbb{Z}/2\mathbb{Z})^0$. □

Proposition 10. *Pour toute extension* $K/\mathbb{Q}$*, le noyau de la flèche de restriction*

$$H^2(X \otimes_{\mathbb{Q}} K, \mathbb{Z}/2\mathbb{Z})^0 \to H^2(V \otimes_{\mathbb{Q}} K, \mathbb{Z}/2\mathbb{Z}) \tag{13.5}$$

est contenu dans $c(\mathrm{Pic}(X \otimes_{\mathbb{Q}} K)_{\mathrm{tors}})$*, où c désigne l'application classe de cycle.*

Preuve. Le lieu de ramification du morphisme $X \to \mathbb{P}^1_{\mathbb{Q}}$, $(y,w) \mapsto w$ est P, qui est de degré 4 sur $\mathbb{Q}$. Par conséquent, le choix d'un point géométrique de X au-dessus de P munit $X \otimes_{\mathbb{Q}} \overline{K}$ d'une structure de courbe elliptique dont le sous-groupe de 2-torsion est $P \otimes_{\mathbb{Q}} \overline{K}$. Les diviseurs de degré 0 sur $X \otimes_{\mathbb{Q}} K$ supportés sur $P \otimes_{\mathbb{Q}} K$ sont donc tous de torsion dans $\mathrm{Pic}(X \otimes_{\mathbb{Q}} K)$. Or leurs classes de cycles dans $H^2(X \otimes_{\mathbb{Q}} K, \mathbb{Z}/2\mathbb{Z})^0$ engendrent le noyau de (13.5), en vertu de la suite exacte de localisation. □

L'image de γ dans $H^2(V \otimes_{\mathbb{Q}} \mathbb{Q}_2, \mathbb{Z}/2\mathbb{Z})$ est nulle puisque 17 est un carré dans $\mathbb{Q}_2$. Les Propositions 8 et 10 entraînent donc que l'hypothèse (T) est satisfaite (avec $\gamma_{2,0} = 0$). Ainsi le Théorème 2 permet-il de conclure la démonstration du Théorème 1.

Remarque 11. Soit X la courbe de Schinzel. Nous avons montré que la suite (∗) n'est pas scindée. À l'instar de l'exemple de [HS09], la courbe de Schinzel vérifie une propriété plus forte : même la suite exacte fondamentale abélianisée

$$1 \longrightarrow \pi_1(\overline{X})^{\mathrm{ab}} \longrightarrow \pi_1(X)^{[\mathrm{ab}]} \longrightarrow \mathrm{Gal}(\overline{\mathbb{Q}}/\mathbb{Q}) \longrightarrow 1,$$

obtenue en poussant $(*)$ le long du morphisme d'abélianisation $\pi_1(\overline{X}) \to \pi_1(\overline{X})^{ab}$, n'est pas scindée. En effet, la suite exacte fondamentale abélianisée est tout aussi fonctorielle que la suite exacte fondamentale, et les deux coïncident dans le cas de la courbe de Reichardt–Lind puisque celle-ci est de genre 1.

Bibliographie

[EW09] H. Esnault and O. Wittenberg. Remarks on cycle classes of sections of the arithmetic fundamental group. *Mosc. Math. J.*, 9(3) :451–467, 2009.

[Gro83] A. Grothendieck. Lettre à Faltings du 27 juin 1983 (en allemand), parue dans : Geometric Galois actions, Vol. 1, London Math. Soc. Lecture Note Ser., vol. 242, Cambridge Univ. Press, Cambridge, 1997.

[HS09] D. Harari and T. Szamuely. Galois sections for abelianized fundamental groups. *Math. Ann.*, 344(4) :779–800, 2009. Avec un appendice par E. V. Flynn.

[Jan88] U. Jannsen. Continuous étale cohomology. *Math. Ann.*, 280(2) :207–245, 1988.

[Lin40] C.-E. Lind. Untersuchungen über die rationalen Punkte der ebenen kubischen Kurven vom Geschlecht Eins. Thèse de doctorat, Uppsala, 1940.

[Rei42] H. Reichardt. Einige im Kleinen überall lösbare, im Grossen unlösbare diophantische Gleichungen. *J. reine angew. Math.*, 184 :12–18, 1942.

[Sai89] S. Saito. Some observations on motivic cohomology of arithmetic schemes. *Invent. Math.*, 98(2) :371–404, 1989.

[Sch84] A. Schinzel. Hasse's principle for systems of ternary quadratic forms and for one biquadratic form. *Studia Math.*, 77(2) :103–109, 1984.

[SGA1] A. Grothendieck. *Revêtements étale et groupe fondamental (SGA 1)*. Séminaire de géométrie algébrique du Bois Marie 1960-61, dirigé par A. Grothendieck, augmenté de deux exposés de Mme M. Raynaud, *Lecture Notes in Math.* 224, Springer-Verlag, Berlin-New York, 1971. Édition recomposée et annotée : *Documents Mathématiques* 3, Société Mathématique de France, Paris, 2003.

[Sti11] J. Stix. The Brauer–Manin obstruction for sections of the fundamental group. *Journal of Pure and Applied Algebra*, 215 (2011), no. 6, 1371-1397.

Chapter 14
On ℓ-adic Iterated Integrals V: Linear Independence, Properties of ℓ-adic Polylogarithms, ℓ-adic Sheaves

Zdzisław Wojtkowiak

Abstract In a series of papers we have introduced and studied ℓ-adic polylogarithms and ℓ-adic iterated integrals which are analogues of the classical complex polylogarithms and iterated integrals in ℓ-adic Galois realizations. In this note we shall show that in the generic case ℓ-adic iterated integrals are linearly independent over $\mathbb{Q}_\ell$. In particular they are non trivial. This result can be viewed as analogous of the statement that the classical iterated integrals from 0 to z of sequences of one forms $\frac{dz}{z}$ and $\frac{dz}{z-1}$ are linearly independent over $\mathbb{Q}$. We also study ramification properties of ℓ-adic polylogarithms and the minimal quotient subgroup of the absolute Galois group $\mathrm{G_K}$ of a number field K on which ℓ-adic polylogarithms are defined. In the final sections of the paper we study ℓ-adic sheaves and their relations with ℓ-adic polylogarithms. We show that if an ℓ-adic sheaf has the same monodromy representation as the classical complex polylogarithms then the action of $\mathrm{G_K}$ in stalks is given by ℓ-adic polylogarithms.

14.1 Introduction

In this paper we study properties of ℓ-adic iterated integrals and ℓ-adic polylogarithms introduced in [Woj04] and [Woj05a]. We describe briefly the main results of the paper, though in the introduction we do not present them in full generality.

Let K be a number field with algebraic closure $\bar{\mathrm{K}}$. Throughout this paper we fix an embedding $\bar{\mathrm{K}} \subset \mathbb{C}$. Let $z \in \mathrm{K} \setminus \{0,1\}$ or let z be a tangential point of

$$\mathbb{P}^1_{\bar{\mathrm{K}}} \setminus \{0,1,\infty\}$$

Z. Wojtkowiak (✉)
Département de Mathématiques, Université de Nice-Sophia Antipolis, Laboratoire Jean Alexandre Dieudonné, U.R.A. au C.N.R.S., No 168, Parc Valrose – B.P.N° 71, 06108 Nice Cedex 2, France
e-mail: wojtkow@unice.fr

J. Stix (ed.), *The Arithmetic of Fundamental Groups*, Contributions in Mathematical and Computational Sciences 2, DOI 10.1007/978-3-642-23905-2_14,

defined over K, and let γ be an ℓ-adic path from $\overrightarrow{01}$ to z on $\mathbb{P}^1_{\bar{K}} \setminus \{0,1,\infty\}$. For any $\sigma \in G_K = \mathrm{Gal}(\bar{K}/K)$ we set

$$f_\gamma(\sigma) := \gamma^{-1} \cdot \sigma(\gamma) \in \pi_1^{\text{ét}}(\mathbb{P}^1_{\bar{K}} \setminus \{0,1,\infty\}, \overrightarrow{01})_{\text{pro}-\ell}\,.$$

Here and later our convention of composing a path α from y to z with a path β from x to y will be that $\alpha \cdot \beta$ is defined as a path from x to z.

Let V be an algebraic variety defined over K and let v be a K-point or a tangential point defined over K. By the comparison homomorphism

$$\pi_1(V(\mathbb{C}), v) \to \pi_1^{\text{ét}}(V_{\bar{K}}, v)_{\text{pro}-\ell}$$

any element of $\pi_1(V(\mathbb{C}), v)$ determines canonically an element of $\pi_1^{\text{ét}}(V_{\bar{K}}, v)_{\text{pro}-\ell}$, and we shall use the same notation for an element of $\pi_1(V(\mathbb{C}), v)$ and its image. In particular, we have the comparison homomorphism

$$\pi_1(U, \overrightarrow{01}) \to \pi_1^{\text{ét}}(\operatorname{Spec} \bar{K}((z)), \overrightarrow{01})_{\text{pro}-\ell}\,,$$

where $U \subset \mathbb{C} \setminus \{0\}$ is a punctured infinitesimal neighbourhood of 0 and $\operatorname{Spec} \bar{K}((z))$ is an algebraic infinitesimal punctured neighbourhood of 0 in $\mathbb{P}^1_{\bar{K}}$. Hence a loop around 0 in $\mathbb{C} \setminus \{0\}$ determines canonically an element of

$$\pi_1^{\text{ét}}(\operatorname{Spec} \bar{K}((z)), \overrightarrow{01})_{\text{pro}-\ell}\,.$$

Similarly we have the comparison map from the torsor of paths from v to z on $V(\mathbb{C})$ to the torsor of ℓ-adic paths from v to z on $V_{\bar{K}}$.

Informally, we define ℓ-adic iterated integrals from $\overrightarrow{01}$ to z as functions

$$l_b(z) = l_b(z)_\gamma : G_K \to \mathbb{Q}_\ell$$

given by coefficients of $f_\gamma(\,)$ indexed by elements b in a Hall basis $\mathcal{B}$ of the free Lie algebra Lie(X, Y) on two generators X and Y. Let $\mathcal{B}_n$ be the set of elements of degree n in $\mathcal{B}$. Let

$$H_n \subset G_{K(\mu_{\ell^\infty})}$$

be the subgroup of $G_{K(\mu_{\ell^\infty})}$ defined by the condition that all $l_b(z)$ and $l_b(\overrightarrow{10})$ vanish on H_n for all $b \in \bigcup_{i<n} \mathcal{B}_i$.

Our first result concerns linear independence of ℓ-adic iterated integrals.

Theorem 1. *Assume that $z \in K \setminus \{0,1\}$ is not a root of any equation of the form $z^p \cdot (1-z)^q = 1$, where p and q are integers such that $p^2 + q^2 > 0$. Then the functions*

$$l_b(z) : H_n \to \mathbb{Q}_\ell$$

for $b \in \mathcal{B}_n$ are linearly independent over $\mathbb{Q}_\ell$.

Our next results concerns ℓ-adic polylogarithms. Hence we recall here their definition, see [Woj05a, Definition 11.0.1.]. Let x and y be the standard generators of

$$\pi_1^{\text{ét}}(\mathbb{P}^1_{\bar{K}} \setminus \{0,1,\infty\}, \overrightarrow{01})_{\text{pro}-\ell}\,,$$

see for example [Woj05a, Picture 1 on page 126]. Let $\mathbb{Q}_\ell\{\{X,Y\}\}$ be the $\mathbb{Q}_\ell$-algebra of non-commutative formal power series in non-commutative variables X and Y. Let

$$E : \pi_1^{\text{ét}}(\mathbb{P}^1_{\bar{K}} \setminus \{0,1,\infty\}, \overrightarrow{01})_{\text{pro}-\ell} \to \mathbb{Q}_\ell\{\{X,Y\}\}$$

be a continuous multiplicative embedding of $\pi_1^{\text{ét}}(\mathbb{P}^1_{\bar{K}} \setminus \{0,1,\infty\}, \overrightarrow{01})_{\text{pro}-\ell}$ into the algebra $\mathbb{Q}_\ell\{\{X,Y\}\}$ given by

$$E(x) = \exp(X)\,,$$

$$E(y) = \exp(Y)\,.$$

The ℓ-adic polylogarithms $l_n(z)$ and the ℓ-adic logarithm $l(z)$ are defined as functions on $\sigma \in G_K$ by the coefficients of the following expansion

$$\log E(f_\gamma(\sigma)) = l(z)(\sigma)X + \sum_{n=1}^{\infty} l_n(z)(\sigma)YX^{n-1} + \ldots,$$

where only relevent terms on the right hand side are written. The ℓ-adic polylogarithms $l_n(z)$ and $l(z)$ depend on a choice of a path γ from $\overrightarrow{01}$ to z. If we want to indicate the dependence on a path γ we shall write $l_n(z)_\gamma$ and $l(z)_\gamma$. The function

$$l(z) : G_K \to \mathbb{Q}_\ell$$

takes its values in $\mathbb{Z}_\ell$ and agrees with the Kummer character $\kappa(z)$ associated to z, see [Woj05b, Proposition 14.1.0.].

Our second result concerns the minimal quotient of G_K, on which the ℓ-adic polylogarithms $l_n(z)$ are defined and their ramification properties. For $z \in K \setminus \{0,1\}$ we consider the fields $K(\mu_{\ell^\infty})$ and $K(\mu_{\ell^\infty}, z^{1/\ell^\infty})$. Let

$$M(K(\mu_{\ell^\infty}, z^{1/\ell^\infty}))^{ab}_{\ell,1-z}$$

be the maximal pro-ℓ abelian extension of $K(\mu_{\ell^\infty}, z^{1/\ell^\infty})$ that is unramified outside ℓ and $1-z$.

Theorem 2. *Assume that $z \in K \setminus \{0,1\}$ is not a root of any equation of the form $z^p \cdot (1-z)^q = 1$, where p and q are integers such that $p^2 + q^2 > 0$. Then we have:*

(1) The ℓ-adic polylogarithm

$$l_n(z) : G_K \to \mathbb{Q}_\ell$$

factors through the group

$$\mathrm{Gal}(M(K(\mu_{\ell^\infty}, z^{1/\ell^\infty}))^{ab}_{\ell,1-z}/K)\,.$$

(2) *The ℓ-adic polylogarithm $l_n(z)$ ramifies only at prime factors of the fractional ideals*

$$(\ell),\ (z),\ (1-z)\,.$$

(3) *The ℓ-adic polylogarithm $l_n(z)$ determines a non-trivial element in the group*

$$\mathrm{Hom}(\mathrm{Gal}(\mathrm{M}(\mathrm{K}(\mu_{\ell^\infty},z^{1/\ell^\infty}))^{ab}_{\ell,1-z}/\mathrm{K}(\mu_{\ell^\infty},z^{1/\ell^\infty})),\mathbb{Q}_\ell)\,.$$

Our third result connects ℓ-adic polylogarithms to non-abelian Iwasawa theory though we are not sure if our terminology of non-abelian Iwasawa theory is not an exaggeration, since the result is quite elementary. Let us set

$$\mathcal{G}=\mathrm{Gal}\,(\mathrm{M}(\mathrm{K}(\mu_{\ell^\infty},z^{1/\ell^\infty}))^{ab}_{\ell,1-z}/\mathrm{K}(\mu_{\ell^\infty},z^{1/\ell^\infty}))$$

and

$$\Phi=\mathrm{Gal}(\mathrm{K}(\mu_{\ell^\infty},z^{1/\ell^\infty})/\mathrm{K})\,.$$

The Galois group $\mathcal{G}$ is a Φ-module, hence it is also a $\mathbb{Z}_\ell[[\Phi]]$-module. Therefore $\mathrm{Hom}(\mathcal{G},\mathbb{Q}_\ell)$ is also a $\mathbb{Z}_\ell[[\Phi]]$-module. If $f\in\mathrm{Hom}(\mathcal{G},\mathbb{Q}_\ell)$ and $\mu\in\mathbb{Z}_\ell[[\Phi]]$ then we denote by f^μ the element f acted (multiplied) by μ.

Let $\chi:\mathrm{G_K}\to\mathbb{Z}_\ell^\times$ denote the ℓ-adic cyclotomic character. Observe that χ and $l(z)$ are continuous functions on Φ, hence we can integrate them against the measure μ.

Theorem 3. *Let z belong to $\mathrm{K}\setminus\{0,1\}$. Then we have*

$$(l_m(z))^\mu=\Big(\int_\Phi\chi^m(x)d\mu\Big)l_m(z)+\sum_{k=1}^{m-1}\Big(\int_\Phi\frac{(-l(z)(x))^k}{k!}\chi^{m-k}(x)d\mu\Big)l_{m-k}(z)$$

for any $\mu\in\mathbb{Z}_\ell[[\Phi]]$.

In the final sections of the paper we study ℓ-adic sheaves. We shall show that if an ℓ-adic sheaf has the same monodromy representation as the classical complex polylogarithm then the Galois action in stalks is given by the ℓ-adic polylogarithms.

We say a few words about our terminology and our notation. The functions $l_n(z)$, $l(z)$ and $l_b(z)$ appear exactly at the same place in our studies as the classical complex polylogarithms, the logarithm and iterated integrals when calculating sections of the universal pro-unipotent connection on $\mathbb{P}^1(\mathbb{C})$ minus a finite number of points. Moreover the equation

$$l(xy)=l(x)+l(y)\,,$$

the fact that $l(z):\mathrm{G_K}\to\mathbb{Z}_\ell(1)$ is a cocycle, functional equations of $l_n(z)$ and $l_b(z)$, the fact that $l_n(\xi)$ is a cocycle for ξ a root of unity, the precise value of

$$l_{2n}(\overrightarrow{10})$$

are proved using geometry, see [Woj05a, Woj05b] and [Woj09]. Geometry is also used to calculate them explicitely, see [NW02] for ℓ-adic polylogarithms and our work in progress for arbitrary ℓ-adic iterated integrals.

Acknowledgements. We would like to thank very much the Max Planck Institute for Mathematics in Bonn for its hospitality where during two summer visits this paper was written. Very warm thanks are also due to Jakob Stix for enormous help in the editorial preparation of the paper and for a lot of very helpful mathematical suggestions.

Notation. The tensor product $\mathrm{M}\otimes_{\mathbb{Z}_\ell}\mathbb{Q}_\ell$ of a $\mathbb{Z}_\ell$-module M is denoted by $\mathrm{M}\otimes\mathbb{Q}_\ell$.

14.2 The Projective Line Minus 0, ∞ and n-th Roots of Unity

In this section we recall some elementary results concerning Galois actions on fundamental groups in the special case of

$$\mathbb{P}^1_{\mathbb{Q}(\mu_n)}\setminus(\{0,\infty\}\cup\mu_n)\,,$$

see [Woj05b] and [DW04]. Let us fix a rational prime ℓ. Let K be a number field containing the group μ_n of n-th roots of unity. We abbreviate

$$\mathrm{V}=\mathbb{P}^1_{\mathrm{K}}\setminus(\{0,\infty\}\cup\mu_n)$$

and denote by

$$\pi_1(\mathrm{V}_{\bar{\mathrm{K}}},\overrightarrow{01})$$

the pro-ℓ completion of the étale fundamental group of $\mathrm{V}_{\bar{\mathrm{K}}}$ based at $\overrightarrow{01}$.

First we describe how to choose generators of $\pi_1(\mathrm{V}_{\bar{\mathrm{K}}},\overrightarrow{01})$. We fix the primitive n-th root of unity

$$\xi:=\exp(\frac{2\pi i}{n})$$

using the fixed embedding $\bar{\mathrm{K}}\subset\mathbb{C}$. Let β be the standard path from $\overrightarrow{01}$ to $\overrightarrow{10}$. Let x be a loop around 0 in the counterclockwise direction based at $\overrightarrow{01}$ in an infinitesimal neighbourhood of 0 and such that the integral along x of the one-form dz/z is $2\pi i$. Let y_0' be a loop around 1 based at $\overrightarrow{10}$ in an infinitesimal neighbourhood of 1 defined in the analogous way. Let s_k be a path from $\overrightarrow{01}$ to $\overrightarrow{0\xi^k}$ in an infinitesimal neibourhood of 0 as in [Woj05b, Picture 2, page 20]. Let $r_k\,:\,\mathrm{V}\to\mathrm{V}$ be the automorphism given by

$$r_k(z)=\xi^k\cdot z\,,$$

and set

$$y_k=\begin{cases}\beta^{-1}\cdot y_0'\cdot\beta & \text{for } k=0\,,\\ s_k^{-1}\cdot r_{k,*}(y_0)\cdot s_k & \text{for } 0<k<n\,.\end{cases}$$

Then the elements

$$x,y_0,y_1,\dots,y_{n-1}$$

are free generators of $\pi_1(V_{\bar{K}}, \overrightarrow{01})$. Observe that

$$s_k^{-1} \cdot r_{k,*}(y_j) \cdot s_k = \begin{cases} y_{j+k} & \text{for } j+k < n\,, \\ x^{-1} \cdot y_{j+k} \cdot x & \text{for } j+k \geqslant n\,. \end{cases} \tag{14.1}$$

Let z be either a K-rational point $z \in V(K)$ or a tangential point defined over K. Let γ be an ℓ-adic path from $\overrightarrow{01}$ to z on $V_{\bar{K}}$. For every $\sigma \in G_K$, the element

$$f_\gamma(\sigma) = \gamma^{-1} \cdot \sigma(\gamma)$$

is a pro-ℓ word in the generators of $\pi_1(V_{\bar{K}}, \overrightarrow{01})$, hence we shall write

$$f_\gamma(\sigma) = f_\gamma(\sigma)(x, y_0, \ldots, y_{n-1})\,.$$

Observe that $r_{k,*}(\gamma) \cdot s_k$ is a path from $\overrightarrow{01}$ to $\xi^k z$ and by (14.1)

$$\begin{aligned} f_{r_{k,*}(\gamma)\cdot s_k}(\sigma) &= s_k^{-1} \cdot r_{k,*}(\gamma^{-1}) \cdot \sigma(r_{k,*}(\gamma) \cdot s_k) = s_k^{-1} \cdot r_{k,*}(f_\gamma(\sigma)) \cdot s_k \cdot f_{s_k}(\sigma) \\ &= f_\gamma(\sigma)(x, y_k, y_{k+1}, \ldots, y_{n-1}, x^{-1} y_0 x, \ldots, x^{-1} y_{k-1} x) \cdot x^{\frac{k(\chi(\sigma)-1)}{n}}\,. \end{aligned} \tag{14.2}$$

Let

$$E : \pi_1(V_{\bar{K}}, \overrightarrow{01}) \to \mathbb{Q}_\ell\{\{X, Y_0, \ldots, Y_{n-1}\}\}$$

be the continuous multiplicative embedding of $\pi_1(V_{\bar{K}}, \overrightarrow{01})$ into the $\mathbb{Q}_\ell$-algebra of non-commutative formal power series $\mathbb{Q}_\ell\{\{X, Y_0, \ldots, Y_{n-1}\}\}$ given by

$$E(x) = \exp(X)\,,$$

$$E(y_j) = \exp(Y_j)$$

for $0 \leqslant j < n$. Let

$$\pi_1(V_{\bar{K}}; z, \overrightarrow{01})$$

be the right $\pi_1(V_{\bar{K}}, \overrightarrow{01})$-torsor of ℓ-adic paths from $\overrightarrow{01}$ to z. The map

$$\delta \to \gamma^{-1} \cdot \delta$$

defines a bijection as right torsors

$$t_\gamma : \pi_1(V_{\bar{K}}; z, \overrightarrow{01}) \to \pi_1(V_{\bar{K}}, \overrightarrow{01})\,.$$

Composing t_γ with the embedding E we get an embedding

$$E_\gamma : \pi_1(V_{\bar{K}}; z, \overrightarrow{01}) \to \mathbb{Q}_\ell\{\{X, Y_0, \ldots, Y_{n-1}\}\}.$$

The Galois group G_K acts on $\pi_1(V_{\bar{K}}, \overrightarrow{01})$ and on $\pi_1(V_{\bar{K}}; z, \overrightarrow{01})$ compatible with the torsor structure. Hence we get two Galois representations

$$\varphi_{\overrightarrow{01}} : G_K \to \mathrm{Aut}(\mathbb{Q}_\ell\{\{X, Y_0, \dots, Y_{n-1}\}\}),$$

$$\psi_\gamma : G_K \to \mathrm{GL}(\mathbb{Q}_\ell\{\{X, Y_0, \dots, Y_{n-1}\}\})$$

deduced from the action of G_K on $\pi_1(V_{\bar{K}}, \overrightarrow{01})$ via the embedding E and on the torsor of paths $\pi_1(V_{\bar{K}}; z, \overrightarrow{01})$ via the embedding E_γ respectively, see [Woj04, Sect. 4] and also [Woj07, Sect. 1].

Before going farther we fix the following notation. The set of **Lie polynomials** in $\mathbb{Q}_\ell\{\{X, Y_0, \dots, Y_{n-1}\}\}$ we denote by

$$\mathrm{Lie}(X, Y_0, \dots, Y_{n-1}).$$

It is a free Lie algebra on $n+1$ generators $X, Y_0, \dots, Y_{n-1}$. The set of **formal Lie power series** in $\mathbb{Q}_\ell\{\{X, Y_0, \dots, Y_{n-1}\}\}$ we denote by

$$L(X, Y_0, \dots, Y_{n-1}).$$

We denote by

$$I_2$$

the closed Lie ideal of $L(X, Y_0, \dots, Y_{n-1})$ generated by Lie brackets with two or more Y's. We shall use the following inductively defined short hand notation

$$[Y_k, X^{(m)}] = \begin{cases} Y_k & \text{if } m = 0 \\ [[Y_k, X^{(m-1)}], X] & \text{for } m > 0 . \end{cases}$$

In an algebra the operator of the left (resp. right) multiplication by a we denote by L_a (resp. R_a).

We recall the definition of **ℓ-adic iterated integrals** from [Woj04]. Let $\mathcal{B}$ be a Hall base of the free Lie algebra $\mathrm{Lie}(X, Y_0, \dots, Y_{n-1})$ on $n+1$ free generators $X, Y_0, \dots, Y_{n-1}$ and let $\mathcal{B}_m$ be the set of elements of degree m in $\mathcal{B}$. For $b \in \mathcal{B}$ we define ℓ-adic iterated integrals

$$l_b(z) : G_{K(\mu_{\ell^\infty})} \to \mathbb{Q}_\ell$$

as follows. For $\sigma \in G_{K(\mu_{\ell^\infty})}$ the expression $(\log \psi_\gamma(\sigma))(1)$ is a Lie element, hence

$$(\log \psi_\gamma(\sigma))(1) = \sum_{b \in \mathcal{B}} l_b(z)(\sigma) \cdot b .$$

More naively, for $\sigma \in G_K$ we define functions

$$li_b(z) : G_K \to \mathbb{Q}_\ell$$

by the equality

$$\log \Lambda_\gamma(\sigma) = \sum_{b \in \mathcal{B}} li_b(z)(\sigma) \cdot b , \tag{14.3}$$

where

$$\Lambda_\gamma(\sigma) := \mathrm{E}(f_\gamma(\sigma)) . \tag{14.4}$$

If $n = 1$ and $\mathrm{Y} = \mathrm{Y}_0$ then the formula from the introduction defining ℓ-adic polylogarithms has the form

$$\log \Lambda_\gamma(\sigma) \equiv l(z)(\sigma)\mathrm{X} + \sum_{n=1}^{\infty} l_n(z)(\sigma)[\mathrm{Y}, \mathrm{X}^{(n-1)}] \mod \mathrm{I}_2 .$$

With the representations $\varphi_{\overrightarrow{01}}$ and ψ_γ there are associated the filtrations

$$\{\mathrm{G}_m = \mathrm{G}_m(\mathrm{V}, \overrightarrow{01})\}_{m \in \mathrm{N}}$$
$$\{\mathrm{H}_m = \mathrm{H}_m(\mathrm{V}; z, \overrightarrow{01})\}_{m \in \mathrm{N}}$$

of G_K, see [Woj04, Sect. 3, pp. 122-124]. We recall that

$$\mathrm{H}_m = \left\{ \sigma \in \mathrm{G}_{\mathrm{K}(\mu_{\ell^\infty})} \;\middle|\; \begin{array}{c} l_b(z)(\sigma) = 0 \text{ and } l_b(\xi^k)(\sigma) = 0 \\ \text{for } 0 \leqslant k < n \text{ and for all } b \in \bigcup_{i<m} \mathcal{B}_i \end{array} \right\}.$$

If $b \in \mathcal{B}_m$ and $\sigma \in \mathrm{H}_m$ then $l_b(z)(\sigma) = li_b(z)(\sigma)$.

Let L be a Lie algebra. The Lie ideals of the lower central series are defined recursively by

$$\Gamma^m\mathrm{L} = \begin{cases} \mathrm{L} & \text{if } m = 1 \\ [\Gamma^{m-1}\mathrm{L}, \mathrm{L}] & \text{for } m > 1. \end{cases}$$

Proposition 4. *For $\sigma \in \mathrm{H}_m(\mathrm{V}; z, \overrightarrow{01})$ we have*

$$(\log \psi_\gamma(\sigma))(1) \equiv \log \Lambda_\gamma(\sigma) \equiv \Lambda_\gamma(\sigma) - 1 \mod \Gamma^{m+1}\mathrm{L}(\mathrm{X}, \mathrm{Y}_0, \ldots, \mathrm{Y}_{n-1}) . \tag{14.5}$$

Proof. It follows from [Woj04, Lemma 1.0.2.] that

$$\psi_\gamma(\sigma) = \mathrm{L}_{\Lambda_\gamma(\sigma)} \circ \varphi_{\overrightarrow{01}}(\sigma) .$$

After taking logarithm and applying the Baker-Campbell-Hausdorff formula we get the first congruence of the proposition. The second congruence is clear. □

Let us set

$$\gamma_k := r_{k,*}(\gamma) \cdot s_k . \tag{14.6}$$

Our next result is a consequence of the formula (14.2).

Proposition 5. *Let $\sigma \in \mathrm{H}_m(\mathrm{V}; z, \overrightarrow{01})$. Then*

$$\log(\Lambda_{\gamma_k}(\sigma)(\mathrm{X}, \mathrm{Y}_0, \ldots, \mathrm{Y}_{n-1})) \equiv \log(\Lambda_\gamma(\sigma)(\mathrm{X}, \mathrm{Y}_k, \ldots, \mathrm{Y}_{n-1}, \mathrm{Y}_0, \ldots, \mathrm{Y}_{k-1}))$$

modulo $\Gamma^{m+1}\mathrm{L}(\mathrm{X}, \mathrm{Y}_0, \ldots, \mathrm{Y}_{n-1})$.

Proof. The proof is the same as the proof of Lemma 15.2.1 in [Woj05b]. □

Corollary 6. *(1) Let* $m > 1$ *and let* $\sigma \in \mathrm{H}_m(\mathrm{V}; z, \overrightarrow{01})$. *Then we have*

$$\log(\Lambda_\gamma(\sigma)(\mathrm{X}, \mathrm{Y}_0, \ldots, \mathrm{Y}_{n-1})) \equiv \sum_{k=0}^{n-1} l_m(\xi^{-k} z)(\sigma)[\mathrm{Y}_k, \mathrm{X}^{(m-1)}]$$

modulo $\Gamma^{m+1}\mathrm{L}(\mathrm{X}, \mathrm{Y}_0, \ldots, \mathrm{Y}_{n-1}) + \mathrm{I}_2$.

(2) Let $\sigma \in \mathrm{G}_{\mathrm{K}(\mu_{\ell^\infty})}$. *Then we have*

$$\log(\Lambda_\gamma(\sigma)(\mathrm{X}, \mathrm{Y}_0, \ldots, \mathrm{Y}_{n-1})) \equiv l(z)(\sigma)\mathrm{X} + \sum_{k=0}^{n-1} l(1 - \xi^{-k} z)(\sigma)\mathrm{Y}_k$$

modulo $\Gamma^2\mathrm{L}(\mathrm{X}, \mathrm{Y}_0, \ldots, \mathrm{Y}_{n-1})$.

Proof. The corollary follows from the very definition of ℓ-adic polylogarithms and from Proposition 5. □

Now we shall define **polylogarithmic quotients** of the representations $\varphi_{\overrightarrow{01}}$ and ψ_γ. Let $\mathcal{I}$ be a closed ideal of $\mathbb{Q}_\ell\{\{\mathrm{X}, \mathrm{Y}_0, \ldots, \mathrm{Y}_{n-1}\}\}$ generated by monomials with any two Y's and by monomials $\mathrm{Y}_k\mathrm{X}$ for $0 \leqslant k \leqslant n-1$. We set

$$\mathrm{Pol}(\mathrm{X}, \mathrm{Y}_0, \ldots, \mathrm{Y}_{n-1}) := \mathbb{Q}_\ell\{\{\mathrm{X}, \mathrm{Y}_0, \ldots, \mathrm{Y}_{n-1}\}\}/\mathcal{I}\,.$$

Observe that the classes

$$\mathrm{X}^i \text{ and } \mathrm{X}^i\mathrm{Y}_k$$

with $i = 0, 1, \ldots$ and $0 \leqslant k \leqslant n-1$ form a topological base of $\mathrm{Pol}(\mathrm{X}, \mathrm{Y}_0, \ldots, \mathrm{Y}_{n-1})$. We denote by

$$\Omega_\gamma(\sigma) \in \mathrm{Pol}(\mathrm{X}, \mathrm{Y}_0, \ldots, \mathrm{Y}_{n-1})$$

the image of the power series $\Lambda_\gamma(\sigma) \in \mathbb{Q}_\ell\{\{\mathrm{X}, \mathrm{Y}_0, \ldots, \mathrm{Y}_{n-1}\}\}$.

Proposition 7. *(1) The representation* $\varphi_{\overrightarrow{01}}$ *induces a representation*

$$\bar{\varphi}_{\overrightarrow{01}} : \mathrm{G}_\mathrm{K} \to \mathrm{Aut}(\mathrm{Pol}(\mathrm{X}, \mathrm{Y}_0, \ldots, \mathrm{Y}_{n-1}))\,.$$

given by

$$\bar{\varphi}_{\overrightarrow{01}}(\sigma)(\mathrm{X}) = \chi(\sigma)\mathrm{X}$$

$$\bar{\varphi}_{\overrightarrow{01}}(\sigma)(\mathrm{Y}_k) = \chi(\sigma)\mathrm{Y}_k + \sum_{i=1}^{\infty} \frac{(-1)^i}{i!}\chi(\sigma)\Big(\frac{k}{n}(\chi(\sigma) - 1)\Big)^i \mathrm{X}^i\mathrm{Y}_k$$

for $k = 0, 1, \ldots, n-1$.

(2) The representation ψ_γ *induces a representation*

$$\bar{\psi}_\gamma : \mathrm{G}_\mathrm{K} \to \mathrm{GL}(\mathrm{Pol}(\mathrm{X}, \mathrm{Y}_0, \ldots, \mathrm{Y}_{n-1}))$$

given by the formula

$$\bar{\psi}_\gamma(\sigma) = \mathrm{L}_{\Omega_\gamma(\sigma)} \circ \bar{\varphi}_{\overrightarrow{01}}(\sigma) .$$

(3) For $n = 1$ we have

$$\log(\Omega_\gamma(\sigma)) = l(z)_\gamma(\sigma)\mathrm{X} + \sum_{i=1}^{\infty}(-1)^{i-1} l_i(z)_\gamma(\sigma)\mathrm{X}^{i-1}\mathrm{Y}_0 .$$

Proof. (1) It follows from [Woj05b, Proposition 15.1.7] that $\varphi_{\overrightarrow{01}}(\mathcal{I}) \subset \mathcal{I}$. Hence $\varphi_{\overrightarrow{01}}$ induces a representation on the quotient space. The explicit formulae also follow from [Woj05b, Proposition 15.1.7].

(2) We recall that $\psi_\gamma(\sigma) = \mathrm{L}_{\Lambda_\gamma(\sigma)} \circ \varphi_{\overrightarrow{01}}(\sigma)$, see [Woj04, Sect. 4], hence the existence of $\bar{\psi}_\gamma$ and the explicit formula. Assertion (3) follows from the definition of ℓ-adic polylogarithms. □

For $\alpha \in \mathbb{Q}_\ell^\times$ we denote by $\tau(\alpha)$ the automorphism of the $\mathbb{Q}_\ell$-algebra Pol(X, Y) such that

$$\tau(\alpha)(\mathrm{X}) = \alpha \cdot \mathrm{X}$$
$$\tau(\alpha)(\mathrm{Y}) = \alpha \cdot \mathrm{Y}$$

and continuous with respect to the topology defined by the powers of the augmentation ideal. For $n = 1$ we have a very simple description of $\bar{\varphi}_{\overrightarrow{01}}$.

Corollary 8. *If $n = 1$ then $\bar{\varphi}_{\overrightarrow{01}}(\sigma) = \tau(\chi(\sigma))$.*

14.3 Linear Independence of ℓ-adic Iterated Integrals

In this section we shall prove linear independence of ℓ-adic polylogarithms in a generic situation. We use the notation of Sect. 14.2.

If $a_1, \dots, a_k$ belong to $\mathrm{K}^\times$ we denote by

$$\langle a_1, \dots, a_k \rangle = \langle a_i \mid 1 \leqslant i \leqslant n \rangle$$

the subgroup of $\mathrm{K}^\times$ generated by $a_1, \dots, a_k$.

Theorem 9. *Suppose that $z \in \mathrm{K}$ is not a root of any equation*

$$z^p \cdot \prod_{k=0}^{n-1} (z - \xi^k)^{q_k} = 1 ,$$

where p and q_k are integers not all equal zero. Suppose that

$$\langle z, 1 - \xi^{-k} z \mid 0 \leqslant k \leqslant n-1 \rangle \cap \langle 1 - \xi^{-k} \mid 1 \leqslant k \leqslant n-1 \rangle \subset \mu_n .$$

Then the homomorphisms

$$l_b(z) : \mathrm{H}_m(\mathrm{V}; z, \overrightarrow{01}) / \mathrm{H}_{m+1}(\mathrm{V}; z, \overrightarrow{01}) \to \mathbb{Q}_\ell$$

for $b \in \mathcal{B}_m$ *are linearly independent over* $\mathbb{Q}_\ell$.

Proof. It follows from the formula

$$\psi_\gamma(\sigma) = \mathrm{L}_{\Lambda_\gamma(\sigma)} \circ \varphi_{\overrightarrow{01}}(\sigma),$$

see [Woj04, page 131], that the morphism

$$\psi_\gamma : \mathrm{G}_\mathrm{K} \to \mathrm{GL}(\mathbb{Q}_\ell\{\{\mathrm{X}, \mathrm{Y}_0, \dots, \mathrm{Y}_{n-1}\}\})$$

induces the morphism of associated graded Lie algebras

$$\Psi_{z,\overrightarrow{01}} : \bigoplus_{m=1}^{\infty} \frac{\mathrm{H}_m(\mathrm{V}; z, \overrightarrow{01})}{\mathrm{H}_{m+1}(\mathrm{V}; z, \overrightarrow{01})} \otimes \mathbb{Q}_\ell \to \mathrm{Lie}(\mathrm{X}, \mathrm{Y}_0, \dots, \mathrm{Y}_{n-1}) \rtimes \mathrm{Der}(\mathrm{Lie}(\mathrm{X}, \mathrm{Y}_0, \dots, \mathrm{Y}_{n-1})).$$

It follows from [Woj05b, Lemma 15.2.5.] that the image of $\Psi_{z,\overrightarrow{01}}$ is contained in

$$\mathrm{Lie}(\mathrm{X}, \mathrm{Y}_0, \dots, \mathrm{Y}_{n-1}) \rtimes \mathrm{Der}^*_{\mathbb{Z}/n}\mathrm{Lie}(\mathrm{X}, \mathrm{Y}_0, \dots, \mathrm{Y}_{n-1}),$$

see [Woj05b, Definition 15.2.4.] for the definition of the right hand factor. The Lie algebra of special derivations

$$\mathrm{Der}^*_{\mathbb{Z}/n}\mathrm{Lie}(\mathrm{X}, \mathrm{Y}_0, \dots, \mathrm{Y}_{n-1})$$

is isomorphic as a vector space to $\mathrm{Lie}(\mathrm{X}, \mathrm{Y}_0, \dots, \mathrm{Y}_{n-1})$ divided by a vector subspace generated by Y_0. The Lie bracket of the Lie algebra of special derivations induces a new bracket denoted by $\{\,\}$ on $\mathrm{Lie}(\mathrm{X}, \mathrm{Y}_0, \dots, \mathrm{Y}_{n-1})$. The obtained Lie algebra we denote by

$$\mathrm{Lie}(\mathrm{X}, \mathrm{Y}_0, \dots, \mathrm{Y}_{n-1})_{\{\,\}},$$

see [Woj05b, page 24 and Lemma 15.2.8.]. To simplify the notation let us set

$$\widetilde{\mathrm{Lie}}_n := \mathrm{Lie}(\mathrm{X}, \mathrm{Y}_0, \dots, \mathrm{Y}_{n-1}) \rtimes \mathrm{Lie}(\mathrm{X}, \mathrm{Y}_0, \dots, \mathrm{Y}_{n-1})_{\{\,\}}$$

Hence, finally, the morphism $\psi_\gamma : \mathrm{G}_\mathrm{K} \to \mathrm{GL}(\mathbb{Q}_\ell\{\{\mathrm{X}, \mathrm{Y}_0, \dots, \mathrm{Y}_{n-1}\}\})$ induces the morphism of graded Lie algebras

$$\Psi_{z,\overrightarrow{01}} : \bigoplus_{m=1}^{\infty} \frac{\mathrm{H}_m(\mathrm{V}; z, \overrightarrow{01})}{\mathrm{H}_{m+1}(\mathrm{V}; z, \overrightarrow{01})} \otimes \mathbb{Q}_\ell \to \widetilde{\mathrm{Lie}}_n .$$

For $\sigma \in \mathrm{H}_m(\mathrm{V}; z, \overrightarrow{01})$ the morphism $\Psi_{z,\overrightarrow{01}}$ is given by the formula

$$\left[\Psi_{z,\overrightarrow{01}}(\sigma)\right]_{\deg m} = \left(\log \Lambda_\gamma(\sigma), \log \Lambda_\beta(\sigma)\right) \quad \mathrm{mod} \quad \Gamma^{m+1}(\widetilde{\mathrm{Lie}}_n),$$

see [Woj07, Sect. 1, page 194]. Hence, it follows from Corollary 6 (2) that the morphism $\Psi_{z,\overrightarrow{01}}$ in degree 1 is given by

$$\left[\Psi_{z,\overrightarrow{01}}(\sigma)\right]_{\deg 1} = \Big(l(z)(\sigma)\mathrm{X} + \sum_{k=0}^{n-1} l(1-\xi^{-k}z)(\sigma)\mathrm{Y}_k, \sum_{k=1}^{n-1} l(1-\xi^{-k})(\sigma)\mathrm{Y}_k\Big).$$

The elements z and $1-\xi^{-k}z$, for $0 \leqslant k < n$ are linearly independent in $\mathrm{K}^\times \otimes \mathbb{Q}$, and the intersection

$$\langle 1-\xi^{-k} \mid 1 \leqslant k \leqslant n-1 \rangle \otimes \mathbb{Q} \cap \langle z, 1-\xi^{-k}z \mid 0 \leqslant k \leqslant n-1 \rangle \otimes \mathbb{Q}$$

is trivial in $\mathrm{K}^\times \otimes \mathbb{Q}$. By Kummer theory we find $\tau \in \mathrm{H}_1 = \mathrm{K}(\mu_{\ell^\infty})$ and $\sigma_k \in \mathrm{H}_1$ for $0 \leqslant k < n$ such that

$$\left[\Psi_{z,\overrightarrow{01}}(\tau)\right]_{\deg 1} = (\mathrm{X}, 0) \quad \text{and} \quad \left[\Psi_{z,\overrightarrow{01}}(\sigma_k)\right]_{\deg 1} = (\mathrm{Y}_k, 0)$$

for $0 \leqslant k < n$. We conclude that the image of $\Psi_{z,\overrightarrow{01}}$ contains the first factor of $\widetilde{\mathrm{Lie}_n}$.

For $m > 1$ and for $\sigma \in \mathrm{H}_m(\mathrm{V}; z, \overrightarrow{01})$ we have

$$\log \Lambda_\gamma(\sigma) \equiv \sum_{b \in \mathcal{B}_m} l_b(z)(\sigma) b \mod \Gamma^{m+1}\mathrm{L}(\mathrm{X}, \mathrm{Y}_0, \ldots, \mathrm{Y}_{n-1}).$$

Hence, it follows that the functions

$$l_b(z) : \mathrm{H}_m(\mathrm{V}_\mathrm{K}; z, \overrightarrow{01}) \to \mathbb{Q}_\ell$$

are linearly independent over $\mathbb{Q}_\ell$. □

Theorem 1 of Sect. 14.1 follows immediately from Theorem 9.

Corollary 10. *The ℓ-adic polylogarithms*

$$l_m(\xi^k z) : \mathrm{H}_m(\mathrm{V}; z, \overrightarrow{01})/\mathrm{H}_{m+1}(\mathrm{V}; z, \overrightarrow{01}) \to \mathbb{Q}_\ell$$

for $k = 0, 1, \ldots, n-1$ are linearly independent over $\mathbb{Q}_\ell$.

Proof. Corollary 10 follows immediately from Theorem 9 and Corollary 6 (1) of Sect. 14.2. □

Remark 11. Theorem 9 is an analogue of the statement, as far as we know unproven, that the iterated integrals indexed by elements of $\mathcal{B}_m$ as in [Woj91] of sequences of length m of 1-forms $\frac{dz}{z}$ and $\frac{dz}{z-\xi^k}$ for $0 \leqslant k \leqslant n-1$ along a fixed path γ from $\overrightarrow{01}$ to a z satisfying the assumption of Theorem 9, are linearly independent over $\mathbb{Q}$.

14.4 Ramification Properties of ℓ-adic Polylogarithms

Let as above K be a number field and z either a rational or a tangential point of

$$\mathbb{P}^1 \setminus \{0,1,\infty\}$$

defined over K. Let γ be an ℓ-adic path on $\mathbb{P}^1_{\bar{K}} \setminus \{0,1,\infty\}$ from $\overrightarrow{01}$ to z.

For an algebraic extension L/K and $z \in \bar{K}$, we denote the maximal pro-ℓ (resp. maximal abelian pro-ℓ) extension of L that is unramified outside ℓ and all prime factors of the fractional ideal (z) by

$$M(L)_{\ell,z} \quad (\text{resp. } M(L)^{ab}_{\ell,z}) .$$

The triple $(\mathbb{P}^1_K \setminus \{0,1,\infty\}, z, \overrightarrow{01})$ has good reduction outside the prime ideals which are factors of the fractional ideals (z) or $(1-z)$. Therefore the action of G_K on the torsor of ℓ-adic paths

$$\pi_1(\mathbb{P}^1_{\bar{K}} \setminus \{0,1,\infty\}; z, \overrightarrow{01})$$

from $\overrightarrow{01}$ to z factors through

$$\mathrm{Gal}(M(K(\mu_{\ell^\infty}))_{\ell,z(1-z)}/K) .$$

Hence the ℓ-adic polylogarithm $l_m(z)_\gamma : G_K \to \mathbb{Q}_\ell$ factors as a map

$$l_m(z)_\gamma \;:\; \mathrm{Gal}\,(M(K(\mu_{\ell^\infty}))_{\ell,z(1-z)}/K) \to \mathbb{Q}_\ell .$$

Let us consider the tower of fields $K \subseteq K(\mu_{\ell^\infty}) \subseteq K(\mu_{\ell^\infty}, z^{1/\ell^\infty})$.

Proposition 12. *The ℓ-adic polylogarithm $l_m(z)_\gamma$ factors through*

$$\mathrm{Gal}\,(M(K(\mu_{\ell^\infty}, z^{1/\ell^\infty}))^{ab}_{\ell,1-z}/K) .$$

Proof. We consider the polylogarithmic quotient $\bar{\psi}_\gamma : G_K \to GL(\mathrm{Pol}(X,Y))$ as in Proposition 7 and restrict to

$$G_{K(\mu_{\ell^\infty}, z^{1/\ell^\infty})} \subset G_K .$$

By Proposition 7 (1) we have $\bar{\varphi}_{\overrightarrow{01}}(\sigma) = \mathrm{id}$ for $\sigma \in G_{K(\mu_{\ell^\infty})}$, so that

$$\bar{\psi}_\gamma(\sigma) = L_{\Omega_\gamma(\sigma)} \in GL(\mathrm{Pol}(X,Y))$$

by Proposition 7 (2). By Proposition 7 (3) for $\sigma \in G_{K(\mu_{\ell^\infty}, z^{1/\ell^\infty})}$ we have

$$\log(\Omega_\gamma(\sigma)) = \sum_{n=1}^{\infty} (-1)^{n-1} l_n(z)_\gamma(\sigma) X^{n-1} Y$$

and

$$\Omega_\gamma(\sigma) = \exp\Big(\sum_{n=1}^{\infty}(-1)^{n-1} l_n(z)_\gamma(\sigma) X^{n-1} Y\Big) = 1 + \sum_{n=1}^{\infty}(-1)^{n-1} l_n(z)_\gamma(\sigma) X^{n-1} Y .$$

Observe that the subgroup in GL(Pol(X,Y)) of automorphisms of the form L_Ω, where $\Omega = 1 + \sum_{n=1}^{\infty} c_n X^{n-1} Y$ is abelian. Hence we deduce a factorization

$$\bar{\psi}_\gamma : \mathrm{Gal}(M(K(\mu_{\ell^\infty}, z^{1/\ell^\infty}))^{ab}_{\ell, z(1-z)}/K) \to \mathrm{GL}(\mathrm{Pol}(X,Y)) .$$

Therefore the ℓ-adic polylogarithm $l_m(z)_\gamma$ factors through the Galois group

$$\mathrm{Gal}(M(K(\mu_{\ell^\infty}, z^{1/\ell^\infty}))^{ab}_{\ell, z(1-z)}/K) .$$

The functions $l_m(z)_\gamma$ are given explicitely by Kummer characters associated to

$$\prod_{i=0}^{\ell^n - 1} (1 - \xi^i_{\ell^n} z^{1/\ell^n})^{i^{m-1}/\ell^n} ,$$

see [NW02]. In fact, the $l_m(z)_\gamma$ factor through $\mathrm{Gal}(M(K(\mu_{\ell^\infty}, z^{1/\ell^\infty}))^{ab}_{\ell,1-z}/K)$, since $1 - \xi^i_{\ell^n} z^{1/\ell^n} \equiv 1$ modulo any prime dividing a prime factor that occurs in (z). □

Corollary 13. *The ℓ-adic polylogarithm $l_m(z)_\gamma$ restricted to the Galois group*

$$\mathrm{Gal}\,(M(K(\mu_{\ell^\infty}, z^{1/\ell^\infty}))^{ab}_{\ell,1-z}/K(\mu_{\ell^\infty}, z^{1/\ell^\infty}))$$

is a homomorphism.

Proof. In the proof of Proposition 12 we have already seen that the representation $\bar{\psi}_\gamma$ restricted to $G_{K(\mu_{\ell^\infty}, z^{1/\ell^\infty})}$ is abelian. □

Combining Proposition 12, Corollary 10 and 13 we get Theorem 2 of Sect. 14.1.

14.5 Iwasawa Action on ℓ-adic Polylogarithms

We keep the notation of Sect. 14.4. Let us consider the tower of fields

$$K \subseteq K(\mu_{\ell^\infty}) \subseteq K(\mu_{\ell^\infty}, z^{1/\ell^\infty}) \subseteq M(K(\mu_{\ell^\infty}, z^{1/\ell^\infty}))^{ab}_{\ell,1-z}$$

with Galois groups

$$\begin{aligned} \mathcal{G} &= \mathrm{Gal}\,(M(K(\mu_{\ell^\infty}, z^{1/\ell^\infty}))^{ab}_{\ell,1-z}/K(\mu_{\ell^\infty}, z^{1/\ell^\infty})) \\ \mathbb{Z}_\ell(1) &= \mathrm{Gal}(K(\mu_{\ell^\infty}, z^{1/\ell^\infty})/K(\mu_{\ell^\infty})) \\ \Gamma &= \mathrm{Gal}(K(\mu_{\ell^\infty})/K) \end{aligned}$$

where $\mathrm{Gal}(\mathrm{K}(\mu_{\ell^\infty}, z^{1/\ell^\infty})/\mathrm{K}(\mu_{\ell^\infty})) = \mathbb{Z}_\ell(1)$ as a Γ-module. We set

$$\Phi = \mathrm{Gal}(\mathrm{K}(\mu_{\ell^\infty}, z^{1/\ell^\infty})/\mathrm{K}) = \mathbb{Z}_\ell(1) \rtimes \Gamma$$

and want to understand the action of Φ and of $\mathbb{Z}_\ell[[\Phi]]$ on $\mathcal{G}$. By Corollary 13, the ℓ-adic polylogarithms $l_n(z)_\gamma$ induce elements of $\mathrm{Hom}(\mathcal{G}, \mathbb{Q}_\ell)$. As our first step to understand $\mathcal{G}$ we shall study the $\mathbb{Z}_\ell[[\Phi]]$-module generated by $l_n(z)_\gamma$ in $\mathrm{Hom}(\mathcal{G}, \mathbb{Q}_\ell)$.

We recall that Φ acts on $\mathcal{G}$ on the left in the following way. Let $\sigma \in \Phi$ and $\tau \in \mathcal{G}$. Let $\tilde{\sigma} \in \mathrm{Gal}(\mathrm{M}(\mathrm{K}(\mu_{\ell^\infty}, z^{1/\ell^\infty}))^{ab}_{\ell,1-z}/\mathrm{K})$ be a lifting of σ. Then the formula

$${}^\sigma\tau := \tilde{\sigma} \cdot \tau \cdot \tilde{\sigma}^{-1}$$

defines the action of Φ on $\mathcal{G}$. The right action of Φ on $\mathrm{Hom}(\mathcal{G}, \mathbb{Q}_\ell)$ is given by

$$f^\sigma(\tau) = f(\tilde{\sigma} \cdot \tau \cdot \tilde{\sigma}^{-1}).$$

Lemma 14. *For any $\alpha, \tau \in \mathrm{G}_\mathrm{K}$ we have in $\mathbb{Q}_\ell\{\{\mathrm{X}, \mathrm{Y}\}\}$*

$$\Lambda_\gamma(\alpha \cdot \tau \cdot \alpha^{-1}) = \Lambda_\gamma(\alpha) \cdot \varphi_{\overrightarrow{01}}(\alpha)(\Lambda_\gamma(\tau)) \cdot \varphi_{\overrightarrow{01}}(\alpha \cdot \tau \cdot \alpha^{-1})(\Lambda_\gamma(\alpha)^{-1}).$$

Proof. This follows from [Woj04, Proposition 1.0.7 and Corollary 1.0.8]. □

We define the product $\star$ on the Lie algebra of formal Lie power series $\mathrm{L}(\mathrm{X}, \mathrm{Y})$ by the Baker-Campbell-Hausdorff formula

$$\mathrm{A} \star \mathrm{B} := \log(\exp(\mathrm{A}) \cdot \exp(\mathrm{B})).$$

Proposition 15. *The action of $\sigma \in \Phi$ on $l_m(z)_\gamma \in \mathrm{Hom}(\mathcal{G}, \mathbb{Q}_\ell)$ is given by*

$$(l_m(z)_\gamma)^\sigma = \chi(\sigma)^m \cdot l_m(z)_\gamma + \sum_{k=1}^{m-1} \frac{(-l(z)_\gamma(\sigma))^k}{k!} \cdot \chi(\sigma)^{m-k} \cdot l_{m-k}(z)_\gamma.$$

Proof. Let $\tau \in \mathcal{G}$ and let $\bar{\sigma}$ and $\bar{\tau}$ be liftings of σ and τ to $\mathrm{Gal}(\bar{\mathrm{K}}/\mathrm{K})$. It follows from Lemma 14 that

$$\log \Lambda_\gamma(\bar{\sigma} \cdot \bar{\tau} \cdot \bar{\sigma}^{-1}) = \log \Lambda_\gamma(\bar{\sigma}) \star \varphi_{\overrightarrow{01}}(\bar{\sigma})(\log \Lambda_\gamma(\bar{\tau})) \star (\varphi_{\overrightarrow{01}}(\bar{\sigma} \cdot \bar{\tau} \cdot \bar{\sigma}^{-1})(-\log \Lambda_\gamma(\bar{\sigma}))).$$

Hence we get modulo I_2 that

$$\sum_{n=1}^{\infty} l_n(z)({}^\sigma\tau)[\mathrm{Y}, \mathrm{X}^{(n-1)}] \equiv \Big(l(z)(\bar{\sigma})\mathrm{X} + \sum_{n=1}^{\infty} l_n(z)(\bar{\sigma})[\mathrm{Y}, \mathrm{X}^{(n-1)}]\Big) \star \Big(\chi(\bar{\sigma}) l(z)(\tau)\mathrm{X}$$
$$+ \sum_{n=1}^{\infty} \chi(\bar{\sigma})^n \cdot l_n(z)(\tau)[\mathrm{Y}, \mathrm{X}^{(n-1)}]\Big) \star \Big(-l(z)(\bar{\sigma})\mathrm{X} - \sum_{n=1}^{\infty} l_n(z)(\bar{\sigma})[\mathrm{Y}, \mathrm{X}^{(n-1)}]\Big).$$

Observe that $l(z)(\bar{\sigma})$ and $\chi(\bar{\sigma})$ depend only on σ. Hence we replace them by $l(z)(\sigma)$ and $\chi(\sigma)$. We get the formula of the proposition by calculating the right hand side of the congruence and comparing coefficients at $[\mathrm{Y},\mathrm{X}^{(n-1)}]$. □

Corollary 16. *Let* $\mu \in \mathbb{Z}_\ell[[\Phi]]$. *Then* $(l_m(z)_\gamma)^\mu$ *equals*

$$\Big(\int_\Phi \chi(x)^m d\mu(x)\Big)\cdot l_m(z)_\gamma + \sum_{k=1}^{m-1}\Big(\int_\Phi \frac{(-l(z)_\gamma(x))^k}{k!}\cdot \chi(x)^{m-k} d\mu(x)\Big)\cdot l_{m-k}(z)_\gamma\,.$$

Proof. This generalization of Proposition 15 is straightforward. □

Observe that we have just proved Theorem 3 from Sect. 14.1.

14.6 Profinite Sheaves

The ℓ-adic polylogarithms and ℓ-adic iterated integrals studied in [Woj04], [Woj05a], [Woj05b] and in [NW02] arise from actions of Galois groups on ℓ-adic paths

$$\pi_1(\mathbb{P}^1_{\bar{\mathbb{Q}}} \setminus \{a_1,\ldots,a_n\}; z, v)\,.$$

On the other side in [BD94], [BL94] and in various other papers motivic polylogarithmic sheaves are studied. Their ℓ-adic realizations are inverse systems of locally constant sheaves of $\mathbb{Z}/\ell^n\mathbb{Z}$-modules in étale topology. Each stalk is equipped with a Galois representation. The relation between parallel transport and the Galois representations in stalks is given by the formula

$$\sigma_t \circ p_* = \sigma(p)_* \circ \sigma_s\,, \tag{14.7}$$

where p_* (resp. $\sigma(p)_*$) is the parallel transport along the path p (resp. $\sigma(p)$) from s to t, σ_s (resp. σ_t) is the action of $\sigma \in \mathrm{G_K}$ in the stalk over s (resp. over t) and $\sigma(p)$ is the image of p by σ in the torsor of paths from s to t.

The formula (14.7) is fundamental to relate ℓ-adic polylogarithms introduced in [Woj05a] with polylogarithmic sheaves, which we discuss next.

Let S be a smooth quasi-projective, geometrically connected algebraic variety over K. We denote by $\mathrm{S_{\acute{e}t}}$ the étale site associated to S. We denote by $\Pi_1(\mathrm{S_{\acute{e}t}})$ the fundamental groupoid on $\mathrm{S_{\acute{e}t}}$, see [SGA1, Exp V].

Definition 17. A **locally constant sheaf** of finite sets on $\mathrm{S_{\acute{e}t}}$ is a functor

$$\Pi_1(\mathrm{S_{\acute{e}t}}) \to \big(\text{category of finite sets}\big)\,.$$

Let $\bar{a}:\ \mathrm{Spec}(\bar{\mathrm{K}}) \to \mathrm{S}$ be a geometric point of S with values in $\bar{\mathrm{K}}$. The category $\Pi_1(\mathrm{S_{\acute{e}t}})$ is equivalent to the category with one object and automorphism group

$\pi_1^{\text{ét}}(S,\bar{a})$. Hence a locally constant sheaf $\mathcal{F}$ of finite sets on $S_{\text{ét}}$ is a finite discrete set $\mathcal{F}_{\bar{a}}$, the stalk of $\mathcal{F}$ in $\bar{a}$, equipped with a continuous action of $\pi_1^{\text{ét}}(S,\bar{a})$.

Let S have a K-point $s : \text{Spec}(K) \to S$ and let $\bar{s} : \text{Spec}(\bar{K}) \to S_{\bar{K}}$, and by abuse of notation also $\bar{s} : \text{Spec}(\bar{K}) \to S$, be the corresponding geometric point of $S_{\bar{K}}$ or S. The structure map $\text{pr} : S \to \text{Spec}(K)$ induces the projection

$$\text{pr}_* : \pi_1^{\text{ét}}(S,\bar{s}) \to \pi_1(\text{Spec}(K),\text{Spec}(\bar{K})) = G_K$$

that is canonically split by

$$s_* : G_K \to \pi_1^{\text{ét}}(S,\bar{s}) .$$

Therefore we have a semidirect product

$$\pi_1^{\text{ét}}(S,\bar{s}) = \pi_1^{\text{ét}}(S_{\bar{K}},\bar{s}) \rtimes G_K ,$$

and for a locally constant sheaf $\mathcal{F}$ both groups $\pi_1^{\text{ét}}(S_{\bar{K}},\bar{s})$ and G_K act on $\mathcal{F}_{\bar{s}}$. If s and t are two K-points of S then the relation between actions of G_K on $\mathcal{F}_{\bar{s}}$ and $\mathcal{F}_{\bar{t}}$ is given by the formula (14.7).

Definition 18. A **profinite sheaf** on S is a projective system of locally constant sheaves of finite sets on $S_{\text{ét}}$.

The category of finite sets we can replace by the category of finite groups, finite abelian groups, finite ℓ-groups, finite ℓ-sets, finite $\mathbb{Z}_\ell$-modules, finite $\mathbb{Z}_\ell$-algebras and so on. Then we speak about a profinite sheaf of groups, abelian groups, and so on. A classical smooth ℓ-adic sheaf provides an example of a profinite sheaf.

Let $\mathcal{F} = \{\mathcal{F}_i\}_{i\in I}$ be a profinite sheaf on S. Then

$$\mathcal{F}_{\bar{s}} := \varprojlim \mathcal{F}_{i,\bar{s}}$$

is the stalk of the profinite sheaf $\mathcal{F}$ over $\bar{s}$. The group $\pi_1^{\text{ét}}(S,\bar{s})$ acts continuously on $\mathcal{F}_{\bar{s}}$. Hence we get two representations:

$$\rho_{\bar{s}} : \pi_1^{\text{ét}}(S_{\bar{K}},\bar{s}) \to \text{Aut}(\mathcal{F}_{\bar{s}}) ,$$

called the monodromy representation in the stalk over $\bar{s}$ and

$$G_K \to \text{Aut}(\mathcal{F}_{\bar{s}}) ,$$

the Galois representation in the stalk over $\bar{s}$. The relation between Galois representations in the stalks over $\bar{s}$ and $\bar{t}$, where t is another K-point of S, are given by the formula

$$\sigma_{\bar{\mathbf{t}}} \circ \mathbf{p}_* = \sigma(\mathbf{p})_* \circ \sigma_{\bar{\mathbf{s}}} , \tag{14.8}$$

where p_* (resp. $\sigma(p)_*$) is the parallel transport along the path p (resp. $\sigma(p)$) from $\bar{s}$ to $\bar{t}$, $\sigma_{\bar{s}}$ (resp. $\sigma_{\bar{t}}$) is the action of $\sigma \in G_K$ in the stalk over $\bar{s}$ (resp. over $\bar{t}$) and $\sigma(p)$ is the image of p by σ in the torsor of paths from $\bar{s}$ to $\bar{t}$.

It is clear from (14.8) that the Galois representation in the stalk over $\bar{s}$ together with the cocycle $\sigma \mapsto p^{-1} \cdot \sigma(p)$ determines uniquely the Galois representation in the stalk over $\bar{t}$. Let us write (14.8) in the form

$$p_*^{-1} \circ \sigma_{\bar{t}} \circ p_* = (p^{-1} \cdot \sigma(p))_* \circ \sigma_{\bar{s}} .$$

Therefore it is crucial to calculate the element

$$f_p(\sigma) := p^{-1} \cdot \sigma(p) \in \pi_1^{\text{ét}}(\mathrm{S}_{\bar{\mathrm{K}}}, \bar{s}) .$$

We started to study $f_p(\sigma)$ in the series of papers on ℓ-adic iterated integrals.

For the next proposition we introduce the following notation

$$\mathrm{F}_{\mathrm{S},\bar{s}}(\mathrm{G}_{\mathrm{K}}) := \{\mathrm{T}^{-1} \cdot \sigma(\mathrm{T}) \in \pi_1^{\text{ét}}(\mathrm{S}_{\bar{\mathrm{K}}}, \bar{s}) \ ; \ \mathrm{T} \in \pi_1^{\text{ét}}(\mathrm{S}_{\bar{\mathrm{K}}}, \bar{s}), \ \sigma \in \mathrm{G}_{\mathrm{K}}\} .$$

Proposition 19. *Let us assume that* $\mathrm{F}_{\mathrm{S},\bar{s}}(\mathrm{G}_{\mathrm{K}})$ *topologically and normally generates* $\pi_1^{\text{ét}}(\mathrm{S}_{\bar{\mathrm{K}}}, \bar{s})$. *Let* $\mathcal{F}$ *be a profinite sheaf on* S *such that the monodromy representation*

$$\rho_{\bar{s}} \ : \ \pi_1^{\text{ét}}(\mathrm{S}_{\bar{\mathrm{K}}}, \bar{s}) \to \mathrm{Aut}(\mathcal{F}_{\bar{s}})$$

is non-trivial. Then the Galois representation $\mathrm{G}_{\mathrm{K}} \to \mathrm{Aut}(\mathcal{F}_{\bar{s}})$ *in the stalk* $\mathcal{F}_{\bar{s}}$ *is also non-trivial.*

Proof. For $\mathrm{T} \in \pi_1^{\text{ét}}(\mathrm{S}_{\bar{\mathrm{K}}}, \bar{s})$ and $\sigma \in \mathrm{G}_{\mathrm{K}}$ formula (14.8) leads to

$$\mathrm{T}_*^{-1} \circ \sigma_{\bar{s}} \circ \mathrm{T}_* \circ (\sigma_{\bar{s}})^{-1} = (\mathrm{T}^{-1} \cdot \sigma(\mathrm{T}))_*.$$

If $\sigma_{\bar{s}} = \mathrm{id}$ for all $\sigma \in \mathrm{G}_{\mathrm{K}}$ then the set $\mathrm{F}_{\mathrm{S},\bar{s}}(\mathrm{G}_{\mathrm{K}})$ lies in the kernel of $\rho_{\bar{s}}$. But the set $\mathrm{F}_{\mathrm{S},\bar{s}}(\mathrm{G}_{\mathrm{K}})$ topologically and normally generates $\pi_1^{\text{ét}}(\mathrm{S}_{\bar{\mathrm{K}}}, \bar{s})$. Hence $\sigma_{\bar{s}}$ cannot be the identity for all $\sigma \in \mathrm{G}_{\mathrm{K}}$. □

14.7 On Profinite Sheaves Related to Bundles of Fundamental Groups

In this section we shall study examples of profinite sheaves for which the monodromy representation determines Galois representations in the stalks. The notation and assumptions are as in Sect. 14.6. We recall only that $\pi_1(\mathrm{S}_{\bar{\mathrm{K}}}, \bar{s})$ is the pro-ℓ completion of $\pi_1^{\text{ét}}(\mathrm{S}_{\bar{\mathrm{K}}}, \bar{s})$.

For $\sigma \in \mathrm{G}_{\mathrm{K}}$ we denote by σ the induced automorphisms of $\pi_1^{\text{ét}}(\mathrm{S}_{\bar{\mathrm{K}}}, \bar{s})$ and of $\pi_1(\mathrm{S}_{\bar{\mathrm{K}}}, \bar{s})$. We have the surjective map $\pi_1^{\text{ét}}(\mathrm{S}_{\bar{\mathrm{K}}}, \bar{s}) \to \pi_1(\mathrm{S}_{\bar{\mathrm{K}}}, \bar{s})$. If $\mathrm{T} \in \pi_1^{\text{ét}}(\mathrm{S}_{\bar{\mathrm{K}}}, \bar{s})$ we denote also by T its image in $\pi_1(\mathrm{S}_{\bar{\mathrm{K}}}, \bar{s})$.

Proposition 20. *Let* S *and* s *be as in Sect. 14.6. We assume that* $\pi_1(\mathrm{S}_{\bar{\mathrm{K}}}, \bar{s})$ *is a free noncommutative pro-*ℓ *group. Let* $\mathcal{P}_{\bar{s}}$ *be a profinite sheaf of* ℓ*-groups on* S *whose stalk over* $\bar{s}$ *is the group* $(\mathcal{P}_{\bar{s}})_{\bar{s}} = \pi_1(\mathrm{S}_{\bar{\mathrm{K}}}, \bar{s})$ *with the monodromy representation*

$$\rho_{\bar{s}} : \pi_1^{\text{ét}}(S_{\bar{K}}, \bar{s}) \to \text{Aut}(\pi_1(S_{\bar{K}}, \bar{s}))$$

given by $\rho_{\bar{s}}(T)(w) = T \cdot w \cdot T^{-1}$. *Then for any* $\sigma \in G_K$ *and any* $w \in \pi_1(S_{\bar{K}}, \bar{s})$ *we have*

$$\sigma_{\bar{s}}(w) = \sigma(w)\,.$$

Proof. Let $\sigma \in G_K$, $T \in \pi_1^{\text{ét}}(S_{\bar{K}}, \bar{s})$ and $w \in \pi_1(S_{\bar{K}}, \bar{s})$. The formula (14.8) implies

$$\sigma_{\bar{s}}(T \cdot w \cdot T^{-1}) = \sigma(T) \cdot \sigma_{\bar{s}}(w) \cdot \sigma(T)^{-1}\,.$$

Let us take T such that its image in $\pi_1(S_{\bar{K}}, \bar{s})$ is w. Then

$$\sigma_{\bar{s}}(w) = \sigma(w) \cdot \sigma_{\bar{s}}(w) \cdot \sigma(w)^{-1}\,.$$

The assumption that $\pi_1(S_{\bar{K}}, \bar{s})$ is a free pro-ℓ group implies that $\sigma_{\bar{s}}(w) = \sigma(w)^{\eta(\sigma,w)}$, where $\eta(\sigma, w) \in \mathbb{Z}_\ell$. Let $w_1, w_2 \in \pi_1(S_{\bar{K}}, \bar{s})$ be two arbitrary noncommuting elements. Then

$$\sigma_{\bar{s}}(w_1 \cdot w_2) = \sigma(w_1 \cdot w_2)^{\eta(\sigma, w_1 \cdot w_2)} = (\sigma(w_1) \cdot \sigma(w_2))^{\eta(\sigma, w_1 \cdot w_2)}$$

and

$$\sigma_{\bar{s}}(w_1) \cdot \sigma_{\bar{s}}(w_2) = \sigma(w_1)^{\eta(\sigma, w_1)} \cdot \sigma(w_2)^{\eta(\sigma, w_2)}\,.$$

Hence we get

$$(\sigma(w_1) \cdot \sigma(w_2))^{\eta(\sigma, w_1 \cdot w_2)} = \sigma(w_1)^{\eta(\sigma, w_1)} \cdot \sigma(w_2)^{\eta(\sigma, w_2)}$$

for two noncommuting elements $\sigma(w_1)$, $\sigma(w_2)$ in the free pro-ℓ group $\pi_1(S_{\bar{K}}, \bar{s})$ and for $\eta(\sigma, w_1 \cdot w_2) \neq 0$, $\eta(\sigma, w_1) \neq 0$ and $\eta(\sigma, w_2) \neq 0$. This implies that $\eta(\sigma, w) = 1$ for all σ and w. □

Proposition 21. *Let* S *and* s *be as above. Let* $\mathcal{P}$ *be a profinite sheaf on* $S \times S$ *whose stalk over* $(\bar{s}, \bar{s})$ *is* $\mathcal{P}_{(\bar{s},\bar{s})} = \pi_1(S_{\bar{K}}, \bar{s})$ *considered as a set. We assume that the monodromy representation*

$$\rho_{(\bar{s},\bar{s})} : \pi_1^{\text{ét}}(S_{\bar{K}}, \bar{s}) \times \pi_1^{\text{ét}}(S_{\bar{K}}, \bar{s}) \to \text{Bijections}(\pi_1(S_{\bar{K}}, \bar{s}))$$

is given by $\rho_{(\bar{s},\bar{s})}(T_1, T_2)(w) = T_1 \cdot w \cdot T_2^{-1}$. *We assume also that the center of the group* $\pi_1(S_{\bar{K}}, \bar{s})$ *is* 1. *Then for any* $\sigma \in G_K$ *and any* $w \in \pi_1(S_{\bar{K}}, \bar{s})$ *we have*

$$\sigma_{(\bar{s},\bar{s})}(w) = \sigma(w)\,.$$

Proof. The formula (14.8) implies

$$\sigma(T_1) \cdot \sigma_{(\bar{s},\bar{s})}(w) \cdot \sigma(T_2)^{-1} = \sigma_{(\bar{s},\bar{s})}(T_1 \cdot w \cdot T_2^{-1})\,. \tag{14.9}$$

Let us take $T_1 = T_2 = T$ and $w = 1$. Then we get

$$\sigma(T) \cdot \sigma_{(\bar{s},\bar{s})}(1) \cdot \sigma(T)^{-1} = \sigma_{(\bar{s},\bar{s})}(1)\,.$$

Hence $\sigma_{(\bar{s},\bar{s})}(1)$ lies in the center of $\pi_1(\mathrm{S}_{\bar{\mathrm{K}}},\bar{s})$ and $\sigma_{(\bar{s},\bar{s})}(1)=1$. Let us take $\mathrm{T}_2=w=1$ in the formula (14.9). Then we get $\sigma(\mathrm{T}_1)=\sigma_{(\bar{s},\bar{s})}(\mathrm{T}_1)$ for any $\mathrm{T}_1\in\pi_1(\mathrm{S}_{\bar{\mathrm{K}}},\bar{s})$. □

14.8 Polylogarithmic Profinite Sheaves and ℓ-adic Polylogarithms

We shall show that if a profinite sheaf of $\mathbb{Z}_\ell$-modules on $\mathbb{P}^1_{\mathrm{K}}\setminus\{0,1,\infty\}$ has the same monodromy representation as the classical complex polylogarithm then the Galois representation in the stalk over $z\in(\mathbb{P}^1_{\mathrm{K}}\setminus\{0,1,\infty\})(\mathrm{K})$ is given by the ℓ-adic polylogarithms evaluated at z.

We start by recalling a result about the monodromy of classical complex polylogarithms. The constant vector bundle with fibre Pol(X, Y)

$$\mathbb{P}^1(\mathbb{C})\setminus\{0,1,\infty\}\times\mathrm{Pol}(\mathrm{X},\mathrm{Y})\to\mathbb{P}^1(\mathbb{C})\setminus\{0,1,\infty\}$$

is endowed with the connection $\nabla=d-\omega$ defined by the 1-form

$$\omega=\frac{1}{2\pi i}\frac{dz}{z}\otimes\mathrm{X}+\frac{1}{2\pi i}\frac{dz}{z-1}\otimes\mathrm{Y}\,.$$

The space of horizontal sections $\Lambda:\ \mathbb{P}^1(\mathbb{C})\setminus\{0,1,\infty\}\to\mathrm{Pol}(\mathrm{X},\mathrm{Y})$ is the solution space of the differential equation

$$d\Lambda(z)-\Big(\frac{1}{2\pi i}\frac{dz}{z}\otimes\mathrm{X}+\frac{1}{2\pi i}\frac{dz}{z-1}\otimes\mathrm{Y}\Big)\cdot\Lambda(z)=0\,.$$

One checks that

$$\Lambda_{\overrightarrow{01}}(z):=\exp\Big(\frac{1}{2\pi i}\log(z)\Big)\mathrm{X}+\frac{1}{2\pi i}\log(1-z)\mathrm{Y}+\sum_{k=2}^{\infty}\frac{-1}{(2\pi i)^k}Li_k(z)\mathrm{X}^{k-1}\mathrm{Y}$$

is locally a horizontal section. The functions $\log(z)$, $\log(1-z)$ and $Li_k(z)$ are calculated along a path α from $\overrightarrow{01}$ to z.

Let x and y be the standard generators of $\pi_1(\mathbb{P}^1(\mathbb{C})\setminus\{0,1,\infty\},\overrightarrow{01})$. To calculate the monodromy of $\Lambda_{\overrightarrow{01}}(z)$ we integrate along the paths $\alpha\cdot x$ and $\alpha\cdot y$. The monodromy transformation of $\Lambda_{\overrightarrow{01}}(z)$ is given by

$$x:\ \Lambda_{\overrightarrow{01}}(z)\to\Lambda_{\overrightarrow{01}}(z)\cdot\exp(\mathrm{X})$$

$$y:\ \Lambda_{\overrightarrow{01}}(z)\to\Lambda_{\overrightarrow{01}}(z)\cdot(1+\mathrm{Y})\,.$$

The group $\pi_1(\mathbb{P}^1(\mathbb{C})\setminus\{0,1,\infty\},z)$ is freely generated by

$${}^{\alpha}x=\alpha\cdot x\cdot\alpha^{-1}\quad\text{and}\quad{}^{\alpha}y=\alpha\cdot y\cdot\alpha^{-1}\,.$$

The monodromy representation of $(\mathrm{Pol}(X,Y),\nabla)$ in z

$$\rho_z : \pi_1(\mathbb{P}^1(\mathbb{C})\setminus\{0,1,\infty\},z)\to \mathrm{GL}(\mathrm{Pol}(X,Y))$$

is given by

$$\rho_z({}^{\alpha}x)=\mathrm{R}_{\exp(X)}$$
$$\rho_z({}^{\alpha}y)=\mathrm{R}_{1+Y}\,.$$

For a word $w({}^{\alpha}x,{}^{\alpha}y)$ in ${}^{\alpha}x$ and ${}^{\alpha}y$ we thus find

$$\rho_z(w({}^{\alpha}x,{}^{\alpha}y))=\mathrm{R}_{w(\exp(X),1+Y)}=\mathrm{R}_{\bar{\mathrm{E}}(w)}\,.$$

Now we shall study the ℓ-adic situation. Let z_0 be a K-point of $\mathbb{P}^1_{\mathrm{K}}\setminus\{0,1,\infty\}$. We start with the description of the action of G_{K} on

$$\pi_1(\mathbb{P}^1_{\bar{\mathrm{K}}}\setminus\{0,1,\infty\},z_0)\,.$$

Let γ be a path from z_0 to $\overrightarrow{01}$ and let p be the standard path from $\overrightarrow{01}$ to $\overrightarrow{10}$. We recall that x and y are the standard generators of $\pi_1(\mathbb{P}^1_{\bar{\mathrm{K}}}\setminus\{0,1,\infty\},\overrightarrow{01})$. Then

$$x_{z_0}=\gamma^{-1}\cdot x\cdot\gamma \quad\text{and}\quad y_{z_0}=\gamma^{-1}\cdot y\cdot\gamma$$

are free generators of the pro-ℓ group $\pi_1(\mathbb{P}^1_{\bar{\mathrm{K}}}\setminus\{0,1,\infty\},z_0)$. The following lemma is a standard exercice.

Lemma 22. *The action of* G_{K} *on* $\pi_1(\mathbb{P}^1_{\bar{\mathrm{K}}}\setminus\{0,1,\infty\},z_0)$ *is given by the formulae*

$$\sigma(x_{z_0}) = f_\gamma(\sigma)^{-1}\cdot x_{z_0}{}^{\chi(\sigma)}\cdot f_\gamma(\sigma)$$
$$\sigma(y_{z_0}) = f_\gamma(\sigma)^{-1}\cdot(\gamma^{-1}\cdot f_p(\sigma)^{-1}\cdot\gamma)\cdot y_{z_0}{}^{\chi(\sigma)}\cdot(\gamma^{-1}\cdot f_p(\sigma)\cdot\gamma)\cdot f_\gamma(\sigma)\,.$$

Let z be another K-point of $\mathbb{P}^1_{\mathrm{K}}\setminus\{0,1,\infty\}$. Let δ be a path from z to z_0. Let us set

$$\gamma_z := \gamma\cdot\delta\,.$$

The following equalities can be found in [Woj04].

$$f_{\gamma\cdot\delta}(\sigma)=\delta^{-1}\cdot f_\gamma(\sigma)\cdot\delta\cdot f_\delta(\sigma) \text{ and } f_{\delta^{-1}}(\sigma)^{-1}=\delta\cdot f_\delta(\sigma)\cdot\delta^{-1}\,. \tag{14.10}$$

Hence we get

$$\delta\cdot f_{\gamma\cdot\delta}(\sigma)\cdot\delta^{-1}=f_\gamma(\sigma)\cdot f_{\delta^{-1}}(\sigma)^{-1}\,. \tag{14.11}$$

The group $\pi_1(\mathrm{P}^1_{\bar{\mathrm{K}}}\setminus\{0,1,\infty\},z)$ is freely generated by the elements

$$x_z := \gamma_z^{-1}\cdot x\cdot\gamma_z \quad\text{and}\quad y_z := \gamma_z^{-1}\cdot x\cdot\gamma_z$$

as a pro-ℓ group. We use the following exponential embedings.

$$E_{\overrightarrow{01}} : \pi_1(\mathbb{P}^1_{\bar{K}} \setminus \{0,1,\infty\}, \overrightarrow{01}) \to \mathbb{Q}_\ell\{\{X,Y\}\}$$

$$E_{\overrightarrow{01}}(x) := \exp(X) \quad \text{and} \quad E_{\overrightarrow{01}}(y) := \exp(Y)\,,$$

$$E_{z_0} : \pi_1(\mathbb{P}^1_{\bar{K}} \setminus \{0,1,\infty\}, z_0) \to \mathbb{Q}_\ell\{\{X,Y\}\}$$

$$E_{z_0}(x_{z_0}) := \exp(X) \quad \text{and} \quad E_{z_0}(y_{z_0}) := \exp(Y)\,,$$

$$E_z : \pi_1(\mathbb{P}^1_{\bar{K}} \setminus \{0,1,\infty\}, z) \to \mathbb{Q}_\ell\{\{X,Y\}\}$$

$$E_z(x_z) := \exp(X) \quad \text{and} \quad E_z(y_z) := \exp(Y)\,.$$

In other words we have trivialized the bundle of fundamental groups along the path γ_z. The action of G_K on $\mathbb{Q}_\ell\{\{X,Y\}\}$ considered over a K-point s is deduced from the action of G_K on $\pi_1(\mathbb{P}^1_{\bar{K}} \setminus \{0,1,\infty\}, s)$ so it depends on s

Using the embeddings E_a, for $a \in \{\overrightarrow{01}, z_0, z\}$, we can define the Λ-series as

$$\Lambda_\delta(\sigma) := E_z(f_\delta(\sigma)) \quad \text{and} \quad \Lambda_\gamma(\sigma) := E_{z_0}(f_\gamma(\sigma))\,.$$

The composition of an embedding E_a with the quotient map $\mathbb{Q}\{\{X,Y\}\} \to \mathrm{Pol}(X,Y)$ we denote by $\bar{E}_a$. We recall that the images of Λ-series by the quotient map are Ω-series. For example we have

$$\Omega_{\delta^{-1}}(\sigma) = \bar{E}_{z_0}(f_{\delta^{-1}}(\sigma))\,.$$

Because of the trivialization of the bundle of fundamental groups we can compare various Λ-series. It follows from (14.10) and (14.11) that

$$\Lambda_{\gamma\cdot\delta}(\sigma) = \Lambda_\gamma(\sigma)\cdot\Lambda_\delta(\sigma)\,,$$

$$(\Lambda_{\delta^{-1}}(\sigma))^{-1} = \Lambda_\delta(\sigma)$$

$$\Lambda_{\gamma\cdot\delta}(\sigma) = \Lambda_\gamma(\sigma)\cdot(\Lambda_{\delta^{-1}}(\sigma))^{-1}\,. \tag{14.12}$$

These equalities imply the analogous equalities for Ω-series.

Theorem 23. *Let z_0 be a K-point of $\mathbb{P}^1_K \setminus \{0,1,\infty\}$, and let $\mathcal{P}$ be a profinite sheaf of $\mathbb{Z}_\ell$-algebras over $\mathbb{P}^1_K \setminus \{0,1,\infty\}$ such that:*

(i) $\mathcal{P}_{z_0} \otimes \mathbb{Q}_\ell = \mathrm{Pol}(X,Y)$ as a $\mathbb{Q}_\ell$-algebra, and
(ii) The monodromy representation of $\mathcal{P}$ on the stalk over z_0 tensored by $\mathbb{Q}_\ell$

$$\rho_{z_0} : \pi_1^{\text{ét}}(\mathbb{P}^1_{\bar{K}} \setminus \{0,1,\infty\}, z_0) \to \mathrm{GL}(\mathrm{Pol}(X,Y))$$

is given by the formula

$$\rho_{z_0}(w) = L_{\bar{E}_{z_0}(w)}\,.$$

Let z be another K-point of $\mathbb{P}^1_K \setminus \{0,1,\infty\}$. Let δ be a path from z to z_0 and let α be a path from $\overrightarrow{01}$ to z. Then

$$\delta_* \circ \sigma_z \circ (\delta_*)^{-1} = \mathrm{L}_{\Omega_\alpha(\sigma)} \circ \mathrm{R}_{\mathrm{B}(\sigma)} \circ \tau(\chi(\sigma)),$$

where $\mathrm{B} : \mathrm{G_K} \to \mathrm{Pol}(\mathrm{X},\mathrm{Y})$ *satisfies* $\mathrm{B}(\sigma \cdot \sigma_1) = (\tau(\chi(\sigma))(\mathrm{B}(\sigma_1))) \cdot \mathrm{B}(\sigma)$ *and*

$$\log(\Omega_\alpha(\sigma)) = l(z)_\alpha(\sigma)\mathrm{X} + \sum_{i=1}^{\infty} (-1)^{i-1} l_i(z)_\alpha(\sigma)\mathrm{X}^{i-1}\mathrm{Y}.$$

Proof. Let us set $\gamma = (\delta \cdot \alpha)^{-1}$. Then γ is a path from z_0 to $\overrightarrow{01}$. Let $\sigma \in \mathrm{G_K}$ and let $w \in \pi_1^{\text{ét}}(\mathbb{P}^1_{\bar{\mathrm{K}}} \setminus \{0,1,\infty\}, z_0)$. It follows from Lemma 22 that

$$\bar{\mathrm{E}}_{z_0}(\sigma(w))) = (\Omega_\gamma(\sigma))^{-1} \cdot (\bar{\mathrm{E}}_{z_0}(w)(\chi(\sigma)\mathrm{X}, \chi(\sigma)\mathrm{Y})) \cdot \Omega_\gamma(\sigma). \tag{14.13}$$

It follows from (14.8) that

$$\sigma_{z_0}(\bar{\mathrm{E}}_{z_0}(w)) = \bar{\mathrm{E}}_{z_0}(\sigma(w)) \cdot \sigma_{z_0}(1).$$

Hence it follows from (14.13) that

$$\sigma_{z_0}(\bar{\mathrm{E}}_{z_0}(w)) = (\Omega_\gamma(\sigma))^{-1} \cdot \bar{\mathrm{E}}_{z_0}(w)(\chi(\sigma)\mathrm{X}, \chi(\sigma)\mathrm{Y}) \cdot \Omega_\gamma(\sigma) \cdot \sigma_{z_0}(1). \tag{14.14}$$

Hence for any $\mathrm{W}(\mathrm{X},\mathrm{Y}) \in \mathrm{Pol}(\mathrm{X},\mathrm{Y})$ we have

$$\sigma_{z_0}(\mathrm{W}(\mathrm{X},\mathrm{Y})) = (\Omega_\gamma(\sigma))^{-1} \cdot \mathrm{W}(\chi(\sigma)\mathrm{X}, \chi(\sigma)\mathrm{Y}) \cdot \Omega_\gamma(\sigma) \cdot \sigma_{z_0}(1). \tag{14.15}$$

We shall calculate the representation of $\mathrm{G_K}$ in $\mathcal{P}_z \otimes \mathbb{Q}_\ell$. It follows from the fundamental formula (14.8) that

$$\delta_* \circ \sigma_z \circ \delta_*^{-1} = \delta_* \circ \sigma(\delta)_*^{-1} \circ \sigma_{z_0}.$$

Observe that

$$\delta_* \circ \sigma(\delta)_*^{-1} = (\delta \cdot \sigma(\delta^{-1}))_* = (f_{\delta^{-1}}(\sigma))_* = \rho_{z_0}(f_{\delta^{-1}}(\sigma)) = \mathrm{L}_{\Omega_{\delta^{-1}}(\sigma)}.$$

Hence we get

$$\delta_* \circ \sigma_z \circ \delta_*^{-1} = \mathrm{L}_{\Omega_{\delta^{-1}}(\sigma)} \circ \sigma_{z_0}.$$

The formula (14.15) implies that

$$\begin{aligned} \mathrm{L}_{\Omega_{\delta^{-1}}(\sigma)} \circ \sigma_{z_0} &= \mathrm{L}_{\Omega_{\delta^{-1}}(\sigma)} \circ \mathrm{L}_{(\Omega_\gamma(\sigma))^{-1}} \circ \mathrm{R}_{\Omega_\gamma(\sigma)\cdot\sigma_{z_0}(1)} \circ \tau(\chi(\sigma)) \\ &= \mathrm{L}_{\Omega_{\delta^{-1}}(\sigma)\cdot(\Omega_\gamma(\sigma))^{-1}} \circ \mathrm{R}_{\Omega_\gamma(\sigma)\cdot\sigma_{z_0}(1)} \circ \tau(\chi(\sigma)). \end{aligned}$$

By (14.12) we deduce that

$$\Omega_{\delta^{-1}}(\sigma) \cdot (\Omega_\gamma(\sigma))^{-1} = (\Omega_{\gamma\cdot\delta}(\sigma))^{-1} = (\Omega_{\alpha^{-1}}(\sigma))^{-1} = \Omega_\alpha(\sigma).$$

With $\mathrm{B}(\sigma) = \Omega_\gamma(\sigma) \cdot \sigma_{z_0}(1)$ we therefore finally get

$$\delta_* \circ \sigma_z \circ \delta_*^{-1} = \mathrm{L}_{\Omega_\alpha(\sigma)} \circ \mathrm{R}_{\mathrm{B}(\sigma)} \circ \tau(\chi(\sigma))\,.$$

The equality $(\tau \cdot \sigma)_z = \tau_z \circ \sigma_z$ implies that $\mathrm{B} : \mathrm{G_K} \to \mathrm{Pol(X,Y)}$ indeed satisfies the formula of the theorem.

Since the path α is from $\overrightarrow{01}$ to z, the formula for $\log(\Omega_\alpha(\sigma))$ follows from the very definition of ℓ-adic polylogarithms. □

Remark 24. (1) We need one more condition to show that $\mathrm{B} = 1$. We can require for example that over $\overrightarrow{01}$ the Galois group $\mathrm{G_K}$ acts on $\mathrm{Pol(X,Y)}$ through $\tau \circ \chi$.

(2) Let us set

$$\Omega_\alpha(\sigma) = 1 + \exp(l(z)(\sigma)\mathrm{X}) + \sum_{n=1}^{\infty} -li_n(z)(\sigma)\mathrm{X}^{n-1}\mathrm{Y}\,.$$

Then the matrice of $\mathrm{L}_{\Omega_\alpha(\sigma)}$ in the base

$$1, \mathrm{Y}, \mathrm{XY}, \mathrm{X}^2\mathrm{Y}, \mathrm{X}^3\mathrm{Y}, \ldots$$

is exactly as the matrice $\mathrm{L}(z)$ expressing the monodromy of polylogarithms in [BD94].

(3) If ξ is a root of unity, then $l(\xi)_\alpha$ vanishes for a suitable choice of a path α and $l_n(\xi)_\alpha$ are cocycles, see [Woj05a, Corollary 11.0.12. and its proof]. Hence the representation of $\mathrm{G_K}$ on $\mathcal{P}_\xi \otimes \mathbb{Q}_\ell$ is an extension of $\mathbb{Q}_\ell(0)$ by $\prod_{n=1}^{\infty} \mathbb{Q}_\ell(n)$ if $\mathrm{B} = 1$.

14.9 Cosimplicial Spaces and Galois Actions

In this last section we will work more generally over a field k, still with a fixed complex embedding $k \subset \mathbb{C}$, but not necessarily a number field. Let V be a smooth quasi-projective, geometrically connected algebraic variety over k and let v be a k-point of V. The étale fundamental group $\pi_1^{\text{ét}}(\mathrm{V}_{\bar{k}}, v)$ and its maximal pro-ℓ quotient $\pi_1(\mathrm{V}_{\bar{k}}, v)$ are equipped with the action of $\mathrm{G}_k = \mathrm{Gal}(\bar{k}/k)$ induced by conjugation and the canonical section v_* as

$$\mathrm{G_K} = \pi_1^{\text{ét}}(\mathrm{Spec}(k), \mathrm{Spec}(\bar{k})) \xrightarrow{v_*} \pi_1^{\text{ét}}(\mathrm{V}, v) \xrightarrow{\gamma \mapsto \gamma(-)\gamma^{-1}} \mathrm{Aut}(\pi_1^{\text{ét}}(\mathrm{V}_{\bar{k}}, v))\,. \tag{14.16}$$

On the other side, given an algebraic variety V and a k-point v there is a cosimplicial algebraic variety $\mathrm{V}^\bullet$, which is a model in algebraic geometry for the loop space based at v, see [Woj93] and [Woj02]. Let $\mathrm{V}(\mathbb{C})$ (resp. $\mathrm{V}^\bullet(\mathbb{C})$) be the set of $\mathbb{C}$-points of V endowed with its natural structure as a (resp. cosimplicial) complex variety. The de Rham cohomology group of complex differential forms

$$\mathrm{H}^0_{\mathrm{DR}}(\mathrm{V}^\bullet(\mathbb{C}))$$

is the algebra of polynomial complex valued functions on the Malcev $\mathbb{Q}$-completion

$$\pi_1(\mathrm{V}(\mathbb{C}),v)\otimes\mathbb{Q}\,.$$

The étale cohomology group

$$\mathrm{H}^0_{\text{ét}}(\mathrm{V}^\bullet_{\bar{k}},\mathbb{Q}_\ell)$$

can be interpreted as the algebra of polynomial $\mathbb{Q}_\ell$-valued functions on the $\mathbb{Q}$-completion, or better on the Malcev $\mathbb{Q}_\ell$-completion, for which we have the comparison isomorphism

$$\pi_1(\mathrm{V}(\mathbb{C}),v)\otimes\mathbb{Q}_\ell=\pi_1(\mathrm{V}_{\bar{k}},v)\otimes\mathbb{Q}_\ell\,,$$

with the Malcev $\mathbb{Q}_\ell$-completion $\pi_1(\mathrm{V}_{\bar{k}},v)\otimes\mathbb{Q}_\ell$ of the pro-ℓ group $\pi_1(\mathrm{V}_{\bar{k}},v)$. The Galois group G_k acts on $\mathrm{H}^0_{\text{ét}}(\mathrm{V}^\bullet_{\bar{k}},\mathbb{Q}_\ell)$. We interpret $\mathrm{H}^0_{\text{ét}}(\mathrm{V}^\bullet_{\bar{k}},\mathbb{Q}_\ell)$ as the algebra of polynomial $\mathbb{Q}_\ell$-valued functions on $\pi_1(\mathrm{V}_{\bar{k}},v)\otimes\mathbb{Q}_\ell$. Therefore G_k acts also on

$$\pi_1(\mathrm{V}_{\bar{k}},v)\otimes\mathbb{Q}_\ell\,.$$

In this section we shall compare these two Galois actions. Our arguments will be very sketchy in some places because of a lot of technical material.

We first fix some notation. The sheaf $\mathrm{A}_{\mathrm{X}_{\text{ét}}}$ (resp. $\mathrm{A}_{\mathrm{X}(\mathbb{C})}$) is the constant sheaf with values in A on the étale site $\mathrm{X}_{\text{ét}}$ (resp. $\mathrm{X}(\mathbb{C})$) for an algebraic variety X. With $\Delta[1]$ we denote the standard simplicial model of the 1-simplex, while $\partial\Delta[1]$ is its boundary. The n-th truncation of a cosimplicial object $\mathrm{X}^\bullet$ will be denoted by $\mathrm{X}^\bullet_{[n]}$.

Let X be a smooth quasi-projective, geometrically connected algebraic variety over k. The inclusion of simplicial sets

$$\partial\Delta[1]\hookrightarrow\Delta[1]$$

induces the morphism of cosimplicial algebraic varieties

$$p^\bullet\,:\,\mathrm{X}^{\Delta[1]}\to\mathrm{X}^{\partial\Delta[1]}\,.$$

Therefore for each n we get the morphism between their n-th truncations

$$p^\bullet_{[n]}\,:\,\mathrm{X}^{\Delta[1]}_{[n]}\longrightarrow\mathrm{X}^{\partial\Delta[1]}_{[n]}\,.$$

For each i, the map

$$p^i\,:\,\mathrm{X}^{\Delta[1]_i}=\mathrm{X}\times\mathrm{X}^i\times\mathrm{X}\to\mathrm{X}^{\partial\Delta[1]_i}=\mathrm{X}\times\mathrm{X}$$

is the projection map on the first and the last factor.

First we shall study the Gauss-Manin connection associated to the morphism $p^\bullet\,:\,\mathrm{X}^{\Delta[1]}\to\mathrm{X}^{\partial\Delta[1]}$. We review briefly the results from [Woj93] in a form suitable for our study here. We apply to the map between the n-th truncations

$$p^{\bullet}_{[n]} : \mathrm{X}^{\Delta[1]}_{[n]} \to \mathrm{X}^{\partial\Delta[1]}_{[n]}$$

the standard construction of the Gauss-Manin connection, see [Woj93]. For each $0 \leqslant i \leqslant n$ the complex of sheaves $\Omega^{\cdot}_{\mathrm{X}^{\Delta[1]_i}}$ is equipped with a canonical filtration

$$\mathrm{F}^j\Omega^{\cdot}_{\mathrm{X}^{\Delta[1]_i}} := Image(\Omega^{\cdot-i}_{\mathrm{X}^{\Delta[1]_i}/\mathrm{X}^{\partial\Delta[1]_i}} \otimes_{\mathcal{O}_{\mathrm{X}^{\Delta[1]_i}}} p^{i,*}\Omega^j_{\mathrm{X}^{\partial\Delta[1]_i}} \to \Omega^{\cdot}_{\mathrm{X}^{\Delta[1]_i}}) .$$

Hence on $\mathrm{X}^{\partial\Delta[1]_i} = \mathrm{X}\times\mathrm{X}$ we have a filtered complex $\mathrm{R}\, p^i_*(\Omega^{\cdot}_{\mathrm{X}^{\Delta[1]_i}})$. We form the total complex

$$\mathrm{Tot}(\mathrm{R}\, p^{\bullet}_{[n],*}(\Omega^{\cdot}_{\mathrm{X}^{\Delta[1]}_{[n]}})) = \bigoplus_{i=0}^{n} \mathrm{R}\, p^i_*(\Omega^{\cdot}_{\mathrm{X}^{\Delta[1]_i}}) .$$

The filtration on each summand on the right hand side induces a filtration on the left hand side. Applying the spectral sequence of a finitely filtered object, we get a spectral sequence converging to the cohomology sheaves

$$\mathcal{H}^j\Big(\mathrm{Tot}(\mathrm{R}\, p^{\bullet}_{[n],*}(\Omega^{\cdot}_{\mathrm{X}^{\Delta[1]}_{[n]}}))\Big)$$

on $\mathrm{X}\times\mathrm{X}$, the E_1-term of which reads

$$\mathrm{E}_1^{p,q} = \Omega^p_{\mathrm{X}\times\mathrm{X}} \otimes_{\mathcal{O}_{\mathrm{X}\times\mathrm{X}}} \mathcal{H}^q\Big(\mathrm{Tot}(\mathrm{R}\, p^{\bullet}_{[n],*}(\Omega^{\cdot}_{\mathrm{X}^{\Delta[1]}_{[n]}/\mathrm{X}^{\partial\Delta[1]}_{[n]}}))\Big) ,$$

where

$$\mathrm{Tot}(\mathrm{R}\, p^{\bullet}_{[n],*}(\Omega^{\cdot}_{\mathrm{X}^{\Delta[1]}_{[n]}/\mathrm{X}^{\partial\Delta[1]}_{[n]}})) = \bigoplus_{i=0}^{n} \mathrm{R}\, p^i_*(\Omega^{\cdot}_{\mathrm{X}^{\Delta[1]_i}/\mathrm{X}^{\partial\Delta[1]_i}}) .$$

Farther we denote the relative de Rham complex $\Omega^{\cdot}_{\mathrm{X}^{\Delta[1]}_{[n]}/\mathrm{X}^{\partial\Delta[1]}_{[n]}}$ on $\mathrm{X}^{\Delta[1]}_{[n]}$ by $\Omega^{\cdot}$ in the algebraic case, by $\Omega^{\cdot}_{hol}$ in the holomorphic case and by $\Omega^{\cdot}_{C^\infty}$ in the smooth complex case. The differential

$$d_1^{0,q} : \mathrm{E}_1^{0,q} \to \mathrm{E}_1^{1,q}$$

is the integrable Gauss-Manin connection on the relative de Rham cohomology sheaves

$$\mathcal{H}^q(\mathrm{Tot}(\mathrm{R}\, p^{\bullet}_{[n],*}\Omega^{\cdot})) .$$

Let x and y be two k-points of X. The fiber of $\mathcal{H}^q(\mathrm{Tot}(\mathrm{R}\, p^{\bullet}_{[n],*}\Omega^{\cdot}))$ over a point $(x,y) \in \mathrm{X}\times\mathrm{X}$ is

$$\mathrm{H}^q_{\mathrm{DR}}((p^{\bullet}_{[n]})^{-1}(x,y)) .$$

Note that if $x = y$ then $(p^{\bullet}_{[n]})^{-1}(x,x)$ is the n-th truncation of the cosimplicial algebraic variety, which is a model in algebraic geometry for the loop space based at x from the very beginning of the section.

Recall that we fixed an embedding $k \subset \mathbb{C}$. Then we get the morphism of cosimplicial complex varieties

$$p(\mathbb{C})^\bullet : \mathrm{X}(\mathbb{C})^{\Delta[1]} \longrightarrow \mathrm{X}(\mathbb{C})^{\partial\Delta[1]}$$

and the maps between the n-th truncations

$$p(\mathbb{C})^\bullet_{[n]} : \mathrm{X}(\mathbb{C})^{\Delta[1]}_{[n]} \longrightarrow \mathrm{X}(\mathbb{C})^{\partial\Delta[1]}_{[n]} .$$

We do the same construction for holomorphic differentials. The holomorphic de Rham sheaf $\Omega^{\cdot}_{\mathrm{X}(\mathbb{C})^{\Delta[1]}_{[n]}}$ is the resolution of the constant sheaf $\mathbb{C}_{\mathrm{X}(\mathbb{C})^{\Delta[1]}_{[n]}}$ on $\mathrm{X}(\mathbb{C})^{\Delta[1]}_{[n]}$. Let us set

$$\mathrm{Tot}(\mathrm{R}\, p(\mathbb{C})^\bullet_{[n],*}(\mathbb{C}_{\mathrm{X}(\mathbb{C})^{\Delta[1]}_{[n]}})) = \bigoplus_{i=0}^{n} \mathrm{R}\, p^i_*(\mathbb{C}_{\mathrm{X}(\mathbb{C})^{\Delta[1]_i}}) .$$

Hence we get that

$$\mathcal{H}^q\Big(\mathrm{Tot}(\mathrm{R}\, p(\mathbb{C})^\bullet_{[n],*}(\mathbb{C}_{\mathrm{X}(\mathbb{C})^{\Delta[1]}_{[n]}}))\Big)$$

is the sheaf of the flat sections of the holomorphic Gauss-Manin connection

$$(d_1^{0,q})_{hol} : \mathcal{H}^q(\mathrm{Tot}(\mathrm{R}\, p(\mathbb{C})^\bullet_{[n],*}\Omega^{\cdot}_{hol})) \to \Omega^1_{\mathrm{X}(\mathbb{C})^2} \otimes_{O_{\mathrm{X}(\mathbb{C})^2}} \mathcal{H}^q(\mathrm{Tot}(\mathrm{R}\, p(\mathbb{C})^\bullet_{[n],*}\Omega^{\cdot}_{hol})) .$$

We shall calculate the monodromy representation of the locally constant sheaf

$$\mathcal{H}^0\Big(\mathrm{Tot}(\mathrm{R}\, p(\mathbb{C})^\bullet_{[n],*}(\mathbb{C}_{\mathrm{X}(\mathbb{C})^{\Delta[1]}_{[n]}}))\Big) .$$

The de Rham complexes of smooth differentials are acyclic for direct image functors. Hence there is a quasi-isomorphism

$$\mathrm{Tot}(\mathrm{R}\, p(\mathbb{C})^\bullet_{[n],*}\Omega^{\cdot}_{hol}) \simeq \mathrm{Tot}(p(\mathbb{C})^\bullet_{[n],*}\Omega^{\cdot}_{C^\infty}) .$$

Let $\omega_1,\dots,\omega_n \in \Omega^1_{C^\infty}(\mathrm{X}(\mathbb{C}))$ be closed one-forms on $\mathrm{X}(\mathbb{C})$. Let us assume that $\omega_i \wedge \omega_{i+1} = 0$ for all i. Then the tensor product $1\otimes\omega_1\otimes\ldots\otimes\omega_n\otimes 1$ defines an element of $O_{\mathrm{X}(\mathbb{C})}\otimes\Omega^{\cdot}_{C^\infty}(\mathrm{X}(\mathbb{C})^n)\otimes O_{\mathrm{X}(\mathbb{C})}$. Hence

$$1\otimes\omega_1\otimes\ldots\otimes\omega_n\otimes 1$$

defines a global section of $\mathcal{H}^0(\mathrm{Tot}(p(\mathbb{C})^\bullet_{[n],*}\Omega^{\cdot}_{C^\infty}))$. We shall calculate the action of $d^0 := (d_1^{0,0})_{C^\infty}$ on the section $1\otimes\omega_1\otimes\ldots\otimes\omega_n\otimes 1$. The connection d^0 is the boundary homomorphism of the long exact sequence associated to the short exact sequence

$$0 \to \mathrm{F}^1/\mathrm{F}^2 \to \mathrm{F}^0/\mathrm{F}^2 \to \mathrm{F}^0/\mathrm{F}^1 \to 0 .$$

We recall that the coface maps

$$\delta^i : \mathrm{X}\times\mathrm{X}^{n-1}\times\mathrm{X} \to \mathrm{X}\times\mathrm{X}^n\times\mathrm{X}$$

are given by

$$\delta^i(x_0,x_1,\dots,x_n) = (x_0,\dots,x_{i-1},x_i,x_i,\dots,x_n)$$

for $0 \leqslant i \leqslant n$. We set

$$\delta_n := \sum_{i=0}^{n} (-1)^{n-i} (\delta^i)^* .$$

The boundary operator of the total complex is given by $\mathrm{D} = \delta_n + (-1)^n d$, where d is the exterior differential of the de Rham complex.

We denote by $\int_a \omega_1, \ldots, \omega_i$ a function defined on a contractible subset of $\mathrm{X}(\mathbb{C})$ containing a and sending z to the iterated integral $\int_a^z \omega_1, \ldots, \omega_i$ along any path contained in this contractible subset. After calculations we get the following result.

Lemma 25. *Let $(a,b) \in \mathrm{X}(\mathbb{C}) \times \mathrm{X}(\mathbb{C})$. We have*

$$\mathrm{D}\Big(\sum_{0 \leqslant i \leqslant j \leqslant n} \int_a \omega_1, \ldots, \omega_i \otimes \omega_{i+1} \otimes \ldots \otimes \omega_j \otimes (-1)^{n-j} \int_b \omega_n, \ldots, \omega_{j+1} \Big) = 0 .$$

We denote by $\pi_1(\mathrm{X}(\mathbb{C}); b, a)$ the right $\pi_1(\mathrm{X}(\mathbb{C}), a)$-torsor of paths from a to b on $\mathrm{X}(\mathbb{C})$ and by

$$\pi_1(\mathrm{X}(\mathbb{C}); b, a) \otimes \mathbb{Q}$$

the induced right $\pi_1(\mathrm{X}(\mathbb{C}), a) \otimes \mathbb{Q}$-torsor. We denote by

$$\mathrm{Alg}_{\mathbb{C}}(\pi_1(\mathrm{X}(\mathbb{C}); b, a) \otimes \mathbb{Q})$$

the algebra of complex valued polynomial functions on $\pi_1(\mathrm{X}(\mathbb{C}); b, a) \otimes \mathbb{Q}$. The fiber of the sheaf $\mathcal{H}^0(\mathrm{Tot}(p(\mathbb{C})^{\bullet}_{*} \Omega^{\cdot}_{C^\infty}))$ over (a,b) is $\mathrm{H}^0_{\mathrm{DR}}((p(\mathbb{C})^{\bullet})^{-1}(a,b))$. The shuffle product defines a multiplication on $\mathrm{H}^0_{\mathrm{DR}}((p(\mathbb{C})^{\bullet})^{-1}(a,b))$, hence the 0-th cohomology group is a $\mathbb{C}$-algebra and if $a = b$ it is a Hopf algebra.

The element $1 \otimes \omega_1 \otimes \ldots \otimes \omega_n \otimes 1$ in $\mathrm{H}^0_{\mathrm{DR}}((p(\mathbb{C})^{\bullet})^{-1}(a,b))$ determines a polynomial complex valued function on the torsor of rational paths $\pi_1(\mathrm{X}(\mathbb{C}); b, a) \otimes \mathbb{Q}$, which to a path γ from a to b associates the iterated integral $\int_\gamma \omega_1 \ldots, \omega_n$. Hence we get an isomorphism of $\mathbb{C}$-algebras

$$\mathrm{H}^0_{\mathrm{DR}}((p(\mathbb{C})^{\bullet})^{-1}(a,b)) \cong Alg_{\mathbb{C}}(\pi_1(\mathrm{X}(\mathbb{C}); b, a) \otimes \mathbb{Q})$$

and if $a = b$ we get an isomorphism of Hopf algebras by the work of Chen. Observe that

$$\varinjlim_n \mathrm{H}^0_{\mathrm{DR}}((p(\mathbb{C})^{\bullet}_{[n]})^{-1}(a,b)) = \mathrm{H}^0_{\mathrm{DR}}((p(\mathbb{C})^{\bullet})^{-1}(a,b)) .$$

The same holds also for cohomology sheaves, considered by us and for the connections. Farther we shall need the following lemma.

Lemma 26. *Let* X *be a smooth quasi-projective, geometrically connected algebraic variety over a field $k \subset \mathbb{C}$. Then there is an affine smooth algebraic curve* S *over k and an algebraic map $f : \mathrm{S} \to \mathrm{X}$ over k such that the induced map*

$$f_* : \mathrm{H}_1(\mathrm{S}(\mathbb{C}), \mathbb{Q}) \to \mathrm{H}_1(\mathrm{X}(\mathbb{C}), \mathbb{Q})$$

is surjective.

Proof. The lemma follows from the successive applications of the Lefschetz hyperplane theorem for quasi-projective varieties, see [GM88, pages 22 and 23]. To assure that S and $f : S \to X$ are over k one takes successive hyperplanes over k. □

Proposition 27. *Let X be a smooth quasi-projective, geometrically connected algebraic variety over a field $k \subset \mathbb{C}$. The monodromy representation of the bundle of flat sections of the Gauss-Manin connection*

$$d^0 = (d_1^{0,0})_{hol} : \mathcal{H}^0(\mathrm{Tot}(\mathrm{R}\, p(\mathbb{C})_*^\bullet \Omega^\cdot_{hol})) \to \Omega^1_{\mathrm{X}(\mathbb{C})^2} \otimes_{\mathcal{O}_{\mathrm{X}(\mathbb{C})^2}} \mathcal{H}^0(\mathrm{Tot}(\mathrm{R}\, p(\mathbb{C})_*^\bullet \Omega^\cdot_{hol}))$$

at a point $(a,b) \in \mathrm{X}(\mathbb{C}) \times \mathrm{X}(\mathbb{C})$

$$\rho_{a,b} : \pi_1(\mathrm{X}(\mathbb{C}),a) \times \pi_1(\mathrm{X}(\mathbb{C}),b) \to \mathrm{Aut}(\mathrm{Alg}_{\mathbb{C}}(\pi_1(\mathrm{X}(\mathbb{C});b,a) \otimes \mathbb{Q}))$$

is given by the formula

$$((\rho_{a,b}(\alpha,\beta))(f))(\gamma) = f(\beta^{-1} \cdot \gamma \cdot \alpha)\,, \tag{14.17}$$

where $(\alpha,\beta) \in \pi_1(\mathrm{X}(\mathbb{C}),a) \times \pi_1(\mathrm{X}(\mathbb{C}),b)$ acts on $f \in \mathrm{Alg}_{\mathbb{C}}(\pi_1(\mathrm{X}(\mathbb{C});b,a) \otimes \mathbb{Q})$ and where $\gamma \in \pi_1(\mathrm{X}(\mathbb{C});b,a) \otimes \mathbb{Q}$.

Proof. First we suppose that X is an affine smooth algebraic curve over a field $k \subset \mathbb{C}$. We can find smooth closed one-forms

$$\eta_1, \ldots, \eta_r \in \Omega^1_{C^\infty}(\mathrm{X}(\mathbb{C}))$$

such that their classes form a base of $\mathrm{H}^1_{\mathrm{DR}}(\mathrm{X}(\mathbb{C}))$ and $\eta_i \wedge \eta_j = 0$ for $1 \leqslant i,j \leqslant r$. Then all possible tensor products $1 \otimes \eta_{i_1} \otimes \ldots \otimes \eta_{i_k} \otimes 1$ form a base of

$$\mathrm{H}^0_{\mathrm{DR}}((p(\mathbb{C})^\bullet)^{-1}(a,b))\,.$$

Let $1 \otimes \omega_1 \otimes \ldots \otimes \omega_n \otimes 1$ be one of such products. The stalk of the locally constant sheaf

$$\mathcal{H}^0(\mathrm{Tot}(\mathrm{R}\, p(\mathbb{C})^\bullet_{[n],*}(\mathbb{C}_{\mathrm{X}(\mathbb{C})^{\Delta[1]}_{[n]}})))$$

over the point (a,b) is equal $\mathrm{H}^0((p(\mathbb{C})^\bullet_{[n]})^{-1}(a,b),\mathbb{C})$, which in turn we calculate using complexes of smooth differential forms. The element $1 \otimes \omega_1 \otimes \ldots \otimes \omega_n \otimes 1$ is considered in the stalk of the sheaf $\mathcal{H}^0(\mathrm{Tot}(\mathrm{R}\, p(\mathbb{C})^\bullet_{[n],*}(\mathbb{C}_{\mathrm{X}(\mathbb{C})^{\Delta[1]}_{[n]}})))$ over the point (a,b). We prolongate $1 \otimes \omega_1 \otimes \ldots \otimes \omega_n \otimes 1$ to a continuous section s of the locally constant sheaf $\mathcal{H}^0(\mathrm{Tot}(\mathrm{R}\, p(\mathbb{C})^\bullet_{[n],*}(\mathbb{C}_{\mathrm{X}(\mathbb{C})^{\Delta[1]}_{[n]}})))$ along

$$(\alpha,\beta) \in \pi_1(\mathrm{X}(\mathbb{C}),a) \times \pi_1(\mathrm{X}(\mathbb{C}),b)\,.$$

We have $s(0) = 1 \otimes \omega_1 \otimes \ldots \otimes \omega_n \otimes 1$. It follows from Lemma 25 that

$$s(1) = \sum_{0 \leqslant i \leqslant j \leqslant n} \Big(\int_\alpha \omega_1, \ldots, \omega_i\Big) \otimes \omega_{i+1} \otimes \ldots \otimes \omega_j \otimes (-1)^{n-j} \Big(\int_\beta \omega_n, \ldots, \omega_{j+1}\Big)$$

as an element of

$$\mathrm{H}^0((p(\mathbb{C})^\bullet)^{-1}(a,b),\mathbb{C}) = \mathrm{H}^0_{\mathrm{DR}}((p(\mathbb{C})^\bullet)^{-1}(a,b)) = \mathrm{Alg}_{\mathbb{C}}(\pi_1(\mathrm{X}(\mathbb{C});b,a)\otimes\mathbb{Q}).$$

Then, for any path γ from a to b, we have

$$s(1)(\gamma) = \sum_{0\leqslant i\leqslant j\leqslant n}(\int_\alpha \omega_1,\dots,\omega_i)\cdot(\int_\gamma \omega_{i+1},\dots,\omega_j)\cdot(-1)^{n-j}(\int_\beta \omega_n,\dots,\omega_{j+1})$$

which by Chen's formulae, see [Che75], equals

$$s(1)(\gamma) = \int_{\beta^{-1}\cdot\gamma\cdot\alpha} \omega_1,\dots,\omega_n\,.$$

Hence the monodromy transformation along (α,β) maps the function

$$f(-) := s(0) \in \mathrm{Alg}_{\mathbb{C}}(\pi_1(\mathrm{X}(\mathbb{C});b,a)\otimes\mathbb{Q})$$

into the function $f(\beta^{-1}\cdot - \cdot\alpha) \in \mathrm{Alg}_{\mathbb{C}}(\pi_1(\mathrm{X}(\mathbb{C});b,a)\otimes\mathbb{Q})$. Therefore the proposition is proved for affine smooth algebraic curves over k.

Now we assume that X is a smooth quasi-projective, geometrically connected algebraic variety over a field $k\subset\mathbb{C}$. It follows from Lemma 26 that there is an affine smooth algebraic curve S over k and an algebraic map $f\,:\,\mathrm{S}\to\mathrm{X}$ over k such that the induced map

$$f_*\,:\,\mathrm{H}_1(\mathrm{S}(\mathbb{C}),\mathbb{Q})\to\mathrm{H}_1(\mathrm{X}(\mathbb{C}),\mathbb{Q})$$

is surjective.

The morphism f induces a morphism of locally constant sheaves

$$f^*\mathcal{H}^0\Big(\mathrm{Tot}\big(\mathrm{R}\,p(\mathbb{C})^\bullet_{[n],*}(\mathbb{C}_{\mathrm{X}(\mathbb{C})^{\Delta[1]}_{[n]}})\big)\Big) \longrightarrow \mathcal{H}^0\Big(\mathrm{Tot}\big(\mathrm{R}\,p(\mathbb{C})^\bullet_{[n],*}(\mathbb{C}_{\mathrm{S}(\mathbb{C})^{\Delta[1]}_{[n]}})\big)\Big)\,.$$

Consider first $(a,b)\in\mathrm{X}(\mathbb{C})\times\mathrm{X}(\mathbb{C})$ which is the image of $(s,t)\in\mathrm{S}(\mathbb{C})\times\mathrm{S}(\mathbb{C})$. Then $\mathrm{H}^0((p(\mathbb{C})^\bullet_{[n]})^{-1}(a,b))$ is a subalgebra of $\mathrm{H}^0((p(\mathbb{C})^\bullet_{[n]})^{-1}(s,t))$. Hence it follows from what we have already proved for smooth affine curves that the monodromy representation of the sheaf

$$\mathcal{H}^0\Big(\mathrm{Tot}\big(\mathrm{R}\,p(\mathbb{C})^\bullet_{[n],*}(\mathbb{C}_{\mathrm{X}(\mathbb{C})^{\Delta[1]}_{[n]}})\big)\Big)$$

at the point (a,b) is given by the formula (14.17). But then it is given by the formula (14.17) at any point of $\mathrm{X}(\mathbb{C})\times\mathrm{X}(\mathbb{C})$. □

We recall that X is a smooth quasi-projective, geometrically connected algebraic variety over the field k with a fixed complex embedding $k\subset\mathbb{C}$. Let us set

$$\mathrm{Tot}\big(\mathrm{R}\,p^\bullet_{[n],*}(\mathbb{Z}/\ell^m\mathbb{Z}_{\mathrm{X}^{\Delta[1]}_{[n],\acute{e}t}})\big) = \bigoplus_{i=0}^{n}\mathrm{R}\,p^i_*(\mathbb{Z}/\ell^m\mathbb{Z}_{\mathrm{X}^{\Delta[1]_i}_{\acute{e}t}})\,,$$

where Tot is the total complex of a bicomplex. Let us define

$$\mathrm{R}^i p^{\bullet}_{[n],*}(\mathbb{Z}/\ell^m\mathbb{Z}_{\mathrm{X}^{\Delta[1]}_{[n],\text{ét}}}) = \mathcal{H}^i\Big(\mathrm{Tot}(\mathrm{R}\, p^{\bullet}_{[n],*}(\mathbb{Z}/\ell^m\mathbb{Z}_{\mathrm{X}^{\Delta[1]}_{[n],\text{ét}}}))\Big). \tag{14.18}$$

Observe that the stalks of the sheaves in (14.18) are equipped with Galois actions.

From now on we assume that the field k is algebraically closed with a fixed complex embedding $k \subset \mathbb{C}$.

Lemma 28. *The cohomology sheaves* $\mathrm{R}^i p^{\bullet}_{[n],*}(\mathbb{Z}/\ell^m\mathbb{Z}_{\mathrm{X}^{\Delta[1]}_{[n],\text{ét}}})$ *are sheaves of finitely generated* $\mathbb{Z}/\ell^m\mathbb{Z}$*-modules on* $(\mathrm{X}\times\mathrm{X})_{\text{ét}}$.

Proof. The spectral sequence of the bicomplex

$$\bigoplus_{i=0}^{n} \mathrm{R}\, p^i_*(\mathbb{Z}/\ell^m\mathbb{Z}_{\mathrm{X}^{\Delta[1]_i}_{\text{ét}}}),$$

converges to the cohomology sheaves in question. The E_1-term reads

$$\mathrm{E}_1^{i,j} = \mathrm{R}^i p^j_*(\mathbb{Z}/\ell^m\mathbb{Z}_{\mathrm{X}^{\Delta[1]_j}_{\text{ét}}})$$

and is a constant sheaf on $(\mathrm{X}\times\mathrm{X})_{\text{ét}}$ with finitely generated $\mathbb{Z}/\ell^m\mathbb{Z}$-modules as stalks. As only finitely many E_1-terms are nonzero, the lemma follows. □

We need to know if the sheaves

$$\mathrm{R}^i p^{\bullet}_{[n],*}(\mathbb{Z}/\ell^m\mathbb{Z}_{\mathrm{X}^{\Delta[1]}_{[n],\text{ét}}})$$

are locally constant and we need to calculate their monodromy representations.

Let Y be a topological space. We denote by Y_{lh} the site of local homeomorphisms on Y. The functors $\mathrm{R}^i p(\mathbb{C})^{\bullet}_{[n],*}$ for the sites $\mathrm{X}(\mathbb{C})_{\mathrm{lh}}$ and $\mathrm{X}(\mathbb{C})$ are defined as $\mathrm{R}^i p^{\bullet}_{[n],*}$ in (14.18) for $\mathrm{X}_{\text{ét}}$. The morphisms of sites

$$\epsilon : (\mathrm{X}\times\mathrm{X})(\mathbb{C})_{\mathrm{lh}} \to (\mathrm{X}\times\mathrm{X})_{\text{ét}} \text{ and } \alpha : (\mathrm{X}\times\mathrm{X})(\mathbb{C})_{\mathrm{lh}} \to (\mathrm{X}\times\mathrm{X})(\mathbb{C})$$

induce the comparison isomorphisms

$$\mathrm{R}^i p^{\bullet}_{[n],*}(\mathbb{Z}/\ell^m\mathbb{Z}_{\mathrm{X}^{\Delta[1]}_{[n],\text{ét}}}) \cong \epsilon_*(\mathrm{R}^i p(\mathbb{C})^{\bullet}_{[n],*}(\mathbb{Z}/\ell^m\mathbb{Z}_{\mathrm{X}(\mathbb{C})^{\Delta[1]}_{[n],\mathrm{lh}}})) \tag{14.19}$$

and

$$\mathrm{R}^i p(\mathbb{C})^{\bullet}_{[n],*}(\mathbb{Z}/\ell^m\mathbb{Z}_{\mathrm{X}(\mathbb{C})^{\Delta[1]}_{[n]}}) \cong \alpha_*(\mathrm{R}^i p(\mathbb{C})^{\bullet}_{[n],*}(\mathbb{Z}/\ell^m\mathbb{Z}_{\mathrm{X}(\mathbb{C})^{\Delta[1]}_{[n],\mathrm{lh}}})). \tag{14.20}$$

We do not know whether the sheaves in (14.19) and (14.20) are locally constant. However we have

$$(\varprojlim_m \mathrm{R}^i p(\mathbb{C})^{\bullet}_{[n],*}(\mathbb{Z}/\ell^m\mathbb{Z}_{\mathrm{X}(\mathbb{C})^{\Delta[1]}_{[n]}}))\otimes\mathbb{Q}_\ell \cong \mathcal{H}^i\Big(\mathrm{Tot}(\mathrm{R}\, p(\mathbb{C})^{\bullet}_{[n],*}(\mathbb{Z}_{\mathrm{X}(\mathbb{C})^{\Delta[1]}_{[n]}}))\Big)\otimes\mathbb{Q}_\ell\,.$$

The sheaf $\mathcal{H}^i\big(\mathrm{Tot}(\mathrm{R}\,p(\mathbb{C})^{\bullet}_{[n],*}(\mathbb{C}_{\mathrm{X}(\mathbb{C})^{\Delta[1]}_{[n]}}))\big)$ is locally constant as the sheaf of flat sections of the integrable connection d^0. Hence the sheaf

$$\mathcal{H}^i\big(\mathrm{Tot}(\mathrm{R}\,p(\mathbb{C})^{\bullet}_{[n],*}(\mathbb{Z}_{\mathrm{X}(\mathbb{C})^{\Delta[1]}_{[n]}}))\big)\otimes\mathbb{Q}$$

is locally constant on $(\mathrm{X}\times\mathrm{X})(\mathbb{C})$. This implies that the sheaf

$$\mathcal{H}^i\big(\mathrm{Tot}(\mathrm{R}\,p(\mathbb{C})^{\bullet}_{[n],*}(\mathbb{Z}_{\mathrm{X}(\mathbb{C})^{\Delta[1]}_{[n],\mathrm{lh}}}))\big)\otimes\mathbb{Q}$$

is locally constant on $(\mathrm{X}\times\mathrm{X})(\mathbb{C})_{\mathrm{lh}}$. Therefore the sheaf

$$(\mathcal{H}^i\big(\mathrm{Tot}(\mathrm{R}\,p(\mathbb{C})^{\bullet}_{[n],*}(\mathbb{Z}_{\mathrm{X}(\mathbb{C})^{\Delta[1]}_{[n],\mathrm{lh}}}))\big))/\mathrm{torsion}$$

is also locally constant on $(\mathrm{X}\times\mathrm{X})(\mathbb{C}))_{\mathrm{lh}}$.

If $\mathcal{A}$ is locally constant sheaf of finite sets on T_{lh}, where T is a topological space, then the stalk of $\mathcal{A}$ in $t\in\mathrm{T}$ can be naturally identified with $\mathcal{A}(\bar{\mathrm{T}})$ for some finite covering $\bar{\mathrm{T}}\to\mathrm{T}$ of T that trivializes $\mathcal{A}$. Moreover if $\bar{\mathrm{T}}\to\mathrm{T}$ is a Galois covering then the finite quotient $\pi_1(\mathrm{T},t)/\pi_1(\bar{\mathrm{T}},t)$ acts on $\mathcal{A}(\bar{\mathrm{T}})$. We apply this to calculate the stalk

$$\Big(\varprojlim_m \mathrm{R}^i\,p(\mathbb{C})^{\bullet}_{[n],*}(\mathbb{Z}/\ell^m\mathbb{Z})_{\mathrm{X}(\mathbb{C})^{\Delta[1]}_{[n],\mathrm{lh}}}\otimes\mathbb{Q}_\ell\Big)_{(a,b)}$$

in $(a,b)\in(\mathrm{X}\times\mathrm{X})(\mathbb{C})$. By the comparison isomorphism (14.19) the same is true for the stalk in the correspinding geometric point (a,b) and the projective system of sheaves

$$\{\mathrm{R}^i\,p^{\bullet}_{[n],*}(\mathbb{Z}/\ell^m\mathbb{Z}_{\mathrm{X}^{\Delta[1]}_{[n],\acute{\mathrm{e}}\mathrm{t}}})\}_{m\in\mathrm{N}}\,.$$

Moreover we get an action of $\pi_1^{\acute{\mathrm{e}}\mathrm{t}}(\mathrm{X}\times\mathrm{X},(a,b))$ on

$$\varprojlim_m\Big(\mathrm{R}^i\,p^{\bullet}_{[n],*}(\mathbb{Z}/\ell^m\mathbb{Z}_{\mathrm{X}^{\Delta[1]}_{[n],\acute{\mathrm{e}}\mathrm{t}}})\Big)_{(a,b)}\otimes\mathbb{Q}_\ell\cong\mathrm{H}^i_{\acute{\mathrm{e}}\mathrm{t}}((p^{\bullet}_{[n]})^{-1}(a,b),\mathbb{Q}_\ell)\,. \tag{14.21}$$

It follows from the work of Chen that

$$\mathrm{H}^0_{\mathrm{DR}}((p(\mathbb{C})^{\bullet})^{-1}(a,b))\cong\mathrm{Alg}_{\mathbb{C}}(\pi_1(\mathrm{X}(\mathbb{C});b,a)\otimes\mathbb{Q})\,.$$

We shall use Sullivan's polynomial differential forms with $\mathbb{Q}$-coefficients, see [Sul77, page 297]. The subscript SDR denotes the corresponding cohomology groups. We get the corresponding isomorphism of $\mathbb{Q}$-algebras

$$\mathrm{H}^0_{\mathrm{SDR}}((p(\mathbb{C})^{\bullet})^{-1}(a,b))\cong\mathrm{Alg}_{\mathbb{Q}}(\pi_1(\mathrm{X}(\mathbb{C});b,a)\otimes\mathbb{Q})$$

where $\mathrm{Alg}_{\mathbb{Q}}$ denotes the algebra of polynomial functions on $\pi_1(\mathrm{X}(\mathbb{C});b,a)\otimes\mathbb{Q}$ with values in $\mathbb{Q}$. If $a=b$ then we get even an isomorphism of Hopf algebras. The comparison isomorphism leads to isomorphisms

$$\mathrm{H}^0_{\text{ét}}((p^\bullet)^{-1}(a,b),\mathbb{Q}_\ell) \cong \mathrm{H}^0((p(\mathbb{C})^\bullet)^{-1}(a,b),\mathbb{Q})\otimes\mathbb{Q}_\ell$$
$$\cong \mathrm{H}^0_{\mathrm{SDR}}((p(\mathbb{C})^\bullet)^{-1}(a,b))\otimes\mathbb{Q}_\ell$$

from étale via singular to de Rham cohomology, the last one calculated using Sullivan polynomial differential forms. We conclude a natural isomorphism

$$\mathrm{H}^0_{\text{ét}}((p^\bullet)^{-1}(a,b),\mathbb{Q}_\ell) \cong \mathrm{Alg}_{\mathbb{Q}_\ell}(\pi_1(\mathrm{X}(\mathbb{C});b,a)\otimes\mathbb{Q})$$

where $\mathrm{Alg}_{\mathbb{Q}_\ell}$ denotes the algebra of polynomials with values in $\mathbb{Q}_\ell$. On the other side we have an isomorphisms of torsors

$$\pi_1(\mathrm{X}(\mathbb{C});b,a)\otimes\mathbb{Q}_\ell \cong \pi_1(\mathrm{X};b,a)\otimes\mathbb{Q}_\ell\,,$$

where $\pi_1(\mathrm{X};b,a)$ is the right $\pi_1(\mathrm{X},a)$-torsor of ℓ-adic paths from a to b on X and $\pi_1(\mathrm{X};b,a)\otimes\mathbb{Q}_\ell$ is the induced right $\pi_1(\mathrm{X},a)\otimes\mathbb{Q}_\ell$-torsor. Therefore we get an isomorphism of $\mathbb{Q}_\ell$-vector spaces

$$\mathrm{H}^0_{\text{ét}}((p^\bullet)^{-1}(a,b),\mathbb{Q}_\ell) \cong \mathrm{Alg}_{\mathbb{Q}_\ell}(\pi_1(\mathrm{X};b,a)\otimes\mathbb{Q}_\ell)\,. \tag{14.22}$$

The shuffle product on $\mathrm{H}^0_{\mathrm{DR}}$ is defined using codegeneracies, and can thus be defined also on $\mathrm{H}^0_{\text{ét}}$. The Hopf algebra structure on $\mathrm{H}^0_{\mathrm{DR}}((p(\mathbb{C})^\bullet)^{-1}(a,a))$ is defined by the maps

$$1\otimes\omega_1\otimes\ldots\otimes\omega_n\otimes 1 \to \sum_{i=0}^{n}(1\otimes\omega_1\otimes\ldots\otimes\omega_i\otimes 1)\otimes(1\otimes\omega_{i+1}\otimes\ldots\otimes\omega_n\otimes 1)\,,$$

hence one can use maps $\mathrm{X}^n \to \mathrm{X}^i\times\mathrm{X}^{n-i}$ to define it. Therefore the isomorphism (14.22) is an isomorphism of $\mathbb{Q}_\ell$-algebras and if $a=b$ it is an isomorphism of Hopf algebras. We get that the monodromy representation on the projective limit of stalks (14.21) for $i=0$ reads

$$\rho_{(a,b)}\;:\;\pi_1^{\text{ét}}(\mathrm{X},a)\times\pi_1^{\text{ét}}(\mathrm{X},b)\longrightarrow \mathrm{Aut}(\,\mathrm{Alg}_{\mathbb{Q}_\ell}(\pi_1(\mathrm{X};b,a)\otimes\mathbb{Q}_\ell))$$

and is given by the formula

$$((\rho_{(a,b)}(\alpha,\beta))(f))(\gamma) = f(\beta^{-1}\cdot\gamma\cdot\alpha)\,. \tag{14.23}$$

Now we suppose that X is defined over a number field K with a complex embedding $\mathrm{K}\subset\mathbb{C}$ and that a and b are two K-points of X. Then the Galois group $\mathrm{G_K}$ acts on $\mathrm{H}^0_{\text{ét}}((p^\bullet)^{-1}(a,b),\mathbb{Q}_\ell)$, where

$$p^\bullet\;:\;\mathrm{X}_{\bar{\mathrm{K}}}^{\Delta[1]}\to\mathrm{X}_{\bar{\mathrm{K}}}^{\partial\Delta[1]}\,.$$

On the other side $\mathrm{G_K}$ acts on $\pi_1(\mathrm{X}_{\bar{\mathrm{K}}};b,a)$, the set of isomorphisms of fiber functors, see [SGA1, Exp V], which in this paper we call the set of ℓ-adic paths, by

$$\sigma(\alpha) = \sigma_{\bar{b}} \circ \alpha \circ \sigma_{\bar{a}}^{-1}\,.$$

If $a = b$ this is the action described in (14.16).

Proposition 29. *Let* X *be a smooth quasi-projective, geometrically connected algebraic variety defined over a number field* K *with a complex embedding* $\mathrm{K} \subset \mathbb{C}$. *Let* a *and* b *be two* K*-points of* X. *Then the isomorphism of* $\mathbb{Q}_\ell$*-algebras*

$$\mathrm{H}^0_{\text{ét}}((p^\bullet)^{-1}(a,b),\mathbb{Q}_\ell) \cong \mathrm{Alg}_{\mathbb{Q}_\ell}(\pi_1(\mathrm{X}_{\bar{\mathrm{K}}};b,a)\otimes\mathbb{Q}_\ell)$$

is an isomorphism of $\mathrm{G_K}$*-modules, where on the left hand side* $\mathrm{G_K}$ *acts on the étale cohomology group and on the right hand side the action of* $\mathrm{G_K}$ *is deduced from the action on* $\pi_1(\mathrm{X}_{\bar{\mathrm{K}}};b,a)$.

Proof. The projective system of sheaves

$$\left\{\mathcal{H}^0\Big(\mathrm{Tot}(\mathrm{R}\,p^\bullet_{[n],*}(\mathbb{Z}/\ell^m\mathbb{Z}_{(\mathrm{X}_{\bar{\mathrm{K}}})^{\Delta[1]}_{[n],\text{ét}}}))\Big)\right\}_{m\in\mathrm{N}}$$

over $\mathrm{X}_{\bar{\mathrm{K}}}\times\mathrm{X}_{\bar{\mathrm{K}}}$ is equipped with Galois action in each stalk. Moreover the projective limit tensored with $\mathbb{Q}_\ell$ is locally constant. The Galois action and parallel transport satisfy the formula (14.8). To see this one need to do all constructions over Spec(K) instead of Spec($\bar{\mathrm{K}}$).

More naively, if $f : \mathrm{X}_{\bar{\mathrm{K}}} \to \mathrm{X}_{\bar{\mathrm{K}}}$ is a morphism of algebraic varieties over $\bar{\mathrm{K}}$, then for any $\sigma \in \mathrm{G_K}$ we have $\sigma\circ f = \sigma(f)\circ\sigma$ on $\mathrm{X}_{\bar{\mathrm{K}}}$. Let $\mathrm{Z} \to \mathrm{X}_{\bar{\mathrm{K}}}\times\mathrm{X}_{\bar{\mathrm{K}}}$ be an étale Galois covering and let $f : \mathrm{Z}\to\mathrm{Z}$ be a covering transformation. Let $\mathcal{A}$ be a sheaf on $(\mathrm{X}_{\bar{\mathrm{K}}}\times\mathrm{X}_{\bar{\mathrm{K}}})_{\text{ét}}$, like the sheaves considered in this paper. Then f induces $f_* : \mathcal{A}(\mathrm{Z})\to\mathcal{A}(\mathrm{Z})$ and once more one have $\sigma\circ f_* = \sigma(f)_*\circ\sigma$ on $\mathcal{A}(\mathrm{Z})$. The map f_* is the monodromy along an element of $\pi_1^{\text{ét}}(\mathrm{X}_{\bar{\mathrm{K}}}\times\mathrm{X}_{\bar{\mathrm{K}}},(a,b))$, which induces f. The equality $\sigma\circ f_* = \sigma(f)_*\circ\sigma$ is the formula (14.8).

For $(\alpha,\beta)\in\pi_1^{\text{ét}}(\mathrm{X}_{\bar{\mathrm{K}}},a)\times\pi_1^{\text{ét}}(\mathrm{X}_{\bar{\mathrm{K}}},a)$, for $\sigma\in\mathrm{G_K}$ and for

$$f \in \mathrm{H}^0_{\text{ét}}((p^\bullet)^{-1}(a,a),\mathbb{Q}_\ell)$$

we have a formula for the Galois action that reads

$$\begin{aligned}\sigma_{(a,a)}((\alpha,\beta)_*(f)) &= (\sigma(\alpha),\sigma(\beta))_*(\sigma_{(a,a)}(f))\\ &= \Big(\gamma\mapsto(\sigma_{(a,a)}(f))(\sigma(\beta)^{-1}\cdot\gamma\cdot\sigma(\alpha))\Big) \qquad (14.24)\end{aligned}$$

by (14.8) and (14.23).

The function $\gamma \to f(\beta^{-1}\cdot\gamma\cdot\alpha)$ is calculated using the Hopf algebra structure on $\mathrm{H}^0_{\text{ét}}((p^\bullet)^{-1}(a,a),\mathbb{Q}_\ell)$. The Galois group $\mathrm{G_K}$ acts on $\mathrm{H}^0_{\text{ét}}((p^\bullet)^{-1}(a,a),\mathbb{Q}_\ell)$ through $\mathbb{Q}_\ell$-isomorphisms. Therefore after applying $\sigma_{(a,a)}$ and setting $\beta = 1$ and $\gamma = 1$ we get that the left hand side of (14.24) is equal $f(\alpha)$.

If we plug $\beta = 1$ and $\gamma = 1$ into the right hand side we get $(\sigma_{(a,a)}(f))(\sigma(\alpha))$. Hence for any $\sigma\in\mathrm{G_K}$ and any $\alpha\in\pi_1(\mathrm{X}_{\bar{\mathrm{K}}},a)$ we have

$$(\sigma_{(a,a)}(f))(\alpha) = f(\sigma^{-1}(\alpha)).$$

Therefore the G_K-modules $H^0_{\text{ét}}((p^\bullet)^{-1}(a,a),\mathbb{Q}_\ell)$ and $\text{Alg}_{\mathbb{Q}_\ell}(\pi_1(X_{\bar{K}},a)\otimes\mathbb{Q})$, where the action of G_K on the second one comes from the action of G_K on $\pi_1(X_{\bar{K}},a)$, are isomorphic. We deduce also a G_K isomorphism

$$H^0_{\text{ét}}((p^\bullet)^{-1}(a,b),\mathbb{Q}_\ell) \cong \text{Alg}_{\mathbb{Q}_\ell}(\pi_1(X_{\bar{K}};b,a)\otimes\mathbb{Q}_\ell)$$

for any pair (a,b). □

References

[BD94] A. A. Beilinson and P. Deligne. Interprétation motivique de la conjecture de Zagier reliant polylogarithmes et régulateurs. In U. Jannsen, S. L. Kleiman, J.-P. Serre, *Motives, Proc. of Sym. in Pure Math.*, pages 97–121, **55**, Part II AMS 1994.

[BL94] A. A. Beilinson and A. Levin. The elliptic polylogarithm. In U. Jannsen, S. L. Kleiman, J.-P. Serre, *Motives, Proc. of Sym. in Pure Math.*, pages 123–190, **55**, Part II AMS 1994.

[Che75] K. T. Chen Iterated integrals, fundamental groups and covering spaces. *Trans. of the Amer. Math. Soc.*, 206:83–98, 1975.

[DW04] J.-C. Douai and Z. Wojtkowiak. On the Galois actions on the fundamental group of $\mathbb{P}^1_{\mathbb{Q}(\mu_n)} \setminus \{0,1,\infty\}$. *Tokyo Journal of Math.*, 27(1):21–34, 2004.

[GM88] M. Goresky and R. MacPherson. Stratified Morse theory. Ergebnisse der Mathematik und ihrer Grenzgebiete (3), vol. **14**, Springer 1988.

[NW02] H. Nakamura and Z. Wojtkowiak. On the explicit formulae for ℓ-adic polylogarithms. In *Arithmetic Fundamental Groups and Noncommutative Algebra, Proc. Symp. Pure Math.*, (AMS) 70:285–294, 2002.

[SGA1] A. Grothendieck. *Revêtements étale et groupe fondamental (SGA 1)*. Séminaire de géométrie algébrique du Bois Marie 1960-61, directed by A. Grothendieck, augmented by two papers by Mme M. Raynaud, *Lecture Notes in Math.* 224, Springer-Verlag, Berlin-New York, 1971. Updated and annotated new edition: *Documents Mathématiques* 3, Société Mathématique de France, Paris, 2003.

[Sul77] D. Sullivan. Infinitesimal computations in topology. *Publications Mathématiques, Institut des Hautes Études Scientifiques*, 47:269–332, 1977.

[Woj91] Z. Wojtkowiak. The basic structure of polylogarithmic functional equations. In L. Lewin, *Structural Properties of Polylogarithms, Mathematical Surveys and Monographs*, pages 205–231, Vol 37, 1991.

[Woj93] Z. Wojtkowiak. Cosimplicial objects in algebraic geometry. In *Algebraic K-theory and Algebraic Topology*, Kluwer Academic Publishers, 407:287–327, 1993.

[Woj02] Z. Wojtkowiak. Mixed Hodge structures and iterated integrals, I. In F. Bogomolov and L. Katzarkov, *Motives, Polylogarithms and Hodge Theory. Part I: Motives and Polylogarithms, International Press Lectures Series* , Vol.3:121–208, 2002.

[Woj04] Z. Wojtkowiak. On ℓ-adic iterated integrals, I. Analog of Zagier conjecture. *Nagoya Math. Journals*, 176:113–158, 2004.

[Woj05a] Z. Wojtkowiak. On ℓ-adic iterated integrals, II. Functional equations and *l*-adic polylogarithms. *Nagoya Math. Journals*, 177:117–153, 2005.

[Woj05b] Z. Wojtkowiak. On ℓ-adic iterated integrals, III. Galois actions on fundamental groups. *Nagoya Math. Journals*, 178:1–36, 2005.

[Woj07] Z. Wojtkowiak. On the Galois actions on torsors of paths I, Descent of Galois representations. *J. Math. Sci. Univ. Tokyo*, 14:177–259, 2007.

[Woj09] Z. Wojtkowiak. On ℓ-adic Galois Periods, relations between coefficients of Galois representations on fundamental groups of a projective line minus a finite number of points. *Publ. Math. de Besançon, Algèbra et Théorie des Nombres*, 155–174, 2007–2009, *Février* 2009.

Workshop Talks

Amnon Besser
Department of Mathematics, Ben-Gurion University of the Negev, Be'er- Sheva, Israel
Coleman integration in families

Suppose that X/S is a smooth family of rigid varieties with good reduction, and that we are given a closed relative 1-form. Then we can try to ask for the corresponding family of Coleman integrals, which is more restrictive, and better behaved than just fiber by fiber Coleman integral. This can be achieved by imposing *equivariance with respect to differentiation in the direction of the base*, a term which needs to be made precise, in addition to the usual Frobenius equivariance. I will explain how one can get this theory, and a more general theory for iterated Coleman integration in families, using new ideas of Ovchinnikov on differential Tannakian theory. This theory adds structure to Tannakian categories so that they become equivalent to representations of differential algebraic groups, rather than ordinary algebraic groups. One interprets the relative Coleman integration theory as saying that there exists a unique path *with zero derivative* and invariant under Frobenius.

Anna Cadoret
Centre Mathématiques Laurent Schwarz, École Polytechnique, Palaiseau, France
Growth of the genus of the generic torsion of abelian schemes over curves

A consequence of the geometric torsion conjecture for abelian varieties over function fields is the following. Let k be an algebraically closed field of characteristic 0. For any integers d, $g \geqslant 1$ there exists an integer $\mathrm{N} := \mathrm{N}(k,d,g) \geqslant 1$ such that for any function field K/k with transcendence degree 1 and genus g and any d-dimensional abelian variety A/K containing no nontrivial k-isotrivial subvariety, the order of torsion in A(K) is bounded by N. In this talk, I will deal with the weak variant of this statement, where the integer $\mathrm{N}(k,d,g)$ is allowed to depend on the abelian variety A/K. More precisely, I will show that if K has genus at least 1 or if K has genus 0 and A/K has semistable reduction over all but possibly one place then, for any integer $g \geqslant 1$, there exists an integer

J. Stix (ed.), *The Arithmetic of Fundamental Groups*, Contributions in Mathematical and Computational Sciences 2, DOI 10.1007/978-3-642-23905-2,

$N := N(A, g) \geqslant 1$ such that for any finite extension L/K with genus $\leqslant g$, the order of torsion in $A(L)$ is bounded by N. Previous joint work with Akio Tamagawa shows that the above holds – without any restriction on K or the reduction type of A/K – for the ℓ-primary torsion with ℓ a fixed prime. So, it is enough to prove that there exists an integer $N := N(A, g) \geqslant 1$ such that for any finite extension L/K with genus $\leqslant g$, the prime divisors of the order of $|A(L)_{tors}|$ are all bounded by N. This is joint work with A. Tamagawa.

Gerd Faltings
Max Planck Institute for Mathematics, Bonn, Germany
Rational points and the motivic polylogarithm

In Kim's new proof of Siegel's theorem one key ingredient is an upper bound for global Galois cohomology. The purpose of this talk is to explain how this might be replaced by motivic cohomology, and to explain some of the difficulties with this approach, and how to resolve them sometimes.

Majid Hadian
Max Planck Institute for Mathematics, Bonn, Germany
Motivic fundamental groups and integral points

By studying the variation of different realizations of motivic path torsors of unirational varieties over number fields and comparing them, we show that S-integral points of such a variety over a totally real number field with highly enough non-abelian fundamental group lie, locally in the p-adic analytic topology, in the zero locus of a non-zero p-adic analytic function.

David Harari
Département de Mathématiques d'Orsay, Université de Paris-Sud, France
The fundamental group and cohomological obstructions to the Hasse principle and weak approximation

Let X be an algebraic variety defined over a number field k. I will explain how the non-triviality of the geometric fundamental group of X can provide obstructions to the existence of a rational point or to the weak approximation property. This is related to a *non-abelian descent* formalism, which gives many interesting examples, as elliptic surfaces and Enriques surfaces. The relationship between *descent obstructions* and Grothendieck's section conjecture will also be discussed.

David Harbater
Department of Mathematics, University of Pennsylvania, Philadelphia, USA
Split covers and local-global principles

This talk will consider curves over complete discretely valued fields, where patching methods can be used. In this context we relate a certain quotient of the fundamental group, which we call the split fundamental group, to local-global principles for structures such as torsors and quadratic forms. This is done via the topological fundamental group of an associated reduction graph. This is joint work with Julia Hartmann and Daniel Krashen.

Yuichiro Hoshi
Research Institute for Mathematical Sciences, Kyoto, Japan
Existence of non-geometric pro-p Galois sections of hyperbolic curves

In this talk, we construct a non-geometric pro-p Galois section of a hyperbolic curve over a number field, as well as over a p-adic local field. Moreover, we observe that there exists a proper hyperbolic curve over a number field which admits infinitely many conjugacy classes of pro-p Galois sections.

Minhyong Kim
Department of Mathematics, University College London, UK
Diophantine geometry and Galois theory 12

In his manuscripts form the 1980's Grothendieck proposed ideas that have been interpreted variously as embedding the theory of schemes into either:

- Group theory and higher-dimensional generalizations,
- or homotopy theory.

It was suggested, moreover, that such a framework would have profound implications for the study of Diophantine problems. In this talk, we will discuss mostly the little bit of progress made on this last point using some mildly non-abelian motives associated to hyperbolic curves.

Guido Kings
Fakultät für Mathematik, Universität Regensburg, Germany
Abelian polylogarithms on curves

Elements in the étale cohomology $\mathrm{H}^1(\mathbb{Z}[1/\mathrm{N}], \mathbb{Z}_p[\![\mathrm{G}]\!](1))$ where G is an almost pro-p-group play a crucial role in the construction of Euler systems and the p-adic study of special values of L-functions. In our talk we discuss how the polylogarithm on curves gives rise to such cohomology classes and explain some of their fundamental properties. In the abelian case we show that the classes are motivic, i. e., in the image of the regulator map from K-theory. According to general conjectures this should imply that these classes are related to special values of L-functions.

Emmanuel Lepage
Institut de Mathématiques de Jussieu, Université de Paris 6, France
The tempered fundamental group of Mumford curves and the metric graph of their stable reduction

The tempered fundamental group of a smooth Berkovich space over a complete non archimedean field is a topological group that classifies étale coverings for the étale topology of Berkovich that become (potentially infinite) topological coverings for the topology of Berkovich after pullback by a finite étale covering. S. Mochizuki proved that one can recover the graph of the stable reduction of a hyperbolic curve from its geometric tempered fundamental group. I show that, for a Mumford cuve, one can in fact also recover the metric of the graph. The proof involves the theory of theta functions of a Mumford curve.

Florian Pop
Department of Mathematics, University of Pennsylvania, Philadelphia, USA
Hints about a "minimalistic" form of BAP

After recalling Bogomolov's Anabelian Program (BAP), I will give some hints about the why and how of a *minimalistic* form of this program (MAP). If true, MAP would bring a completely new quality in birational anabelian geometry.

Jonathan P. Pridham
Department of Pure Mathematics and Mathematical Statistics, University of Cambridge, UK
Pro-algebraic ℓ-adic fundamental groups

For smooth varieties in finite and mixed characteristics, the ℓ-adic pro-algebraic fundamental group is largely determined as a Galois representation by cohomology of semisimple local systems, and the same holds for big Malcev completions. There are also a pro-algebraic crystalline fundamental group and a non-abelian étale-crystalline comparison theorem, but these cannot be recovered from cohomology alone.

Gereon Quick
Mathematisches Institut, Westfälische Wilhelms-Universität Münster, Germany
Torsion algebraic cycles and étale cobordism

We show that over an algebraically closed field of positive characteristic the classical integral cycle class map from algebraic cycles to étale cohomology factors through a quotient of ℓ-adic étale cobordism. This implies that there is a strong topological obstruction for cohomology classes to be algebraic and that examples of Atiyah and Hirzebruch for non-algebraic integral cohomology classes and of Totaro for non-trivial elements in the Griffiths group also work in positive characteristic.

João Pedro P. dos Santos
Institut de Mathématiques de Jussieu, Université de Paris 6, France
A new characterization of essentially finite bundles and some applications to fundamental groups of rationally connected varieties

The category of essentially finite bundles (Nori 1976) over a projective variety X/k is a Tannakian category such that, the k-affine group scheme associated to it by Tannaka duality classifies torsors $P \to X$ with finite structure group. This group scheme is then a generalization of the algebraic fundamental group. I will show that an essentially finite vector bundle E over X is simply a vector bundle which, when pulled back by a suitable proper and surjective morphism $f : Y \to X$ becomes trivial. This can be appreciated in two ways:

- It gives a technically simpler characterization of essentially finite bundles.
- It generalizes a well known result, due to Lange and Stuhler, characterizing k-representations of the fundamental group as bundles trivialized by étale morphisms.

This property will allow us to conclude that *all* fundamental group schemes associated to a smooth rationally connected variety are trivial. This is joint work with I. Biswas.

Annette Werner
Institut für Mathematik, Goethe-Universität Frankfurt, Germany
Vector bundles on p-adic curves and parallel transport

We define functorial isomorphisms of parallel transport along étale paths for a class of vector bundles on a p-adic curve. All bundles of degree zero with potentially strongly semistable reduction belong to this class. In particular, they give rise to representations of the algebraic fundamental group of the curve.

Kirsten Wickelgren
AIM and Harvard University, Cambdridge MA, USA
Étale π_1 obstructions to rational points

Grothendieck's section conjecture says that for a hyperbolic curve over a number field, rational points are in bijection with sections of étale π_1 of the structure map. We use cohomological obstructions of Jordan Ellenberg to study such sections. We will relate Ellenberg's obstructions to Massey products, and explicitly compute versions of the first and second for $\mathbb{P}^1_{\mathbb{Q}} - \{0, 1, \infty\}$. Over $\mathbb{R}$, we show the first obstruction alone determines the connected components of real points of the curve from those of the Jacobian.

Olivier Wittenberg
CNRS and Département de Mathématiques et Appl.,École Normale Supérieure, Paris, France
Around cycle classes of sections

To every section of the arithmetic fundamental group of a smooth $K(\pi, 1)$ variety X over a field k, one may associate a cycle class in the étale cohomology with compact supports of X. We discuss various applications of this construction. By investigating the algebraicity of this cycle class, we prove that Schinzel's quartic curve has no section over the rationals, as predicted by the section conjecture. This is partly joint work with Hélène Esnault.

Zdzisław Wojtkowiak
Département de Mathématiques, Université Nice Sophia-Antipolis, France

Periods of mixed Tate motives, examples, ℓ-adic Galois side

The iterated integrals on $\mathbb{P}^1(\mathbb{C}) - \{0, 1, \infty\}$ from 0 to 1 suitably normalized in sequences of 1-forms dz/z and $dz/(z-1)$ are periods of mixed Tate motives over Spec($\mathbb{Z}$). The natural question is: do we obtain in this way all periods of mixed Tate motives over Spec($\mathbb{Z}$)? We will discuss an ℓ-adic version of this problem. In ℓ-adic setting analogs of periods are coefficients and the analogs of iterated integrals are geometric coefficients. We will show that all coefficients of mixed Tate motives over Spec($\mathbb{Z}$) can be expressed as some suitable sums of coefficients of the natural representation of $\mathrm{Gal}_{\mathbb{Q}}$ on the pro-ℓ completion of

$$\pi_1(\mathbb{P}^1_{\bar{\mathbb{Q}}} - \{0, 1, -1, \infty\}, \overrightarrow{01}),$$

and these coefficients are geometric coefficients. We shall also show that all Tate motives over Spec($\mathbb{Z}[1/p]$) can be expressed as some suitable sums of coefficients of the natural representation of $\mathrm{Gal}_{\mathbb{Q}(\mu_p)}$ on the pro-ℓ completion of

$$\pi_1(\mathbb{P}^1_{\bar{\mathbb{Q}}} - (\{0, \infty\} \cup \mu_p), \overrightarrow{01}).$$

9783642239045